# Interdisciplinary Applied Mathematics

Volumes published are listed at the end of this book.

**Springer**

*New York*
*Berlin*
*Heidelberg*
*Hong Kong*
*London*
*Milan*
*Paris*
*Tokyo*

# Interdisciplinary Applied Mathematics

## Volume 20

Problems in engineering, computational science, and the physical and biological sciences are using increasingly sophisticated mathematical techniques. Thus, the bridge between the mathematical sciences and other disciplines is heavily traveled. The correspondingly increased dialog between the disciplines has led to the establishment of the series: *Interdisciplinary Applied Mathematics.*

The purpose of this series is to meet the current and future needs for the interaction between various science and technology areas on the one hand and mathematics on the other. This is done, firstly, by encouraging the ways that mathematics may be applied in traditional areas, as well as point towards new and innovative areas of applications; and, secondly, by encouraging other scientific disciplines to engage in a dialog with mathematicians outlining their problems to both access new methods and suggest innovative developments within mathematics itself.

The series will consist of monographs and high-level texts from researchers working on the interplay between mathematics and other fields of science and technology.

Christopher P. Fall     Eric S. Marland
John M. Wagner          John J. Tyson
Editors

# Computational Cell Biology

With 210 Illustrations

 Springer

Christopher P. Fall
Center for Neural Science
New York University
New York, NY 10003
USA
fall@cns.nyu.edu

Eric S. Marland
Department of Mathematical Sciences
Appalachian State University
Boone, NC 28608
USA
marlandes@appstate.edu

John M. Wagner
IBM TJ Watson Research Center
Yorktown Heights, NY 10598
USA
wagnerjo@us.ibm.com

John J. Tyson
Department of Biology
Virginia Polytechnic Institute
   and State University
Blacksburg, VA 24061
USA
tyson@vt.edu

*Editors:*

S.S. Antman
Department of Mathematics
   *and*
Institute for Physical Science and Technology
University of Maryland
College Park, MD 20742
USA

L. Sirovich
Laboratory of Applied Mathematics
Department of Biomathematical Sciences
Mount Sinai School of Medicine
One Gustave L. Levy Place
New York, NY 10029-6574
USA

J.E. Marsden
Control and Dynamical Systems
Mail Code 107-81
California Institute of Technology
Pasadena, CA 91125
USA

S. Wiggins
Control and Dynamical Systems
Mail Code 107-81
California Institute of Technology
Pasadena, CA 91125
USA

Mathematics Subject Classification (2000): 92-01, 92BXX, 92C30, 92C20

Library of Congress Cataloging-in-Publication Data
Computational cell biology / editors, Christopher P. Fall . . . [et al.].
      p. cm. — (Interdisciplinary applied mathematics)
      Includes bibliographical references and index.

      1. Cytology—Computer simulation.   2. Cytology—Mathematical models.   I. Fall, Christopher P.
II. Series.
QH585.5.D38 C65 2002
571.6′01′13—dc21                                          2001054912

ISBN  978-1-4419-2975-4          e-ISBN  978-0-387-22459-6

Printed in the United States of America.      (BPR/HAM)

9  8  7  6  5  4  3      Corrected third printing, 2005.

springeronline.com

Joel Edward Keizer 1942–1999

Joel Keizer's thirty years of scientific work set a standard for collaborative research in theoretical chemistry and biology. Joel served the University of California at Davis for 28 years, as a Professor in the Departments of Chemistry and of Neurobiology, Physiology and Behavior, and as founder and Director of the Institute for Theoretical Dynamics. Working at the boundary between experiment and theory, Joel built networks of collaborations and friendships that continue to grow and produce results. This book evolved from a textbook that Joel began but was not able to finish. The general outline and goals of the book were laid out by Joel, on the basis of his many years of teaching and research in computational cell biology. Those of us who helped to finish the project—as authors and editors—are happy to dedicate our labors to the memory of our friend and colleague, Joel Edward Keizer. All royalties from this book are to be directed to the Joel E. Keizer Memorial Fund for collaborative interdisciplinary research in the life sciences.

# Preface

This text is an introduction to dynamical modeling in cell biology. It is not meant as a complete overview of modeling or of particular models in cell biology. Rather, we use selected biological examples to motivate the concepts and techniques used in computational cell biology. This is done through a progression of increasingly more complex cellular functions modeled with increasingly complex mathematical and computational techniques. There are other excellent sources for material on mathematical cell biology, and so the focus here truly is computer modeling. This does not mean that there are no mathematical techniques introduced, because some of them are absolutely vital, but it does mean that much of the mathematics is explained in a more intuitive fashion, while we allow the computer to do most of the work.

The target audience for this text is mathematically sophisticated cell biology or neuroscience students or mathematics students who wish to learn about modeling in cell biology. The ideal class would comprise both biology and applied math students, who might be encouraged to collaborate on exercises or class projects. We assume as little mathematical and biological background as we feel we can get away with, and we proceed fairly slowly. The techniques and approaches covered in the first half of the book will form a basis for some elementary modeling or as a lead in to more advanced topics covered in the second half of the book. Our goal for this text is to encourage mathematics students to consider collaboration with experimentalists and to provide students in cell biology and neuroscience with the tools necessary to access the modeling literature and appreciate the value of theoretical approaches.

The core of this book is a set of notes for a textbook written by Joel Keizer before his death in 1999. In addition to many other accomplishments as a scientist, Joel founded and directed the Institute of Theoretical Dynamics at the University of California, Davis. It is currently the home of a training program for young scientists studying nonlinear dynamics in biology. As a part of this training program Joel taught a course entitled "Computational Models of Cellular Signaling," which covered much of the material in the first half of this book.

Joel took palpable joy from interaction with his colleagues, and in addition to his truly notable accomplishments as a theorist in both chemistry and biology, perhaps his greatest skill was his ability to bring diverse people together in successful collaboration. It is in recognition of this gift that Joel's friends and colleagues have brought this text to completion. We have expanded the scope, but at the core, you will still find Joel's hand in the approach, methodology, and commitment to the interdisciplinary and collaborative nature of the field. The royalties from the book will be donated to the Joel E. Keizer foundation at the University of California at Davis, which promotes interdisciplinary collaboration between mathematics, the physical sciences, and biology.

*Audience*: We have aimed this text at an advanced undergraduate or beginning graduate audience in either mathematics or biology.

*Prerequisites*: We assume that students have taken full–year courses in calculus and biology. Introductory courses in differential equations and molecular cell biology are desirable but not absolutely necessary. Students with more substantial background in either biology or mathematics will benefit all the more from this text, especially the second half. No former programming experience is needed, but a working knowledge of using computers will make the learning curve much more pleasant. Note that we often point students to other textbooks and monographs, both because they are important references for later use and because they might be a better source for the material. Instructors may want to have these sources available for students to borrow or consult.

*Organization*: We consider the first six chapters, through intercellular communication, to be the core of the text. They cover the basic elements of compartmental modeling, and they should be accessible to anyone with a minimum background in cell biology and calculus. The remainder of the chapters cover more specialized topics that can be selected from, based on the focus of the course. Chapters 7 and 8 introduce spatial modeling, Chapters 9 and 10 discuss biochemical oscillations and the cell cycle, and Chapters 11–13 cover stochastic methods and models. These chapters are of varying degrees of difficulty.

Finally, in the first appendix, some of the mathematical and computational concepts brought up throughout the book are covered in more detail. This appendix is meant to be a reference and a learning tool. Sections of it may be integrated into the chapters as the topics are introduced. The second appendix contains an introduction to the XPPAUT ODE package discussed below. The final appendix contains psuedocode versions of the code used to create some of the data figures in the text.

*Internet Resources*: This book will have its own web page which will contain a variety of resources. We will maintain a list of the inevitable mistakes and typos and will make available actual code for the figures in the book. The web address is `http://www.ComputationalCellBiology.net`.

*Software*: We designed the text to be independent of any particular software, but have included appendices in support of the XPPAUT package. XPPAUT has been developed by Bard Ermentrout at the University of Pittsburgh, and it is currently available free of charge. XPPAUT numerically solves and plots the solutions of ordinary differential equations. It also incorporates a numerical bifurcation software and some methods for stochastic equations. Versions are currently available for Windows, Linux, and Unix systems. Recent changes in the Macintosh platform (OSX) make it possible to use XPP there as well. Ermentrout has recently published an excellent user's manual available through SIAM (Ermentrout 2002).

There are a large number of other software packages available that can accomplish many of the same things as XPPAUT can, such as MATLAB, MapleV, Mathematica, and Berkeley Madonna. Programming in C or Fortran is also possible. However XPPAUT is easy to use, requires minimal programming skills, has an excellent online tutorial, and is distributed without charge. The aspect of XPPAUT which is available in very few other places is the bifurcation software AUTO, originally developed by E.J. Doedel. The bifurcation tools in XPPAUT are necessary only for selected problems, so many of the other packages will suffice for most of the book. The the book and web site contain code that will reproduce many of the figures in the book. As students solve the exercises and replicate the simulations using other packages, we would encourage the submission of the code to the editors. We will incorporate this code into the web site and possibly into future editions of the book.

There are many people to thank for their help with this project. Of course, we are deeply indebted to the contributors, who first completed or wrote from scratch the chapters and then dealt with the numerous revisions necessary to homogenize the book to a reasonable level. Carla Wofsy and Byron Goldstein, as well as Albert Goldbeter, encouraged us to go forward with the project and provided valuable suggestions. We thank Chris Dugaw and David Quinonez for their assistance with typesetting several of the chapters, and Randy Szeto for his work with the graphic design of the book. We thank James Sneyd for many helpful comments on the manuscript, and also Tim Lewis for commenting on several of the chapters. Carol Lucas generously provided many corrections for the first half of the text. C.F., J.W., and E.M. were supported in part by the Institute of Theoretical Dynamics at UC Davis during some of the preparation of the manuscript.

We suspect that Joel, for a start, would have thanked Lee Segel, Jim Murray, Leah Edelstein-Keshet and others whose pioneering textbooks in mathematical biology certainly informed this one. We *know* that Joel would have thanked many friends and colleagues for contributing to the true excitement he felt in his "second career" studying biology. While we have dedicated this work to the memory of Joel, Joel's dedication might well have been to his wife, Susan; his daughter, Sarah; his son and daughter-in-law, Sidney and Noelle; and his grandson, Justin Joel.

We hope you enjoy this text, and we look forward to your comments and suggestions. We strongly believe that a textbook such as this might serve to help to develop the field of computational cell biology by introducing students to the subject. This textbook will be more successful in helping to forge a community if it represents what most of us agree is necessary to teach beginning students. This is only a first step, and we truly look forward both to input about the material already presented and to suggestions and contributions of additional material and topics for future editions.

*For further information please visit*
`http://www.ComputationalCellBiology.net`.

# Contributors

Timothy C. Elston
North Carolina State University
Department of Statistics

G. Bard Ermentrout
University of Pittsburgh
Department of Mathematics

Christopher P. Fall
New York University
Center for Neural Science

James P. Keener
University of Utah
Department of Mathematics

Joel E. Keizer
University of California at Davis
Institute of Theoretical Dynamics

Yue-Xian Li
University of British Columbia
Department of Mathematics

Eric S. Marland
Appalachian State University
Department of Mathematical Sciences

Alexander Mogilner
University of California at Davis
Department of Mathematics

Béla Novák
Budapest University of Technology and
Economics
Department of Agricultural Chemical
Technology

George Oster
University of California at Berkeley
Departments of Molecular and Cellular Biology
and ESPM

John E. Pearson
Los Alamos National Laboratory
Applied Theoretical and Computational Physics

John Rinzel
New York University
Center for Neural Science and
Courant Institute of Mathematical Sciences

Arthur S. Sherman
National Institutes of Health
Mathematical Research Branch
National Institute of Diabetes and
Digestive and Kidney Diseases

Gregory D. Smith
College of William and Mary
Department of Applied Science

John J. Tyson
Virginia Polytechnic Institute
and State University
Department of Biology

John M. Wagner
IBM TJ Watson Research Center
Functional Genomics and Systems Biology

Hongyun Wang
University of California at Santa Cruz
Department of Applied Mathematics
and Statistics

Graphic design by
Randy Szeto

# Contents

# PART I

# Introductory Course

# CHAPTER 1

# Dynamic Phenomena in Cells

**Christopher P. Fall and Joel E. Keizer**

Over the past several decades, progress in the measurement of rates and interactions of molecular and cellular processes has initiated a revolution in our understanding of dynamic phenomena in cells. Spikes or bursts of plasma membrane electrical activity or intracellular signaling via receptors, second messengers, or other networked biochemical pathways in single cells, or more complex processes that involve small clusters of cells, organelles, or groups of neurons, are examples of the complex behaviors that we know take place on the cellular scale. The vast amount of quantitative information uncovered in recent years, leveraged by the intricate mechanisms already shown to exist, results in an array of possibilities that makes it quite hard to evaluate new hypotheses on an intuitive basis. Using mathematical analysis and computer simulation we can show that some seemingly reasonable hypotheses are not possible. Analysis and simulations that confirm that a given hypothesis is reasonable can often result in quantitative predictions for further experimental exploration. Rapid advances in computer hardware and software technology combined with pioneering work giving structure to the interface between mathematics and biology have put the ability to test hypotheses and evaluate mechanisms with simulations within the reach of all cell biologists and neuroscientists.

## 1.1   Scope of Cellular Dynamics

Generally speaking, the phrase *dynamic phenomenon* refers to any process or observable that changes over time. Living cells are inherently dynamic. Indeed, to sustain

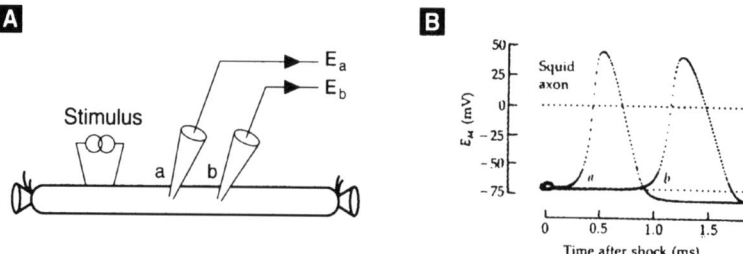

**Figure 1.1**  (A) Schematic diagram of the recording electrodes (a and b) used to detect action potentials following a stimulus shock in an isolated giant axon from squid. Adapted from Hille (2001). (B) The membrane potential recorded at electrodes a and b in the upper panel following a depolarizing shock. Reprinted from delCastillo and Moore (1959).

the characteristic features of life such as growth, cell division, intercellular communication, movement, and responsiveness to their environment, cells must continually extract energy from their surroundings. This requires that cells function thermodynamically as open systems that are far from static thermal equilibrium. Much energy is utilized by cells in the maintenance of gradients of ions and metabolites necessary for proper function. These processes are inherently dynamic due to the continuous movement of ionic and molecular species across the cell membrane.

Electrical activity of excitable cells is a widely studied example of cellular dynamics. The classical behavior of an action potential in the squid giant axon is shown in Figure 1.1. This single spike of electrical activity, initiated by a small positive current applied by an external electrode, propagates as a traveling pulse along the axonal membrane. Hodgkin and Huxley were the first to propose a satisfactory explanation for action potentials that incorporated experimental measurements of the response of the squid axon to depolarizations of the membrane potential. We will describe voltage gated ion channel models in Chapter 2.

Membrane transporters allow cells to take up glucose from the blood plasma. Cells then use glycolytic enzymes to convert energy from carbon and oxygen bonds to phosphorylate adenosine diphosphate (ADP) and produce the triphosphate ATP. ATP, in turn, is utilized to pump $Ca^{2+}$ and $Na^+$ ions from the cell and $K^+$ ions back into the cell, in order to maintain the osmotic balance that helps give red cells the characteristic shape shown in Figure 1.2. ATP is also used to maintain the concentration of 2,3-diphosphoglycerate, an intermediary metabolite that regulates the oxygen binding conformation of hemoglobin. The final products of glucose metabolism in red cells are pyruvate and lactate, which move passively out of the cell down a concentration gradient through specific transporters in the plasma membrane. Because red cells possess neither a nucleus nor mitochondria, they are not capable of reproduction or energetically demanding processes. Nonetheless, by continually extracting energy from the transformation of glucose to lactate, red blood cells maintain the capacity to shuttle oxygen and carbon dioxide between the lungs and the capillaries. Remarkably, this

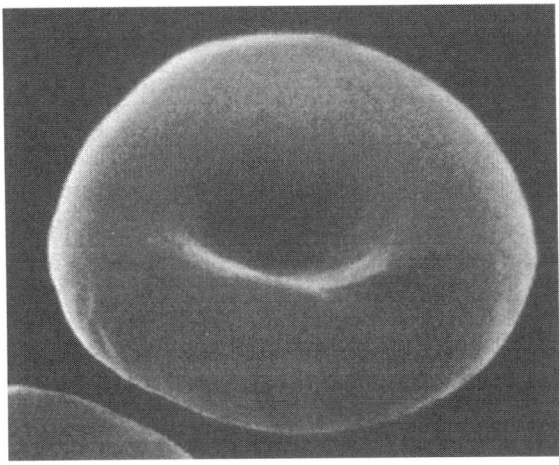

**Figure 1.2** A normal red blood cell with its characteristic discoid shape. The cell is approximately 5 $\mu$m in diameter. Reprinted from Grimes (1980).

efficient biochemical factory is only 5 $\mu$m across, with a volume of less than $10^{-14}$ L. In later Chapter 3 we will discuss models for the transport of glucose across the cell membrane.

Electrical activity and membrane transport are coupled cellular mechanisms. Experimental measurements of the membrane potential in pancreatic cells have revealed regular bursts of electrical activity corresponding to insulin secretion that is stimulated by increases in blood glucose levels. These oscillations occur at physiological levels of glucose, as shown in the microelectrode recordings from a pancreatic beta cell in an anesthetized rat in Figure 1.3. Recent work in vitro has shown that the rapid spikes of electrical activity, known as *action potentials*, are caused by rapid influx of $Ca^{2+}$ from the exterior of the beta cell followed by a slower efflux of $K^+$. The periods of rapid spiking are referred to as *active phases* of the burst, which are separated by intervals referred to as *silent phases*. A variety of mechanisms have been proposed to explain bursting behavior, and computer models of bursting were the first to predict that oscillations of $Ca^{2+}$ within the cytoplasm should occur in phase with the electrical activity. Oscillations in $Ca^{2+}$ were recorded for the first time in vitro eight years after they were predicted by a theoretical model (Chay and Keizer 1983). These oscillations are important physiologically, because cytoplasmic $Ca^{2+}$ plays a major role in triggering insulin secretion. This topic will be revisited in Chapter 5.

The control of cellular processes by interlocking molecular mechanisms can also produce spatiotemporal oscillatory $Ca^{2+}$ signals that are independent of electrical activity (Lechleiter and Clapham 1992). Figure 1.4 shows the spiral pattern of cytoplasmic $Ca^{2+}$ oscillations that occurs when an immature *Xenopus leavis* egg (an *oocyte*) is stimulated by a microinjection of inositol 1,4,5-trisphosphate (IP$_3$). IP$_3$ is a phospholipid membrane metabolite that is widely involved in signaling by receptors in the plasma membrane and that triggers release of $Ca^{2+}$ from the *endoplasmic reticulum* (ER). The

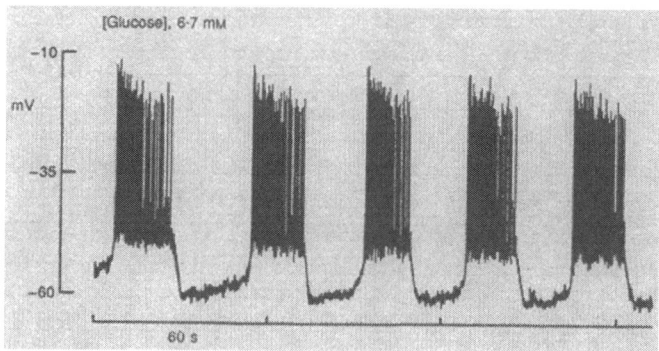

**Figure 1.3** Periodic bursts of electrical activity recorded in vivo from a pancreatic beta cell from the intact pancreas of an anesthetized rat. Reprinted from Sanchez-Andres et al. (1995).

ER is an intracellular compartment that functions, among other purposes, as a store for $Ca^{2+}$. The ER maintains an internal $Ca^{2+}$ concentration ($[Ca^{2+}]_{ER}$) that is comparable to that of the external medium (ca. 5 mM), whereas the cytoplasmic $Ca^{2+}$ concentration ($[Ca^{2+}]_i$) is typically 1000-fold smaller. The spiral waves of $Ca^{2+}$ in *Xenopus* oocytes can be explained quantitatively by kinetic models of the feedback mechanisms responsible for uptake and release of $Ca^{2+}$ from the ER. Simple models of regenerative $Ca^{2+}$ release that are solved on a spatial domain provide insight into the processes of *self organization* that result in spiral waves (Winfree 1987). These models are discussed in Chapter 8.

Circadian rhythms are regular changes in cellular processes that have a period of about 24 hours (from the Latin *circa*, about, and *dies*, day) and represent another dynamical phenomenon that is widely observed in cells. A great deal about the mechanisms of circadian rhythms has been uncovered in recent years, and circadian biology offers a rich source of unsolved modeling problems. Internal clocks provide an organism with the ability to predict changes in the environment. For example, flower opening and insect-egg hatching occur in advance of the rising sun (Winfree 1987). Cell division in *Euglenids* may also synchronize to light–dark cycles, as shown in Figure 1.5. The dark/light bands correspond to periods of absence and presence of light that simulate the normal dark/light cycle during a day. As shown in Figure 1.5, the growth rate of *Euglena* is temperature dependent, and cell division sychronizes to a 24–hour dark/light cycle only when the temperature is in the range found in its natural environment. At this temperature the population doubling time is close to 24 hours. Recent experiments with the fruit fly *Drosophila* and other organisms suggest that circadian rhythms like this are controlled by oscillations in gene transcription. Further consideration of circadian rhythms will be given in Chapter 9.

**Figure 1.4** A spiral wave of $Ca^{2+}$ ions detected as the bright fluorescence from an indicator dye after microinjection of $IP_3$ into an immature frog egg, the *Xenopus laevis* oocyte. Provided by Drs. James Lechleiter and Patricia Camacho, University of Texas Health Sciences Center.

The cell division cycle is the process by which a cell grows and divides into two daughter cells that can repeat the process. The eukaryotic cell cycle consists of a regular sequence of events as shown in Figure 1.6: chromosome replication during a restricted period of the cycle (S phase), chromosome segregation during metaphase and anaphase (M phase, or *mitosis*), and finally *cytokinesis*, in which two daughter cells separate. This cycle involves a cascade of molecular events that center on the proteins Cdc2 and cyclin, which make up a complex known as M phase promoting factor, or MPF. This complex has been shown to oscillate in synchrony with cell division and to be regulated by a series of phosphorylation and dephosphorylation reactions. Related dynamical changes occur during *meiosis*, in which germ line cells produce eggs and sperm. We will discuss models of the cell cycle progression in Chapter 10.

After DNA replication is complete, each chromosome consists of two "sister chromatids," which must be separated during mitosis so that each daugher nucleus gets one and only one copy of each chromosome. Segregation of sister chromatids during mitosis is another complex dynamical process that involves self-organizing structures in the cell that pull sister chromatids apart. This wonderfully coordinated dynamical behavior is just one of many examples of motile cellular processes. Other important examples include muscle contraction, cell movement, and projections of cell membrane called *pseudopodia*. Molecular motors will discussed at length in Chapter 12 and Chapter 13.

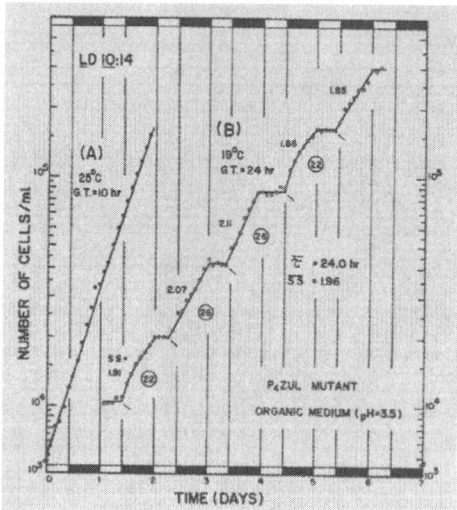

**Figure 1.5** Population growth of a mutant *Euglena* strain shows log growth at 25°C independent of the light/dark cycle, indicated by the alternating light/dark bands on the time axis. At 19°C a circadian (approximatly 24 hour) growth rhythm develops that entrains with the light/dark cycle. Reprinted from Edwards, Jr. (1988).

## 1.2 Computational Modeling in Biology

Even the simplest of the dynamic phenomena described in the previous section are exceedingly complex, and computer models have proven to be an important tool in helping to dissect the molecular processes that control their evolution in time. In the physical sciences, theoretical methods in combination with experimental measurements have for many years provided rich insights into dynamical phenomena. The abundance of quantitative experimental data now available at the cellular level has opened the door to similar collaborations in neurobiology and cell physiology.

### 1.2.1 Cartoons, Mechanisms, and Models

The interplay of experiment, theory, and computation follows a conceptual framework similar to that which has proven successful in the physical sciences:

1. Taking clues from experimental work, the first step is to sort through possible molecular mechanisms and focus on the most plausible ones. In most cases, this step requires close consultation with experimentalists working on the problem.
2. The selection of mechanisms defines the basis for a schematic representation, or *cartoon*, that depicts the overall model. To be useful, the cartoon should be explicit enough to be translated into a series of elementary steps representing the individual mechanisms.
3. Next, the basic laws of physics and chemistry can be used to translate the elementary steps of the mechanism into mathematical expressions.

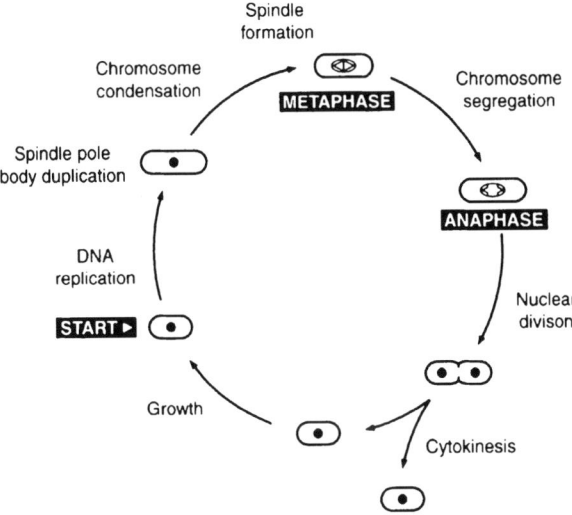

**Figure 1.6** The cell division cycle in fission yeast. Spindle formation initiates metaphase, where condensed chromosomes pair up for segregation into daughter cells. The cycle is completed when cytokinesis cleaves the dividing cell.

4. These expressions are then combined into time dependent differential equations that quantify the changes described by the whole model.
5. Finally, careful study of the differential equations reveals whether or not the cartoon is a successful model of the biological system.

The challenge of the theorist in biology then becomes similar to that in astrophysics or quantum mechanics: to analyze the equations, simplify them if possible, solve them, and, most importantly, make predictions that can be tested by further experiment. Further experiments may uncover inconsistencies in a model that will require changes. The process that we have outlined above and will revisit in later chapters is an iterative cycle of ever improving approximation where the mathematical or computer model plays the role of a quantitative hypothesis.

## 1.2.2  The Role of Computation

Advances in computer hardware and numerical analysis have made the solution of complicated systems of ordinary differential equations fast, accurate, and relatively easy. Indeed, the role of computation is critical because the differential equations describing biological processes nearly always involve control mechanisms that have nonlinear components. Simple linear differential equations often can be solved analytically, which means that we can obtain an exact solution using traditional mathematical methods. Nonlinearities often make it difficult or impossible to obtain an exact solution; however, we can obtain quite good estimates using numerical methods implemented on computers. Spatial variation is often an important feature in cellular mechanisms, so one is confronted with analyzing and solving spatially explicit *partial differential equations*, which can be still more complicated and less analytically tractable than ordinary differential equations.

Computer models permit one to test conditions that may at present be difficult to attain in the laboratory or that simply have not yet been examined by experimentalists. Each numerical solution of the differential equations can therefore provide a *simulation* of a real or potential laboratory experiment. These simulations can be used to help assess parameters, such as diffusion constants or kinetic constants, that may be difficult to measure experimentally. Numerical simulations can test how intervention by pharmacological agents might affect a process. With simulations one can test specific hypotheses about the role of individual mechanistic components or make predictions that can be tested in the laboratory. Often the most important result of a simulation is negative: A well–crafted model can rule out a particular mechanism as a possible explanation for experimental observations.

## 1.2.3   The Role of Mathematics

The scope of mathematical techniques employed to investigate problems in mathematical biology spans almost all of applied mathematics. The modeling of processes is discussed in detail here, but only the basics of the mathematics and the elementary tools for the analysis of these models are introduced. Rigorous mathematics plays at least three important roles in the computational modeling of cell biology. One role is in the development of the techniques and algorithms that make up the tools of *numerical analysis*. In its essence, the computation of solutions to mathematical problems on computers is fundamentally a process of estimation, and the accuracy and efficiency of these methods of estimation are the subjects of much study. We will introduce briefly a few ideas from numerical analysis at the end of this chapter and in appropriate places throughout the rest of this book.

The process of developing model mechanisms that we described above is also fundamentally a process of approximation due to the simplifications that must be introduced to produce a useful model. Not only must these simplifications make sense in terms of the physical process being studied, but they must also be valid from a mathematical standpoint. We will learn the basics about more mathematical concepts such as the reduction of scale and stochastic methods in later chapters.

It is one thing to solve the differential equations that result from the formulation of a model, but another thing to understand why a model behaves as it does. Mathematicians have developed techniques and tools for the analysis of systems of differential equations that describe complex interrelated processes. These tools reveal the structure, properties, and dynamical behavior of the system much as anatomical, physiological, and molecular biological techniques reveal the physical basis of the model. In particular, such analysis reveals details about behaviors in a model such as the os-

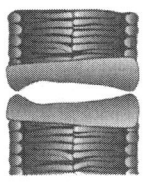
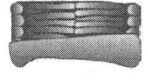
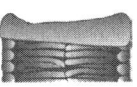

Closed          Open

**Figure 1.7** Mechanistic cartoon of a gated ionic channel showing an aqueous pore that is selective to particular types of ions. The portion of the transmembrane protein that forms the "gate" is sensitive to membrane potential, allowing the pore to be in an open or closed state.

cillations and other complex behaviors that often motivate study of such biological phenomena. These techniques are covered in several chapters and in the appendices.

Rigorous analysis of complicated differential equations requires specialized training, because there are many subtleties that are appreciated only with experience. Similarly, choosing proper numerical methods and selecting valid simplifications requires caution. While the creation and manipulation of simple models is within the reach of all cell biologists, the careful scientist will seek collaborations with experienced mathematicians, particularly for the valid simplification of complicated models into more tractable ones. In the middle ground between established disciplines such as biology and mathematics, fruitful scientific work can be done, and all parties gain valuable insight from the interdisciplinary experience.

## 1.3 A Simple Molecular Switch

In this section we illustrate, with a simple model of channel conformation, the kinds of phenomena that are investigated in detail in subsequent chapters. We introduce some of the methods underlying the analysis of these models and also try to demonstrate the basic conceptual modeling framework utilized throughout the book.

We begin with a simple channel because it is an intuitively clear example of transition between different molecular states corresponding to different conformations of a macromolecule. Let us be clear that we are modeling only proteins that are switching between an "open" state and a "closed" state and nothing more at this point. The simplest cartoon of gating is a channel with two states, one with the pore open and the other with it closed, corresponding to the mechanism shown in Figure 1.7. This kinetic "cartoon" is easily translated into a conventional kinetic model of the sort often employed in biochemistry.

The model takes the form of the diagram in (1.1). Diagrams like this, which will be used extensively in this book, represent molecular states or entities by symbols and transitions between states by solid lines or arrows:

$$C \underset{k^-}{\overset{k^+}{\rightleftharpoons}} O. \tag{1.1}$$

The $C$ in (1.1) corresponds to the closed state, the $O$ corresponds to the open state of the channel, and the arrows represent elementary molecular processes. These states represent a complex set of underlying molecular conformations. The transitions between $C$ and $O$ are *unimolecular* processes because they involve only the channel molecule (*bimolecular* processes will be introduced in Chapter 3). An important aspect of transitions between molecular states is that they are reversible, which is a consequence of microscopic reversibility of molecular processes.

Using special techniques, transitions between closed and open states can be measured for single channel molecules. However, here the focus is on the average change for a collection of channels. Because it is not unusual to have several thousand channels of a given type in the plasma membrane of a cell, the average behavior of the entire ensemble of channels is often what determines the cellular dynamics.

The rate of an elementary process in a kinetic diagram is determined by the so-called *law of mass action*. (Despite the name, mass action is not technically a physical law, but rather is a constitutive relation that holds as a very good approximation for any well–mixed system.) This "law," which dates back to the early studies of chemical kinetics, states that the rate of a process is proportional to the product of the concentrations of the molecular species involved in the process. If we define $k^+$ as the proportionality constant or the *rate constant*, the rate of the transition from state $C$ to state $O$ is given by $J_+ = k^+[C]$, where the square brackets denote concentration, with $[C]$ representing the concentration of channel molecules in state $C$. In this case $k^+$ is unimolecular, with practical units of $s^{-1}$. Similarly, the rate of the reverse reaction, $C \leftarrow O$, is given by $J_- = k^-[O]$ with the unimolecular rate constant $k^-$.

To translate the mechanism in (1.1) into an equation, the law of mass action is applied to the concentration of channels in states $C$ and $O$. For cellular mechanisms, a variety of measures of "concentration" can be used. For example, if the channels are in intact cells, concentration is often expressed in terms of total cell volume. Another measure in common use involves total weight of protein in a sample. The total number of transporters is useful for single cells. Here we choose the latter to define concentration so that $[O] = f_O = N_O/N$, where $f_O$ will refer to the fraction of open channels, and $N$ and $N_O$ the total number of channels and open channels, respectively (similarly, $f_C$ and $N_C$ refer to the closed state).

Because the kinetic model involves only interconversion of channel states, the total number of channels should be preserved. This introduces the idea of a conservation law, $N_C + N_O = N$, which states mathematically that channels are neither created nor destroyed. Using conservation relations, one of the dependent variables can be eliminated because $N_C = N - N_O$. The differential equation for $N_C$ therefore becomes redundant, and the number of differential equations to be solved is reduced to only one along with the algebraic equation for $N_C$. The fraction of channels in the closed state is therefore $1 - f_O$.

Having established the correspondence of the diagram with rate expressions, it is easy to write down the differential equations that the diagram represents. To do so one must keep track of the change that each elementary process in the diagram makes for each state, which we refer to as a *flux*. Thus the process connecting states $O$ and $C$ causes a loss of state $O$ in the reverse direction and again in the forward direction. These small whole numbers that correspond to losses or gains of a state (e.g., $-1$ for state $C$ in the process $C \to O$) are called the *stoichiometric coefficients* for the mechanism. Using the coeficients in conjunction with the kinetic diagram, the ordinary differential equations follow for the rate of change in the states:

$$\text{flux } O \to C = j_- = k^- f_O, \tag{1.2}$$

$$\text{flux } C \to O = j_+ = k^+(1 - f_O). \tag{1.3}$$

The difference between the two fluxes represents the change in $f_O$ over time:

$$\begin{aligned}
\frac{df_O}{dt} &= j_+ - j_- \\
&= -k^- f_O + k^+(1 - f_O) \\
&= -\left(k^- + k^+\right)\left(f_O - \frac{k^+}{(k^- + k^+)}\right).
\end{aligned} \tag{1.4}$$

Defining $\tau = 1/(k^- + k^+)$ and $f_\infty = k^+/(k^- + k^+)$, we have

$$\frac{df_O}{dt} = \frac{-(f_O - f_\infty)}{\tau}. \tag{1.5}$$

The example that we have presented here shows the basic framework under which all of the models in this book are developed and understood. With the formulation of the model equations, the first steps in the modeling process are completed. What remains is the analysis of the equations.

## 1.4  Solving and Analyzing Differential Equations

Many students have worked with differential equations in their studies of the physical sciences or elementary mathematics, and they may have been introduced to solution

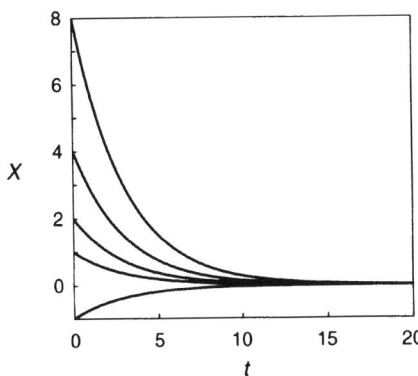

**Figure 1.8** A selection from the family of solutions to (1.6) for $\tau = 2$.

techniques explicitly in an advanced calculus course. In general, the differential equations that arise for the rate of change in cellular properties will be complicated and difficult or impossible to solve exactly using analytical techniques. One example that can be solved is the simple linear equation

$$\frac{dX}{dt} = \frac{-X}{\tau}. \tag{1.6}$$

In order to solve (1.6), we seek a function $X(t)$ whose derivative is proportional to itself with proportionality constant $1/\tau$. Remembering that the derivative of the exponential function is still an exponential function, we guess that the exact solution to (1.6) is

$$X = Ce^{-t/\tau}. \tag{1.7}$$

It can be verified that (1.7) is indeed a solution of (1.6) by differentiating the solution with respect to $t$, thereby recovering our original equation

$$\frac{dX}{dt} = \frac{d}{dt}(Ce^{-t/\tau}) = -\frac{1}{\tau}(Ce^{-t/\tau}) = \frac{-X}{\tau}. \tag{1.8}$$

Because a differential equation describes how a variable changes for all time, the solution for a differential equation has to prescribe both how a variable evolves and at what value it begins. Thus there is a family of solutions to a differential equation, and the correct one is chosen by specifying an *initial condition*. This is an important concept that is particularly relevant to the numerical solution of differential equations. We choose a *particular solution* to (1.6) from the *general solution* given by (1.7) by adjusting the constant $C$ such that the initial condition $X = X(0)$ is satisfied at time $t = 0$, or

$$X(0) = Ce^{-0/\tau}, \tag{1.9}$$

which means that $C = X(0)$. Together this gives the particular solution

$$X = X(0)e^{-t/\tau} \tag{1.10}$$

for the initial condition $X(0)$. Representatives from the family of exact solutions for different values of $X(0)$ are shown in Figure 1.8 for $\tau = 2$.

A deep understanding of the rules of differentiation and integration from calculus is not required for the level of modeling and mathematics that will be encountered in the first half of this book. The exponential function is encountered frequently, and so a review of its properties as given in elementary calculus textbooks is advisable. It is also helpful when analyzing experimental literature to understand the difference between the time constant $\tau$ and the half-time $t_{1/2}$ in the context of the exponential function. The half-time is the time at which the value of the function decays to $\frac{1}{2}$ of the initital value, or the solution of

$$X(0)e^{-t/\tau} = \frac{X(0)}{2}, \tag{1.11}$$

or

$$e^{-t/\tau} = \frac{1}{2}, \tag{1.12}$$

which is $t = \tau \ln 2$. In contrast, after one time constant or at time $t = \tau$, the value of $X$ is given by

$$X = X(0)e^{-\tau/\tau} = X(0)e^{-1} \approx X(0) \cdot 0.37 \tag{1.13}$$

Note that these principles apply to growing as well as decaying functions.

The solution $X(t)$ always "relaxes" or decays to zero at long times. (Note that to mathematicians an equation *decays* to steady state whether it approaches a value greater or smaller than the initial value.) An equation related to the simple exponential decay equation given in (1.6) is the exponential approach to a steady state other than zero. The single channel model given in (1.5) describes exponential decay (or growth) to a steady state fraction of open channels, $f_\infty$. We rewrite it here to aid the presentation:

$$\frac{df_O}{dt} = \frac{-(f_O - f_\infty)}{\tau}. \tag{1.14}$$

It is easy to solve (1.14) if we make a *variable substitution*. If we create a new variable $Z$ and define it as

$$Z = f_O - f_\infty, \tag{1.15}$$

then because $f_\infty$ is a constant, and therefore

$$\frac{dZ}{dt} = \frac{df_O}{dt}, \tag{1.16}$$

we can rewrite (1.14) as

$$\frac{dZ}{dt} = \frac{Z}{\tau}. \tag{1.17}$$

We saw in (1.10) that the solution to this equation is

$$Z = Z(0)e^{-t/\tau}. \tag{1.18}$$

Recalling that $Z = f_O - f_\infty$ (and therefore that $Z(0) = f_O(0) - f_\infty$, we can once again exchange the variables and arrive at the analytic solution to (1.14) for a given initial condition $f_O(0)$:

$$f_O(t) = f_\infty + (f_O(0) - f_\infty)e^{-t/\tau}. \tag{1.19}$$

## 1.4.1 Numerical Integration of Differential Equations

Even if (1.14) were more complicated and could not be solved exactly, a numerical approximation could still be calculated. The simplest and perhaps the oldest method of numerical solution goes back to the mathematician Euler and is easy to understand. The method is called the *forward Euler* method and it is a prototype for all other

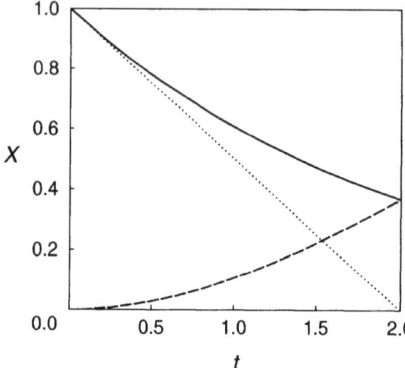

**Figure 1.9** The forward Euler method of numerical integration relies upon a series of short linear approximations using the derivative at the old time point. The solution to (1.6) with $\tau = 2$ is shown in solid, the linear approximation using the derivative at $t = 0$ ($-0.5$) is shown in dots, and the difference between these two curves is shown in dashes. Note that by $t = 2$, the magnitude of error between the actual and approximate functions is equal to the value of the actual function.

methods of solving ODEs numerically. Consider an approximation to the derivative in (1.6)

$$\frac{dX}{dt} \approx \frac{\Delta X}{\Delta t} = \frac{X(t + \Delta t) - X(t)}{\Delta t}, \tag{1.20}$$

where $\Delta X$ and $\Delta t$ are small, but not infinitesimal like the differentials $dX$ and $dt$. If this approximation to the derivative is substituted into the differential equation, the resulting equation can be solved for $X(t + \Delta t)$, giving

$$\frac{dX}{dt} = \frac{-X(t)}{\tau},$$

$$\frac{X(t + \Delta t) - X(t)}{\Delta t} = \frac{-X(t)}{\tau},$$

$$X(t + \Delta t) = X(t) - \frac{X(t)\Delta t}{\tau}. \tag{1.21}$$

The smaller $\Delta t$ is, the better the Euler approximation of the derivative will be. Also, because the Euler approximation gives a piecewise linear estimate of the solution, the further from linear the problem is, the smaller $\Delta t$ must become to give an accurate solution (see Figure 1.9). The essence of numerical integration is that we start at some value and crawl along the solution in increments of $\Delta t$ by estimating the change over that interval. If $\Delta t$ is very small, our estimate of the rate of change is good and our solution is accurate, but it may take a very long time to compute the solution. This is termed *computationally expensive*, because it either requires a faster computer or a longer time to run.

Two solutions to (1.19) obtained by integrating the equation using the Euler method are shown in Figure 1.10A. The time step was chosen to be $dt = 0.03$, and two different initial conditions were used, $f_0(0) = 1$ and $f_0(0) = 0$. Independent of the initial condition, $f_0(t)$ relaxes to its *steady state* value $f_\infty = 0.5$. It corresponds to the point where the rate vanishes, as can be seen graphically in Figure 1.10A. The rate at which the steady state is approached depends on the value of $\tau$, which is 3 in these simulations.

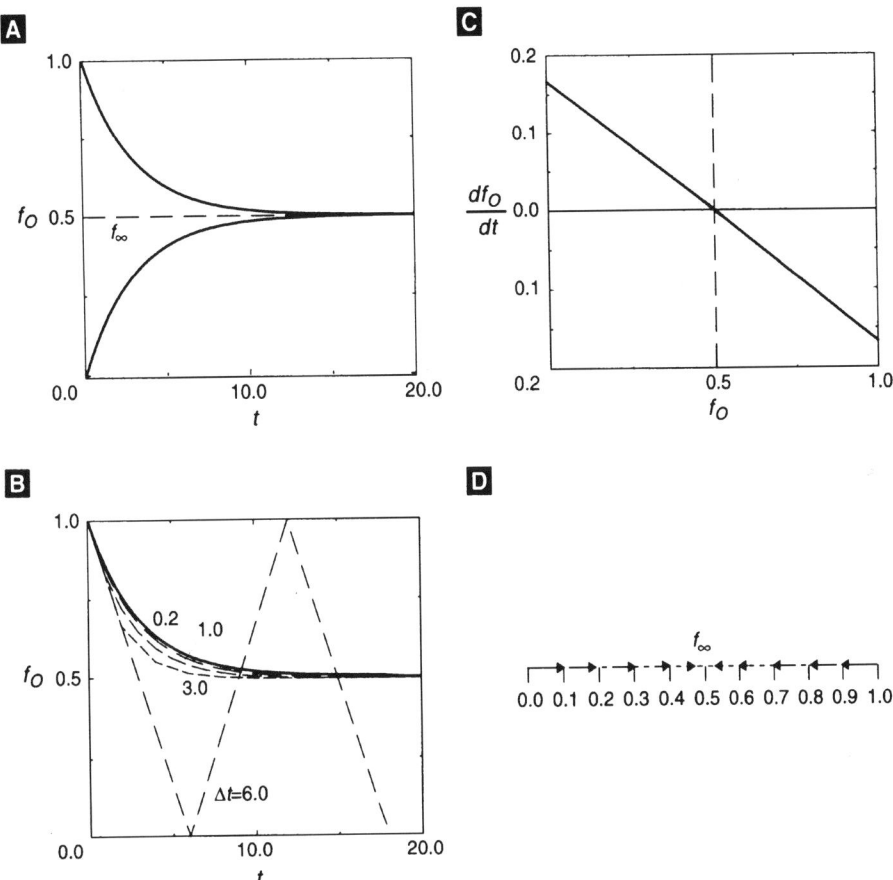

**Figure 1.10** (A) The exponential decay of the open fraction of channels. Initial conditions at either 1 or 0 both decay to the steady–state value of 0.5. (B) The effect of the step size in the Euler method for the simulation in panel A starting with the initial condition $f_O(0) = 1$. The exact solution, $f_O(t) = 0.5(1 + \exp(-t/3))$, is given by the full line. (C) Plot of the rate of change of $n$ as a function of $n$. (D) Phase portrait with the arrows representing the direction and relative magnitude of the rate for each value of $f_O$. All the arrows point toward the steady state, $f_\infty = 0.5$.

The solution to the equation in the Euler method depends on step size as shown in Figure 1.10B. Only step sizes that are more than an order of magnitude smaller than the value of $\tau$ do a reasonable job of approximating the exact exponential solution, which for the parameter values used is $f_O(t) = 0.5(1 + \exp(-t/3))$. Unreasonably large step sizes like $\Delta t = 6$ give approximations that are not even close to the exact solution. In fact, the numerical method has become *unstable*, and the computed solution oscillates around the true solution.

There are many other methods of numerical integration that give better approximations to the derivative. These methods are generally more complicated, but have fewer restrictions on $\Delta t$. These more complicated methods also address some other important problems such as *numerical stability*. There are many fine texts on numerical analysis that discuss these issues and explain the various advantages and disadvantages of each method. As mentioned above, however, the best way to ensure an optimal and valid means of solution is to collaborate with a mathematician who has experience in scientific computation or numerical analysis.

There is a different way to plot the results of solving the differential equation that frequently gives insight into the properties of the solution. This is demonstrated in Figure 1.10C, where the function $df_O/dt = -(f_O - f_\infty)/\tau$ is plotted versus the value of $f_O$ for the two initial conditions in Figure 1.10A. Since $f_O$ is restricted on physical grounds to be between 0 and 1, the plot shows that $f_\infty$ is the unique steady state by making it clear that when $f_O > 0.5$, $n$ decreases with time because $df_O/dt < 0$, and when $f_O < 0.5$, $f_O(t)$ increases with time.

To further emphasize this concept, arrows in Figure 1.10D show the direction in which $f_O$ is changing. This type of plot is called a *phase portrait*, in one dimension. Phase portraits are particularly useful for analyzing ODEs with two variables, where they are typically called *phase plane* diagrams. Phase plane diagrams are discussed in Appendix A. Because phase portrait diagrams will be used extensively in the remainder of the text, it would be a useful digression to review that material now.

## 1.4.2  Introduction to Numerical Packages

While it is important to understand the limitations of whichever numerical algorithm is used for the solution of a problem, fortunately it is not necessary to face the task of implementing these algorithms on a computer from scratch. Several excellent software programs have been developed that not only solve ODEs, but represent solutions graphically and allow their dynamical properties to be analyzed. These packages include very sophisticated commercial mathematical packages such as Matlab and Mathematica, which can often be obtained at reduced cost in the form of student versions. There are also myriad university produced packages such as Berkeley Madonna that are designed to solve ODEs. A new direction for computational cell biology is the creation of several highly sophisticated packages such as The Virtual Cell, which is an integrated database and computational system expressly designed for cell biology modeling.

Among the best for our purposes here is a public domain package, XPPAUT, that has been developed by Bard Ermentrout at the University of Pittsburgh (Ermentrout 2002). The name of the program evolved from a DOS version that was called *Phase-Plane*, refering to the program's ability to carry out phase plane analysis. A version that ran in X-windows under Unix or linux was then developed and was called X-PhasePlane (or XPP for short). Finally, when the automatic bifurcation tool AUTO developed by E. Doedel, was added it became X-PhasePlane-Auto, or XPPAUT. XPPAUT is an excellent tool for solving and analyzing ordinary differential equations, one that provides very sophisticated tools without automating the process too much to be useful for our didactic objectives. Moreover, XPPAUT is free and available for both Unix/Linux, Windows, and MacOSX. We adopt it as the basic software program for use with this text.

# Suggestions for Further Reading

At the end of each chapter will be listed several sources for further reading, together with short descriptions. Because Chapter 1 serves as an introduction to the whole book, we have listed here several sources that might serve as companions to the whole book. Because of their applicability to more than one chapter, those listed here may be listed again later.

- *Modeling Dynamic Phenomena in Molecular and Cellular Biology*, Lee Segel. A great place to start for a more mathematical treatment of some of the contents of this book (Segel 1984).
- *Mathematical Models in Biology*, Leah Edelstein-Keshet. A classic introductory textbook for general mathematical biology, and a good source for a different perspective on topics such as time scales, phase plane analysis, and elementary numerical analysis as applied to biological problems (Edelstein-Keshet 1988).
- *Understanding Nonlinear Dynamics*, Daniel Kaplan and Leon Glass, and *Nonlinear Dynamics and Chaos*, Steven Strogatz. Extremely readable entry–level books on nonlinear dynamics, including sections on chaos, fractals, and data analysis (Kaplan and Glass 1995; Strogatz 1994).
- *Mathematical Physiology*, James Keener and James Sneyd. Keener and Sneyd treat many of the topics presented in this book from a more analytic perspective as opposed to the computational focus presented here (Keener and Sneyd 1998).
- *Computer Methods for Ordinary Differential Equations and Differential-Algebraic Equations*, U.M Acher and L.R. Petzold (Asher and Petzold 1998).
- *Cellular Biophysics*, Volumes 1 and 2, Thomas Weiss. These two volumes cover in more detail the biophysics of transport processes and electrical properties in cells (Weiss 1996).
- *Simulating, Analyzing, and Animating Dynamical Systems: A Guide to XPPAUT for Researchers and Students*, by Bard Ermentrout. A complete user's manual for the public domain ODE package XPP (Ermentrout 2002).
- *Mathematical Biology*, James Murray. While not especially didactic, this volume is recognized as an essential handbook describing models throughout mathematical biology (Murray 1989).

## 1.5  EXERCISES

1. Find the XPPAUT site on the World Wide Web (see Appendix B). Install XPPAUT on your computer and follow the web-based tutorial.

2. Verify by differentiation (or by integration) that (1.7) is the solution to (1.6) and that (1.19) is the solution to (1.14).

3. Create a program that is suitable for solving the ODE in (1.14) and explore what happens to the solution when the intial condition and characteristic time $\tau$ are changed. Compare to the analytical solution.

4. Suggest three processes in biology not listed here that might be modeled on the computer.

5. Explain how the list of steps in Section 1.2.1 is carried out in the example begun in Section 1.3.

# CHAPTER 2

# Voltage Gated Ionic Currents

## Christopher P. Fall and Joel E. Keizer

In Chapter 1 we introduced models of simple channel behavior but ignored the idea that something might flow through such a channel. In this chapter we will learn how to model current flowing through ion channels and how to include the feedback regulation of the channel behavior by voltage. But let us be very clear about what we are modeling in this chapter. Only in the most idealized sense are we modeling the behavior of excitable cells like neurons. In fact, we will be creating *point models* of patches of membrane, with only implicit consideration through the use of ion concentrations of the idea that the membrane encloses a compartment. We do not consider at all that a neuron is a highly complicated cell that integrates signals within *dendrites* and a *soma* before producing *action potentials* in an excitable *axon*. Some of the diverse neuronal shapes are shown in Figure 2.1.

Electrophysiology is the study of ionic currents and electrical activity in cells and tissues. Because this field has its roots in classical physics, traditionally it has been the most quantitative field in cell physiology. The groundbreaking work of the physiologists Hodgkin and Huxley and others in elucidating the mechanism of action potentials in the squid giant axon before and after the Second World War was the first major breakthrough of dynamical modeling in physiology. In the latter half of the twentieth century, the introduction of the patch-clamp technique established firmly that ionic currents are carried by proteins that act as gated ionic pores. More recently, genetic engineering techniques have been employed to clone, modify, and characterize the gating mechanisms of many types of channels (Hille 2001).

Vagal Motoneuron            Olivary Neuron            Layer 2/3 Pyramidal Cell

Layer V Pyramidal Cell        Purkinje Cell                    α-motoneuron

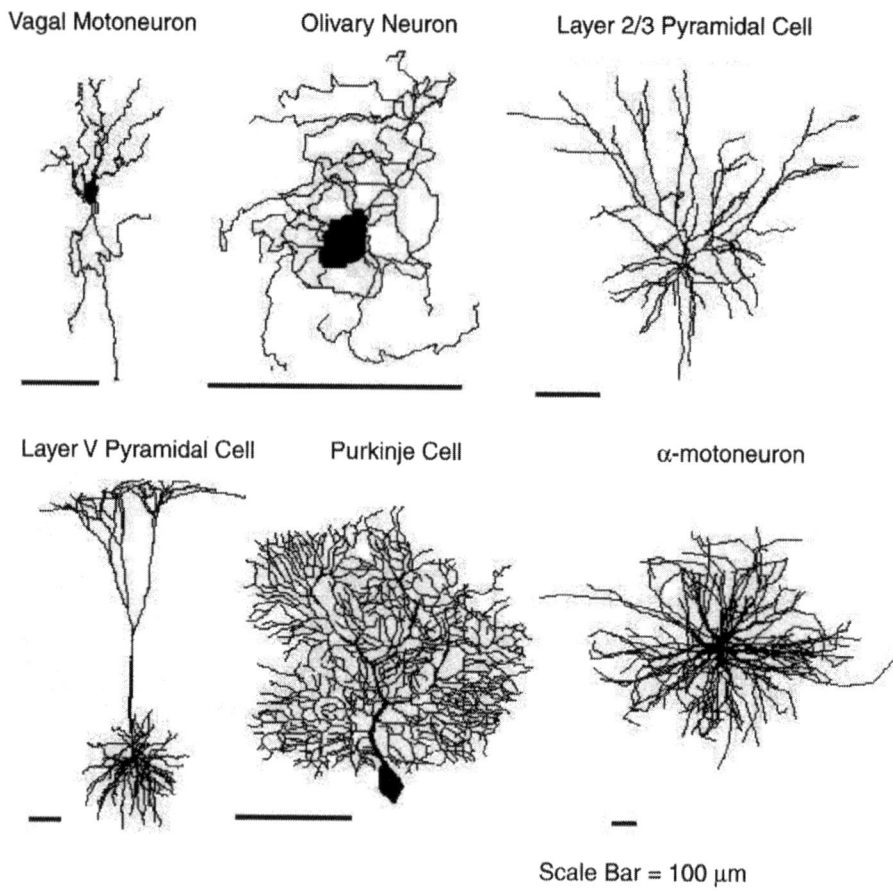

Scale Bar = 100 μm

**Figure 2.1**  Examples of the diverse shapes of mammalian neurons. Reprinted from Koch and Segev (2000).

In this chapter we focus on voltage gated ionic currents. We begin by reviewing the basic concepts of electrical behavior in cells. Next, we describe classical activation and inactivation kinetics and how the voltage clamp technique can be used to study these currents. We use the Morris–Lecar model for action potentials in the giant barnacle muscle to illustrate how voltage gated channels can interact to produce oscillations and action potentials. The Morris–Lecar model is nonlinear but involves only two variables. With only two variables, we can analyze the dynamics of the equations for this model using phase plane techniques. For completeness, we close with brief introductions to the Hodgkin–Huxley model of the squid giant axon and FitzHugh–Nagumo models.

**Table 2.1   Important Definitions in Electrophysiology**

| Definition | Abbreviation | Value |
|---|---|---|
| Avogadro's number | N | $6.02 \cdot 10^{23}$ /mol |
| Faraday's constant | F | $9.648 \cdot 10^4$ C/mol |
| elementary charge | e | $1.602 \cdot 10^{-19}$ C |
| gas constant | R | 8.315 J/(mol · K) |
| joule | J | $1 \text{ V} \cdot \text{C}$ |
| volt | V | 1 J/C |
| ampere | A | 1 C/s |

## 2.1   Basis of the Ionic Battery

The electrical behavior of cells is based upon the transfer and storage of charge. We are used to thinking about electricity as the movement of electrons, but current can be carried by any charged particle, including ions such as $K^+$, $Na^+$, and $Ca^{2+}$ in solution. As we will see, the ability of cells to generate electrical signals is entirely dependent on the evolution of ion–specific pumps and pores that allow the transfer of charge up and down gradients. Ion pumps use energy in the form of ATP to transport ions against a concentration gradient.

Before we begin our discussion of the ionic battery, it is usefull to recall several important definitions from elementary physics listed in Table 2.1. An ion's *valence* is the number of charges, plus or minus, that it carries. A given number of *divalent* $Ca^{2+}$ ions would carry twice the amount of charge as the same number of *univalent* $K^+$ ions.

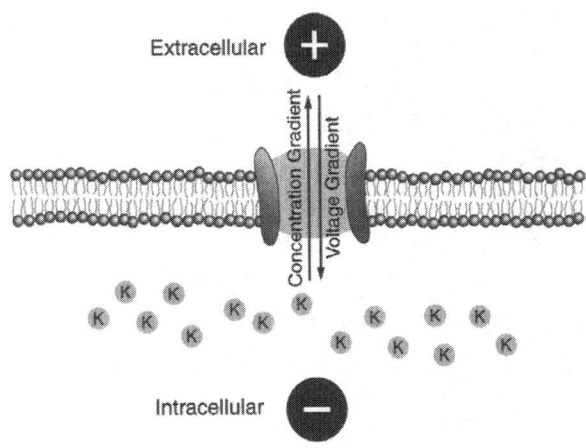

**Figure 2.2**   The basis of the ionic battery. Adapted from McCormick (1999).

Using these relationships, we see that the transfer of 1 mole of K$^+$ ions in a period of one second would carry a current equal to Faraday's constant. It requires one joule of energy (i.e. ATP consumed by ion pumps) to separate one coulomb of charge across one volt of potential (which is the definition of volt).

## 2.1.1  The Nernst Potential: Charge Balances Concentration

Biological fluids such as cytoplasm and extracellular fluid contain numerous ions. Consider the case where the two ions K$^+$ and any monovalent anion A$^-$ are in solution such that the concentration is different across the impermeable membrane but the two ions are equal in concentration on the same side of the membrane. As shown in Figure 2.3A, before we make any changes, there is no potential difference across the membrane because the charge between the K$^+$ ions and the A$^-$ ions is balanced on each side due to the equivalent concentrations. As shown in Figure 2.3B, if we insert a nonselective pore into the membrane, concentration and charge equilibrate such that there are equal concentrations of each ion on both sides of the membrane, and the voltage across the membrane is again zero.

It is when we insert into the membrane an ion-selective pore that allows only the passage of K$^+$ that the phenomenon shown in Figure 2.3C occurs. Because [K$^+$] is greater on one side of the membrane, K$^+$ ions diffuse through the K$^+$ pore down the concentration gradient. Because the membrane is not permeable to the anion A$^-$, each K$^+$ ion that passes down the concentration gradient carries a positive charge that is not balanced by an accompanying A$^-$. Because the transfer of these charges establishes

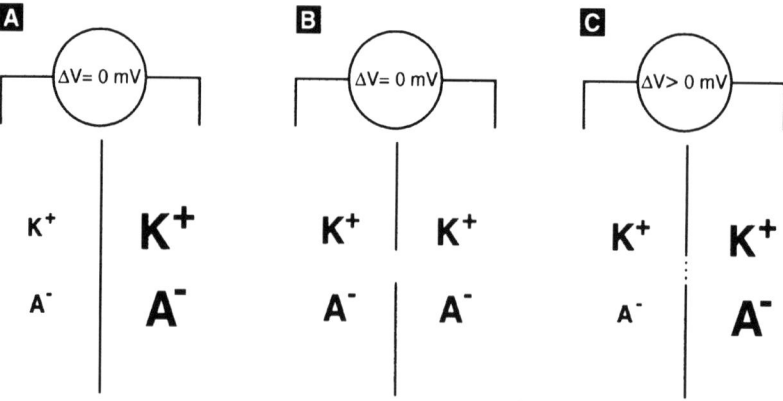

**Figure 2.3**  (A) Concentration and charge are balanced on each side of the membrane, so there is no $\Delta V$ across the membrane. (B) Due to the nonselective pore, charge and concentration are balanced everywhere, and so there is no $\Delta V$ across the membrane. (C) A K$^+$ selective pore allows K$^+$ but not A$^-$ to pass through the membrane. K$^+$ moves to equilibrate concentration until counterbalanced by the accumulating negative charge, because A$^-$ cannot move. The "voltmeter" is seeing excess + charge on the left side and excess − charge on the right side.

an electrical potential gradient, $K^+$ ions continue to move from high concentration to low concentration until the growing force due to the electrical potential difference is balanced by the (opposite) force generated by the concentration difference.

The *equilibrium potential*, where the electrical and osmotic forces are balanced, is given by the *Nernst equation*. The Nernst equation is derived from the expression for the change in Gibbs free energy when one mole of an ion of valence $z$ is moved across a membrane:

$$\Delta G = -RT \ln \frac{[\text{ion}]_{\text{out}}}{[\text{ion}]_{\text{in}}} + \Delta V F z. \tag{2.1}$$

At equilibrium, $\Delta G$ is zero. Rearranging gives us the Nernst potential:

$$\Delta V = V_{\text{Nernst}} = \frac{RT}{zF} \ln \frac{[\text{ion}]_{\text{out}}}{[\text{ion}]_{\text{in}}}$$
$$= 2.303 \frac{RT}{zF} \log_{10} \frac{[\text{ion}]_{\text{out}}}{[\text{ion}]_{\text{in}}}$$
$$= \frac{61.5}{z} \log_{10} \frac{[\text{ion}]_{\text{out}}}{[\text{ion}]_{\text{in}}} \ (\text{at } 37°C)$$
$$\tag{2.2}$$

where $R$ and $F$ are given in Table 2.1, $T$ is temperature (in kelvin), and $z$ is the valence of the ion as previously defined. At body temperature, $RT/F$ is approximately 60 mV. Therefore, a 10–fold difference in the concentration of a monovalent ion like $K^+$ would result in approximately -60 mV of potential difference across a membrane ($[K^+]$ is greater inside a neuron). Because the Nernst potential represents the equilibrium of the thermodynamic system, the potential difference evolves to that given by the Nernst equation regardless of the initial starting potential. This tendency for the system to move toward the equilibrium potential is the basis of the *ionic battery* used in the modeling of electrophysiological phenomena. In electrophysiology, the equilibrium potential is called the *reversal potential*, because departure from that point of zero current flux results in the positive or negative flow of ions .

---

**Table 2.2   Resting Ion Concentrations. From McCormick (1999).**

| Ion | Cytoplasmic Concentration (mM) | Extracellular Concentration (mM) |
|---|---|---|
| **Squid Giant Axon** | | |
| $K^+$ | 400 | 20 |
| $Na^+$ | 50 | 440 |
| $Cl^-$ | 40 | 560 |
| **Mammalian Neuron** | | |
| $K^+$ | 135 | 3 |
| $Na^+$ | 18 | 145 |
| $Cl^-$ | 7 | 120 |

## 2.1.2   The Resting Membrane Potential

The Nernst potential is the equilibrium potential for one permeant ion. In reality, no channel is perfectly selective for a given ion, and there are various channels selective for various ions in a given cell as well. The Goldman–Hodgkin–Katz (GHK) equation is related to the Nernst equation, but considers the case where there are multiple conductances. The GHK equation determines the resting membrane potential of a cell from a weighted sum of the various conductances:

$$V_{GHK} = \frac{RT}{F} \ln \frac{P_K[K^+]_{out} + P_{Na}[Na^+]_{out} + P_{Cl}[Cl^-]_{in}}{P_K[K^+]_{in} + P_{Na}[Na^+]_{in} + P_{Cl}[Cl^-]_{out}} \tag{2.3}$$

where $P_i$ is the relative permeability for ion $i$, which must be determined experimentally. While it looks like a straightforward extension of the Nernst equation, the GHK equation requires assumptions about both the interaction of ions and their ability to diffuse within channels.

We are not going to deal with these complicating details here, and in addition, we will assume perfect selectivity for our ion channels. If we define the *conductance, g* = $1/R$ as the reciprocal of the resistance, we can use a similar weighted–sum formalism for calculating the membrane potential of a cell:

$$V_m = \frac{\sum_i (V_i \cdot g_i)}{\sum_i g_i}, \tag{2.4}$$

where $V_i$ is the Nernst–equation–derived reversal potential for ion $i$ calculated using (2.2). For example, the membrane potential for a cell containing $Na^+$, $K^+$, and $Cl^-$ ions would be

$$V_m = \frac{(V_{Na} \cdot g_{Na}) + (V_K \cdot g_K) + (V_{Cl} \cdot g_{Cl})}{g_{Na} + g_K + g_{Cl}}. \tag{2.5}$$

This is the expression for the resting membrane potential that we will use throughout the remainder of the book. Because the resting membrane potential is the weighted average of Nernst potentials for the various ions, the ion with the greatest permeability contributes the most (see Exercise 1c).

The Nernst potential represents an equilibrium between electrical and osmotic forces on an ion across a membrane: It is what the membrane potential would be if a particular ion were at equilibrium across a membrane. Each ion has its own characteristic Nernst potential given by equation (2.2), which is determined by its concentration gradient across the membrane. The actual membrane potential is some "average" of the Nernst potentials of the several ions that are able to cross the membrane; each ion contributing to the actual membrane potential according to the permeability (or conductance) of the membrane for that particular ion. The resting potential across a membrane given by (2.5) is a steady state in which the currents carried by different ions across the membrane cancel each other out. The GHK potential is a steady state solution of a different kinetic equation for total current, based on different assumptions about how ions cross the membrane.

## 2.2 The Membrane Model

We know from Ohm's law that current flows down a voltage gradient in proportion to the resistance in the circuit. Current is therefore expressed as

$$I = \frac{V}{R} = gV. \tag{2.6}$$

To a first approximation, our task in modeling electophysiological phenomena is to describe how the conductance of the membrane to various ions changes with time and then to keep track of the changes in current and voltage that result.

The conceptual idea behind contemporary electrophysiological models originates in the work of K.S. Cole, who pioneered the notion that cell membranes could be likened to an electronic circuit (Cole 1968). Cole's basic circuit elements are (1) the phospholipid bilayer, which acts as a capacitor in that it accumulates ionic charge as the electrical potential across the membrane changes; (2) the ionic permeabilities of the membrane, which act as resistors in an electronic circuit; and (3) the electrochemical driving forces, which act as batteries driving the ionic currents. These ionic and capacitive currents are arranged in a parallel circuit, as shown in Figure 2.4. This analogy to electrical circuits is now widely relied upon for developing models of electrical activity in membranes.

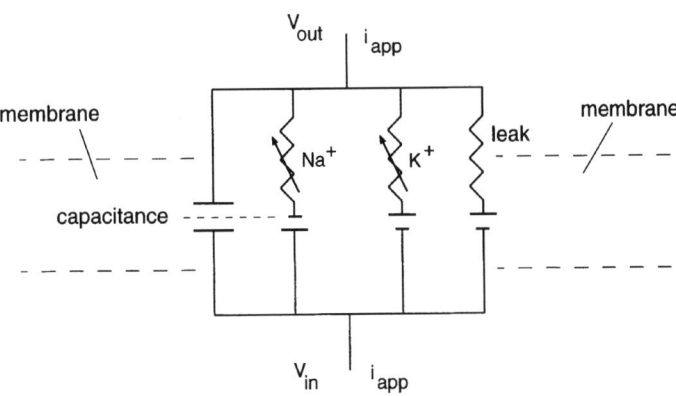

**Figure 2.4** The equivalent electrical circuit for an electrically active membrane. The capacitance is due to the phospholipid bilayer separating the ions on the inside and the outside of the cell. The three ionic currents, one for $Na^+$, one for $K^+$, and one for a non–specific leak, are indicated by resistances. The conductances of the $Na^+$ and $K^+$ currents are voltage dependent, as indicated by the variable resistances. The driving force for the ions is indicated by the symbol for the electromotive force, which is given in the model by the difference between the membrane potential $V = V_{in} - V_{out}$ and the reversal potential.

## 2.2.1  Equations for Membrane Electrical Behavior

Given the several conductances and their reversal potentials, we calculate using (2.5) what the membrane potential will be after the system has stabilized. This equations for the resting membrane potential tell us nothing about how the system evolves to the steady state. Because we are interested in the time course of the membrane voltage, we have to study the dynamics of the various currents that flow in and out of the cell. We can approximate the current flow through a single $K^+$ channel using Ohm's law and an assumption that the reversal potential stays constant:

$$I_K = -g_K(V - V_K). \tag{2.7}$$

Here $g_K$ is the conductance of the $K^+$ channel and the leading negative sign is necessary because of our definitions for "in" and "out." $V_K$ is the $K^+$ reversal potential determined by the Nernst equation, and $V - V_K$ represents the driving force across the membrane provided by the ionic battery. We assume that the reversal potential for a given ion remains constant, which is equivalent to assuming that restorative mechanisms such as ionic pumps can keep pace with electrical activity on a time scale that prevents the ionic battery from running down. This is a reasonable assumption for a large cell, which would have a small surface area to volume ratio. In a small cell, with a large surface to volume ratio, the ion transfer necessary to change the membrane potential might have a large effect on the intracellular ionic concentration and thus the strength of the ionic battery (see Exercise 2e).

Of course, numerous ions are responsible for the electrical behavior in a cell, and the total current is the sum of the individual ionic currents.

$$I_{ion} = \sum I_i = \sum -g_i(V - V_i) = -g_K(V - V_K) - g_{Na}(V - V_{Na}) - \cdots. \tag{2.8}$$

To translate the electric circuit diagram into ODEs, we use the traditional interpretation of each circuit element along with Kirchhoff's law. Assuming that the membrane acts as a capacitor, the capacitive current across the membrane can be written

$$I_{cap} = C\frac{dV}{dt}, \tag{2.9}$$

where $C$ is the capacitance of the membrane and $V$ is the membrane potential, defined as the electrical potential difference between the inside and outside of the cell. To establish the differential equation satisfied by the voltage $V$, Kirchhoff's law of charge conservation is applied to the circuit in Figure 2.4. Kirchhoff's law dictates that capacitive current must balance with the ionic current and any currents that might be applied, say, through experimental manipulation. This implies that

$$I_{cap} = I_{ion} + I_{app}, \tag{2.10}$$

where the sum is over all the ionic currents. Using the expressions in (2.8)–(2.10) this can be rewritten

$$C\frac{dV}{dt} = -\sum_i g_i(V - V_i) + I_{app}. \tag{2.11}$$

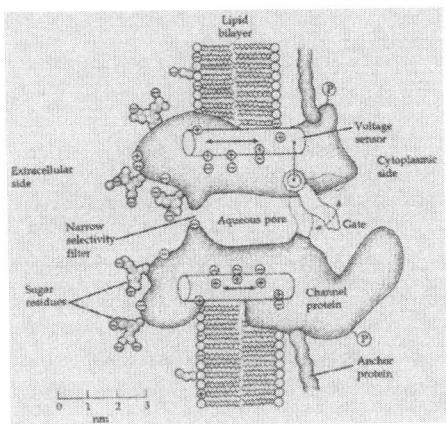

**Figure 2.5** Mechanistic cartoon of a gated ionic channel showing an aqueous pore that is selective to particular types of ions. The portion of the transmembrane protein that forms the "gate" is sensitive to membrane potential, allowing the pore to be in an open or closed state. Reprinted from Hille (2001).

To solve this differential equation for voltage, we must know how the gated conductances $g_i$ depend on $V$ (and possibly time). In general, the $g_i$ will not be linear functions of $V$, and therefore the problem is to find the time and possible voltage dependence of the various conductances.

## 2.3 Activation and Inactivation Gates

Channels can be thought to have *gates* that regulate the permeability of the pore to ions, as illustrated schematically in Figure 2.5. These gates can be controlled by membrane potential, producing *voltage gated* channels; by chemical ligands, producing *ligand gated* channels; or by a combination of factors. In a series of experiments, Alan Hodgkin, Andrew Huxley, and others established experimentally the voltage dependence of ion conductances in the electrically excitable membrane of the squid giant axon (Hodgkin and Huxley 1939). Hodgkin and Huxley shared a Nobel Prize with John Eccles in 1963 for work on nerve signaling.

### 2.3.1 Models of Voltage–Dependent Gating

The mathematical description of voltage–dependent activation and inactivation gates is based on the mechanism

$$C \underset{k^-}{\overset{k^+}{\rightleftharpoons}} O, \tag{2.12}$$

which was presented in Chapter 1. What distinguishes a voltage–dependent gating mechanism from a passive mechanism is the voltage dependence of the rate constants.

Recall from (1.1)–(1.5) that the fraction of open channels $f_O$ satisfies the differential equation

$$\frac{df_O}{dt} = \frac{-(f_O - f_\infty)}{\tau}, \tag{2.13}$$

where

$$f_\infty = \frac{k^+}{k^+ + k^-} \quad \text{and} \quad \tau = \frac{1}{k^+ + k^-}. \tag{2.14}$$

Because ionic channels are composed of proteins with charged amino acid side chains, the potential difference across the membrane can influence the rate at which the transitions from the open to closed state occur. According to the Arrhenius expression for the rate constants, the membrane potential $V$ contributes to the energy barrier for these transitions:

$$k^+ \propto \exp\left(\frac{-\Delta V^+}{RT}\right) \quad \text{and} \quad k^- \propto \exp\left(\frac{-\Delta V^-}{RT}\right). \tag{2.15}$$

The rate constants will have the form

$$k^+ = k_o^+ \exp(-\alpha V) \quad \text{and} \quad k^- = k_o^- \exp(-\beta V), \tag{2.16}$$

where $k_o^+$ and $k_o^-$ are independent of $V$. Substituting the relationships in (2.16) into the expressions for $f_\infty$ and $\tau$ and rearranging, we obtain

$$f_\infty = \frac{1}{1 + k_o^-/k_o^+ \exp((\alpha - \beta)V)}, \tag{2.17}$$

$$\tau = \frac{1}{k_o^+ \exp(-\alpha V)} \cdot \frac{1}{1 + k_o^-/k_o^+ \exp((\alpha - \beta)V)}. \tag{2.18}$$

We can define

$$S_o = \frac{1}{\beta - \alpha} \tag{2.19}$$

and

$$V_o = \frac{\ln(k_o^-/k_o^+)}{\beta - \alpha}. \tag{2.20}$$

If we substitute $S_o$ and $V_o$ into (2.17) and (2.18) (see Exercise 4), we have

$$f_\infty = \frac{1}{1 + \exp(-(V - V_o)/S_o)} \tag{2.21}$$

$$\tau = \frac{\exp(\alpha V)}{k_o^+} \cdot \frac{1}{1 + \exp(-(V - V_o)/S_o)}. \tag{2.22}$$

Finally, both of these expressions can be rewritten in terms of hyperbolic functions (Exercise 5):

$$f_\infty = 0.5(1 + \tanh((V - V_o)/2S_o)), \tag{2.23}$$

$$\tau = \frac{\exp(V(\alpha + \beta)/2)}{2\sqrt{k_o^+ k_o^-} \cosh((V - V_o)/2S_o)}. \tag{2.24}$$

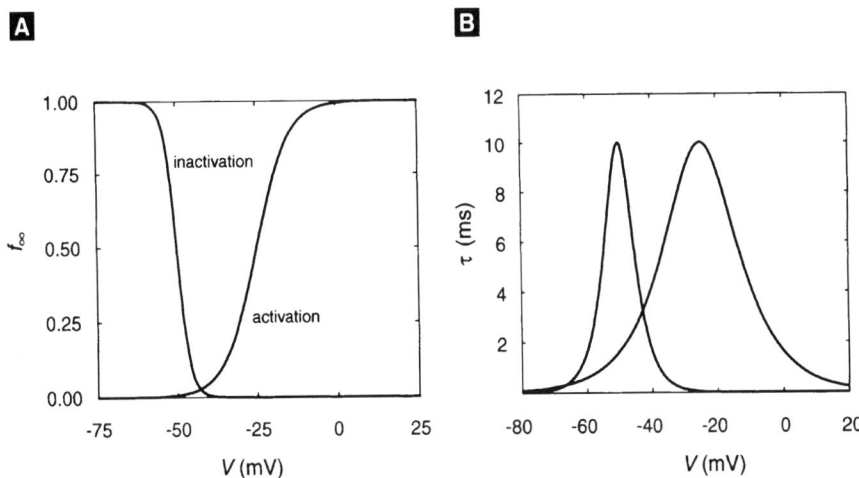

**Figure 2.6** (A) Equilibrium open fractions ($f_\infty$) for an inactivation gate ($V_o = -50$ mV and $S_o = -2$ mV) and activation gate ($V_o = -25$ mV and $S_o = 5$ mV) as a function of voltage. (B) The characteristic relaxation times $\tau$ for the activation and inactivation gates in (A) as a function of voltage, which are peaked around the values of $V_o$ and have a width determined by $S_o$.

Recall that $f_\infty$ gives the fraction of channels open at equilibrium at the membrane potential $V$. Thus for a fixed value of $V$ it gives the open fraction after transient changes in $f_O$ have damped out with a characteristic time $\tau$.

An activation gate tends to open and an inactivation gate tends to close when the membrane is depolarized. Whether a gate activates or inactivates with depolarization is determined by the sign of $S_o$: a positive sign implies activation and a negative sign inactivation. This is illustrated in Figure 2.6A, where the dependence of $f_\infty$ on $V$ has been plotted for an activation gate with $V_o = -25$ mV and $S_o = 5$ mV and an inactivation gate with $V_o = -50$ mV and $S_o = -2$ mV.

Notice that the magnitude of $S_o$ determines the steepness of the dependence of $f_\infty$ on $V$, whereas the value of $V_o$ determines the voltage at which half of the channels are open. The dependence of $\tau$ on $V$ for these activation and inactivation gates is illustrated in Figure 2.6B (assuming that $\alpha = -\beta$ and $2\sqrt{k_o^+ k_o^-} = 0.2$ ms$^{-1}$). When $\alpha = -\beta$ and $\phi = 1/\left(2\sqrt{k_o^+ k_o^-}\right)$,

$$\tau = \frac{\phi}{\cosh((V - V_o)/2S_o)}. \tag{2.25}$$

## 2.3.2 The Voltage Clamp

In order to measure the voltage across a cell membrane or the current flowing through a membrane, microelectrodes are inserted into cells. These electrodes can be used both

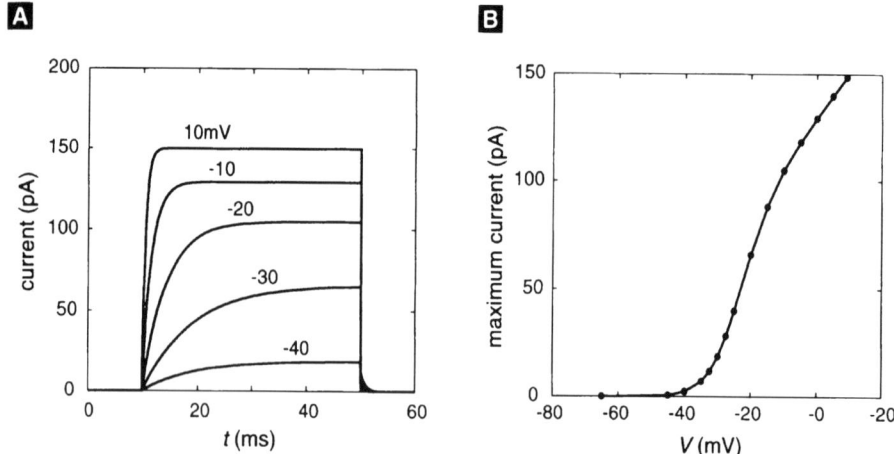

**Figure 2.7**   Simulation of voltage clamp experiment using (2.28) and (2.29). (A) Current records resulting from 40 ms depolarizations from the holding potential of −60 mV to the indicated test potentials. (B) the maximum (steady state) current as a function of test potential taken from records like those in (A).

to measure current and voltage and to apply external current. In order to measure the voltage dependence of the activation and inactivation of ion conductances, a technique called the *voltage clamp* is used. This is an electronic feedback device that adjusts the applied current $I_{app}$ to match and counter the membrane currents such that the membrane voltage is held constant. To see what this accomplishes, consider a membrane with a single gated ionic current. If we assume that the total conductance is the result of the activation of many channels, the conductance $g$ that we have used above can be defined as the product of the maximum possible conductance $\bar{g}$ and the fraction of open channels $f_O$ that we have already encountered.

$$g = f_O \bar{g}. \tag{2.26}$$

**Table 2.3   Consistent Electrical Units**

| Name (Symbol) | Units | Abbreviation |
|---|---|---|
| voltage ($V$) | $10^{-3}$ volt | mV |
| time ($t$) | $10^{-3}$ second | ms |
| conductance ($g$) | $10^{-9}$ siemens | nS |
| capacitance ($C$) | $10^{-12}$ farad | pF |
| current ($I$) | $10^{-12}$ ampere | pA |

We can include this new relationship in the differential equation for membrane potential that we have already seen:

$$C\frac{dV}{dt} = -f_O\bar{g}(V - V_{rev}) + I_{app},\tag{2.27}$$

where $V_{rev}$ is the reversal potential given by the Nernst equation. If we can apply a current that is equal and opposite to the current flowing through the membrane,

$$I_{app} = \bar{g}\xi(V - V_{rev}),\tag{2.28}$$

then the right–hand side of (2.27) is zero and the voltage must be constant. Because $V$ is constant, $\xi = f_O$ and the time dependence of the applied current comes only from the dependence of $f_O$ on $t$ as determined by the gating equation covered in Chapter 1:

$$\frac{d\xi}{dt} = \frac{-(f_O - f_\infty)}{\tau}.\tag{2.29}$$

Thus the time dependence of the applied current provides a direct measurement of the gated current at a fixed voltage. Note that throughout the remainder of the text we will drop the overbar with the understanding that conductances $g_i$ refer to maximum conductances to be scaled by gating variables.

To carry out a voltage clamp measurement like this it is necessary to block all but a single type of current. While this is not always possible, specific toxins and pharmacological agents have proven useful. For example, tetrodotoxin (TTX) from the puffer fish selectively blocks voltage gated $Na^+$ currents.

It is not difficult to simulate a voltage clamp measurement using (2.28)–(2.29). However, to carry out either an experimental measurement or a simulation, a consistent set of electrical units must be used. As we have seen, the standard unit for membrane potential is millivolts (mV), and because the characteristic times for voltage–dependent gates $\tau$ are in milliseconds (ms), this is taken as the standard unit of time. Currents are typically expressed in $\mu A/cm^2$ and capacitances as $\mu F/cm^2$. For a typical cell of area $10^{-6}$ cm$^2$, this translates to a whole–cell current of picoamperes (1 pA = $10^{-12}$ A) and a whole cell capacitance of picofarads (1 pF = $10^{-12}$ F). Cellular dimensions are usually reported in micrometers ("microns"), and there are $10^{-8}$ square centimeters per square micron. Because most biological channels have a conductance $g$ on the order of 1 to 150 pS, whole–cell conductances of nanosiemens are usually expressed in mS/cm$^2$, because the units of $V - V_{rev}$ are mV. This standard set of units is summarized in Table 2.3. An alternative consistent set of units uses current in femtoamperes (1 fA = $10^{-15}$ A), conductance in pS, and capacitance in fF.

To simulate a voltage clamp experiment we have solved (2.29) and plotted the resulting current given by (2.28). The simulation shown in Figure 2.7A represents a typical set of experiments in which the membrane potential is clamped at a holding potential (−60 mV in Figure 2.7), then changed to various test potentials for a fixed interval (40 ms), and finally returned to the holding potential. The value of the holding potential generally is chosen so that there is little or no current through the channel. This greatly simplifies the interpretation of the current at the test voltages. Figure 2.7A shows the

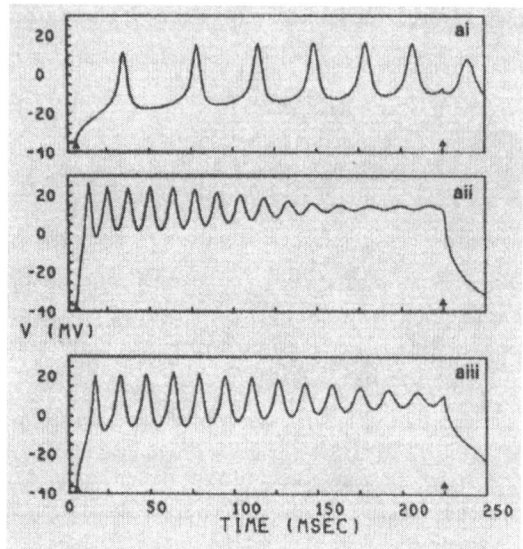

**Figure 2.8** Depolarization-induced electrical activity in giant barnacle muscle fibers; the arrows indicated the start and end of the depolarizating currents. Reprinted from Morris and Lecar (1981).

current that develops during this protocol for 5 test voltages $V_{test}$. The increase in current when the potential is clamped at the test values is governed by the exponential increase in $f_O$ with characteristic time $\tau(V_{test})$. When the potential is clamped again at the holding potential, the resulting current is called the *tail current*. Its decline is also exponential, but because $V = -60$ mV during this period, the characteristic time is now $\tau(-60$ mV$)$.

Figure 2.7B gives a plot of the steady–state current as a function of the test voltage. According to (2.28) it can be expressed as $I = gf_\infty(V - V_{rev})$. Thus for an activating current like that in the simulations, when $V$ is large enough, $f_\infty \approx 1$ and the current is a linear function of $V$. The curvature in Figure 2.7B at lower voltages is caused by the shape of the activation function $f_\infty$. In the jargon of circuit theory, currents like this are said to *rectify*. The delay in the onset of the maximum current, which is determined by the value of $\tau$, has led to channels like the one simulated in Figure 2.7 being referred to as *delayed rectifiers*.

# 2.4   Interacting Ion Channels: The Morris–Lecar Model

Application of a depolarizing current to barnacle muscle fibers produces a broad range of electrical activity. Figure 2.8 illustrates the sort of oscillations that are induced by current injections of 180, 540, and 900 $\mu$A/cm$^{-2}$ into these fibers. Careful experimental work by a number of research groups has indicated that the giant barnacle muscle fiber contains primarily voltage gated K$^+$ and Ca$^{2+}$ currents along with a K$^+$ current that is activated by intracellular $Ca^{2+}$, a so-called $K_{Ca}^+$ current. Neither of the voltage gated currents shows significant inactivation in voltage clamp experiments. The trains of depolarization-induced action potentials in Figure 2.8 must occur via a mechanism different from that proposed by Hodgkin and Huxley for the squid giant axon, which, as we will see later, include channel inactivation.

Morris and Lecar proposed a simple model to explain the observed electrical behavior of the barnacle muscle fiber (Morris and Lecar 1981). Their model involves only a fast activating $Ca^{2+}$ current, a delayed rectifier $K^+$ current, and a passive leak. They tested the model against a number of experimental conditions in which the interior of the fiber was perfused with the $Ca^{2+}$ chelator EGTA in order to reduce activation of the $K_{Ca}$ current. Their simulations provide a good explanation of their experimental measurements. The model translates into two equations:

$$C\frac{dV}{dt} = -g_{Ca}m_\infty(V - V_{Ca}) - g_K w(V - V_K) - g_L(V - V_L) + I_{app}, \tag{2.30}$$

$$\frac{dw}{dt} = \frac{\phi(w_\infty - w)}{\tau}. \tag{2.31}$$

Here $m_\infty$ is the fraction of voltage-dependent $Ca^{2+}$ channels open, and this is a function of voltage but not time. Furthermore, $w$ is the fraction of open channels for the delayed rectifier $K^+$ channels, and the conductances $g_L$, $g_{Ca}$, and $g_K$ are for the leak, $Ca^{2+}$, and $K^+$ currents, respectively. We use $w$ rather than the previously used $f_O$ for the fraction of open channels for historical reasons. The functions

$$m_\infty = 0.5[1 + \tanh((V - v_1)/v_2)], \tag{2.32}$$

$$w_\infty = 0.5[1 + \tanh((V - v_3)/v_4)], \tag{2.33}$$

$$\tau = 1/\cosh((V - v_3)/(2 \cdot v_4)), \tag{2.34}$$

**Table 2.4  Morris–Lecar Oscillator Parameters (Type II)**

| Parameter | Value |
|---|---|
| $C$ | $20\mu F/cm^2$ |
| $V_K$ | $-84$ mV |
| $g_K$ | $8$ mS/cm$^2$ |
| $V_{Ca}$ | $120$ mV |
| $g_{Ca}$ | $4.4$ mS/cm$^2$ |
| $V_{leak}$ | $-60$ mV |
| $g_{leak}$ | $2$ mS/cm$^2$ |
| $v1$ | $-1.2$ mV |
| $v2$ | $18$ mV |
| $v3$ | $2$ mV |
| $v4$ | $30$ mV |
| $\phi$ | $0.04$/ms |

are the equilibrium open fractions for the $Ca^{2+}$ current and the $K^+$ current, and the activation time constant for the delayed rectifier. Representative parameters are given in Table 2.4. Again note that $m$ is not a dynamic variable. The reason for this is that we have assumed that the time constant for $m$ is short enough that $m$ is always in steady state, $m = m_\infty$. The idea of fast and slow processes is arguably one of the most important concepts in modeling. Although we make the assumption without argument here, its implications in modeling are addressed in a more mathematical context in Chapter 4 and Appendix A.

We have solved the Morris–Lecar equations for four values of the applied current $I_{app}$ and plotted the time series for $V$ in Figure 2.9A. Current in the Morris–Lecar model is specified in $\mu A/cm^2$, however to simplify notation somewhat we will assume that the cell has a total surface area of $10^{-6}$ $cm^2$, so that $\mu A/cm^2$ corresponds to 1 pA of total current. In the absence of applied current the equations have a stable steady state near $-60$ mV. Although increasing $I_{app}$ to 60 pA produces a brief transient action potential, the effect of the depolarization simply produces a steady state near $-35$ mV. Depolarization with a current of 150 pA, on the other hand, produces a steady train of action potentials reminiscent of those observed experimentally in Figure 2.8. In the presence of depolarizing currents much greater than this, the simulated barnacle cell can no longer sustain continuous spiking, as shown at $I_{app} = 300$ pA.

## 2.4.1  Phase Plane Analysis

The mechanistic features underlying continuous spiking and action potentials can be understood easily using phase plane analysis. Because of the ease of representation in two dimensions, two variable models such as the Morris–Lecar model are particularly amenable to this technique. Phase plane analysis is a powerful way to determine how the behavior of a system will change with changes in the various parameters in a system. Several types of plots are utilized as part of what is generically called phase plane analysis:

- a *phase portrait* consists of the variables describing a system plotted against each other rather than as a function of time to produce a *trajectory* in *phase space*. A phase portrait tells us how the variables interact for a given set of parameters.
- A *vector field* shows us the direction in which a system will evolve from any location in phase space.
- *Nullclines* are plotted in phase space, and show us the values of a pair of variables at which one of the variables does not change. In other words, for a coupled system of equations $X(x, y, t)$ and $Y(x, y, t)$ nullclines are the solutions to the equations

$$\frac{dX}{dt} = 0, \quad \frac{dY}{dt} = 0 \tag{2.35}$$

Note that there is a nullcline for each variable. The points of intersection of two nullclines are particularly interesting points in phase space that we will discuss below.

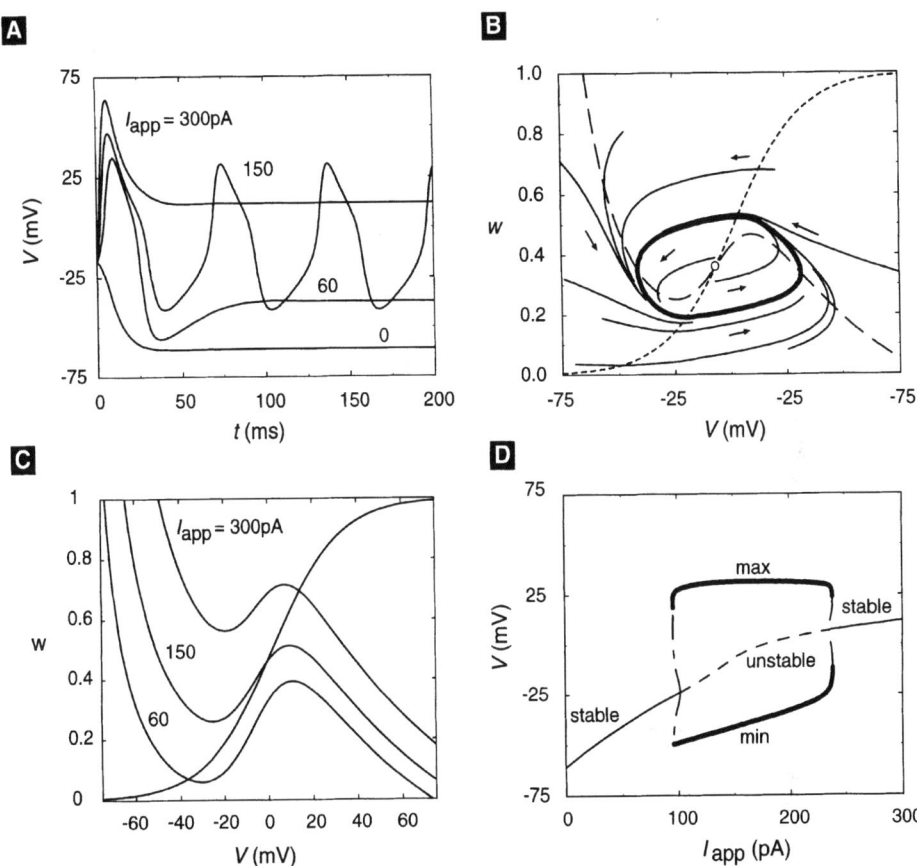

**Figure 2.9** (A) Voltage simulation for the Morris–Lecar equations using the indicated applied currents and parameters as in Table 2.4. Oscillations occur at $I_{app} = 150$ pA. (B) The phase plane for the Morris-Lecar model for $I_{app} = 150$ pA. The heavy line is the limit cycle corresponding to the oscillation in (A), and the lighter lines are short trajectories that circulate in the counter-clockwise direction toward the stable limit cycle. The short arrows indicate the vector field. The $V$-nullcline is the long-dashed line and the $w$-nullcline is the short-dashed line. (C) Nullclines for several values of $I_{app}$. (D) A bifurcation diagram that catalogues the dynamical states of the Morris–Lecar model as a function of $I_{app}$ with the other parameters fixed. The maximum and minimum of $V$ on the limit cycle are represented by the heavy lines. Compare the values for $I_{app} = 150$ pA (long-dashed line) with the voltage record in (A).

Phase plane analysis plots for the Morris–Lecar model are shown in Figure 2.9B and Figure 2.9C. Figure 2.9B shows trajectories, nullclines, and the vector field for $I_{app} = 150$ pA, which leads to the pattern of repetitive spiking. The phase portrait consists of a number of representative trajectories together with a unique, closed trajectory indicated by the heavy line. This trajectory is called a *stable limit cycle* because it is

the cyclic curve to which all the neighboring trajectories converge no matter where
in phase space they originate. The trajectories circulate around the steady state in a
counterclockwise direction as indicated by the velocity vector field, which is shown as
small arrows.

The nullclines are indicated by the lighter segmented lines. The $V$-nullcline, which
has the inverted "N" shape, is given by the long-dashed line. It is the solution to

$$0 = -g_{Ca}m_\infty(V - V_{Ca}) - g_K w(V - V_K) - g_L(V - V_L) + I_{app} \tag{2.36}$$

for each value of $w$.

The $w$-nullcline is the solution to

$$0 = \frac{\phi(w_\infty - w)}{\tau} \tag{2.37}$$

for each value of $V$.

Note that the limit cycle in Figure 2.9B circulates around the intersection of the
two nullclines. We saw in Figure 2.9A that different values of $I_{app}$ lead to different
behaviors and that $I_{app} = 150$ pA was the only value of the four tested that results in
oscillations. Figure 2.9C shows how changing the value of $I_{app}$ affects the nullclines
of the system. We see that increasing $I_{app}$ from 60 pA to 150 pA and then to 300 pA
raises the $V$ nullcline but leaves the $w$ nullcline unaltered. If we examine Figure 2.9A
and Figure 2.9C carefully, we notice something interesting: As we saw above, the limit
cycle obtained when $I_{app} = 150$ pA circulates around the intersection of the nullclines
at that parameter value. At $I_{app} = 300$ pA the system evolves to a steady state that
corresponds to the intersection of the nullclines obtained for that $I_{app}$. The same is true
for $I_{app} = 60$ pA. We know that a point on a nullcline corresponds to a point at which
the variable of interest is not changing, and so it makes sense that the intersection of
nullclines represents a combination of variables for which the system as a whole does
not change. The intersection of the nullclines for $I_{app} = 60$ pA and $I_{app} = 300$ pA are
examples of *fixed points*, and represent *stable steady states*. What is different about the
intersection of nullclines obtained when $I_{app} = 150$ pA?

## 2.4.2  Stability Analysis

Fixed points can be either stable or unstable and the intersection of nullclines obtained
when $I_{app} = 150$ pA is an unstable fixed point. Notice that the two trajectories that start
nearest to the intersection of the nullclines in Figure 2.9B diverge away from the intersection and toward the limit cycle. The trajectory of the system is thus driven away
from the intersection of the nullclines while at the same time being constrained to orbit
around it. The existence of a stable limit cycle should be no more surprising than the
existence of stable steady states. It corresponds to a closed trajectory to which all neighboring trajectories converge. The tools provided for phase plane analysis and stability
analysis in some ODE packages make it easy to determine steady states, limit cycles,
and stability, and therefore how a system like the Morris–Lecar model will behave.

Appendix A reviews how the *eigenvalues* determine the global stability properties of a linear system. Eigenvalues describe the solution of the linear system: Negative real eigenvalues indicate a stable solution, positive real eigenvalues indicate an unstable solution, and complex eigenvalues indicate the presence of oscillations in the solution. Stability analysis for nonlinear systems is not quite so straightforward, in that we can only determine the stability in such a system in a very small region around the a fixed point. In essence we linearize the system, either analytically (with a *Taylor series* as described in Appendix A) or numerically (with a software package). We then use the same tools that we used for the linear system in the small linearized region around the fixed point in the nonlinear system.

This book emphasizes the use of computational tools rather than mathematical analysis. An intuitive understanding of the concept of stability can be obtained without mathematical rigor. However, it must be stressed that there are myriad subtleties that can be appreciated only with an understanding of the underlying mathematics. The topic of stability analysis of linear and nonlinear equations is covered in more depth in Appendix A, and the student should be familiar with the analytical techniques discussed there in order to appreciate the output of computational tools. Other excellent sources covering this material on a reasonably introductory level are available (Edelstein-Keshet 1988; Kaplan and Glass 1995), and a particularly detailed analysis of the firing proterties of the Morris–Lecar system has been published (Rinzel and Ermentrout 1998).

The simulations in Figure 2.9A show that a stable limit cycle occurs for the Morris–Lecar model only for certain values of $I_{app}$, such as $I_{app} = 150$ pA. We can determine the stability of the fixed points corresponding to different values of $I_{app}$ using a numerical package. The intersection of the nullclines for $I_{app} = 150$ pA indicated by the open circle in Figure 2.9B is an unstable steady state at $V = -.460$ and $w = 0.459$, a fixed point with eigenvalues $\lambda_{\pm} = 0.264$ and $0.033$ (values are rounded). It is unstable because the eigenvalues are positive real numbers, and yet these parameters result in stable oscillations. This is an excellent example of how the local stability of a fixed point does not tell the whole story for a nonlinear system. Unlike the case with linear systems (covered in Appendix A), here positive eigenvalues can be associated with stable oscillations in phase space around the unstable fixed point.

**Table 2.5   Fixed Points and Their Eigenvalues**

| $I_{app}$ | V | w | Eigenvalue | Eigenvalue |
|---|---|---|---|---|
| 0 pA | −60.855 V | 0.015 | −0.037 | −0.096 |
| 60 pA | −37.755 V | 0.070 | −0.055 + i0.063 | −0.055 − i0.063 |
| 110 pA | −19.219 V | 0.196 | 0.055 + i0.045 | 0.055 − i0.045 |
| 150 pA | −0.460 V | 0.459 | 0.264 | 0.033 |
| 180 pA | 6.656 V | 0.577 | 0.025 + i0.139 | 0.025 − i0.139 |
| 300 pA | 14.302 V | 0.694 | −0.137 + i0.117 | −0.137 − i0.117 |

Using a systematic stepping procedure, we can test the stability of the system over a wide range of parameter values. Table 2.5 shows the eigenvalues for the values of $I_{app}$ that we have examined plus some additional intermediate values. As $I_{app}$ is increased from $I_{app} = 0$, the eigenvalues are seen to change from negative (stable node) to complex with negative real parts (stable focus), to complex with positive real parts (unstable focus) as oscillations emerge. The oscillations at $I_{app} = 150$ correspond to eigenvalues that are both real and positive (unstable node). As $I_{app}$ is increased further, the eigenvalues become complex again with positive real parts, and eventually become complex with negative real parts as oscillations cease. Fixed points at which the qualitative character of the solution changes with a change in a parameter are called *bifurcation* points. The mathematical properties of bifurcations are also treated in more detail in Appendix A. In particular, Appendix A discusses the *Hopf bifurcation theorem*, which establishes the conditions that result in the limit cycle oscillations that we see here.

By making even smaller changes in $I_{app}$ and testing the stability of the steady state, it is possible to locate the specific values at which the stability of the steady state changes. This procedure could become very laborious, and so numerical algorithms have been developed that allow investigation of bifurcations to be done automatically. In the bifurcation diagram shown in Figure 2.9D, the characteristic values of the membrane potential are plotted on the ordinate as a function of $I_{app}$. The thin full and dashed lines are the steady–state values of $V$ for each value of $I_{app}$, with the full lines representing stable steady states and the dashed lines unstable states. The points $I_{app} \approx 94$ pA and 212 pA are bifurcation points where the stability of the steady state changes. Near these points two new dynamical features appear: a stable limit cycle and an unstable limit cycle. This type of bifurcation is called a *subcritical Hopf bifurcation*, as discussed in Appendix A. Note that while the detail is not apparent in Figure 2.9D, the unstable limit cycle turns back and then coalesces at a turning point bifurcation with the stable limit cycle at a value of $I_{app}$ that is smaller than the bifurcation point, and therefore there is a small region of bistability. The bifurcation diagram in Figure 2.9D also records the maximum and minimum values of $V$ on the limit cycles with heavy full lines. For example, at $I_{app} = 150$ pA the points on the heavy line correspond to the maximum and minimum of the spikes in Figure 2.9A.

### 2.4.3  Why Do Oscillations Occur?

If the following three conditions on the Morris–Lecar equations hold, then oscillations will occur:

- the $V$ nullcline has the inverted "N" shape like that in Figure 2.9B;
- a single intersection of the $V$- and $w$-nullclines occurs between the maximum and minimum of the "N;"
- the rate of change of $V$ is much greater than $w$.

All three conditions are met for the parameter values in Table 2.4, giving rise to oscillations in Figure 2.9A. The importance of the slow change in $w$, i.e., the "delay" of

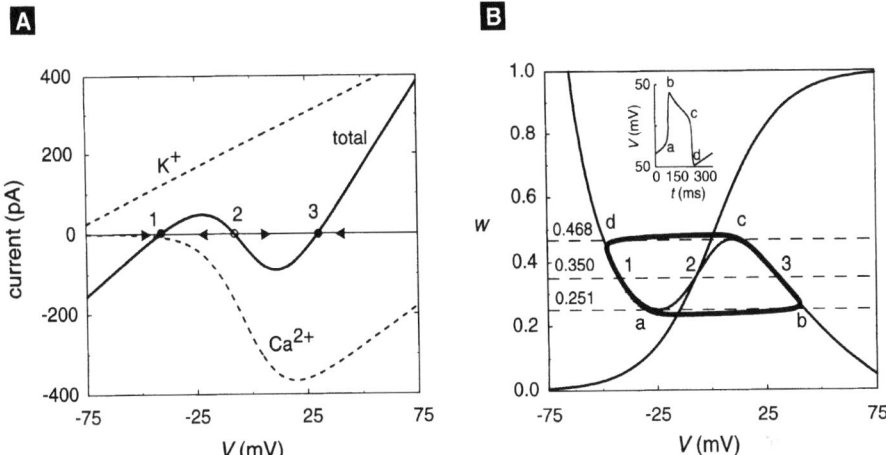

**Figure 2.10** (A) The K$^+$, Ca$^{2+}$, and total current ($I_{Ca} + I_K + I_{leak} - I_{app}$) when $w = 0.35$. States 1 and 3 are stable steady states, and state 2 is unstable as indicated by the velocity vectors. (B) The phase plane for the Morris–Lecar model for $I_{app} = 150$ pA, except that $\tau(V)$ has been increased by a factor of ten. The values $w = 0.468$ and $0.251$ correspond to the maximum and minimum of the $V$-nullcline. The points 1, 2, and 3 at $w = 0.350$ are the steady states in (A). Inset shows the voltage record for a single spike.

the delayed rectifier, can be seen by examining the trajectories in Figure 2.9B. If the rate of change of $w$ were large with respect to $V$, then the trajectories would not depolarize and hyperpolarize rapidly as they do in Figure 2.9A. This is tested by decreasing the value of the parameter $\phi$, thereby increasing the value of the characteristic time for relaxation of $w$.

By altering the time scale we see why oscillations occur when the rate of change of $V$ is very much faster than $w$ and the nullclines have the shape in Figure 2.9. In this case, we can treat changes in $V$ under the assumption that $w$ is constant, and we need only to consider the voltage equation with $w$ fixed:

$$C\frac{dV}{dt} = -g_{Ca}m_\infty(V - V_{Ca}) - g_K w(V - V_K) - g_L(V - V_L) + I_{app}. \qquad (2.38)$$

Because only the voltage is changing on this time scale, we can examine its dynamical behavior using the one–dimensional phase portrait, rather than a phase plane. This is shown in Figure 2.10A, where the total current, which is proportional to the rate of change of $V$, is plotted along with the Ca$^{2+}$ and K$^+$ currents for $w = 0.35$ and $I_{app} = 150$ pA. The total current vanishes at three points, which are the steady states for the voltage when $w$ is fixed. States 1 and 3 are stable, and state 2 is unstable, as indicated by the velocity vectors in the figure. Note that at state 1 the membrane is polarized and the outward K$^+$ current dominates the Ca$^{2+}$ inward current, whereas the opposite holds true at state 3. As we shall see, the oscillations can be thought of as transitions back and forth between the polarized and depolarized states, driven by slow changes in activation of the delayed rectifier current.

Figure 2.10A also explains why the $V$-nullcline in the Morris–Lecar model has the inverted "N" shape: The $K^+$ and leak currents exceed the inward $Ca^{2+}$ current at polarized voltages between states 1 and 2, whereas the $Ca^{2+}$ current exceeds the other currents between states 2 and 3. This voltage-dependent competition between inward and outward currents leads to a maximum and minimum in total current and therefore the inverted "N" shape for the nullcline.

The three steady states of the voltage also can be found graphically in the phase plane of Figure 2.10B by locating the intersection of the line $w = 0.35$ with the $V$-nullcline. It is clear from the figure that if $w$ exceeds 0.468 (the maximum on the right branch of the $V$-nullcline), then the voltage has only a single polarized steady state (intersection with $w$) on the far left branch of the $V$-nullcline. Similarly, if $w$ is smaller than 0.251 (the minimum on the left branch of the $V$-nullcline), then the voltage has only a single depolarized steady state on the right branch. For $0.251 \leq w \leq 0.468$ two stable steady states and one unstable state occur, and the voltage is said to be *bistable*.

To understand how bistability on the fast time scale (neglecting $w$) leads to oscillations in the complete system, we need to understand how $w$ changes on the longer time scale. Assume that initially, the membrane is polarized at state 1 in Figure 2.10B, with $w = 0.35$. Because $w = 0.35$ is above the $w$-nullcline at point 1, $w$ will decrease. As $w$ decreases, $V$ will stay close to the $V$-nullcline because it relaxes rapidly to the closest steady–state value. The trajectory thus follows the polarized branch, as indicated by the heavy line, until the minimum at $w = 0.251$ is reached. Beyond the minimum (near a), stable polarized states no longer exist, and $V$ rapidly relaxes to the only remaining steady state (near b), which is on the depolarized, far right branch of the $V$-nullcline. During the depolarization, however, the $w$-nullcline is crossed. Thus on the depolarized branch, $w$ increases and tracks the $V$-nullcline upward until the maximum at $w = 0.468$ is reached (near c) and the membrane rapidly repolarizes to the polarized branch (near d).

The abrupt transitions from the polarized to depolarized branch and back again have led to the name *relaxation oscillator* for systems of equations that have well-separated time scales. For the parameters in Figure 2.9, the Morris–Lecar model is not a relaxation oscillator. However, when the characteristic time for $w$ is increased by a factor of 10 (by decreasing $\phi$), the limit cycle (heavy line in Figure 2.10B) closely approximates that for a relaxation oscillator. If the characteristic time were increased sufficiently, or the characteristic time for $V$ were decreased sufficiently by altering the capacitance, then the trajectory would coincide with the bistable portions of the $V$-nullclines, and the rapid excursions of the voltage would occur precisely at $w = 0.251$ and 0.468. The inset in Figure 2.10B shows a single voltage spike that illustrates the rapid upstroke and downstroke for the limit cycle. The inset also illustrates that the shape of the depolarized and polarized portions of the spike reflects the shape of the two branches of the $V$-nullcline.

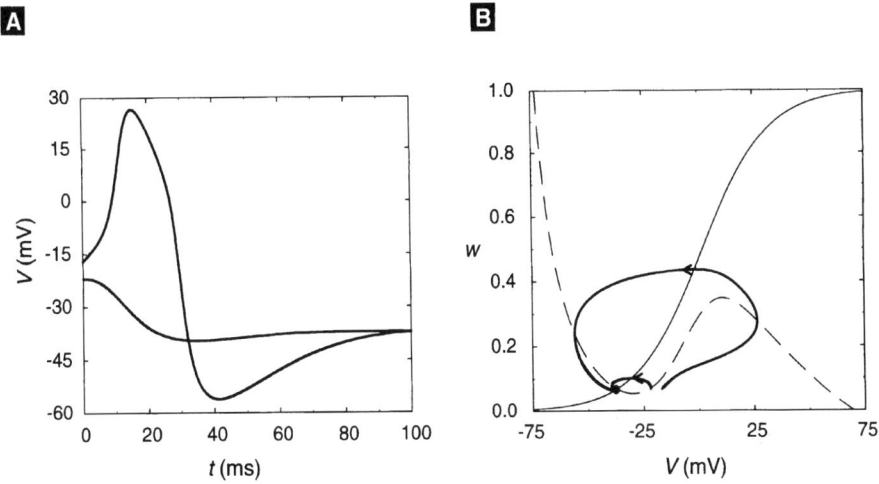

**Figure 2.11** Excitability in the Morris–Lecar model for $I_{app} = 60$ pA. (A) An initial deviation of the voltage to $-22$ mV relaxes rapidly to the steady–state voltage, whereas a deviation to $-17$ mV produces an action potential. (B) The trajectories in (A) represented in the phase plane. When $V$ changes much faster than $w$, the location of the $V$-nullcline (long-dashed line) sets the threshold for action potential spikes.

### 2.4.4 Excitability and Action Potentials

Another dynamical feature of the Morris–Lecar model is *excitability*. A useful working definition of excitability is that a system is excitable when small perturbations return to the steady state, but larger (i.e., above a threshold) perturbations cause large transient deviations away from the steady state. An example of this is shown in Figure 2.11A, where the time course of the voltage with an applied current of 60 pA and initial conditions of $w(0) = w^{ss} = 0.070$ and either $V(0) = -22$ or $-17$ mV are used. For subthreshold $V(0)$ such as $-22$ mV, the voltage increases only slightly before it decreases monotonically to its steady state–value of about $-37$ mV. For suprathreshold deviations, like that for $V(0) = -17$ mV, the voltage increases dramatically, producing an action potential spike before returning to steady–state.

The explanation for excitability can be understood most easily in the phase plane. In Figure 2.11B we have plotted the trajectories for the two initial conditions in Figure 2.11A along with the two nullclines. The trajectories for subthreshold initial conditions like $V(0) \leq -22$ mV start at points in phase space *above* the $V$-nullcline and *below* the $w$ nullcline. This implies that the initial velocity vector, and therefore the initial trajectory, points in the direction of smaller potentials and larger values of the activation of the delayed rectifier $K^+$ current. Thus the voltage begins to decrease and continues to do so because the $K^+$ current activates as $w$ increases. This contrasts with the trajectory that starts at $V(0) = -17$ mV, for which the initial velocity vector points in the direction of

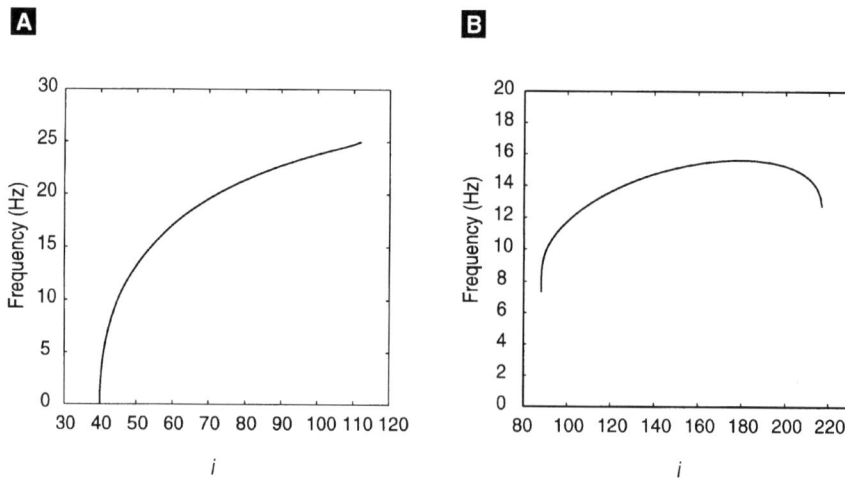

**Figure 2.12** Examples of voltage–freqency plots for the Morris–Lecar model with parameters that result in (A) Type I dynamics and (B) Type II dynamics. Note that the axes are somewhat different. Plotted according to Rinzel and Ermentrout (1998).

increasing voltage. Even though $w$ is increasing in this region, which is also below the $w$-nullcline, the rate of increase of voltage exceeds that of $w$, and the trajectory moves to higher voltages until it crosses the $V$-nullcline and begins to decrease.

The threshold value of $V(0)$ above which action potentials occur depends on the shape of the nullclines and the rate of activation of $w$ (altered by $\phi$). However, it is close to the point where a line drawn parallel to the $V$ axis at $w^{ss}$ crosses the $V$-nullcline. It is not hard to check that it will approach this point as $w$ is made to change more slowly than $V$.

## 2.4.5   Type I and Type II Spiking

Many investigators are now interested in applying biophysical models of neurons and other spike–generating mechanisms to the study of information transfer using information–theoretic measures. It is particularly important, therefore, that the dynamical behavior of a model be characterized over the space of parameters and that the characteristics of the spike generating model match the intended use. Both the Hodgkin–Huxley model, described below, and the Morris–Lecar model produce trains of action potentials with commonly used parameter sets and when sufficiently depolarized. Muscle fiber and axons innervating muscle fiber are examples of situations where a strong, consistent signal is required. In our investigations of the Morris–Lecar model, we have seen that depolarization beyond threshold results in oscillations that begin at a characteristic nonzero frequency. This behavior has been defined as a Type II oscillator (Rinzel and Ermentrout 1998).

Many neurons exhibit firing properties that are fundamentally different from those of the Type II oscillator. Models of these cells can produce arbitrarily low frequencies of oscillations. These models are classified as Type I oscillators (Rinzel and Ermentrout 1998). The Morris–Lecar model can exhibit Type I or Type II behavior, depending on the parameters that are chosen, and the Hodgkin–Huxley model also can be modified to show this behavior. See Table 2.6 for the Morris–Lecar Type I parameters that differ from Type II parameters. Examples of voltage–frequency plots for the Morris–Lecar model in these two regimes are shown in Figure 2.12. The Morris–Lecar Type II oscillator might not be appropriate for the study of subtle aspects of information transfer or intra–neuronal coupling. We have seen that Type II spiking results from a subcritical Hopf bifurcation as input current is increased. Type I spiking results from a saddle-node bifurcation. The difference between these two bifurcations is discussed in Appendix A. We also note that there are other levels of complexity for spiking and oscillating models. Models of bursting cells, which exhibit trains of oscillations separated by periods of quiescence, require additional slow variables. These models will be discussed in detail in Chapter 5.

## 2.5 The Hodgkin–Huxley Model

Ordinarily, a discussion of membrane electrophysiology might begin with the famous model of the squid giant axon developed by Hodgkin and Huxley (the HH model) (Hodgkin and Huxley 1952). Our goal here has been to facilitate a clear mathematical understanding of oscillatory behavior in systems of ion channels. While it remains a seminal accomplishment in the history of physiology, the HH model is complicated and not amenable to the phase plane methods of analysis that we have used to understand dynamical electrical behavior in cells. The student should be familiar with the HH model, for historical reasons and because it is still widely used. We cannot cover the work of Hodgkin and Huxley in the detail it deserves, but the concepts learned in the study of the Morris–Lecar oscillator will aid the student in understanding the HH model. We recommend that the student consult Bertil Hille's *Ionic Channels of Excitable Membranes* (Hille 2001), which contains one of the best treatments of the HH model and the history behind it. "Hille" is also the best consolidated source of information

**Table 2.6  Morris–Lecar Type I Oscillator Parameters**

| Parameter | Value |
|---|---|
| $g_{Ca}$ | 4 mS/cm$^2$ |
| $v3$ | 12 mV |
| $v4$ | 17.4 mV |
| $\phi$ | 0.066/ms |

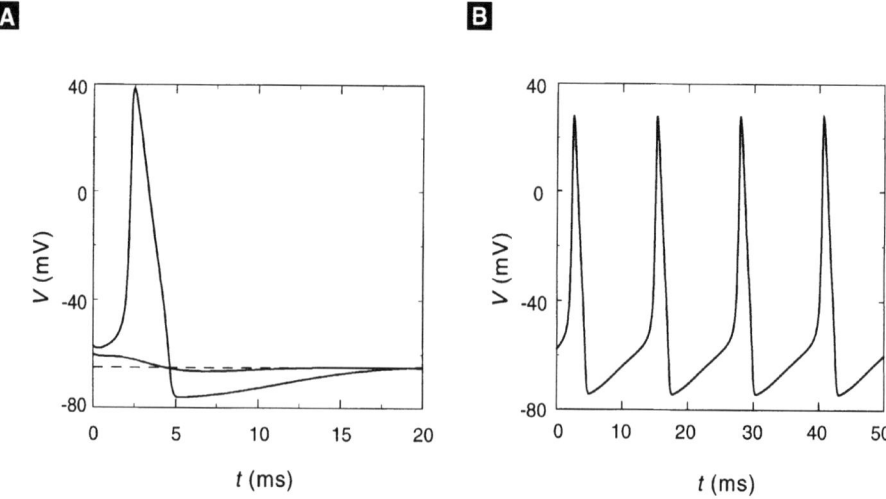

**Figure 2.13** (A) The solution of the Hodgkin-Huxley equations for three different initial values of the membrane potential and no applied current. When the initial value exceeds ca. −59 mV, an action potential is produced. (B) Continuous spiking occurs under the same conditions with an applied current $I_{app} = 15$.

about the biophysics of ion channels, and it should be familiar to any person attempting to model electrical behavior.

The HH model is empirical (Hodgkin and Huxley 1952). Many voltage clamp experiments were performed by Hodgkin and Huxley, and their data were fit to expressions that they incorporated into the model without consideration of an underlying mechanism for the channel gates. One of the remarkable aspects of Hodgkin and Huxley's work is that their model was developed without a molecular understanding of the mechanism. In fact, it required almost thirty years of intensive research after their work to formulate a realistic cartoon of the mechanisms underlying the ionic currents in cells. Although Hodgkin and Huxley justified these expressions on empirical grounds, it is possible to derive the gating expression used in the Hodgkin–Huxley model using mechanistic models.

From their voltage clamp and other measurements, Hodgkin and Huxley deduced that the sodium conductance involved two voltage-dependent "gates," an *activation* gate and an *inactivation*, gate and that the potassium conductance had a single activation gate. Note that the presence of an inactivation gate differs from the Morris–Lecar model. To account for these facts they represented the ionic conductances in the following form:

$$g_{Na} = \bar{g}_{Na} m^3 h, \tag{2.39}$$

$$g_K = \bar{g}_K n^4, \tag{2.40}$$

where the terms $\bar{g}$ represent maximal conductances, and $m$ and $n$ are the activation gating variables and $h$ the inactivation. The exponents on $m$ and $n$ were chosen for the best fit to experimental data. These gating variables were postulated to satisfy linear differential equations where the variables "relax" to voltage–dependent values, e.g., $m_\infty$, that vary between zero and one with voltage-dependent time constants, e.g., $\tau_m$.

Putting the ODEs for the gating variables together with (2.11) gives the primary equations for the Hodgkin–Huxley model:

$$CdV/dt = -\bar{g}_{Na}m^3h(V - V_{Na}) - \bar{g}_K n^4(V - V_K),$$
$$- \bar{g}_{leak}(V - V_{leak}) + I_{app}, \tag{2.41}$$
$$dm/dt = -(m - m_\infty)/\tau_m, \tag{2.42}$$
$$dh/dt = -(h - h_\infty)/\tau_h, \tag{2.43}$$
$$dn/dt = -(n - n_\infty)/\tau_n. \tag{2.44}$$

Hodgkin and Huxley added the third conductance $\bar{g}_{leak}$ to their voltage equation to account for a small voltage-independent conductance that they attributed to a "leak" in the membrane, possibly through their microelectrode. The nonlinear terms in (2.41) are obvious in the activation and inactivation gates. Not so obvious are the nonlinearities in (2.42)–(2.44). However, all the voltage-dependent terms in those equations are nonlinear functions of $V$ as well.

The student should solve the HH equations to provide a basis for comparison and further exploration (Exercise 9b). The equations for the Hodgkin–Huxley spike generator are readily available from other sources, and it is not difficult to run simulations with a package for solving ODEs. One must be careful to choose a suitable numerical method of solution due to the nonlinear equations. Figure 2.13 shows several simulations that can be made with the HH equations. The *Runge–Kutta* method has been used with a time step of 0.05, which is appropriate because reducing the time step to 0.01 gives no noticeable change in the results.

Figure 2.13A shows calculations with $I_{app} = 0$, but with $V(0) = -65, -60,$ and $-57$ mV. The steady state for the voltage with these parameters is clearly $-65$ mV; however, if the initial value of $V$ exceeds about $-59$ mV, then the equations produce an action potential spike, and we have another example of a Type II oscillator. The Hodgkin-Huxley equations can also produce repetitive firing of action potentials. This is illustrated in Figure 2.13B. The parameters in that simulation are identical to those in Figure 2.13A, except that $I_{app} = 15$.

## 2.6  FitzHugh–Nagumo Class Models

We have seen that a fast variable with a cubic nullcline for voltage and a slower variable with a monotonically increasing nullcline for channel opening are sufficient to produce oscillations. In some cases, and particularly for mathematical analysis, it may be beneficial to completely abstract the mathematical equations for an oscillator from the underlying physical processes in order to understand general properties of these systems. Because two–variable systems are useful for the phase plane methods we have discussed, we can contrive such a system.

The two differential equations will be functions of both state variables:

$$\frac{dv}{dt} = f(v, w),$$
$$\frac{dw}{dt} = g(v, w).$$

(2.45)

By definition, the nullcline is obtained by setting the right–hand side of the equation to zero and plotting the function in the $(v, w)$ phase plane. We can create the behavior we desire by specifying the shapes of $f(v, w)$ and $g(v, w)$. The slow variable, $w$, is easiest, because we want $g(v, w)$ to be similar to a line describing the increase of $v$ with $w$. Note that $v$ here is the dependent variable for the equation for a line:

$$g(v, w) : v = \gamma w,$$

(2.46)

and therefore

$$\frac{dw}{dt} = v - \gamma w$$

(2.47)

where $\gamma$ is the slope parameter.

The fast variable $v$ will have a form similar to (2.47) with $w$ as the dependent variable:

$$f(v, w) : w = f(v),$$

(2.48)

and therefore

$$\frac{dv}{dt} = f(v) - w.$$

(2.49)

Because we know that $w$ needs to be a cubic function of $v$, let us try

$$f(v) = Bv(v - \beta)(\delta - v),$$

(2.50)

where $A$ is a parameter to scale the amplitude of the curve. The equations for the complete system are therefore

$$\frac{dv}{dt} = Av(v - \beta)(\delta - v) - Cw,$$
$$\frac{dw}{dt} = \epsilon(v - \gamma w).$$

(2.51)

The parameter $\epsilon$ has been added to more easily control the speed of one variable relative to the other, and the parameter $C$ affects the coupling strength. For an appropriate choice of all of the parameters, we would expect that this system of abstracted differential equations would produce oscillations (Exercise 10b).

This dynamical paradigm appeared first in cellular neuroscience in simplifications of the Hodgkin–Huxley equations by FitzHugh (1961), and later as part of an independent development by Nagumo et al. (1962). In fact, the various parameters we have contrived have physical interpretations in the context of a simplification of the Hodgkin–Huxley circuit diagram we discussed earlier. This type of system, involving a linear nullcline for the slow variable and a cubic nullcline that has the inverted "N" shape for the fast variable, are given the generic name *FitzHugh–Nagumo* (or FH-N) models. The FH-N model is discussed in more detail in Section 7.5.2 and in Appendix B.

## 2.7  Summary

Voltage gated ion channels and the currents that flow through them underlie much of the electrical behavior of cells. We derived a mechanism–based model for ion channels and found that we could produce oscillatory behavior in electrical membranes with only two variables. Therefore, we could analyze the underlying dynamics with phase plane techniques and other methods of dynamical systems analysis. The dynamical features of "slow" variables coupled to "fast" variables with either "N"-shaped or inverted "N"-shaped nullclines are characteristic of FitzHugh–Nagumo type oscillators common to many biological mechanisms at the cellular level. In addition to producing oscillations in the barnacle muscle, the same dynamical structures will appear in mechanistic models of insulin secretion (Chapter 4) and $Ca^{2+}$ oscillations (Chapter 5).

## Suggestions for Further Reading

Because of the introductory nature of this material, there are several excellent books to serve as resources for further study and a different perspective:

- *Principles of Neural Science*, edited by Eric Kandel, James Schwartz, and Thomas Jessel, and *Fundamental Neuroscience*, edited by Michael Zigmond, Floyd Bloom, Story Landis, James Roberts, and Larry Squire. These are general introductory textbooks on neuroscience, including chapters on ion channels and electrical behavior (Kandel et al. 2000; Zigmond et al. 1999).
- *Ionic Channels of Excitable Membranes*, Bertil Hille. As discussed in the text above, this is the "bible" of ion channels, with a particularly complete treatment of the Hodgkin–Huxley axon model and the work behind it by those and other scientists (Hille 2001).
- *Cellular Biophysics*, Volume 2, Thomas Weiss. Volume two of this two–volume set is a more expanded discussion of the electrical properties of cells, and also contains a chapter on the Hodgkin–Huxley model (Weiss 1996).

- *Foundations of Cellular Neurophysiology*, Daniel Johnston and Samuel Wu. Covers the introductory material discussed here as well as advanced topics such as transmitter release, plasticity, elementary networks, and extracellular electrical behavior (Johnston and Wu 1995).
- *Methods in Neuronal Modeling*, Christof Koch and Idan Segev, editors. This is a compilation of chapters from various authors on a wide variety of topics related to neuronal modeling. Particularly relevant is the chapter by John Rinzel and Bard Ermentrout on the "Analysis of Neural Excitability and Oscillations," which describes phase plane methods in the context of the Morris–Lecar model (Koch and Segev 1998).
- *Mathematical Models in Biology*, Leah Edelstein-Keshet. This is a great introductory textbook on general mathematical biology. Chapter 8 contains material on the FH-N oscillator and a general treatment of oscillations and phase plane analysis (Edelstein-Keshet 1988).
- *Mathematical Physiology*, James Keener and James Sneyd. Keener and Sneyd treat the topics presented in this chapter, as well as many other topics in physiology, from a more analytic perspective as opposed to the computational focus presented here (Keener and Sneyd 1998).

## 2.8  EXERCISES

1. The following problems explore calculation of the Nernst potential and the resting membrane potential.

   (a) Fill in the missing values in the following table:

   | Ion | Cytoplasmic Concentration (mM) | Extracellular Concentration (mM) | Equilibrium Potential (mV) |
   |-----|--------------------------------|----------------------------------|----------------------------|
   | $K^+$ | 400 | 20 | |
   | $Na^+$ | 50 | 440 | |
   | $Cl^-$ | 40 | 560 | |

   (b) Calculate the resting membrane potential given the following conductances:
   $$g_{Na} = 0.5 \cdot 10^{-6} \text{ S},$$
   $$g_K = 10 \cdot 10^{-6} \text{ S},$$
   $$g_{Cl} = 2.5 \cdot 10^{-6} \text{ S}.$$

   (c) Fill in the missing information in the following table:

   | Conductance Change | Ion Action | Membrane Potential Change |
   |--------------------|------------|---------------------------|
   | large $g_K$ increase | | |
   | | $Na^+$ enters | |
   | | | $V$ moves to $V_{Cl}$ |

2. Assume that a proto-cell has a membrane capacitance of 1 $\mu$F/cm$^2$ and contains only $K^+$ selective channels. The initial intracellular and extracellular concentrations for $K^+$ are given in the table above, and there are intracellular and extracellular anions $A^-$ equal in concentration to $K^+$.

(a) If the membrane $K^+$ channels are opened, what is the steady–state voltage across the membrane assuming that ion concentrations do not change?

(b) How many $K^+$ ions must move out of the cell per $cm^2$ to achieve a voltage difference equal to that in part (a)?

(c) Assuming that our proto-cell is perfectly round and that the extracellular $K^+$ concentration does not change, plot the final intracellular $K^+$ concentration as a function of the diameter of the cell. Use a range of realistic diameter values and the number of ions transfered that you calculated in part (b).

(d) Assuming that the extracellular $K^+$ concentration remains constant, what would the reversal potential be for the range of intracellular $K^+$ concentrations calculated in part (c).

(e) How do you reconcile parts (a) and (d)?

3. Derive (2.5) from (2.8). *Hint:* assume steady state and $I_{app} = 0$.

4. Verify the steps between (2.16) and (2.22).

5. Use the results in Exercise 4 to verify that the expressions in (2.23)–(2.24) are correct. Show that a special case of (2.24) when $\alpha = -\beta$ is

$$\tau = \frac{1}{2\sqrt{k_o^+ k_o^-}\cosh((V - V_o)/2S_o)}. \tag{2.52}$$

[*Hint:* Recall that $\tanh(x) = (e^x - e^{-x})/(e^x + e^{-x})$ and $\cosh(x) = (e^x + e^{-x})/2$.]

6. It is possible to estimate a characteristic time for the relaxation of the membrane potential using (2.27). When the channels are completely open, show that the equation can be written

$$dV/dt = -(V - V_{rev})/\hat{\tau} + I_{app}/C, \tag{2.53}$$

where $\hat{\tau} = C/\bar{g}$. Show that $\hat{\tau}$ is the characteristic time for relaxation of the voltage to a steady–state value $V^{ss} = V_{rev} + I_{app}/\bar{g}$.

7. For biological membranes a typical capacitance per unit area is $1\ \mu F/cm^2$, whereas conductances per unit area are in the range $10^{-4}$ to $10^{-3}$ $S/cm^2$. Using these estimates, show that typical relaxation times for the membrane potential are in the range of 1 to 10 ms.

8. Write a program suitable for simulating the Morris–Lecar equations (2.30)–(2.34) for the parameter values given in Table 2.4 and the initial conditions $V(0) = -60$ and $w(0) = 0.01$.

(a) Use the program to solve the Morris–Lecar model for the four values of $I_{app}$ given in Figure 2.9A.

(b) Plot the nullclines and simulate this model in the phase plane to verify that there is a unique stable limit cycle. Locate the steady states and calculate the eigenvalues for representative values of $I_{app}$.

(c) Show that the time constant of the delayed rectifier in the Morris–Lecar model determines whether or not the steady state at $I_{app} = 150$ pA is stable or unstable and also determines the maximum and minimum values of the voltage on the limit cycle.

(d) For $I_{app} = 60$ pA check that the value of $V(0)$ above which action potentials occur is close to the point where a line drawn parallel to the $V$ axis at $w^{ss}$ crosses the $V$-nullcline. Also verify that it will be exactly at that point in the limit that $w$ changes much more slowly than the voltage. [Hint: Increasing the parameter $\phi$ increases the time constant for $w$.]

(e) Explain how you could locate the unstable limit cycles near the bifurcation point in the Morris–Lecar model in Figure 2.9 by integrating the equations *backwards* in time. Use a plot to demonstrate this.

9. Exploration of the Hodgkin–Huxley equations.

(a) Read Chapter 2 of Hille's *Ionic Channels of Excitable Membranes* (Hille 2001). Write a paragraph explaining in words how the HH model works, and include an explanatory diagram.

(b) Research the voltage–dependent expressions for time constants and the asymptotic limits in (2.42)–(2.44), together with parameters. Replicate Figure 2.13. [*Hint*: See Hodgkin and Huxley (1952) or another source such as Keener and Sneyd (1998).]

10. System (2.51) constitutes a FitzHugh–Nagumo model. Select parameters for these equations that are likely to result in oscillatory behavior.

(a) Sketch or calculate the nullclines.

(b) Add a parameter to apply external current, and simulate action potentials by solving the equations in time.

# CHAPTER 3

# Transporters and Pumps

## Eric S. Marland and Joel E. Keizer

Ionic channels are not the only mechanism that cells use to transport impermeant species across membranes. Cells have developed a great variety of transport proteins for moving both ions and molecules from one cellular compartment to another. For example, to maintain the concentration imbalance of $Na^+$, $K^+$, and $Ca^{2+}$ across the plasma membrane it is necessary to pump ions against significant concentration gradients. In the case of $Ca^{2+}$ ions the ratio of concentrations outside to inside is greater than four orders of magnitude (ca. 2 mM outside and 0.1 $\mu$M inside). In addition to pumps, there are numerous specific cotransporters and exchangers that allow ions and small molecules to be transported selectively into internal compartments or out of the cell. Unlike ionic channels, for which the driving force is a passive combination of electrical potential and ionic concentration differences, most transporters and pumps expend considerable energy. In many animal cells, for example, it has been estimated that nearly 25% of the ATP that is utilized is devoted to maintaining low cytoplasmic $Na^+$ and high cytoplasmic $K^+$ concentrations via $Na^+/K^+$ pumps (Cooper 1997).

In this chapter we provide an overview of some of the mechanisms, other than ionic channels, that cells use to pump and transport small molecules and ions. We first introduce the ideas behind passive transport, using a passive glucose transporter (GLUT) as an example. We then introduce analytic, diagrammatic, and numerical methods for calculating rates of transport and apply them to a simplfied model of the GLUT transporter. Using the cotransport of glucose and $Na^+$ as an example, we then discuss how to create models of transporters and how this transporter functions physiologically in intestinal epithelial cells. A great deal is known about the kinetic steps involved in the pumping of $Ca^{2+}$ by $Ca^{2+}$ pumps in internal stores, and we use this mechanism to illustrate how phosphorylation by ATP drives the pumping mechanism. In the final section of the chapter we focus on the cyclic nature of these mechanisms and describe additional transporters that operate via comparable kinetic cycles.

# 3.1 Passive Transport

We must be a little careful about what we call passive transport, because there is no free lunch. The following example is not without its energy costs, but it uses background thermal energy rather than the explicit energy in a particular molecule such as ATP.

Glucose is a six–carbon sugar that is a major fuel for intermediary metabolism in animals. It is derived from carbohydrates in the gut and is transported through epithelial cells in the intestines into the blood stream and thence to the brain, pancreas, liver, muscle, and other organs. There glucose is taken up and metabolized via glycolytic enzymes. Although uptake from the gut involves active cotransport of glucose with sodium ions, peripheral tissues such as fat, muscle, and liver transport glucose via a class of passive membrane transporters referred to as GLUT transporters. Because glucose is a major energy source for cells, understanding the rate of glucose transport into cells via GLUT is important physiologically.

At least six isotypes of GLUT transporters have been isolated, GLUT1 through GLUT6, each of which is prevalent in one or more types of tissue (Bell et al. 1993). GLUT2, for example, is found in glucose-sensing pancreatic beta cells that secrete insulin from pancreatic tissue, as well as in liver cells. Extensive kinetic experiments have lead to a cartoon description of the steps involved in the tranport processes (Whitesell et al. 1991). The transitions of the transporter itself, facing the inside of the cell to facing the outside of the cell, is driven by heat or thermal fluctuation. Thus the heat of the system and the concentration gradient of glucose are the only driving forces in the system. Because no other energy is needed, it is called passive transport. It is similar in this sense to diffusion, which we discuss in Chapter 7.

Figure 3.1A shows four different states of the transporter. State $S_1$ has an empty binding site for glucose exposed to the exterior of the cell. When glucose binds to this state, the transporter makes a transition to state $S_2$, with glucose bound and facing the exterior. In $S_2$, a glucose molecule is bound to the transporter, which is still facing the exterior. State $S_3$ is the state with the transporter then facing the interior. When glucose dissociates from GLUT and ends up inside the cell, the transporter is left in state $S_4$. Finally, the cycle can repeat if $S_4$ makes the conformational transition to $S_1$. All of these processes are reversible.

This kinetic "cartoon" is easily translated into a conventional kinetic model of the sort often employed in biochemistry (Hill 1977). Here the model takes the form of the diagram in Figure 3.1B. The labeled corners in Figure 3.1B correspond to the states $S_1$–$S_4$ of a GLUT transporter described in the previous paragraph, and the lines represent elementary molecular processes. The transition from $S_1$ to $S_2$ is a *bimolecular* process, because it requires the interaction of a glucose molecule (indicated as $[G]_{out}$

**A**

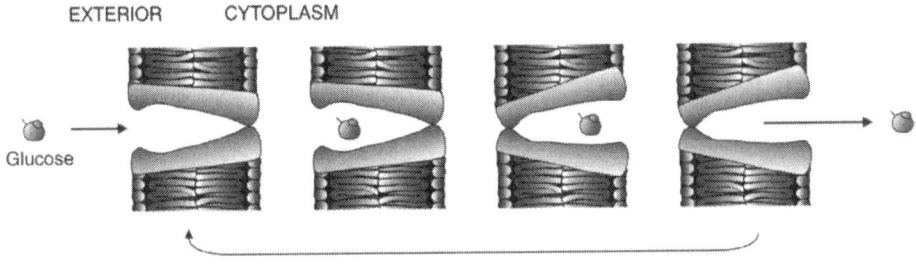

EXTERIOR    CYTOPLASM

Glucose

**B**

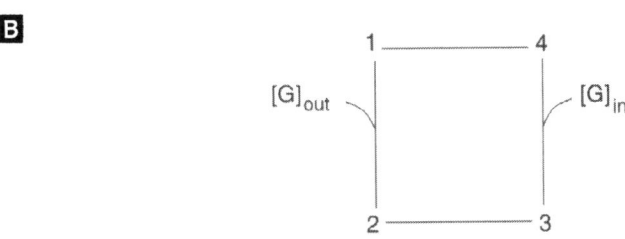

**Figure 3.1** (A) Cartoon of four states of a GLUT transporter, showing the empty pore facing the exterior of the cell, glucose bound facing the exterior, glucose bound facing the interior, and the open pore facing the interior of the cell. Adapted from Leinhard et al. (1992). (B) Four–state kinetic diagram of a GLUT transporter based on the cartoon in (A).

in the diagram) and the GLUT molecule in $S_1$. The transition from $S_2$ to $S_1$, on the other hand, involves only the GLUT molecule and is therefore unimolecular: Thus only $S_2$ appears at the end of the line connecting $S_2$ to $S_1$. This illustrates an important aspect of transitions between molecular states: They are reversible, corresponding to the property of microscopic reversibility of molecular processes.

The rates of the elementary processes depicted in the kinetic diagram are determined again by the *law of mass action* as we discussed in Chapter 1. Thus the rate of the transition from $S_1$ to $S_2$ is given by $J_{12} = k_{12}[G]_{out}x_1$, where the square brackets denote concentration, and $x_1$ represents the fraction of the GLUT molecules in $S_1$, $x_1 = N_1/N$, where $N$ is the total number of transporters. The factor $k_{12}$ is the rate constant, in this case bimolecular, with practical units of $s^{-1}mM^{-1}$. Similarly the rate of the reverse reaction, $2 \rightarrow 1$, is given by $J_{21} = k_{21}x_2$ with $k_{21}$ a unimolecular rate constant (with units $s^{-1}$). Table 3.1 lists the forward and reverse rate expressions for all of the processes in the kinetic diagram.

**Table 3.1   Rate Expressions for Glut Transporter**

| Forward Process | Rate | Reverse Process | Rate |
|---|---|---|---|
| $S_1 \to S_2$ | $k_{12}[G]_{out}\, x_1$ | $S_2 \to S_1$ | $k_{21}\, x_2$ |
| $S_2 \to S_3$ | $k_{23}\, x_2$ | $S_3 \to S_2$ | $k_{32}\, x_3$ |
| $S_3 \to S_4$ | $k_{34}\, x_3$ | $S_4 \to S_3$ | $k_{43}[G]_{in}\, x_4$ |
| $S_4 \to S_1$ | $k_{41}\, x_4$ | $S_1 \to S_4$ | $k_{14}\, x_1$ |

Having established the correspondence of the diagram with rate expressions, we can write down the differential equations that the diagram represents. To do so we must keep track of the change that each elementary process in the diagram makes for each state. Thus the fraction $x_1$ of transporters in $S_1$ decreases with the transition to $S_2$ or $S_4$, and increases with transitions from $S_1$ or $S_4$. Using this idea, in conjunction with the kinetic diagram and Table 3.1, the ordinary differential equations follow for the rate of change in the number of states:

$$dx_1/dt = -k_{12}[G]_{out}\, x_1 + k_{21}x_2 + k_{41}x_4 - k_{14}x_1,$$
$$dx_2/dt = k_{12}[G]_{out}\, x_1 - k_{21}x_2 - k_{23}x_2 + k_{32}x_3,$$
$$dx_3/dt = k_{23}x_2 - k_{32}x_3 - k_{34}x_3 + k_{43}[G]_{in}\, x_4,$$
$$dx_4/dt = k_{34}x_3 - k_{43}[G]_{in}\, x_4 - k_{41}x_4 + k_{14}x_1. \tag{3.1}$$

Because the kinetic model involves only interconversion of GLUT states, the total number of transporters should be preserved. Again we use a conservation law, $N_1 + N_2 + N_3 + N_4 = N$ or $x_1 + x_2 + x_3 + x_4 = 1$, to ensure that transporters are neither created nor destroyed. This condition can be checked by adding together the expressions on the right–hand side of (3.1). It is easily verified that all of the terms cancel, leading to the result $d(x_1 + x_2 + x_3 + x_4)/dt = 0$, which shows that the sum of the fractions of transporters in different states does not change. That is, $x_1 + x_2 + x_3 + x_4$ has a constant value, which in this case is 1.

We see again that one of the dependent variables can be eliminated using the conservation law by writing $x_4 = 1 - x_1 - x_2 - x_3$. The differential equation for $x_4$ becomes redundant and the number of differential equations to be solved is reduced to only three along with the algebraic equation for $x_4$:

$$dx_1/dt = m_{11}x_1 + m_{12}x_2 + m_{13}x_3 + k_{41},$$
$$dx_2/dt = m_{21}x_1 + m_{22}x_2 + m_{23}x_3,$$
$$dx_3/dt = m_{31}x_1 + m_{32}x_2 + m_{33}x_3 + k_{43}[G]_{in},$$
$$x_4 = 1 - x_1 - x_2 - x_3. \tag{3.2}$$

where the $3 \times 3$ array $m_{ij}$ is a matrix whose elements can be found by substituting $x_4 = 1 - x_1 - x_2 - x_3$ into the first three equations of (3.1). We leave it to the reader to verify, for example, that $m_{11} = -(k_{12}[G]_{out} + k_{14})$ and $m_{33} = -(k_{32} + k_{34} + k_{43}[G]_{in})$.

With the formulation of the model equations, the first steps in the modeling process are completed. What remains is the analysis of the equations. We will leave the analysis of this model for an exercise, and we will use a reduced model to demonstrate several ways to analyze the transport rates for transporters.

## 3.2 Transporter Rates

In the previous section we described a four-state model of a GLUT-type glucose transporter. Like all models of transporters, the "states" are distinct molecular arrangements of the transporter protein and the small molecules (*ligands*) that interact with it. These states are the basic kinetic ingredients of the kinetic mechanism, and we identify the states by numbering them $S_1, S_2, \ldots, S_M$, where $M$ is the total number. For the GLUT transporter, we summarize the mechanism by a kinetic diagram with lines between the states representing possible transitions. Each line stands for a forward and reverse step that can be either unimolecular or bimolecular. Which states are connected together and the nature of the transitions connecting the states must be determined experimentally, and the resulting diagrams summarize succinctly a great deal of kinetic information. As we have seen, these state diagrams can easily be translated into differential equations that describe how the number of transporters in each state changes with time.

What is important physiologically is not the rate at which states of a transporter change with time but rather the rate at which the ions or molecules that are transported get across the membrane. This can be determined from the kinetic mechanism but requires additional analysis, because the transport rate is a property of the entire mechanism rather than an individual step. To make this distinction clear, consider the simplified three-state version of the mechanism for a GLUT transporter in Figure 3.2. In this simplification, $S_1$ and $S_4$ in the four-state diagram in Figure 3.1 have been treated as a single state. The reasons why we can make this type of simplification of a mechanism are described more fully in Chapter 4. The three transitions represented are the binding of glucose to the transporter from the exterior ($S_1$–$S_2$) and interior ($S_1$–$S_3$) of the cell and the conformation change in which the glucose moves from the exterior to the interior ($S_2$–$S_3$). Using $x_i$ to represent the fraction of the total number

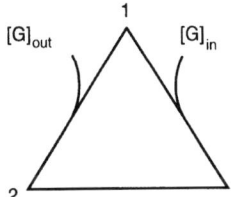

**Figure 3.2** A simplified three-state diagram for the GLUT transporter. $S_1$ and $S_4$ in the four-state diagram in Figure 3.1 have been treated as a single state.

of transporters in state $S_i = S_1, S_2, S_3$, the kinetic equations for this diagram are

$$dx_1/dt = -J_{12} + J_{31},$$
$$dx_2/dt = J_{12} - J_{23},$$
$$dx_3/dt = J_{23} - J_{31}. \tag{3.3}$$

Here the $J_{ij}$ are the *net rates* for the indicated transition, with the $i \to j$ direction taken as positive. Thus,

$$J_{12} = k_{12}^*[G]_{\text{out}}x_1 - k_{21}x_2, \tag{3.4}$$
$$J_{23} = k_{23}x_2 - k_{32}x_3, \tag{3.5}$$
$$J_{31} = k_{31}x_3 - k_{13}^*[G]_{\text{in}}x_1. \tag{3.6}$$

In these equations we use a superscript $*$ to indicate a bimolecular rate constant. Net rates often appear in analyzing transport rates, and we will refer to the $J_{ij}$ as *fluxes*. We can see by adding equations (3.3) together that the total number of tranporters is conserved: $x_1 + x_2 + x_3 = 1$ is already satisfied.

Experimentally, the rate at which glucose is transported into the cell is determined by the rate at which the concentration of glucose accumulates inside the cell in the absence of metabolism. Because $[G]_{\text{in}}$ is measured in units of millimolar, its rate of change of can be determined from the diagram and the rate equations (3.3) to be

$$\frac{d[G]_{\text{in}}}{dt} = \frac{\text{millimoles of transporter}}{\text{cellular volume}} \cdot J_{31} = \left(\frac{10^3 N}{V_{\text{in}} A}\right) \cdot J_{31} = R_{\text{in}}, \tag{3.7}$$

where A is Avogadro's number, $V_{\text{in}}$ is the cellular volume in liters, and the factor $10^3 N/A$ converts the total number of transporters $N$ to millimoles. A related measure of the transport rate is the rate of change of $[G]_{\text{out}}$, which give a measure of the transport rate $R_{\text{out}}$ based on the loss of glucose from outside the cell. In analogy to (3.7), $R_{\text{out}}$ is easily seen to be given by

$$-\frac{d[G]_{\text{out}}}{dt} = \left(\frac{10^3 N}{V_{\text{out}} A}\right) \cdot J_{12} = R_{\text{out}}. \tag{3.8}$$

In order to simplify the interpretation of experiments, $V_{\text{out}}$ is usually chosen to be much greater than $V_{\text{in}}$, and so to a good approximation $[G]_{\text{out}}$ can be taken as a constant.

These two measures of the rate of transport of glucose are generally not proportional to one another, because $J_{12} \neq J_{31}$. Inspection of (3.3) shows, however, that the two fluxes are equal at steady–state, in which case

$$J_{12}^{\text{ss}} = J_{23}^{\text{ss}} = J_{31}^{\text{ss}} = J^{\text{ss}}. \tag{3.9}$$

This state condition can be inferred directly from the diagram in Figure 3.2 by noting that the total flux into and out of each state must vanish for the number of transporters in each state to be steady. Thus at steady–state $J^{\text{ss}}$ provides a unique measure of the transport rate, which can be written

$$R^{ss} = \left( \frac{10^3 N}{V_{in} A} \right) \cdot J^{ss}. \tag{3.10}$$

This rate is achieved, however, only after a transient period during which the states of the transporter come to steady–state.

To calculate the transport rate using (3.10) it is necessary to calculate the value of $J^{ss}$, and this, in turn, requires the steady–state values of the $x_i$. There are three ways that this can be done: numerically, by solving the differential equations; with linear algebra, which gives an analytical expression for the $x_i$; or using diagrammatic methods. The first method is explored in the exercises.

## 3.2.1 Algebraic Method

We can obtain the steady–state value of the transport rate by solving the linear equations for the $x_i$. Substituting the expressions for the fluxes given in (3.4)–(3.6) into (3.3) and eliminating $x_3$ using the conservation condition $x_3 = 1 - x_1 - x_2$ gives the $2 \times 2$ linear equations

$$dx/dt = \hat{A}x + y \tag{3.11}$$

with

$$\hat{A} = \begin{pmatrix} -(k_{12} + k_{13} + k_{31}) & k_{21} - k_{31} \\ k_{12} - k_{32} & -(k_{21} + k_{23} + k_{32}) \end{pmatrix} \text{ and } y = \begin{pmatrix} k_{31} \\ k_{32} \end{pmatrix}. \tag{3.12}$$

To simplify notation we have introduced the *pseudo–unimolecular* rate constants

$$k_{12} = k_{12}^*[G]_{out} \text{ and } k_{13} = k_{13}^*[G]_{in}. \tag{3.13}$$

The steady state of (3.11) is determined by the algebraic equation

$$\hat{A}x^{ss} = -y. \tag{3.14}$$

This equation has the solution

$$x^{ss} = -\hat{A}^{-1}y, \tag{3.15}$$

with $\hat{A}^{-1}$ the inverse matrix of $\hat{A}$. Using the explicit expression for the $\hat{A}^{-1}$ given in Appendix A, (3.15) gives

$$x_1^{ss} = (a_{22}y_1 - a_{12}y_2) \det \hat{A},$$
$$x_2^{ss} = (-a_{21}y_1 + a_{11}y_2) \det \hat{A},$$
$$x_3^{ss} = 1 - x_1^{ss} - x_2^{ss}. \tag{3.16}$$

To evaluate the transport rate using the algebraic method we need to substitute the expressions in (3.16) into one of the expressions in (3.9) for $J^{ss}$. For example, using the

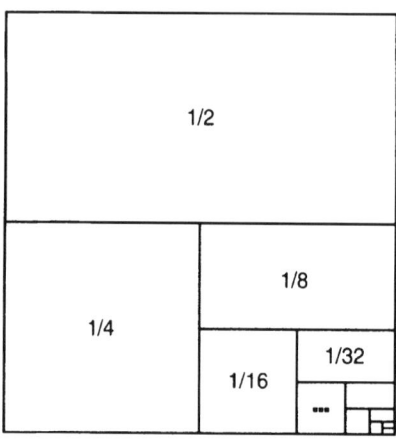

**Figure 3.3** A diagram representing the geometric series in (3.18).

first expression gives

$$J_{12}^{ss} = \frac{1}{\det \hat{A}}(k_{12}(a_{22}y_1 - a_{12}y_2) - k_{21}(-a_{21}y_1 + a_{11}y_2)).  \tag{3.17}$$

To get an explicit expression in terms of the rate constants $k_{ij}$ using (3.17) we need to substitute the expressions for the matrix elements $a_{ij}$ given in (3.12) and then calculate the determinant of $\hat{a}$. The resulting expression is messy and offers numerous opportunities for making algebraic mistakes. An alternative is to use the diagrammatic method, which circumvents all of these algebraic difficulties.

## 3.2.2   Diagrammatic Method

Diagrams that represent mathematical expressions are in common use in both quantum electrodynamics ("Feynman diagrams") and statistical mechanics but are significantly less familiar in biology. Although the diagrammatic method for obtaining $J^{ss}$ involves a few new ideas, it leads to vastly simpler, more transparent expressions for the fluxes than those provided by the algebraic method. To help motivate the use of diagrams, consider the following infinite sum:

$$\frac{1}{2} + \frac{1}{4} + \cdots + \frac{1}{2^n} + \cdots = \sum_{n=1}^{\infty} \frac{1}{2^n}.  \tag{3.18}$$

Even if the reader has previously encountered this geometric series, few probably remember that the sum converges exactly to the value one. On the other hand, a simple glance at the diagram in Figure 3.3 makes the answer clear immediately.

Diagrams of the sort that are used in solving for the fluxes for the three-state GLUT transporter are shown in Figure 3.4. In general, a diagram is a set of vertices (representing the states) and lines representing unimolecular (or pseudo–unimolecular) transitions between states. A *complete diagram* for the GLUT transporter, which in-

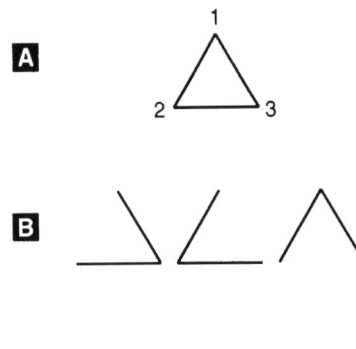

**Figure 3.4** (A) The complete diagram, (B) partial diagrams, and (C) directed diagrams for state $S_1$ of the 3-state GLUT transporter model.

cludes all of the lines and all of the vertices in the model, is shown in Figure 3.4A. Note that it differs from the kinetic diagram in Figure 3.1, because the glucose concentrations have been absorbed into the pseudounimolecular rate constants as in (3.13). The three *partial diagrams* for the model (with the vertices unlabeled) are shown in Figure 3.4B. Partial diagrams are obtained from the complete diagram by removing lines, and they have the maximum number of lines possible without forming a cycle. A third type of diagram, called a , *directional diagram* can be constructed from the partial diagrams. Directional diagrams have arrowheads attached to the lines such that all of the lines "flow" into a single vertex. The directional diagrams for state $S_1$ are given in Figure 3.4C. Note that for ease in writing, the arrowhead on a line will be dropped whenever its direction is obvious, as in the first and second directional diagrams in Figure 3.4C. Three comparable directional diagrams can be drawn for states $S_2$ and $S_3$.

There are diagrams represent algebraic expressions just as the areas in the diagram for the geometric series in Figure 3.3 represent fractions. For these diagrams each line with an arrowhead (a *directed line*) represents the unimolecular or pseudounimolecular rate constant for the indicated transition. For example, the two lines in the third diagram of Figure 3.4C represent $k_{21}$ and $k_{31}$. Diagrams with several directed lines represent the product of all the indicated rate constants. Thus the the directional diagrams stand for products of two rate constants. For example, the first directional diagram in Figure 3.4C is shorthand notation for the product $k_{23}k_{31}$, whereas the second and third diagrams represent $k_{21}k_{32}$ and $k_{21}k_{31}$, respectively.

There is a general theorem that connects the directional diagrams with the fractional occupancy of states in the kinetic diagram. In particular, the fractional occupancy of state $S_i$ is given by the expression

$$x_i^{\text{ss}} = \frac{\text{sum of all state } S_i \text{ directional diagrams}}{\text{sum of all directional diagrams}}. \tag{3.19}$$

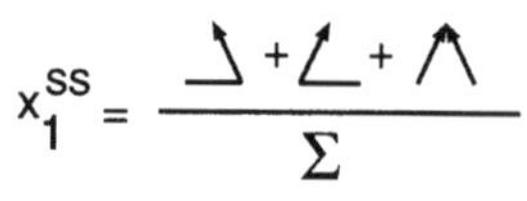

$$x_1^{ss} = \frac{\diagdown + \diagup + \diagup\diagdown}{\Sigma}$$

$$= \frac{k_{23}k_{31} + k_{32}k_{21} + k_{21}k_{31}}{\Sigma}$$

**Figure 3.5** A diagrammatic expression for the fractional occupancy of GLUT transporters in state $S_1$ at steady state.

For $x_1^{ss}$ this yields the expressions given in Figure 3.5, where we have adopted the symbol $\Sigma$ to represent the sum of all directional diagrams. The division by $\Sigma$ ensures that $\sum_i x_i^{ss} = 1$.

Although the expression for $x_i^{ss}$ in (3.19) can be proven for any mechanism that can be represented by a kinetic diagram, we give a proof only for the three-state GLUT model. Because the steady–state solution for this model is unique, we need only show that the expression in (3.19) leads to the equality of all the fluxes $J_{ij}$ at steady–state. Rather than write out the algebra, we use the diagrams themselves to complete the proof. This is illustrated in Figure 3.6, where $J_{12}^{ss}$ is calculated. The second equality uses (3.19), and in the third we have used the directed lines corresponding to the rate constants $k_{12}$ and $k_{21}$ to add extra directed lines to the diagram. Two pairs of terms cancel to give the final equality, in which only the difference of the two *cyclic diagrams* appears. A cyclic diagram is derived from a partial diagram with one additional flux added to produce a cycle. Similar manipulations show that the third equality also holds for $J_{23}^{ss}$ and $J_{31}^{ss}$ (Exercise 5). Thus $J_{12}^{ss} = J_{23}^{ss} = J_{31}^{ss}$, which is the condition for steady state.

The final equality in Figure 3.6 is a corollary that can be generalized for any kinetic diagram, i.e.,

$$J_{ij}^{ss} = \frac{\text{sum of differences of cyclic diagrams with } i, j \text{ in the cycle}}{\Sigma}. \tag{3.20}$$

Thus the steady–state flux for the three-state model is given by the difference of the two cyclic fluxes (counterclockwise − clockwise) divided by the sum over all partial diagrams for the complete diagram.

### 3.2.3  Rate of the GLUT Transporter

We need one more key fact about cyclic diagrams in order to simplify the expression for the transport rate: The product of the bimolecular and unimolecular rate constants in the counterclockwise direction of a cycle equals that of those in the clockwise direction. This is called the *thermodynamic restriction* on the rate constants because it is a consequence of the laws of chemical thermodynamics. To see why the thermodynamic restriction is true, consider the situation in which no transport occurs, i.e., $[G]_{out} = [G]_{in}$. If we revert to the notation for chemical reactions, then the three steps

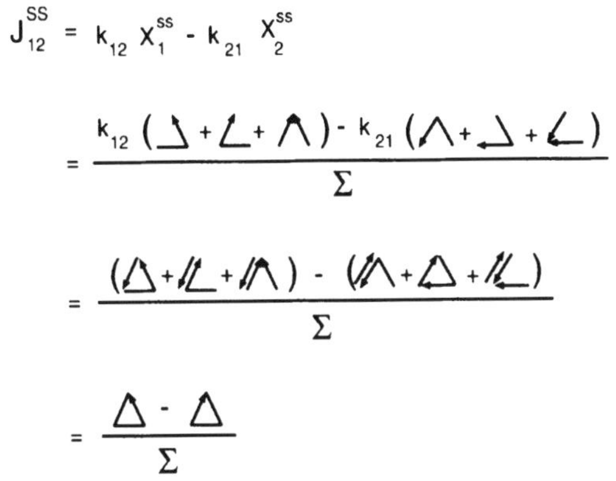

$$J^{ss}_{12} = k_{12}\, X^{ss}_1 - k_{21}\, X^{ss}_2$$

$$= \frac{k_{12}\,(\lrcorner + \angle + \wedge) - k_{21}\,(\wedge + \lrcorner + \measuredangle)}{\Sigma}$$

$$= \frac{(\triangle + \measuredangle + \wedge) - (\wedge + \triangle + \measuredangle)}{\Sigma}$$

$$= \frac{\triangle - \triangle}{\Sigma}$$

**Figure 3.6** Calculation of $J^{ss}_{12}$ using diagrams.

in the cycle for the three-state GLUT transporter can be written

$$G_{out} + S_1 = S_2 \text{ with equilibrium constant } K_{12} = k_{21}/k^*_{12},$$
$$S_2 = S_3 \text{ with equilibrium constant } K_{23} = k_{32}/k_{23},$$
$$S_3 = S_1 + G_{in} \text{ with equilibrium constant } K_{31} = k^*_{13}/k_{31}. \qquad (3.21)$$

where $S_1$, $S_2$, and $S_3$ represent the three states of the transporter. It is easy to show that the equilibrium constants $K_{ij}$ for the "reactions" are the ratios of the rate constants, as indicated next to each reaction in (3.21) (see Exercise 6). If we add these three chemical reactions together we get the net reaction

$$G_{out} = G_{in}. \qquad (3.22)$$

A basic property of equilibrium constants is that when reactions are added, the equilibrium constants are multiplied. Therefore, the equilibrium constant for the net reaction (3.22) is

$$K_{net} = K_{12}K_{23}K_{31} = \frac{k_{21}k_{32}k^*_{13}}{k^*_{12}k_{23}k_{31}}. \qquad (3.23)$$

But at chemical equilibrium the concentrations of product ($G_{in}$) and reactant ($G_{out}$) in (3.22) are equal, so that $K_{net} = [G]^{eq}_{in}/[G]^{eq}_{out} = 1$. Using this fact in (3.23) and rearranging gives the following thermodynamic restriction on the rate constants:

$$k^*_{13}k_{32}k_{21} = k^*_{12}k_{23}k_{31}. \qquad (3.24)$$

*When constructing models, it is essential that the thermodynamic restriction on rate constants be satisfied for all cycles. Otherwise, the model will violate the second law of thermodynamics.*

Using the results in the previous sections we can write an explicit expression for the rate of the three-state GLUT transporter. Combining (3.10) with the final equation in Figure 3.6, we obtain

$$R^{ss} = \frac{10^3 N}{V_{in} A} \cdot \frac{k_{12}^*[G]_{out} k_{23} k_{31} - k_{13}^*[G]_{in} k_{32} k_{21}}{\Sigma} = \frac{10^3 N k_{12}^* k_{23} k_{31}([G]_{out} - [G]_{in})}{V_{in} A \Sigma}, \quad (3.25)$$

where in the second equality we have used (3.24), the thermodynamic restriction on the rate coefficients. Again, $V_{in}$ is the volume. According to (3.25), the steady–state transport rate is positive when the concentration of glucose outside of the cell exceeds that inside, and vanishes when the two concentrations are the same. This is a consequence of the thermodynamic restriction on the rate constants and is just what is expected for a passive transport mechanism. In the next section we consider the $Na^+$/glucose cotransporter, which utilizes a gradient of $Na^+$ to transport glucose from a low concentration to a higher concentration.

For a *symmetric* transporter there is no difference between the kinetic steps occurring inside and outside of the cell. This means that the rate constants for the transitions $2 \to 3$ and $2 \leftarrow 3$ are the same and that the association and dissociation rate constants are the same inside and outside as well. In this case there are only three different rate constants:

$$k_{12}^* = k_{13}^* = k^+ \text{ (glucose association)},$$
$$k_{21} = k_{31} = k^- \text{ (glucose dissociation)},$$
$$k_{23} = k_{32} = k \text{ (transport)}. \quad (3.26)$$

It is not difficult to evaluate the sum $\Sigma$ of the directed diagrams explicitly in this case, which is

$$\Sigma = k^+(2k + k^-)(K + [G]_{out} + [G]_{in}), \quad (3.27)$$

where we have written the dissociation constant $K = (k^-/k^+)$. Thus for the symmetric GLUT transporter model the transport rate can be written

$$R^{ss} = \frac{R_{max}([G]_{out} - [G]_{in}))}{K + ([G]_{out} + [G]_{in})}, \quad (3.28)$$

where the maximal rate is

$$R_{max} = \frac{10^3 N k k^-}{(2k + k^-)V_{in} A}. \quad (3.29)$$

Equations (3.28) and (3.29) provide explicit expressions for the transport rate for the symmetric transporter in terms of the rate constants for the model.

Glucose uptake can be measured experimentally using 3-O-methyl glucose, a non-metabolizable analogue of glucose. This further simplifies the expressions, because the concentration of the analogue is initially zero, $[G]_{in} = 0$, inside the cell. As a practical matter experiments involve large numbers of cells rather than a single cell. However, both $N$ and $V_{in}$ increase in proportion to the number of cells, so that the value of $R_{max}$ is still characteristic of a single cell. So for this type of experiment the rate expression

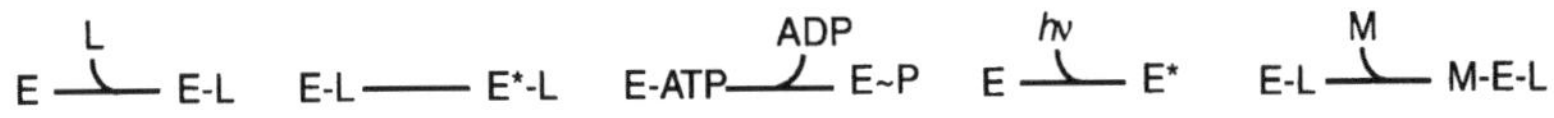

$$E \xrightarrow{\;\;L\;\;} E\text{-}L \qquad E\text{-}L \xrightarrow{\hspace{1cm}} E^*\text{-}L \qquad E\text{-}ATP \xrightarrow{\;\;ADP\;\;} E\text{\textasciitilde}P \qquad E \xrightarrow{\;\;h\nu\;\;} E^* \qquad E\text{-}L \xrightarrow{\;\;M\;\;} M\text{-}E\text{-}L$$

**Figure 3.7** Elementary kinetic processes for transporters representing ligand (*L*) binding, ligand transport, phosphorylation, light excitation, and multiple ligand binding.

in (3.28) can be written

$$R^{ss} = \frac{R_{\max}[G]_{out}}{K + [G]_{out}}. \tag{3.30}$$

One way to analyze the experimental rate of glucose uptake is using an Eadie–Hofstee plot. The Eadie–Hofstee plot is a graph of the experimental rate of glucose uptake $R^{ss}$ versus $R^{ss}/[G]_{out}$ for a range of values of $[G]_{out}$. According to (3.30) this plot should give a straight line with $y$-intercept equal to $R_{\max}$ and slope equal to $K$. This can be seen by first rearranging (3.30) to get

$$R^{ss}/[G]_{out} = \frac{R_{\max}}{K + [G]_{out}}. \tag{3.31}$$

Then we multiply both sides by $K+[G]_{out}$ and divide by $R^{ss}/[G]_{out}$, rearranging to obtain:

$$[G]_{out} = \frac{R_{\max}}{R^{ss}/[G]_{out}} - K. \tag{3.32}$$

If this expression for $[G]_{out}$ is substituted in the second factor in the identity

$$R^{ss} = \frac{R^{ss}}{[G]_{out}} \cdot [G]_{out}, \tag{3.33}$$

we obtain

$$R^{ss} = R_{\max} - \frac{R^{ss}}{[G]_{out}} \cdot K, \tag{3.34}$$

which is the Eadie–Hofstee expression for the rate. Exercise 9 illustrates how transport rates can be simulated for the four-state model of a GLUT transport and how to analyze the results using an Eadie–Hofstee plot.

## 3.3 The Na+/Glucose Cotransporter

A great variety of specialized proteins have evolved to transport specific substances across membranes in cells. Whereas the mechanisms of these transporters differ in detail, they also share a number of common features. For example, all of the known transporters bind the ligand or ligands that they transport, and of course, they must dissociate them as well. These steps must occur on both sides of the membrane for transport to occur, so there must be a process or processes in which the ligands are transported across the membrane. Figure 3.7 illustrates some of the elementary kinetic processes that are found for transporters, including chemical modification of the

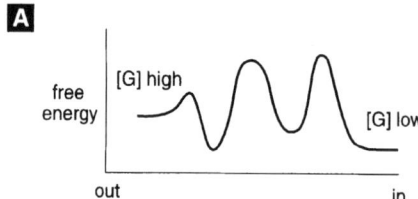

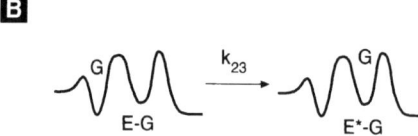

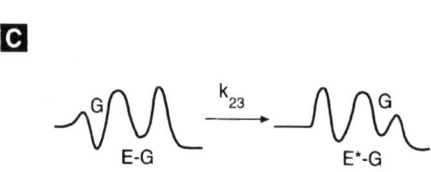

**Figure 3.8** (A) Schematic representation of the free energy profile across a membrane for a transporter protein. Two possibilities for state changes for the transport step 2 to 3 for the GLUT transporter are indicated: (B) simple barrier crossing; and (C), barrier crossing via a conformational transition.

transporter by phosphorylation, light-induced conformational changes, and multiple ligand binding.

From the point of view of chemical physics, the transport step can be viewed as energetic rearrangements that involve both the transporter protein and the ligand being transported. Figure 3.8A is a schematic representation of what the energy profile across a GLUT transporter might look like when the glucose concentration outside is high and inside is low. The energy profile is the Gibbs free energy, rather than the potential energy: For the average kinetic events that we are considering the influence of entropy effects must be taken into account. The reason that the free energy of glucose is higher outside is simply that the concentration of glucose is higher outside the cell than inside. The peaks of the free energy represent barriers to the movement of glucose across the transporter. Two possibilities for the transition from state 2 to state 3 are shown in Figure 3.8Band Figure 3.8C. The first represents a barrier crossing in which the transition $2 \rightarrow 3$ does not influence the shape of the energy profile. In the second, on the other hand, the energy profile is different after the transition, as might be the case if the transition involved a conformational change.

Although understanding the transport step is an important feature of building a model of a transporter, it describes neither how a transporter works nor the rate of transport, which was seen in Section 3.2 to be a property of the complete model, not a single step. To illustrate how a complete model of a transporter is created, we consider the $Na^+$/glucose cotransporter from intestinal epithelial cells. This transporter utilizes a concentration gradient of $Na^+$ to transport glucose from the intestine into the epithelial

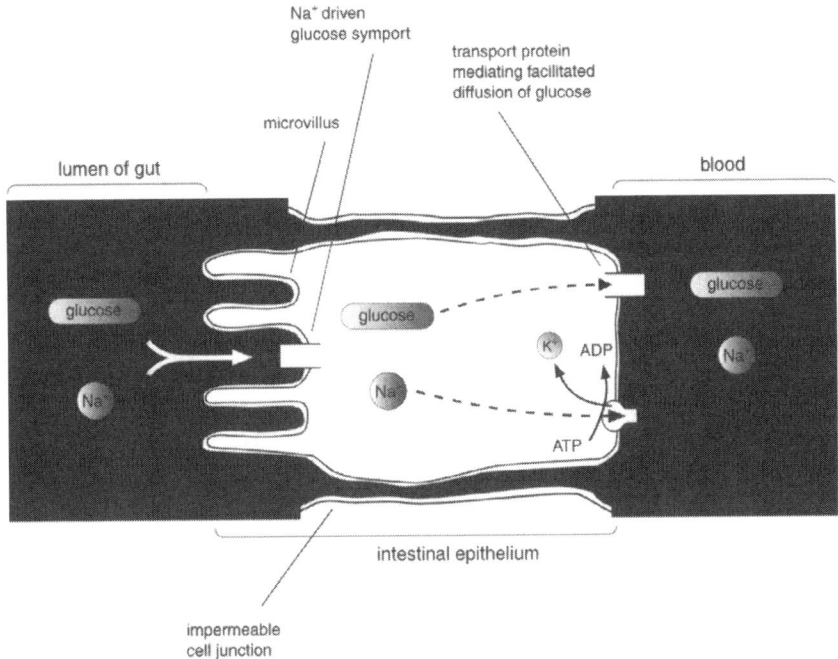

**Figure 3.9**  A cartoon representing the cotransport of glucose and Na⁺ into intestinal epithelium, followed by the passive transport of glucose into the blood. Energy stored in the gradient of Na⁺ (higher in the lumen of the gut) is utilized to transport glucose from a high concentration to a low concentration. Na⁺ that accumulates in the epithelial cells is removed by active transport into the blood by the Na⁺/K⁺ ATPase. Adapted from Alberts et al. (1994).

cells that line the gut. This is "uphill" transport, because the concentration of glucose in the epithelial cells exceeds that in the intestine. As shown schematically in Figure 3.9, the cotransporter works in concert with a Na⁺/K⁺ ATPase and passive transport of glucose by GLUT transporters, both at the basolateral side of the epithelium, to move glucose from the intestine to the blood stream. The Na⁺/K⁺ ATPase helps eliminate the Na⁺ that accompanies glucose uptake during cotransport, thereby maintaining a low concentration of Na⁺ inside of the cell.

A model for any transporter must incorporate a number of basic experimental facts. One of these is *stoichiometry*, which for the cotransporter is the number of Na⁺ ions transported per glucose molecule. Experimental measurements on the Na⁺/glucose cotransporter from intestine yield a stoichiometry of 2 Na⁺ to 1 glucose. Another important fact about the cotransporter is the absolute requirement for Na⁺. If Na⁺ is absent from the external medium, glucose is not transported. In addition, the cotransporter

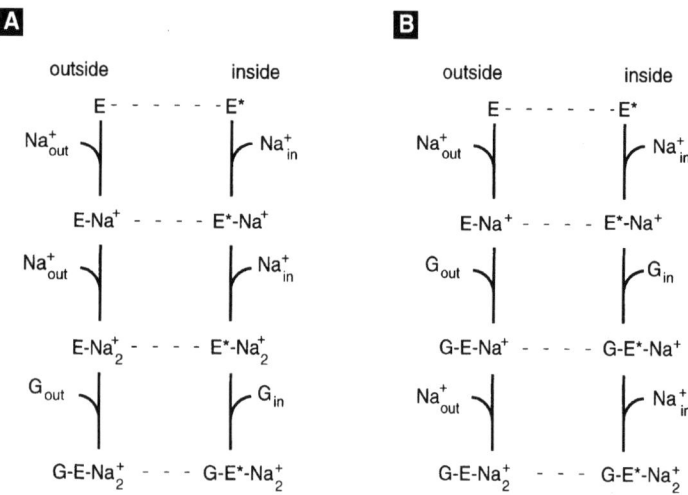

**Figure 3.10** Two possible skeleton diagrams representing binding of $Na^+$ and glucose for the $Na^+$/glucose cotransporter. Only (A) agrees with the experimental stoichiometry.

is *electrogenic*, because transport generates an electrical current due to the transport of $Na^+$.

These observations require 2 $Na^+$ and 1 glucose association steps on each side of the membrane. A partial skeleton for the cotransporter might contain the kinetic steps connected together as shown in Figure 3.10A. It is also possible that the second sodium binds *after* the glucose as in Figure 3.10B. However, this can be ruled out if the states with $Na^+$ and glucose bound from the outside are connected by conformational transitions to comparable states inside (as indicated by the dashed lines). In that case the six state cycle in Figure 3.10B (E to E-$Na^+$ to G-E-$Na^+$ to G-E*-$Na^+$ to E*-$Na^+$ to E* to E) would transport only a single $Na^+$ for every glucose molecule, implying a stoichiometry at steady state less than 2:1. The third possibility, not shown in Figure 3.10, is that glucose binds first. This is ruled out, however, by the experimental observation that the cotransporter supports $Na^+$ currents even in the absence of glucose. For details see (Parent et al. 1992a). Thus we are left with the ordered binding of ligands indicated on the left in Figure 3.10.

If we number the eight states on the left in Figure 3.10 sequentially $S_1$ through $S_8$, starting at "E" and moving counterclockwise, there are a number of possibilities for conformational changes connecting the left and right sides of the diagram. Figure 3.11 illustrates six alternatives. Alternative (B) is easily eliminated, because it does not transport glucose. Although diagram (C) does transport glucose ($S_4$ to $S_5$), it does not includes steps that transport only $Na^+$, and therefore conflicts with the fact that the transporter produces a $Na^+$ current in the absense of glucose. Diagram (D) can be ruled out because it has the wrong stoichiometry (1 $Na^+$:1 glucose). This leaves as possible

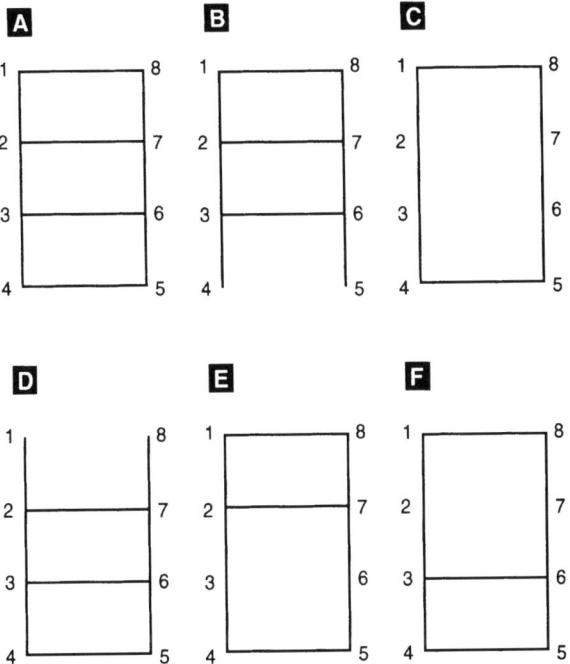

**Figure 3.11** Six possible diagrams for the Na$^+$/glucose transporter with transport steps included. Only (A), (E), and (F) are compatible with experiment.

mechanisms diagram (A), which is fully connected, and diagrams (E) and (F), each of which is missing Na$^+$ transport steps.

All three of these diagrams are compatible with the experimental evidence, and all three can be "reduced" to a diagram with the 6-state skeleton given in Figure 3.12A. This method of reducing diagrams uses the *rapid equilibrium approximation* that applies to steps for which the forward and reverse rates are rapid with respect to other steps in the diagram. The details of how this method works are explained in Chapter 4, although the basic idea can be seen by comparing Figure 3.12A and Figure 3.12B. The experimental values of rate constants for the six–state model have been assigned by Parent and colleagues (Parent et al. 1992b). Step $S_4$ to $S_5$ in the six–state model is the dissociation of glucose inside the cell, and this step is extremely fast. This permits the two states to be approximated as a single combined state (state $S_{4,5}$ in Figure 3.12B and reduces the diagram to five states as shown. We must be careful in doing so to readjust the rates to account for the reduction. The details of the process for doing this is given in Chapter 4, but it is not difficult. In short, only a portion of the combined state $S_{4,5}$ reacts to the other states. The portion to be used in each reaction is determined as a result of the reduction using simple algebra.

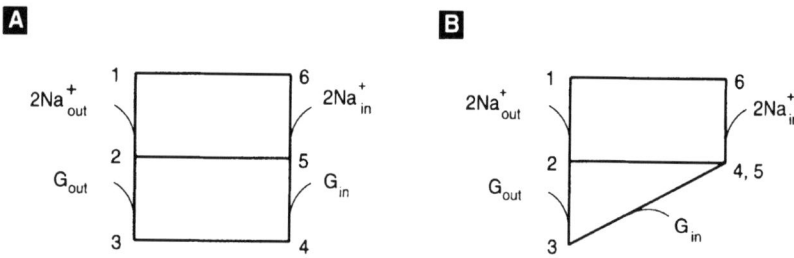

**Figure 3.12** (A) Six-state and (B) five-state simplification of the eight-state diagram for the Na$^+$/glucose cotransporter.

It is possible to write diagrammatic expressions for the transport rate for either the five-state or six-state models in Figure 3.12. However, the number of directed diagrams and cyclic diagrams increases quickly with the complexity of cycles in the complete diagram. For example, for the five-state model there are 6 pairs of cyclic diagrams and 55 directed diagrams. Nonetheless, the general expressions in (3.19) and (3.20) remain valid and can be used to obtain explicit expressions for the steady–state fluxes. The diagrammatic method does not, however, provide information about the transient time-dependence of the fluxes. This is most conveniently obtained by numerical integration of the equations.

## 3.4  SERCA Pumps

The Ca$^{2+}$-ATPase that is found in the endoplasmic reticulum (ER) and sarcoplasmic reticulum (SR) of muscle is typical of transporters that utilize the chemical energy stored in ATP to pump ions against a gradient. Typical free cytosolic Ca$^{2+}$ concentrations [Ca$^{2+}$]$_i$ are of the order of 0.1 $\mu$M, whereas Ca$^{2+}$ concentrations in the ER and SR are in the range of 0.1 to 1 mM. These pumps, which are abbreviated SERCA for "Sarco-Endoplasmic Reticulum Ca$^{2+}$ ATPase," therefore have to surmount a 3 to 4 order of magnitude concentration difference. Topologically, the SR and ER are equivalent to the "outside" of the cell, and both compartments function to store Ca$^{2+}$ for a variety of cellular processes. In muscle, Ca$^{2+}$ release from the SR is involved in triggering muscle contraction, whereas Ca$^{2+}$ release from the ER is involved in stimulating hormone secretion and other intracellular signaling cascades. Pumping of Ca$^{2+}$ by SERCA is the primary mechanism by which SR and ER Ca$^{2+}$ stores are maintained. A different type of Ca$^{2+}$ pump (PMCA), which is found in the plasma membrane, functions to pump Ca$^{2+}$ out of the cell.

Although there are several isotypes of SERCA found in different tissues, the rate at which they pump Ca$^{2+}$ has a simple dependence on [Ca$^{2+}$]$_i$. The pumping rate can be measured using vesicles prepared from either SR or ER membranes. The rate of vesicle

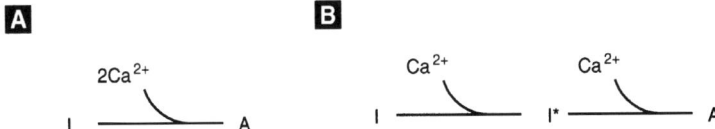

**Figure 3.13** Two simplified models explaining the $Ca^{2+}$ dependence of the experimental SERCA pump rate. (A) A single binding event versus (B) two sequential binding events.

accumulation of $^{45}Ca^{2+}$, a radioactive isotope of $Ca^{2+}$, can then be used to determine the pumping rate. Experimentally, the rate has a sigmoidal dependence on $[Ca^{2+}]_i$ with a Hill coefficient close to two, i.e.,

$$R = \frac{R_{max}[Ca^{2+}]_i^2}{K^2 + [Ca^{2+}]_i^2}. \tag{3.35}$$

The Hill coefficient is related to the stoichiometry of the SERCA pump, which is known to be $2Ca^{2+}$:1ATP. X-ray diffraction of SERCA pumps in bilayers has produced a low-resolution structure with three segments: a stalk region just outside the bilayer close to the binding sites for $Ca^{2+}$, a head region that contains the ATP binding site, and a large transmembrane region through which the $Ca^{2+}$ is transported. Binding experiments have revealed two binding sites for $Ca^{2+}$.

A simple model that is consistent with the $Ca^{2+}$ dependence of the pump rate can be constructed using only two states: an inactive state $I$ and an active state $A$ connected by the mechanism shown in Figure 3.13A. This model leads to the rate expression in (3.35) if we assume that the two states rapidly equilibrate and that only the active state transports $Ca^{2+}$. Rapid equilibration implies the balance of the forward and reverse rates in Figure 3.13A. This leads to the equilibrium condition

$$k^-/k^+ = K_{eq} = \frac{[Ca^{2+}]_i^2[I]}{[A]}, \tag{3.36}$$

where [I] and [A] are the per unit area concentrations of SERCA pumps in the two states, and the equilibrium constant $K_{eq}$ is the ratio of the rate constants. Solving (3.36) for [I], substituting that expression into the conservation condition $[I] + [A] = N$, and then solving for [A] gives the concentration of active SERCAs:

$$[A] = \frac{N[Ca^{2+}]_i^2}{K^2 + [Ca^{2+}]_i^2}. \tag{3.37}$$

Here we have defined $K = \sqrt{K_{eq}}$, which has the units of concentration. As can be seen from (3.37), the numerical value of $K$ equals the concentration of $Ca^{2+}$ at which half of the SERCAs are in the active state. A small value of $K$ is said to correspond to a *high affinity* binding site, and a large value to a *low affinity* site. If the rate constant for the

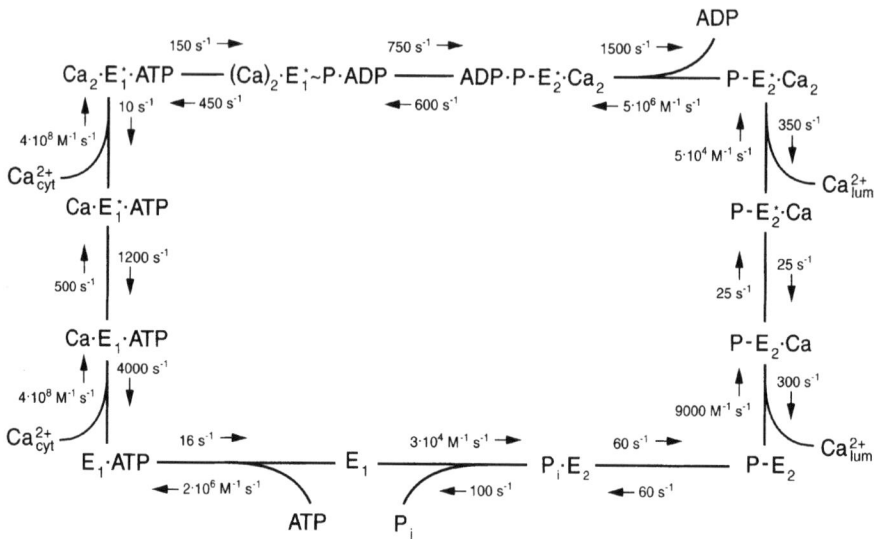

**Figure 3.14** A twelve–state model of the SERCA pump. Note the two sequential $Ca^{2+}$ binding steps on the left–hand side. Although the cycle is driven by the hydrolysis of ATP, all of the steps in the diagram contribute to the steady–state rate. Redrawn from Läuger (1991).

active state to transport $Ca^{2+}$ is $k$, then (3.37) gives the transport rate

$$R = k[A] = \frac{R_{max}[Ca^{2+}]_i^2}{K^2 + [Ca^{2+}]_i^2} \tag{3.38}$$

with $R_{max} = kN$.

Although this mechanism agrees with the measured transport rate and provides an expression for $R_{max}$, there are several things wrong with it. First, it assumes the simultaneous binding of two $Ca^{2+}$, which is highly improbable. Second, it doesn't provide an explanation for the transport rate constant $k$. Third, it doesn't explain how ATP might be involved. We can eliminate the first criticism by expanding the model to include sequential binding of two $Ca^{2+}$, as indicated in Figure 3.13B. For this mechanism there are two simultaneous binding equilibria:

$$K_1 = \frac{[Ca^{2+}]_i[I]}{[I^*]} \quad \text{and} \quad K_2 = \frac{[Ca^{2+}]_i[I^*]}{[A]}. \tag{3.39}$$

In analogy to what was done for the previous model, these equations can be combined with the conservation condition $[I] + [I^*] + [A] = N$ to obtain

$$[A] = \frac{N}{1 + \left(K_1 K_2 / [Ca^{2+}]_i^2\right) + K_2 / [Ca^{2+}]_i}. \tag{3.40}$$

This agrees with the rate expression in (3.35) under the condition that $K_2 \ll$ $[Ca^{2+}]_i$ *and* $K_2 \ll K_1$. In this case [A] is approximately given by (3.37) with $K = \sqrt{K_1 K_2}$, the geometric mean of the two dissociation constants. Although this does not explain the pumping rate, it does suggest that the binding of $Ca^{2+}$ might be sequential with the first site of much lower affinity than the second, i.e., $K_1 \gg K_2$.

Constructing a complete kinetic model of the SERCA pump requires more experimental information than is contained in the pumping rate. In fact, a great deal is known about the other steps involved in the transport cycle. Figure 3.14 gives a 12-state model that includes rate constants for all of the steps indicated. The two conformations of the transporter, $E_1$ and $E_2$, correspond to the bound $Ca^{2+}$ facing the cytosol and inside (*lumen*) of the ER, respectively. The cycle is initiated by ATP binding to $E_1$, followed by the binding of two $Ca^{2+}$ from the cytosol. Using the fact that the $Ca^{2+}$ binding steps are fast, the equilibrium constants can be calculated from the forward and reverse rate constants ($K = k^-/k^+$) to be $K_1 = 1 \cdot 10^{-5}$ M and $K_2 = 2.5 \cdot 10^{-8}$ M. So in agreement with the two–state binding model, this model involves sequential binding of $Ca^{2+}$, with the first step having much lower affinity than the second. In fact, using the expression $K = \sqrt{K_1 K_2}$ gives $K = 5 \cdot 10^{-7}$ M, which is close to the experimental value obtained from rate measurements in vesicles. The cycle in Figure 3.14 is driven by the phosphorylation of SERCA, which facilitates the conformational transition that exposes bound $Ca^{2+}$ to the lumen.

## 3.5 Transport Cycles

Like enzymes, transporters are unchanged by the transport process. Indeed, transporters can be thought of as enzymes whose primary purpose is to alter the location of a molecule rather than its chemical state. If we take the more general point of view suggested by nonequilibrium thermodynamics, an enzyme and a transporter are simply different classes of the same generic type of protein that catalyze a change in free energy. As we noted in the previous section, a transporter accomplishes this by altering the concentration that the molecule experiences. An enzyme, on the other hand, alters the chemical bonds in the molecule.

The catalytic nature of a transporter is apparent in the cyclic structure of the transport mechanism. The GLUT transporter, the $Na^+$/glucose cotransporter, and the SERCA pump described in Section 3.1–Section 3.4 all function in cycles that leave the transporter unchanged. Three additional examples of transport cycles that have been used to explain experimental transport rates are given in Figure 3.15 and Figure 3.16: a P-type proton pump, the adenine nucleotide transporter from mitochondria, and bacteriorhodopsin, a light-driven proton pump.

In a transport cycle a ligand is moved from one cellular compartment to another by some type of driving force. In passive transport, like that for the GLUT transporter, the driving force is simply the concentration difference of ligand. For transporters that involve more than one ligand, such as the $Na^+$/glucose cotransporter or the adenine

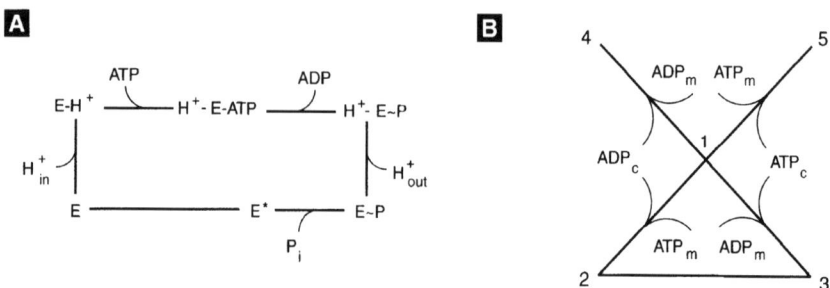

**Figure 3.15** (A) A six–state cycle for a P-type proton ATPase. The phosphorylation step precedes and facilitates dissociation of the proton outside of the cell. (B) Five–state cycle for the adenine nucleotide translocator from mitochondria. The cycle 1-2-3-1 exchanges an $ADP^{3-}$ from the cytoplasm for an $ATP^{4-}$ from the mitochondria. Because the two ions have different charges, the translocator is electrogenic. Steps 1-4 and 1-5 slow the translocation rate by tying up the translocator in states not involved in the cycle. Redrawn from (A) Pietrobon and Caplan (1985) and (B) Bohnensack (1982).

nucleotide transporter in Figure 3.14B, the driving force is a combination of ligand concentration differences. For ATP-dependent pumps, on the other hand, the driving force is the chemical energy stored in the terminal phosphate bond of ATP, which is transferred in a phosphorylation step (cf. Figure 3.14 and Figure 3.15A) to the transporter. Phosphorylation maintains the high–energy state, and the high–energy phosphate bond facilitates conformational transitions that lead to the transport and dissociation of the transported ion. A third form of driving force is light, as indicated in the transport cycle for bacteriorhodopsin in Figure 3.16. In this case energy from a photon excites a state of the transporter causing a trans to cis conformational change in the structure of retinal that is otherwise inaccessible to thermal motion, thereby releasing a proton at the exterior face of the membrane.

As long as we are interested in the average properties of a transporter, it is correct to picture transport cycles as occurring in a fixed direction governed by the driving forces. Dynamic changes in an individual transporter molecule, on the other hand, are stochastic. This is a result of the microscopic reversibility of the kinetic steps in a cycle. Although it is more probable that an individual GLUT transporter will move glucose from a high concentration to a low concentration, the reverse will occur with nonvanishing probability. In fact, any step in a transport cycle can and will occur in the opposite direction to the average transport rate. This reversibility has been demonstrated experimentally for a number of transporters, one of the most convincing being the reversal of SERCA pumps to produce ATP by reversing the $Ca^{2+}$ gradient. Thus the dynamic changes in an individual transporter molecule consist of a series of random positive and negative steps around the cycle that over time lead to an average transport rate in the direction dictated by the driving forces.

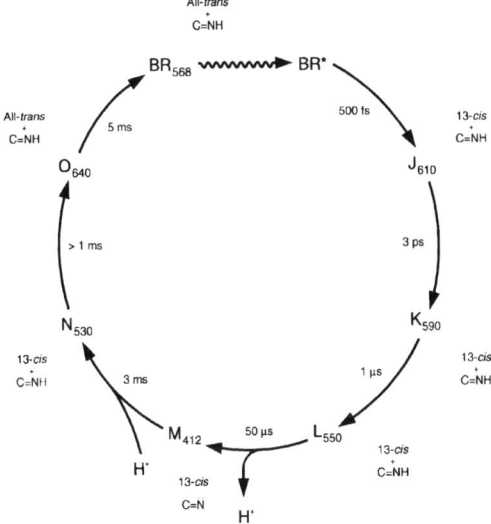

**Figure 3.16** An irreversible diagram for the light–driven proton pump, bacteriorhodopsin, a 27 kD protein from the salt tolerant bacterium, *Halobacterium halobium*. Light absorbed in the step $BR_{568}$-$BR^*$ drives the trans to cis conformational change in retinal, leading to the release of a proton. Redrawn from Läuger (1991).

Cycles are a common feature of other cellular processes with some of the most complex cycles governing muscle contraction, the rotary motion of flagella, and other so-called molecular motors. In metabolism the operation of the $F_0F_1$ ATPase, which is responsible for converting proton gradients and ADP into ATP, is governed by a combined catalytic-transport cycle.

# Suggestions for Further Reading

The material introduced here is relatively straightforward and appears in many fine texts. Some that may be particularly useful at an elementary level are worth mentioning specifically:

- *Free Energy Transduction and Biochemical Cycle Kinetics*, Terrell L. Hill. There are many books that introduce the basic concepts of transport rates. However, this book stands out in explaining both the basic concepts and the diagrammatic method in a concise and complete fashion. Hill also has numerous more advanced books on the subject (Hill 1977).
- *The Fluctuating Enzyme*, edited by G. Rickey Welch. For a somewhat higher level discussion of transporter kinetics, see the chapter "Not Just Catalysts: Molecular Machines in Bioenergetics" by G. Rickey Welch and Douglas B Kell (Welch and Kell 1986).
- *Electrogenic Ion Pumps*, Peter Läuger. This chapter gives a development of the basic physical chemical basis behind many of the electrogenic ion pumps covered here (Läuger 1991).

## 3.6  EXERCISES

1. Verify that the fraction of transporters in states 1 and 2 for the three-state GLUT model satisfies (3.11) and (3.12).

2. Carry out the matrix multiplication indicated in (3.15) to verify that the expressions for the fractional occupany in (3.16) are correct.

3. Write down the directional diagrams for states 2 and 3 for the three-state GLUT transporter model shown in Figure 3.2 and the algebraic expressions that they represent.

4. There are 9 directional diagrams for the three-state GLUT transporter model. Write down the algebraic expression for their sum, which is the denominator $\Sigma$ in (3.20) and Figure 3.6.

5. Show that manipulations like those in Figure 3.6 give the same final expression for $J_{23}^{ss}$ and $J_{31}^{ss}$ as a difference between directed cycle diagrams.

6. The equilibrium constant for a chemical reaction is given by the ratio of the product of the concentrations of "products" divided by the product of concentrations of "reactants." (By convention reactants are the chemical species on the left–hand side of the equation and the products are on the right–hand side.) Show for the three reactions in (3.21) that the equilibrium constants are the ratio of the rate constants as indicated in the equations. [Hint: Recall that "equilibrium" for a reaction occurs when the rate of the forward reaction equals the rate of the reverse reaction.]

7. Using the definition of the equilibrium constant in Exercise 6, show that the equilibrium constant for the sum of two chemical reactions is the product of the equilibrium constants for the individual reactions.

8. Verify that the expression for the sum of the directed diagrams, $\Sigma$, for the symmetric three-state GLUT transporter model is given by (3.27).

9. Write a program to simulate the 4-state model of a GLUT transporter shown in Figure 3.1B. Using this program and the values of the rate constants given below simulate data for an Eadie–Hofstee plot of the rate of transport of glucose when $[G]_{in} = 0$ mM. Make an Eadie–Hofstee plot of the rate using a plotting program and determine $R_{max}$ and $K$. Use the rate constants $k_{12} = k_{43} = 2.4$ mM$^{-1}$min$^{-1}$, $k_{21} = k_{34} = 42$ min$^{-1}$, $k_{14} = k_{14} = k_{41} = k_{23} = k_{32} = 1000$ min$^{-1}$. The transport rate is given by $R = c \cdot J_{34}^{ss}$, where $c = 2$ mM is the concentration of GLUT transporters per unit volume of cells. [Hints: Because of the size of the rate constant $k_{14}$, etc. you will need to use a small step size (try 0.0001min). For the same reason you only will need to integrate for about 0.3min.]

10. Write down the 6 pairs of cyclic diagrams and the 11 directed diagrams for state 1 for the five-state diagram in Figure 3.12B. What are the directed diagrams for state 4,5?

11. The cardiac form of the Na$^+$/Ca$^{2+}$ exchanger is electrogenic with a stoichiometry of 3Na$^+$:1Ca$^{2+}$. Assuming that the 3 Na$^+$ bind sequentially to sites of decreasing affinity, how do you anticipate that the transport rate will depend on $[Na^+]_{out}$? Prove your answer.

# CHAPTER 4

# Fast and Slow Time Scales

**James P. Keener and Joel E. Keizer**

One of the hallmarks of cellular processes is their complexity. For example, in Chapter 3 we described a detailed model for the SERCA pump that might require 11 ODEs and 22 kinetic constants for its analysis. Similarly, the Hodgkin–Huxley model, which includes only three currents in the squid giant axon, involves 4 differential equations and information about three voltage gated currents. As complex as these processes are, they do not begin to represent the true complexity of cellular processes like muscle contraction or insulin secretion, which depend on the coupling of numerous dynamic components. In Chapter 5 we describe some examples of "whole-cell" modeling that attempt to deal with these larger issues.

With a view toward this more complex type of modeling, here we describe several techniques that can be used to simplify the molecular mechanisms that make up these models. These techniques rely on the separation of variables into ones that are "fast" and ones that are "slow." This type of separation has already been used implicitly to simplify previous models. In the Morris-Lecar model (Section 2.4), for example, the rate of activation of the $Ca^{2+}$ current was assumed to be instantaneous, and in Section 3.2 transient behavior was ignored in defining transport rates. In this chapter we introduce two important methods for *time–scale analysis* to simplify molecular models: the *rapid equilibrium approximation* and the *quasi-steady state approximation*. These methods are closely related, and both can be used to eliminate variables and simplify the analysis of the differential equations. A more detailed treatment of time scale analysis using perturbation theory is given in Section A.6. First, however, we motivate the discussion with the following example.

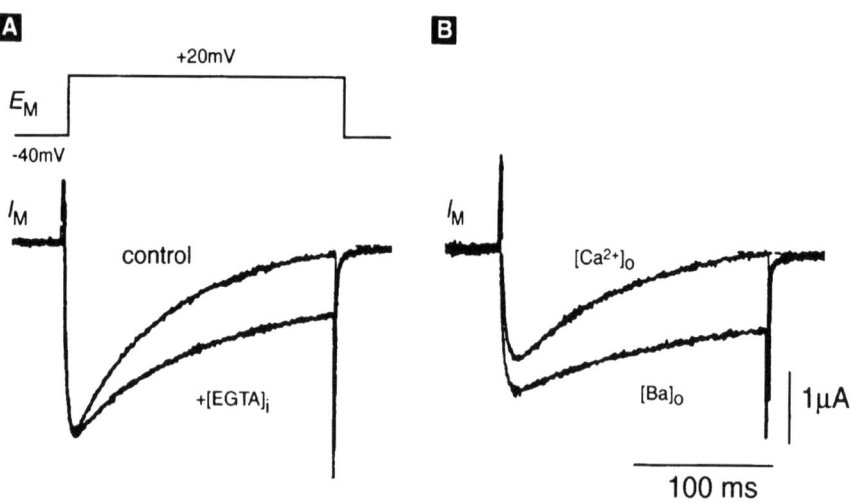

**Figure 4.1**  Whole-cell currents for L-type channels from a neuron in the sea hare *Aplysia*. From Eckert and Tillotson (1981) as reproduced in Hille (2001).

## 4.1  The Rapid Equilibrium Approximation

The voltage clamp measurements in Figure 4.1 show typical whole-cell $Ca^{2+}$ currents for L-type channels from a neuron in the sea hare *Aplysia*. The control curve in Figure 4.1A shows that these channels rapidly activate and then slowly inactivate when the cell is depolarized to 20 mV. The $Ca^{2+}$ dependence of the inactivation step is illustrated by the slowing of inactivation when the mobile $Ca^{2+}$ chelator EGTA is injected into the cell. The $Ca^{2+}$ dependence of inactivation is confirmed by the experiment in Figure 4.1B, in which $Ba^{2+}$ replaces $Ca^{2+}$ outside the cell.

A cartoon for the mechanism underlying $Ca^{2+}$ inactivation of L-type channels is given in Figure 4.2A. The cartoon illustrates the formation of a *microdomain* of elevated $Ca^{2+}$ at the cytoplasmic face of an open $Ca^{2+}$ channel (i.e., a small localized region in the vicinity of the channel in which $Ca^{2+}$ concentration can be quite high). Domains like this have been predicted to form within a few microseconds of the opening of a channel due to the combined effects of high $Ca^{2+}$ concentrations outside the cell (ca. 2 mM) and low basal concentrations in the cytosol (ca. 0.1 $\mu$M). When this is combined with slow diffusion of $Ca^{2+}$ within the cell, calculations predict $Ca^{2+}$ domains with peak values approaching 200–500 $\mu$M within nanometers of the channel. The slow rate of diffusion of $Ca^{2+}$ is caused by tight binding of $Ca^{2+}$ to numerous buffering sites in the cytoplasm, which greatly retards its ability to diffuse. The high concentration of $Ca^{2+}$ in a domain suggests that an open channel may be subject to direct block of the open state by binding of a $Ca^{2+}$ ion at the cytoplasmic face of the channel. A simple mechanism that accounts for this is given in Figure 4.2B.

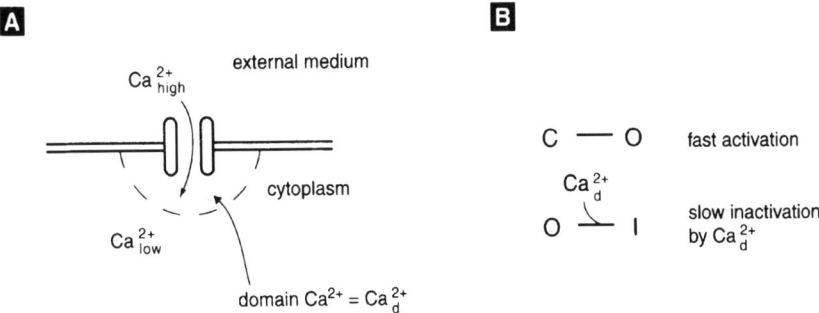

**Figure 4.2** (A) Cartoon of domain calcium. (B) State diagram for $Ca^{2+}$ channel.

The three states $C$, $O$, and $I$ represent closed, open, and inactivated states of the channel. Step 1 is the activation step, whereas step 2 represents the binding of domain $Ca^{2+}$, written as $Ca_d^{+2}$. The mechanism postulates a low affinity site for $Ca^{2+}$ binding, which means that the inactivated state can be reached only when the channel is open and the $Ca^{2+}$ domain has formed. Since simulations show that the peak concentration in a domain falls rapidly when a $Ca^{2+}$ channel closes, it is possible to associate a unique value of domain $Ca^{2+}$ with the open state, whose value depends only on the current through the open channel.

This mechanism is easily translated into a mathematical model. All the steps are unimolecular, except for the bimolecular binding of domain $Ca^{2+}$. If we represent the fractions of channels in the three states by $x_C$, $x_O$, and $x_I$, then the kinetic equations for the model can be written

$$dx_C/dt = -V_1, \tag{4.1}$$
$$dx_O/dt = V_1 - V_2, \tag{4.2}$$

with $x_I = 1 - x_C - x_O$ and with

$$V_1 = k_1^+ x_C - k_1^- x_O, \tag{4.3}$$
$$V_2 = k_2^+ [Ca^{2+}]_d x_O - k_2^- (1 - x_C - x_O) \tag{4.4}$$

the rates of steps 1 and 2. Because the concentration of $Ca^{2+}$ in the domain, $[Ca^{2+}]_d$, depends only on the current, it is a function of the electrical driving force and the single–channel conductance. (In general, the value of $[Ca^{2+}]_d$ depends on the external $Ca^{2+}$ concentration and the membrane potential and is proportional to the single channel conductance; cf. Exercise 5. Specific values for the rate constants are given in Exercise 2. Figure 4.3 shows a simulation with the model that depicts the $Ca^{2+}$ current for a cell that is depolarized at $t = 10$ ms to a voltage where the channel is open. Simulations like this have been used to duplicate the time course of voltage clamp measurements for L-type $Ca^{2+}$ currents in pancreatic beta cells. Key evidence that supports the domain model has come from recent experiments with genetically engineered L-type channels,

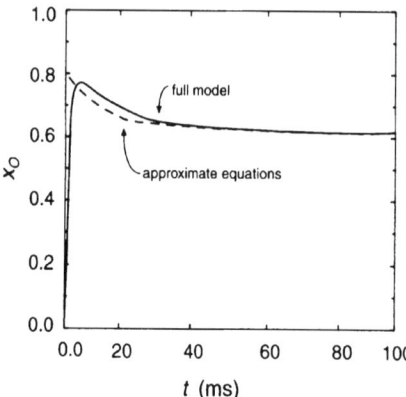

**Figure 4.3** Computed solution and approximate solution for open fraction of L-type calcium channels.

and it now seems certain that the essential ideas of the model are correct.

In both the experiments and the simulations, activation of the channel is fast compared to inactivation. In the model this is due to the fact that both the rate constants for step 1 are much larger than those for step 2. For example, the forward rate for $V_1$ is about 47 times faster than that for $V_2$. Because of this, step 1 rapidly "equilibrates" the states $C$ and $O$.

To see how this equilibration takes place, it is a good approximation to ignore $V_2$, at least at first. Assuming that $x_C(0) = 1$ (all channels are initially closed), it follows that $x_I = 0$ and $x_C + x_O = 1$. We use this to calculate that

$$dx_O/dt = - \left( x_O - \frac{1}{1 + K_1} \right) / \tau_{\text{act}} \tag{4.5}$$

with $K_1 = k_1^-/k_1^+$ and $\tau_{act} = 1/(k_1^+ + k_1^-)$. The number $\tau_{act}$ is the time constant for activation. Because this process is fast, within a few milliseconds $V_1 \approx 0$. This condition continues to hold even as the fractional occupancies $x_C$, $x_O$, and $x_I$ change.

The rapid equilibrium approximation is a method to exploit this observation that some kinetic steps are "fast." By "fast" we mean "faster than the time scales of physiological interest," i.e., faster than the slowest times scales in the process. Here the fast process is the process $V_1$, and for most times after the short initial phase, $V_1 \approx 0$. According to (4.3), the condition $V_1 = 0$ implies that

$$x_C = \left( k_1^-/k_1^+ \right) x_O = K_1 x_O, \tag{4.6}$$

which is the equilibrium condition for step 1 in the mechanism.

Now, it might be tempting to set $V_1 = 0$ in (4.2), but this is the wrong thing to do. Instead, we recognize that since $x_C$ and $x_O$ are in equilibrium, the quantity of interest is the total number of channels in the states $C$ and $O$. Notice that by adding (4.1) and (4.2) together, $V_1$ is eliminated, and we find the rate of change of the combined state

$y = x_C + x_O$ to be

$$d(x_C + x_O)/dt = -V_2. \tag{4.7}$$

Using the equilibrium condition (4.6), we find that

$$x_O = \frac{1}{1 + K_1} y \quad \text{and} \quad x_C = \frac{K_1}{1 + K_1} y,$$

so that

$$dy/dt = k_2^-(1 - y) - \frac{k_2^+[Ca^{2+}]_d}{1 + K_1} y. \tag{4.8}$$

This ODE can be rearranged into the familiar form

$$dy/dt = \frac{y_\infty - y}{\tau}, \tag{4.9}$$

where

$$\tau = \frac{1 + K_1}{k_2^+[Ca^{2+}]_d + k_2^-(K_1 + 1)}, \tag{4.10}$$

$$y_\infty = k_2^- \tau. \tag{4.11}$$

Written this way, (4.9) has a form similar to (2.13) for a voltage gated channel, except that now $[Ca^{2+}]_d$ replaces the voltage.

The only tricky part remaining about the solution to (4.9), which is an exponential, is finding the correct initial condition. Since we have assumed that step 1 is fast, the initial condition for (4.9) must take into account the rapid initial equilibration of $C$ and $O$. As in Figure 4.1 the initial condition typically is at a hyperpolarized potential where all the channels are closed and $x_C(0) = 1$. After the initial equilibration, some of the channels will have moved to state $O$, so that $y = x_C + x_O = 1$. This gives the equilibrated initial condition for (4.9) as

$$y(0) = 1. \tag{4.12}$$

Using this initial condition, the solution to the rapid equilibrium approximation for the model is plotted as the dashed line in Figure 4.3. Two things are notable in comparing the approximation to the complete solution. First, by neglecting the rapid activation of the channel, the approximation slightly overestimates the peak current, which is given analytically using the initial condition in (4.12) as $I_{peak} = g(1/(1 + K_1))(V - V_{Ca})$ (see Exercise 5). Second, the exponential decline in current predicted by (4.9) does a good job of approximating the rate of inactivation of the current. The approximation works well because the time scale for the fast process (ca. 1 ms) is much faster than that for inactivation (ca. 45 ms). As long as the time constant for the fast process is at least an order of magnitude faster than the remaining processes, the rapid equilibrium approximation provides a reasonable approximation to the complete model.

The rapid equilibrium approximation is a useful way to reduce the complexity of models. For the domain $Ca^{2+}$ inactivation model, the simplification is not really necessary, because the full model involves only two linear differential equations that can be analyzed by the matrix methods in Appendix A. However, the fact that the equation for the simplified model resembles that for voltage–gated channels provides a conceptual bridge to the properties of ligand–gating of channels.

## 4.2   Asymptotic Analysis of Time Scales

Mathematicians have developed a more systematic method of dealing with fast and slow variables called *asymptotic analysis*. Although we do not work through all the intricacies of this technique here, the central idea of asymptotics is easy to understand, and it can be applied profitably to many modeling problems in cell and neurobiology. The idea is to define the fast and slow time scales in the model, and then to rescale time so that on this new scale only the slow, or alternatively only the fast, variables are changing.

To illustrate the idea we reexamine the model of $Ca^{2+}$ currents for L-type channels introduced in the previous section. As we saw in that analysis, the two time scales are the time scale of activation $\tau_{act} = 1/(k_1^+ + k_1^-)$ and the time scale of inactivation $\tau([Ca^{2+}]_d)$.

Having identified the fast and slow time scales, we proceed to *nondimensionalize* all of the variables in (4.1)–(4.4), including time. Nondimensionalization is simply eliminating the units of a variable by dividing by a parameter in the model that has the same units as the variable. In this case the variables $x_C$, $x_O$, and $x_I$ are percentages, and so are already nondimensional. The only variable having dimensions is $t$. The choice of a nondimensional time determines whether our analysis focuses on the fast or the slow time scale.

To nondimensionalize time using the slow time scale, we could set $\hat{t} = \tau([Ca^{2+}]_d)t$. However, it is somewhat easier and essentially equivalent to set $\hat{t} = k_2^- t$. In terms of this new time scale, (4.1) and (4.2) become

$$\epsilon dx_C/d\hat{t} = -\frac{k_1^+}{k_1^-}x_C + x_O, \tag{4.13}$$

$$\epsilon dx_O/d\hat{t} = \frac{k_1^+}{k_1^-}x_C - x_O - \epsilon\frac{k_2^+}{k_2^-}[Ca^{2+}]_d x_O + \epsilon(1 - x_C - x_O), \tag{4.14}$$

where $\epsilon = k_2^-/k_1^-$ is a small number.

The important observation is that since $\epsilon$ is small, unless the right–hand side of (4.13) is small, $x_C$ changes rapidly. However, if we add the two equations (4.13) and (4.14) together and divide by $\epsilon$ we find an equation that is independent of $\epsilon$,

$$d(x_C + x_O)/d\hat{t} = -\frac{k_2^+}{k_2^-}[Ca^{2+}]_d x_O + (1 - x_C - x_O). \tag{4.15}$$

The basic idea of asymptotic analysis is to treat $\epsilon$ not as a fixed number, but as a parameter that can be varied. In the asymptotic limit that $\epsilon \to 0$, we find that the right hand–side of (4.13) is zero. This is the lowest–order solution in the asymptotic analysis on the slow time scale, and is exactly the same as (4.6). This approximation is sometimes called the *quasi-steady state* approximation, where "quasi" emphasizes that $k_1^+ x_C - k_1^- x_O$ is nearly, but not exactly, zero. We complete the slow time scale analysis by using (4.13) to find $x_C$ and $x_O$ in terms of $y = x_C + x_O$, and then using (4.15) to find an equation describing the evolution of $y$. This equation turns out to be exactly (4.8).

The analysis on the fast time scale is similar. This time we choose a nondimensional time $\tilde{t} = k_1^- t$, and write equations (4.1) and (4.15) as

$$dx_C/d\tilde{t} = -\frac{k_1^+}{k_1^-} x_C + x_O, \tag{4.16}$$

$$d(x_C + x_O)/d\tilde{t} = \epsilon(-\frac{k_2^+}{k_2^-}[\text{Ca}^{2+}]_d x_O + (1 - x_C - x_O)). \tag{4.17}$$

This time, in the asymptotic limit that $\epsilon \to 0$, equation (4.17) reduces to

$$d(x_C + x_O)/d\tilde{t} = 0, \tag{4.18}$$

so that $x_C + x_O = 1$ and

$$dx_C/d\tilde{t} = -\frac{k_1^+}{k_1^-} x_C + 1 - x_C, \tag{4.19}$$

at least for a short time.

This type of time–scale analysis can be summarized by five steps, as follows:

- Analyze the parameters of the model to assess whether there are time scales that can be separated into "fast" and "slow."
- Define time constants for each time domain whose ratios define a small parameter $\epsilon$.
- Select appropriate parameters in the model to nondimensionalize the dependent variables.
- Nondimensionalize the differential equations in each time domain and see which terms can be neglected as $\epsilon \to 0$.
- Analyze the simplified equations, which represent the behavior of the variables on the two time scales.

In the next sections we describe how to carry out this analysis for several different arrangements of fast and slow reactions.

## 4.3   Glucose–Dependent Insulin Secretion

Insulin is secreted from $\beta$-cells in the pancreas in an oscillatory fashion. Glucose must be metabolized by the $\beta$-cell to stimulate insulin secretion, and the insulin, which is prepackaged in secretory vesicles, is secreted from the $\beta$-cell into the capillary system by exocytosis. However, the secreted insulin affects the transport of glucose into the cell

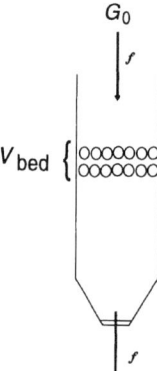

**Figure 4.4**  Flow system for experimental study of insulin secretion. $V_{bed}$ is the volume of the reaction bed, $f$ is the volume flow rate, and $G_0$ is the inflow concentration of glucose. Redrawn from Maki and Keizer (1995).

by activating GLUT1 transporters and inactivating GLUT2 transporters. Thus, there is both positive and negative feedback, necessary ingredients for sustained oscillations.

These oscillations have been studied experimentally in a flow system depicted in Figure 4.4. A thin layer of insulin-secreting $\beta$-cells is sandwiched between beads and exposed to a steady flow of solution. By collecting the solution exiting the bed, one can determine how the rate of insulin release from the cells in the bed depends on the composition and flow rate of the influx solution.

A mathematical model of this process must involve (at least) three variables: the extracellular glucose and insulin concentrations and the intracellular glucose concentration. We assume that the volume of islets behaves like a well stirred chemical flow reactor so that the concentration of any quantity is uniform throughout the bed (justification of this assumption is complicated and goes beyond the level of the discussion given here). Thus, the rate of change of the average concentration of a species in the flow is the rate of change due to production or uptake by cells in the bed plus the rate of change due to the flow. Note that we use a nonbracketed notation for concentration in order to correspond to the original paper that forms the basis of this section (Maki and Keizer 1995). For glucose, $G$, the relation is

$$\frac{dG}{dt} = -R_1 - R_2 - k_0(G - G_0), \qquad (4.20)$$

where $R_1$ and $R_2$ are the uptake (or release) rates for glucose through GLUT1 and GLUT2 receptors, respectively, and $G_0$ is the glucose concentration in the inflow. For insulin, $I$,

$$\frac{dI}{dt} = R_s - k_0(I - I_0), \qquad (4.21)$$

where $R_s$ is the rate of insulin secretion and $I_0$ is the insulin concentration in the inflow. In both cases, $k_0 = f/V_{bed}$ is the inverse of the residence time for this reactor bed. (A schematic diagram of a secreting cell is shown in Figure 4.5).

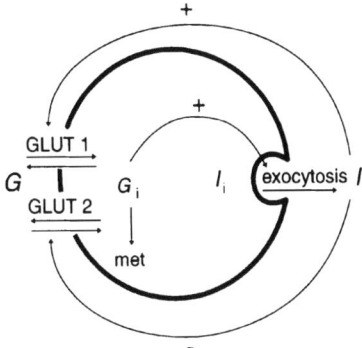

**Figure 4.5** Schematic diagram of an insulin secreting cell. The plus sign represents positive feedback and the negative sign represents negative feedback. Insulin positively affects both uptake and release for GLUT1. Insulin negatively affects only uptake and has no effect on release for GLUT2. Internal insulin, $I_i$, remains approximately constant. Redrawn from Maki and Keizer (1995).

Since there is no flow associated with the intracellular space, the intracellular glucose concentration $G_i$ is found by keeping track of productions rates, via

$$\frac{dG_i}{dt} = R_1 + R_2 - R_m \qquad (4.22)$$

where $R_m$ is the rate of glucose usage through metabolism.

Now we must specify the transfer rates. As described in Chapter 3, the GLUT transporters can be described by four states, two glucose free states and two glucose bound states. If we assume that transporter is always in quasi-equilibrium, then the flux through the transporter is given by

$$R_{GLUT} = \frac{K_G V_{max}(G - G_i)}{(K_G + G)(K_G + G_i)}. \qquad (4.23)$$

The specific transporters GLUT1 and GLUT2 have fluxes of this form with the modification that GLUT1 is activated by insulin, and GLUT2 is inhibited by insulin. Thus, for the GLUT1 transporter we take

$$R_1 = \frac{K_1 V_{max1}(G - G_i)}{(K_1 + G)(K_1 + G_i)} \cdot \frac{I^n}{K_i^n + I^n}, \qquad (4.24)$$

and for the GLUT2 transporter

$$R_2 = \frac{K_2 V_{max2}(GJ^m - G_i)}{(K_2 + G)(K_2 + G_i)}, \qquad (4.25)$$

where n = 1 and m = 2. The variable $J$ is a (phenomenological) inhibition variable, related to $I$. The variable $J$ acts to shut down the inward flux of glucose when $I$ is large, and satisfies the differential equation

$$dJ/dt = \frac{J^\infty - J}{\tau}, \qquad (4.26)$$

where $J^\infty(I) = 1/(I + K_I)$, and $\tau$ is the time constant associated with this time–dependent process.

The metabolism of internal glucose is an enzymatic process, the first step of which is that glucose is phosphorylated by glucokinase, in a reaction that is Michaelis–Menten-like, and so is well described by the Michaelis–Menten rate law

$$R_m = \frac{V_m G_i}{K_m + G_i}.$$ (4.27)

(A derivation of rate laws of this type is given at the end of this chapter.)

The detailed mechanism underlying the secretion of insulin is not fully understood, so a phenomenological equation describing the rate of secretion as a function of the rate of metabolism is used:

$$R_s = \frac{V_s \left( R_m^4 + L^4 \right)}{K_s^4 + R_m^4 + L^4}.$$ (4.28)

Notice that with $L \neq 0$ there is secretion of insulin even when $R_m = 0$. Now that the model is complete, we can begin an analysis of it. To do so we need to know something about the parameters. In Table 4.1 are listed the parameters that are fixed by the experiment, experimentally variable, and adjustable.

To reduce the complexity of the model we would like to determine whether there are any (relatively) fast or slow variables. One way to do this is to numerically simulate the full system of equations using typical parameter values and observe whether there are some variables that change much faster than others. However, we suspect that there are differences in time scales here for the simple reason that the ratio of the two time constants $k_0$ and $\tau$ in Table 4.1 is large, being $\tau k_0 = 8 \times 10^3$. This suggests that the flow processes (involving $G$ and $I$) are fast compared to inhibition through $J$. To be sure that this is correct, we introduce dimensionless parameters and variables $\hat{G} = G/K_m$, $\hat{G}_i = G_i/K_m$, $\hat{G}_0 = G_0/K_m$, $\hat{I} = I/K_I$, and $\hat{t} = t/\tau$, and obtain the four dimensionless

**Table 4.1**  Standard Dimensional Parameters

| | | |
|---|---|---|
| Fixed by experiment | $V_m$ | 0.24 mM/min |
| | $K_m$ | 9.8 mM |
| | $V_s$ | 0.034 mM/min |
| | $K_s$ | 0.13 mM/min |
| | $V_{max2}$ | 32.0 mM/min |
| | $K_2$ | 17.0 mM |
| | $V_{max1}$ | 120.0 mM/min |
| | $K_1$ | 1.4 mM |
| | $L$ | 0.01 mM/min |
| Experimentally variable | $k_0$ | 400.0/min |
| | $I_0$ | 0.0 mM |
| | $G_0$ | 8-22 mM |
| Adjustable in the model | $K_I$ | $1 \times 10^{-6}$ mM |
| | $K_i$ | $4.0 \times 10^{-5}$ mM |
| | $\tau$ | 20.0 min |

differential equations

$$d\hat{G}/d\hat{t} = -\hat{R}_1 - \hat{R}_2 - \hat{k}_0(\hat{G} - \hat{G}_0), \tag{4.29}$$

$$d\hat{G}_i/d\hat{t} = \hat{R}_1 + \hat{R}_2 - \hat{R}_m, \tag{4.30}$$

$$d\hat{I}/d\hat{t} = \hat{R}_s - \hat{k}_0\hat{I}, \tag{4.31}$$

$$dJ/d\hat{t} = \hat{J}^\infty(I) - J, \tag{4.32}$$

where

$$\hat{R}_1 = \frac{\hat{K}_1\hat{V}_{\max1}(\hat{G} - \hat{G}_i)}{(\hat{K}_1 + \hat{G})(\hat{K}_1 + \hat{G}_i)} \cdot \frac{\hat{I}^n}{\hat{K}_i^n + \hat{I}^n}, \tag{4.33}$$

$$\hat{R}_2 = \frac{\hat{K}_2\hat{V}_{\max2}(\hat{G}J^m - \hat{G}_i)}{(\hat{K}_2 + \hat{G})(\hat{K}_2 + \hat{G}_i)}, \tag{4.34}$$

$$\hat{R}_m = \frac{\hat{V}_m\hat{G}_i}{1 + \hat{G}_i}, \tag{4.35}$$

$$\hat{R}_s = \frac{\hat{V}_s(\hat{R}_m^4 + \hat{L}^4)}{\hat{K}_s^4 + \hat{R}_m^4 + \hat{L}^4}, \tag{4.36}$$

$$\hat{J}^\infty = \frac{1}{1 + \hat{I}}. \tag{4.37}$$

The definitions of the dimensionless parameters and their values are given in Table 4.2.
The first noticeable feature from Table 4.2 is that there are two numbers, $\hat{V}_s$ and $\hat{k}_0$, that are quite large. Since these parameters both occur in (4.31) for $I$, this implies that $\hat{I}$ is a fast variable, so that $\hat{I}$ changes rapidly in order to bring the right–hand side of (4.31) close to zero. Thus, our first quasi-steady-state approximation is to take

$$\hat{I} = \frac{1}{\hat{k}_0}\hat{R}_s = \frac{\hat{V}_s}{\hat{k}_0}\left(\frac{\hat{R}_m^4 + \hat{L}^4}{\hat{K}_s^4 + \hat{R}_m^4 + \hat{L}^4}\right). \tag{4.38}$$

**Table 4.2**   Standard Dimensionless Parameters

| Dimensionless parameter | Dimensional definition | Standard value |
|---|---|---|
| $\hat{V}_m$ | $\tau V_m/K_m$ | 0.50 |
| $\hat{V}_s$ | $\tau V_s/K_I$ | $6.8 \times 10^5$ |
| $\hat{K}_s$ | $\tau K_s/K_m$ | 0.27 |
| $\hat{V}_{\max2}$ | $\tau V_{\max2}/K_m$ | 65.3 |
| $\hat{K}_2$ | $K_2/K_m$ | 1.7 |
| $\hat{V}_{\max1}$ | $\tau V_{\max1}/K_m$ | 245.0 |
| $\hat{K}_2$ | $K_1/K_m$ | 0.14 |
| $\hat{L}$ | $\tau L/K_m$ | 0.02 |
| $\hat{k}_0$ | $\tau k_0$ | $8 \times 10^3$ |
| $\hat{G}_0$ | $G_0/K_m$ | 0.8-2.2 |
| $\hat{K}_i$ | $K_i/K_I$ | 40.0 |

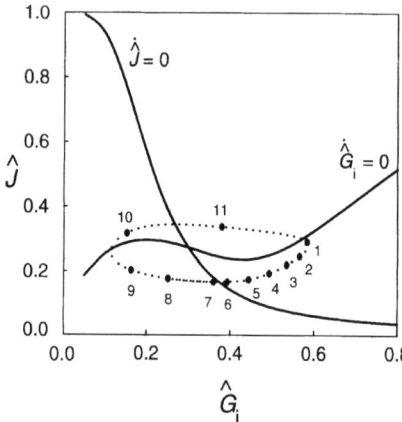

**Figure 4.6**  Phase portrait of glucose oscillations in the phase plane. Redrawn from Maki and Keizer (1995).

Our second observation is that because $\hat{k}_0$ is a large number, $\hat{G}$, governed by (4.29), is also a rapidly equilibrating variable, so we take the quasi-steady-state approximation

$$\hat{G} = \hat{G}_0 - \frac{1}{\hat{k}_0}\hat{R}_1 - \frac{1}{\hat{k}_0}\hat{R}_2. \tag{4.39}$$

However, since $\hat{V}_{max1}/\hat{k}_0 = 3.06 \times 10^{-2}$ and $\hat{V}_{max2}/\hat{k}_0 = 8.2 \times 10^{-3}$ are small, it is legitimate to ignore $\hat{R}_1/\hat{k}_0$ and $\hat{R}_2/\hat{k}_0$ in (4.39) and take

$$\hat{G} = \hat{G}_0. \tag{4.40}$$

With these simplifications we are left with a two–variable model that we can readily analyze and simulate. We leave this simulation as an exercise, to verify that indeed there are glucose oscillations with a period of about 1 (about 20 minutes in dimensional time). So that you can check your answer, in Figure 4.6 is shown the phase portrait for this oscillation.

## 4.4  Ligand Gated Channels

The $Ca^{2+}$-activated potassium channel shown in Figure 4.7 provides another example in which some transitions are much faster than others, and this can be used to derive simplified kinetics. This channel has two open states and two closed states. The channel has two binding sites for $Ca^{2+}$ and opens when one of the sites is occupied. Thus, the closed state may have zero or one $Ca^{2+}$ ion bound, and the open state may have one or two ions bound. The binding process is considered a fast process, while the transition between open and closed states is slow.

To describe this process mathematically, we let the variables $x_1, x_2, x_3$, and $x_4$ denote the fractional occupancies of states $C_1, C_2, O_1$, and $O_2$. Then the transition rates between

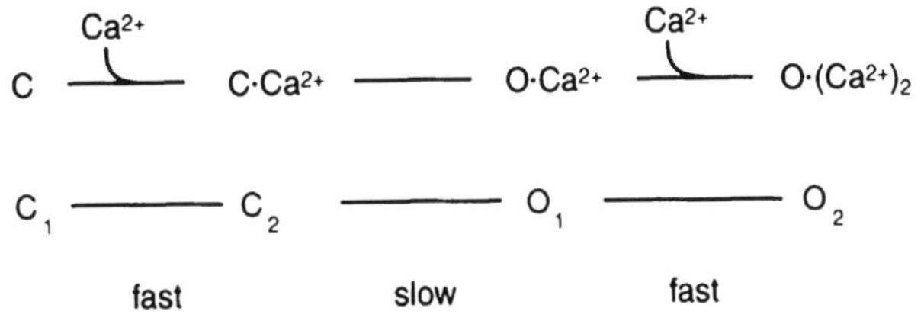

$$C \xrightarrow{\overset{Ca^{2+}}{\frown}} C{\cdot}Ca^{2+} \longrightarrow O{\cdot}Ca^{2+} \xrightarrow{\overset{Ca^{2+}}{\frown}} O{\cdot}(Ca^{2+})_2$$

$$C_1 \longrightarrow C_2 \longrightarrow O_1 \longrightarrow O_2$$

$$\text{fast} \qquad\qquad \text{slow} \qquad\qquad \text{fast}$$

**Figure 4.7** Diagram for the calcium-activated potassium channel. Opening of the channel occurs after binding one calcium ion in either of two binding sites.

these states are $V_a$, $V_b$, and $V_c$, where

$$V_a = k_a^+[Ca^{2+}]_i x_1 - k_a^- x_2, \tag{4.41}$$

$$V_b = k_b^+ x_2 - k_b^- x_3, \tag{4.42}$$

$$V_c = k_c^+[Ca^{2+}]_i x_3 - k_c^- x_4. \tag{4.43}$$

Of course, $x_1 + x_2 + x_3 + x_4 = 1$.

Since we assume that $V_a$ and $V_c$ (the association/dissociation steps for binding $Ca^{2+}$) are fast, we use the rapid equilibrium assumption to set $V_a = V_c = 0$. This implies that

$$x_1 = \left(\frac{K_a}{[Ca^{2+}]_i}\right) x_2, \tag{4.44}$$

$$x_3 = \left(\frac{K_c}{[Ca^{2+}]_i}\right) x_4, \tag{4.45}$$

with dissociation constants

$$K_a = k_a^-/k_a^+, \qquad K_c = k_c^-/k_c^+. \tag{4.46}$$

Next, to find the evolution of the slow states, we combine states $x_1$ and $x_2$ into the closed state $x_C$ and combine states $x_3$ and $x_4$ into the open state $x_O$, with

$$x_1 + x_2 = x_C, \qquad x_3 + x_4 = x_O. \tag{4.47}$$

Of course, because of our rapid equilibrium assumption,

$$x_2 = \frac{x_C}{1 + K_a/[Ca^{2+}]_i}, \tag{4.48}$$

$$x_4 = \frac{x_O}{1 + K_c/[Ca^{2+}]_i}. \tag{4.49}$$

The reduced equations follow from

$$\frac{dx_C}{dt} = -V_a + V_a - V_b = -V_b \tag{4.50}$$

and, of course, $x_C + x_O = 1$. Therefore,

$$\frac{dx_O}{dt} = V_b = k_b^+ x_2 - k_b^- x_3 = \frac{k_b^+ x_C}{1 + K_a/[Ca^{2+}]_i} - \frac{(K_c/[Ca^{2+}]_i)k_b^- x_O}{1 + K_c/[Ca^{2+}]_i} = k^+ x_C - k^- x_O,$$

where

$$k^+ = \frac{k_b^+}{1 + K_a/[Ca^{2+}]_i}, \quad k^- = \frac{(K_c/[Ca^{2+}]_i)k_b^-}{1 + K_c/[Ca^{2+}]_i}.$$

In other words, this process is equivalent to the two–state process

$$C \overset{k^+}{\underset{k^-}{\rightleftharpoons}} O.$$

Parameters are known from experiments to be $k_b^+ = 480/s$, $k_b^- = 280/s$, $K_a = k_a^-/k_a^+ = 180e^{-V/15.5}\mu M$, and $K_c = k_c^-/k_c^+ = 11e^{-V/13}\mu M$, where $V$, the voltage, is in units of mV.

## 4.5   The Neuromuscular Junction

A similar analysis works to find a model of the acetylcholine receptor (AchR) in the postsynaptic membrane of neuromuscular junctions. Neuromuscular junctions consist of a presynaptic cell and a postsynaptic cell that are separated by a small synaptic cleft, as depicted in Figure 4.8. When an action potential reaches the nerve terminal, several processes lead to the release of a chemical neurotransmitter, such as acetylcholine, from the presynaptic cell into the synaptic cleft. The neurotransmitter binds to receptors on the postsynaptic membrane that act as channels for some ion, such as sodium or potassium.

A model for the opening and closing of the Ach receptor (see Figure 4.9) is based on the idea that a receptor is a four-state device, with three closed states and one open state. The three closed states have 0, 1, or 2 Ach molecules bound to the receptor, and

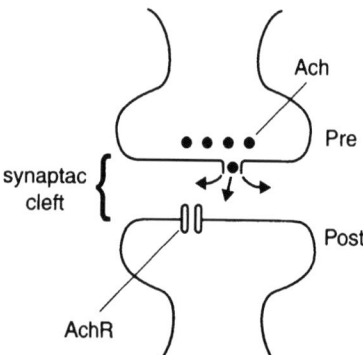

**Figure 4.8**   Diagram of a neuromuscular junction with acetylcholine as the neurotransmitter.

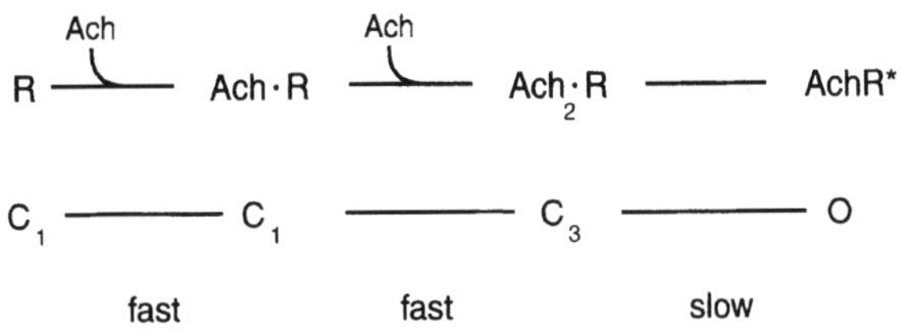

**Figure 4.9** Four-state model for the Ach receptor. The channel opens after two binding events.

the open state requires that two Ach molecules be bound. We assume that the binding of Ach is a fast process and the opening and closing of the channel is slow in contrast.

Using the rapid equilibrium approxmation, it is a straightforward exercise to combine the three closed states into one and to reduce this process to an equivalent two–state process. In fact, it is readily found that if $x_O$ is the proportion of receptors in the open state, then

$$dx_O/dt = k_{34}\frac{k_{23}[\text{Ach}]}{k_{32}}\frac{k_{12}[\text{Ach}]}{k_{32}\left(k_{21}+k_{12}[\text{Ach}]+k_{12}[\text{Ach}]k_{23}[\text{Ach}]\right)}(1-x_O)-k_{43}x_O. \quad (4.51)$$

# 4.6  The Inositol Trisphosphate (IP$_3$) receptor

The IP$_3$ receptor is a $Ca^{2+}$ channel located in the endoplasmic reticulum that is regulated both by IP$_3$ and by $Ca^{2+}$. Each receptor consists of three independent subunits, each of which must be in the open state for the channel to be open. Each subunit has one binding site for IP$_3$ and two binding sites for calcium. Thus there are eight possible states for the subunit. Binding with IP$_3$ "potentiates" the subunit. The two calcium binding sites activate and inactivate the subunit, and a subunit is in the open state when IP$_3$ and the activating calcium site are bound but the inactivating site is unbound.

To make a mathematical model for this receptor we must first give names to the eight different states. We denote by $x_i$ the proportion of receptors in state $i$, where $i$ is the label of the vertex on the cube in Figure 4.10. Thus, for example, $x_1$ represents the state in which all binding sites are unbound, and $x_6$ is the open state of the subunit.

For these eight different states, there are 24 different rate constants $k_{ij}$. However, since each cycle must satisfy the thermodynamic constraint on kinetic constants, there are six restrictions on these 24 parameters. Notice that the rate constants that involve binding of a substrate must be proportional to the concentration of that substrate. Thus, $k_{15}, k_{26}, k_{37}$, and $k_{48}$ are proportional to the concentration of IP$_3$, while the eight rate constants $k_{12}, k_{14}, k_{23}, k_{43}, k_{58}, k_{56}, k_{67}$, and $k_{87}$ are all proportional to the $Ca^{2+}$ concentration. Now we make two simplifying assumptions. First we assume that the rate constants are independent of whether activating $Ca^{2+}$ is bound or not, and second we assume that the kinetics of $Ca^{2+}$ activation are independent of IP$_3$ binding and $Ca^{2+}$

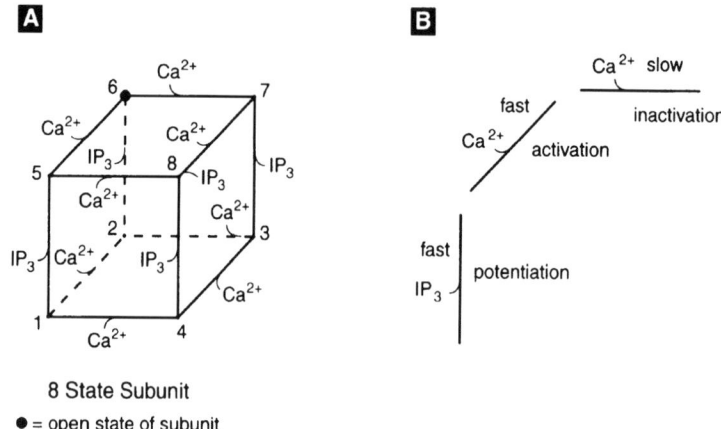

Figure 4.10   (A) Transition diagram for the eight-state IP$_3$ receptor. (B) Relative rates of reaction for transitions in (A).

inactivation. The first of these implies that $k_{14} = k_{23}, k_{15} = k_{26}, k_{58} = k_{67}$, and $k_{48} = k_{37}$, as well as for the reverse reactions. The second implies that $k_{12} = k_{56} = k_{87} = k_{43}$ and similarly for the reverse reactions. Thus we are left with *only* 10 rate constants. The parameter values that were used by DeYoung and Keizer are displayed in Table 4.3.

An examination of the rate constants in Table 4.3 reveals that some processes are much faster than others. In fact, according to this table, the binding (and unbinding) of inactivating calcium is a slow process compared to the binding of IP$_3$ and activating calcium. For example, $k_{15}$ and $k_{48}$ are much larger than $k_{58}, k_{14}$, and $k_{12}$. Therefore, we lump the eight variables into two groups with

$$y = x_1 + x_2 + x_5 + x_6, \tag{4.52}$$
$$1 - y = x_3 + x_4 + x_7 + x_8. \tag{4.53}$$

Next, we assume that all the fast processes are in quasi-equilibrium. That is, we assume that the processes 1-5, 2-6, 3-7, 4-8, in which IP$_3$ is bound, and the processes 5-6, 8-7, 1-2, 4-3, in which activating calcium is bound, are in quasi-equilibrium.

**Table 4.3**  Rate constants for the eight–state IP$_3$ receptor (DeYoung and Keizer 1992).

| | |
|---|---|
| $k_{15} = 400[IP_3] \ \mu M^{-1}s^{-1}$ | $k_{51} = 52 \ s^{-1}$ |
| $k_{58} = 0.2[Ca^{2+}]_i \ \mu M^{-1}s^{-1}$ | $k_{85} = 0.21 \ s^{-1}$ |
| $k_{48} = 400[IP_3] \ \mu M^{-1}s^{-1}$ | $k_{84} = 377.2 \ s^{-1}$ |
| $k_{14} = 0.2[Ca^{2+}]_i \ \mu M^{-1}s^{-1}$ | $k_{41} = 0.029 \ s^{-1}$ |
| $k_{12} = 20[Ca^{2+}]_i \ \mu M^{-1}s^{-1}$ | $k_{21} = 1.64 \ s^{-1}$ |

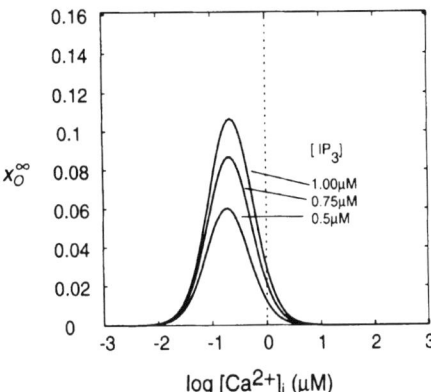

Figure 4.11 Open probability for IP₃ receptor for different concentrations of IP₃.

This gives us a set of algebraic relationships that can be solved to find the variables $x_i$ in terms of $y$. For example,

$$x_6 = \frac{k_{15}k_{12}}{(k_{51} + k_{15})(k_{21} + k_{12})} y. \tag{4.54}$$

We substitute these into the differential equation for $y$,

$$\frac{dy}{dt} = k_{41}x_4 + k_{85}x_8 + k_{76}x_7 + k_{32}x_3 - (k_{14}x_1 + k_{58}x_5 + k_{67}x_6 + k_{23}x_2) \tag{4.55}$$

and obtain the equation

$$\frac{dy}{dt} = \left(\frac{d_1 + k_{85}[\text{IP}_3]}{d_2 + [\text{IP}_3]}\right)(1 - y) - d_3[\text{Ca}^{2+}]_i\, y, \tag{4.56}$$

where the $d_i$ are combinations of the rate coefficients. Using the parameter values given in Table 4.3, we find that $d_1 = 0.027\ \mu\text{M s}^{-1}$, $d_2 = 0.94\ \mu\text{M}$, $d_3 = 0.2\ \mu\text{M}^{-1}\text{s}^{-1}$. This equation is readily converted (see Exercise 8a) to an equation of the form

$$\frac{dy}{dt} = -\frac{y - y^\infty([\text{Ca}^{2+}]_i, [\text{IP}_3])}{\tau(\text{Ca}^{2+}, [\text{IP}_3])} \tag{4.57}$$

with an open probability

$$x_O = x_6^3 = \left(\frac{[\text{IP}_3][\text{Ca}^{2+}]_i y}{(K_1 + [\text{IP}_3])(K_2 + [\text{Ca}^{2+}]_i)}\right)^3. \tag{4.58}$$

The relationship of constants $K_1$ and $K_2$ to the rate coefficients are determined in Exercise 8a. Of course, the equilibrium open probability is

$$x_O^\infty = \left(\frac{[\text{IP}_3][\text{Ca}^{2+}]_i y^\infty}{(K_1 + [\text{IP}_3])(K_2 + [\text{Ca}^{2+}]_i)}\right)^3. \tag{4.59}$$

The equilibrium open probability $x_O^\infty$ is plotted in Figure 4.11 for several values of [IP₃]. Notice that the maximum open probability increases with increasing [IP₃], and the peak of the curve shifts to the right with increasing [IP$_3$]. Both of these features are observed experimentally.

# 4.7 Michaelis–Menten Kinetics

One of the most important reductions of a complex reaction is associated with the names of Michaelis and Menten, and in fact, this reduction is used in several places in this book. In many texts this reduction is used to introduce the ideas and methods of time scale reduction. However, the reduction of Michaelis and Menten is slightly different from the reductions discussed so far in this chapter, because it does not rely on large differences in reaction rates. Instead, the difference in time scale of reaction comes from an entirely different source, which we now discuss.

Working in the early part of the twentieth century, Michaelis and Menten set about to explain several key experimental facts regarding the conversion of substrate to product catalyzed by simple enzymes. Figure 4.12 illustrates their results for the enzyme invertase, which converts sucrose to glucose and fructose. These graphs give the time course of accumulation of product (measured as the change in optical rotation of the solution) for a fixed total enzyme concentration [E]$_T$. In the three curves the initial concentration of sucrose, [S]$_T$, is increased from 5.2 mM to 10.4 mM to 20.8 mM. The initial rate of increase of product, given by the slope of the three curves ($V(0)$), is plotted for these and similiar experiments in Figure 4.12. As the concentration of sucrose increases, the initial rate saturates at the value $V_{max}$. This function is *hyperbolic* and can be fit with the expression:

$$V(0) = V_{max}[S]_T/([S]_T + K_m). \qquad (4.60)$$

The concentration of substrate at which $V(0) = V_{max}/2$ is called the *Michaelis constant* or $K_m$ for the enzyme. In further experiments Michaelis and Menten established that for a fixed concentration of sucrose, $V(0)$ increased linearly with [E]$_T$. This hyperbolic dependence of the rate on [S]$_T$ and linear dependence on [E]$_T$ has been established for a number of enzymes, and Michaelis and Menten proposed a kinetic model to explain these facts.

Their work was motivated by Victor Henri, who had suggested earlier that enzyme and substrate might form a complex. Using the notation E to represent the free enzyme in solution, S the substrate, P the product, and ES the enzyme–substrate complex, Michaelis and Menten proposed that the kinetics could be described by the two chemical reactions

$$E + S \underset{k_1^-}{\overset{k_1^+}{\rightleftharpoons}} ES,$$

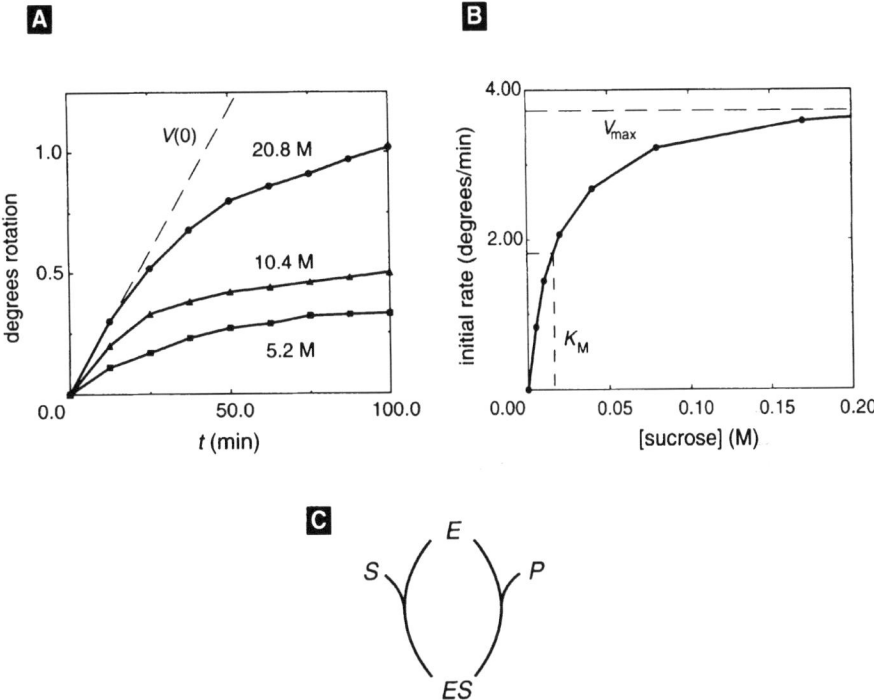

**Figure 4.12** (A) Experimental rate of loss of optical activity of sucrose for three initial concentrations of sucrose and fixed concentrations of the enzyme. Data of Michaelis and Menten, replotted from Wong (1975). (B) The initial rate $V(0)$ in (A) of the invertase catalyzed reaction plotted as a function of sucrose concentration. (C) Two–state diagram for the Michaelis–Menten model.

$$ES \xrightarrow{k_2^+} P + E. \tag{4.61}$$

This scheme can be recast easily as a diagram for enzyme states like that used in Figure 3.2 for the GLUT transporter. This is done in Figure 4.12C, where the two states of the enzyme are the free form (E) and the complex (ES). This two–state model actually involves four variables, because S and P also vary. However, the total concentration of enzyme ($[E]_T = [E] + [ES]$) and of substrate ($[S]_T = [S] + [ES] + P_C$) are conserved, so that only two of the concentrations change independently.

In this analysis we choose the concentration of substrate [S] and enzyme–substrate complex [ES] as variables and eliminate the concentration of enzyme using the conservation condition $[E] = [E]_T - [ES]$. Because the catalytic process is irreversible, the concentration of product $P_C$ does not appear in the kinetic equations for [S] and [ES], which can be obtained from the mass action laws applied to (4.61):

$$d[S]/dt = -k_1^+[E]_T[S] + (k_1^- + k_1^+[S])[ES], \qquad (4.62)$$
$$d[ES]/dt = k_1^+[E]_T[S] - (k_1^- + k_2^+ + k_1^+[S])[ES]. \qquad (4.63)$$

The two important time scales in the Michaelis–Menten model are the time that it takes for substrate to be converted into product, and the time scale on which enzyme–substrate complex forms. Thus, the important rates are $k_1^+[E]_T$ and $k_1^+[S]$. Now if we assume, as did Michaelis and Menten, that there is very little enzyme compared to substrate, then we also expect there to be very little complex compared to substrate. This means that the rate $k_1^+[E]_T$ is much smaller than the rate $k_1^+[S]$, at least initially before a lot of product has been made. Thus, it is the small ratio of the concentration of catalyst to total concentration of substrate ($[S]_T$), i.e., $\epsilon = [E]_T/[S]_T$, that makes the two time scales widely different. The two natural time scales in the model are therefore $\tau_s = 1/k_1^+[E]_T$ (slow) and $\tau_f = 1/k_1^+[S]_T$ (fast).

Having identified the fast and slow time scales, we proceed to *nondimensionalize* all of the variables in (4.62) and (4.63), including the time. We define the nondimensional dependent variables $\hat{s} = [S]/[S]_T$ and $\hat{es} = [ES]/[E]_T$. The choice of a nondimensional time then determines whether our analysis focuses on the fast or the slow time scale.

If we nondimensionalize time using the slow time scale, then $\hat{t} = t/\tau_s = tk_1^+[E]_T$, and we restrict ourselves to slow changes. Substituting these nondimensional variables into (4.62) and (4.63) gives

$$d\hat{s}/d\hat{t} = -\hat{s} + \left( \frac{k_1^-}{k_1^+[S]_T} + \hat{s} \right) \hat{es}, \qquad (4.64)$$

$$\epsilon d\hat{es}/d\hat{t} = \hat{s} - \left( \frac{k_1^- + k_2^+}{k_1^+[S]_T} + \hat{s} \right) \hat{es}. \qquad (4.65)$$

The small dimensionless parameter $\epsilon$ appears as a multiplicative factor on the left–hand side of (4.65). If we think of the asymptotic limit as $\epsilon \to 0$, then we can set the right–hand side of (4.64) equal to zero. This is the lowest–order term in the asymptotic analysis on the slow time scale, which allows us to solve explicitly for $\hat{es}$ and obtain

$$\hat{es} = \frac{\hat{s}}{(k_1^- + k_2^+)/(k_1^+[S]_T) + \hat{s}}. \qquad (4.66)$$

Substituting this expression into the right–hand side of (4.64) gives the rate of change of substrate

$$d\hat{s}/d\hat{t} = -\frac{\hat{V}_{\max}\hat{s}}{\hat{s} + \hat{K}_m}, \qquad (4.67)$$

with

$$\hat{V}_{\max} = \frac{k_2^+}{k_1^+[S]_T}, \qquad \hat{K}_m = \frac{k_1^- + k_2^+}{k_1^+[S]_T}. \qquad (4.68)$$

Equation (4.67) can be written in terms of the original dimensional variables as

$$d[S]/dt = -\frac{V_{max}[S]}{[S] + K_{m}},$$ 

(4.69)

with $V_{max} = k_2^+[E]_T$ and $K_m = (k_1^- + k_2^+)/k_1^+$. This is identical to the expression obtained by Michaelis and Menten for the initial rate of product formation (4.60). However, time–scale analysis suggests that this expression is valid on the slow time–scale $\tau_s = 1/k_1^+[E]_T$ and is not restricted to the initial period of the catalytic process.

To see how the model simplifies on the fast time–scale $\tau_f = 1/k_1^+[S]_T$, we nondimensionalize the dependent variables in the same fashion, but now introduce the rescaled time $\tilde{t} = tk_1^+[S]_T$. This leads to the equations

$$d\hat{s}/d\tilde{t} = -\epsilon\left(\hat{s} - \left(\frac{k_1^-}{k_1^+[S]_T} + \hat{s}\right)\hat{e}\hat{s}\right),$$ 

(4.70)

$$d\hat{e}\hat{s}/d\tilde{t} = \hat{s} + \left(\frac{k_1^- + k_2^+}{k_1^+[S]_T} + \hat{s}\right)\hat{e}\hat{s}.$$ 

(4.71)

On this time scale when $\epsilon \to 0$, the left–hand side of (4.70) vanishes. Thus, $d\hat{s}/d\tilde{t} = 0$, and $\hat{s} = [S]/[S]_T$ is constant. Using this result in (4.71) and reverting to dimensional variables gives, after a bit of algebra,

$$d[ES]/dt = k_1^+[S]([E]_T - [ES]) - (k_1^- + k_2^+)[ES].$$ 

(4.72)

This equation describes the exponential increase of the enzyme–substrate complex concentration to its steady–state value $[ES]^{ss} = [E]_T[S]/(K_m + [S])$. This is precisely the dimensional form of the expression for $\hat{e}\hat{s}$ on the slow time scale in (4.66). For this reason, this approximation is often referred to as the *quasi-steady-state* approximation, where "quasi" emphasizes that the value of $[ES]^{ss}$ changes in time, but only on a slow time scale.

## Suggestions for Further Reading

- *Mathematical Models in Biology*, Leah Edelstein-Keshet. This is a great introductory textbook on general mathematical biology. Chapter 7 contains material on reduction of scale and molecular events (Edelstein-Keshet 1988).
- *Principles of Applied Mathematics*, James Keener. This book contains a good treatment of perturbation theory as well as other approximation techniques (Keener 1999).
- *Applied Mathematics*, J. David Logan. This book also contains a good treatment of perturbation theory (Logan 1997).

## 4.8  Exercises

1.  Consider the system of linear differential equations

$$dx/dt = -3x + y, \qquad dy/dt = 100(2x - y),$$ (4.73)

subject to initial conditions $y(0) = 0$, $x(0) = 1$.

(a)  For this system, which is the fast variable and which is the slow variable?

(b)  What is the quasi-steady-state approximation for this system?

(c)  Use the quasi-steady-state approximation to eliminate $y$ and then show that the solution of the simplified system is $x(t) = \exp(-t), y(t) = 2\exp(-t)$.

(d)  Show that the exact solution of this system is

$$x(t) = 0.02\exp(-102.02t) + 0.98\exp(-0.98t),$$ (4.74)

$$y(t) = -1.98\exp(-102.02t) + 1.98\exp(-0.98t).$$ (4.75)

(e)  Compare the approximate solution with the exact solution. When do they agree and when do they disagree?

2.  Create a program for solving the domain $Ca^{2+}$ inactivation mechanism in Figure 4.2. Use the parameter values $k_1^+ = 0.7 ms^{-1}$, $k_1^- = 0.2\ ms^{-1}$, $k_2^+ = 0.05\ mM^{-1}\ ms^{-1}$, $k_2^- = 0.005\ ms^{-1}$, and $[Ca^{2+}]_d = 0.3\ mM$. Include in your model an auxiliary variable for calculating whole cell $Ca^{2+}$ currents using $I_{Ca} = gx_O(V - V_{Ca})$, with $g = 5\ nS$, $V = 20\ mV$, and $V_{Ca} = 60\ mV$. Use the results to confirm the results in Figure 4.3.

3.  Use the data in Exercise 2 to create the simplified version of the domain $Ca^{2+}$ inactivation model based on the rapid equilibrium approximation. Using the initial condition in (4.12), solve the ODE and compare your result for the $Ca^{2+}$ current ($I_{Ca} = gx_O(V - V_{Ca})$) to that in Figure 4.3. Use (4.9), (4.10) and (4.11) to calculate the numerical value of the relaxation time $\tau$ and the $Ca^{2+}$ current after inactivation is complete. Verify that (4.12) is correct and use it to show that the peak current is given by the expression $I_{peak} = g(1/(1 + K_1))(V - V_{Ca})$.

4.  On the fast time scale it is a good approximation to neglect step 2 in the domain $Ca^{2+}$ inactivation model in Figure 4.2.

(a)  Assuming that $x_C(0) = 1$, use the approximation to verify that on the fast time scale

$$dx_O/dt = -\left(x_O - \frac{1}{1 + K_1}\right)/\tau_{act},$$ (4.76)

with $K_1 = k_1^-/k_1^+$ and $\tau_{act} = 1/(k_1^+ + k_1^-)$, the time constant for activation.

(b)  Using the parameters in Exercise 2, evaluate $\tau_{act}$.

5.  A model that simulates the voltage dependence of domain $Ca^{2+}$ activation of L-type $Ca^{2+}$ channels in the pancreatic beta cell of mouse was developed by Sherman, Keizer, and Rinzel using the mechanism in Figure 4.2. They fit data to experimental voltage clamp records using the voltage-dependent kinetic constants: $k_1^+(V) = 0.78/(1 + \exp[-(3 + V)/10])\ ms^{-1}$, $k_1^- = (0.78 - k_1^+(V))\ ms^{-1}$, $k_2^+[Ca^{2+}]_d = 7.56 \cdot 10^{-4}[Ca^{2+}]_{out}V/(1 - \exp(V/13.4))\ mM^{-1}\ ms^{-1}$, $k_2^- = 0.002\ ms^{-1}$, with the external $Ca^{2+}$ concentration $[Ca^{2+}]_{out} = 3\ mM$ and $V$ in mV.

(a)  Create a program for this model with the current as an auxiliary variable using the expression $I_{Ca} = gx_O[Ca^{2+}]_{out}V/(1 - \exp(V/13.4))$ (This is a special case of the Goldman-Hodgkin-Katz expression $[ion]_{out}^z V/(1 - \exp(zFV/RT))$ for the driving force, which is based on consideration of the rate at which an ion of charge $z$ can diffuse through a pore with a linear gradient of electrical potential. This generalization of the Nernst expression has proven particularly useful for modeling $Ca^{2+}$ currents.)

(b) Use the program to simulate voltage clamp currents in which the voltage is increased from a holding potential of $-100$ mV to depolarized test potentials with a duration of 250 msec. Record the peak and final equilibrium currents and plot them as a function of the test potential. Explain the inverted bell-shaped I versus V curves.

6. Verify (4.51) as a model for the Ach receptor.

7. What are the six thermodynamic constraints that must hold for the kinetic parameters of the full $IP_3$receptor model? Verify that these constraints hold for the parameter values in Table 4.3.

8. (a) Verify the graph in Figure 4.11 by finding $y^\infty$, $K_1$, and $K_2$ and graphing $x_O$ in (4.59).

   (b) Use (4.56) to calculate $y^\infty([Ca^{2+}]_i, [IP_3])$. Plot the equilibrium open probability for several values of $IP_3$.

9. (a) Derive a model for ryanodine (RyR) receptors in cardiac cells. Assume that each sub-unit of a receptor has two binding sites for $Ca^{2+}$, one that activates the subunit when $Ca^{2+}$ is bound and one that inactivates the subunit when $Ca^{2+}$ is bound. Assume fur-ther that the binding of activating $Ca^{2+}$ is independent of inactivating $Ca^{2+}$, and that the binding of inactivating $Ca^{2+}$ is independent of activating $Ca^{2+}$. Show that as a re-sult there are four independent rate constants, and that the thermodynamic constraint is automatically satisfied.

   (b) Let $x_0$ be the fraction of receptors with no $Ca^{2+}$ bound, $x_1$ the fraction of receptors with only activating $Ca^{2+}$ bound, $x_3$ the fraction of receptors with only inactivating $Ca^{2+}$ bound, and $x_4$ the fraction of receptors with both binding sites bound. Draw a diagram of this reaction mechanism. Which of the rate constants are proportional to the $Ca^{2+}$ concentration?

   (c) Using the rate constants $k_{12} = 15[Ca^{2+}]_i \; \mu M^{-1}s^{-1}$, $k_{13} = 0.8[Ca^{2+}]_i \; \mu M^{-1}s^{-1}$, $k_{21} = 7.6$ $s^{-1}$, $k_{31} = 0.84 \; s^{-1}$, identify which processes are fast and which are slow. Use the quasi-steady-state approximation to compress this into a two-state model. What are the effective rate constants for this compressed two–state model?

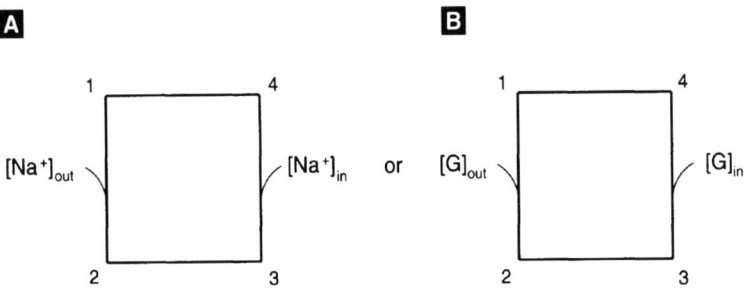

**Figure 4.13** Transition diagrams for sodium and GLUT four-state transporters.

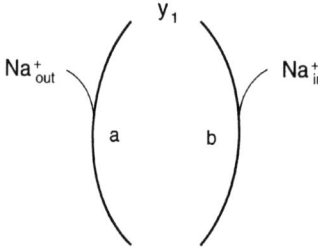

**Figure 4.14** Transition diagram for the reduced four-state sodium transporter.

10. Find the reduced slow time scale equations for a four-state transporter, such as the GLUT transporter described in Chapter 3 (see Figure 4.13). Assume that the 1-4 and 2-3 transitions are rapid compared to other transitions by lumping states 1 and 4 and states 2 and 3 into two variables,

$$x_1 + x_4 = y_1, \qquad x_2 + x_3 = y_2, \tag{4.77}$$

where $y_1 + y_2 = 1$, and show that

$$dy_1/dt = -k_{12}x_1 + k_{21}x_2 + k_{34}x_3 - k_{43}x_4. \tag{4.78}$$

Solve the equilibrium conditions for $x_1$ and $x_4$ in terms of $y_1$ and for $x_2$ and $x_3$ in terms of $y_2$. Show that

$$dy_1/dt = -\frac{1}{\tau}\left(y_1 - y_1^{ss}\right), \tag{4.79}$$

where

$$y_1^{ss} = \frac{k_{21}^{(a)} + k_{21}^{(b)}}{k_{12}^{(a)} + k_{21}^{(a)} + k_{12}^{(b)} + k_{21}^{(b)}},$$

$$\tau = \frac{1}{k_{12}^{(a)} + k_{21}^{(a)} + k_{12}^{(b)} + k_{21}^{(b)}}.$$

$$k_{12}^{(a)} = \frac{k_{12}}{1 + K'}, \qquad k_{21}^{(a)} = \frac{k_{21}}{1 + K}, \qquad k_{12}^{(b)} = \frac{k_{43}K'}{1 + K'}, \qquad k_{21}^{(b)} = \frac{k_{34}K}{1 + K}. \tag{4.80}$$

Notice that this equation describes the transition between states 1 and 2 by two different processes, depicted in Figure 4.14 as processes "a" and "b."

11. Verify (4.69) and (4.72) by expressing (4.67) and (4.71) in dimensional variables.

# CHAPTER 5

# Whole–Cell Models

## Arthur S. Sherman, Yue-Xian Li, and Joel E. Keizer

In modeling whole cells we try to understand complex properties of cells by combining interlocking transport and regulatory mechanisms. We use a modular approach and develop models of each individual process separately using available experimental data. We then construct progressively more complete models by combining components to understand how they work together. Sometimes, we proceed in the opposite order, beginning with a comprehensive model, which we simplify in order to determine the minimal essential elements. One particularly useful simplification technique is to exploit separation of time scales to set fast processes to equilibrium as described in Chapter 4.

Three model systems are investigated in this chapter; the bullfrog sympathetic ganglion neuron (Figure 5.2), the pituitary gonadotroph (Figure 5.11), and the pancreatic $\beta$-cell (Figure 5.18). The order chosen is pedagogical and is in fact opposite to the historical order in which the models were developed. These models focus on the consequences of integrating plasma membrane (PM) and endoplasmic reticulum (ER) membrane fluxes into a coordinated system for control of membrane potential and $Ca^{2+}$ concentrations. The concentration of cytoplasmic $Ca^{2+}$, $[Ca^{2+}]_i$, is of great importance for the life and death of cells. It is a key regulator of many cell processes, such as secretion, gene transcription, and apoptosis. We will restrict our attention to models with a small number of discrete, well-mixed compartments; diffusion in spatially extended compartments will be treated in Chapter 7. Other models for these particular systems can be found (Keener and Sneyd 1998; Tang et al. 1996).

These simplified models incorporate ion pumps and voltage and ligand gated channels like those discussed in previous chapters, and provide illustrations of the modular

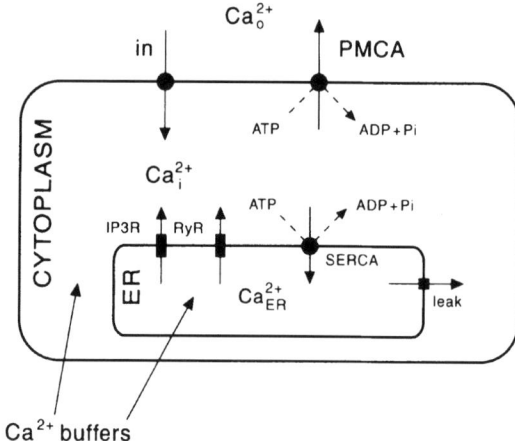

**Figure 5.1**  Basic elements involved in $Ca^{2+}$ signaling in cells.

approach. A common dynamical theme in all three models is oscillations, some driven by the ER, some by the plasma membrane, and some by interactions between the two. We will see, as previously in Chapter 2, that the combination of fast positive feedback with slow negative feedback is a ubiquitous way for cells to produce oscillations.

## 5.1   Models of ER and PM Calcium Handling

The most striking feature of $Ca^{2+}$ in cells is its low concentration, approximately 0.1 $\mu$M at rest, and only 1 $\mu$M or so at its peak. In contrast, $Na^+$ and $K^+$ are found in millimolar concentrations. Cells need to keep $[Ca^{2+}]_i$ so low because $Ca^{2+}$ binds to many proteins and modifies their enzymatic properties. Thus, rises in $Ca^{2+}$ must be kept brief and highly localized to avoid runaway activation of enzymatic cascades. These needs are met by two basic mechanisms, buffering and sequestration. Buffers are specialized $Ca^{2+}$-binding proteins that soak up 95–99% of the $Ca^{2+}$ in the cytosol. $Ca^{2+}$ is sequestered in internal stores—the sarcoplasmic reticulum (SR) in muscle cells, and the endoplasmic reticulum (ER) in other cells—by proteins that hydrolyze ATP to move $Ca^{2+}$ against steep concentration gradients. These proteins are called sarco/endoplasmic $Ca^{2+}$-ATPase (SERCA) pumps. Other pumps, plasma membrane $Ca^{2+}$-ATPases (PMCA), remove $Ca^{2+}$ from the cell. ER and SR membranes contain ion channels, analogous to but different from plasma membrane channels, that allow $Ca^{2+}$ back out of the stores into the cytoplasm when needed. See Figure 5.1. Whereas all cells have $Ca^{2+}$ pumps for negative feedback and homeostasis, some cells have evolved ion channels such as the ryanodine receptor (RyR) and the $IP_3$-Receptor, which are both activated and inhibited by $Ca^{2+}$ and provide both positive and negative feedback. The large ratios of bound to free $Ca^{2+}$ and sequestered to cytoplasmic $Ca^{2+}$ mean that brief channel openings can lead to large excursions of free cytoplasmic $Ca^{2+}$ or oscillations that can be exploited for signaling.

Mitochondria and vesicles may also act as $Ca^{2+}$ reservoirs and have their own specialized uptake and release mechanisms for $Ca^{2+}$ and other ions. These have not been modeled as extensively, and will be treated only in passing here.

$Ca^{2+}$ indicator dyes, which are exogenous buffers that fluoresce differentially depending on $[Ca^{2+}]_i$ are the most common way to measure the concentration of free $Ca^{2+}$. As we will see below, buffers change (generally, slow down) the dynamics of $Ca^{2+}$ in cells. The optical measurement of $Ca^{2+}$ is discussed further in Chapter 8. Some cells possess $Ca^{2+}$-dependent $K^+$ or $Cl^-$ channels, and one can use the currents generated by these channels as an alternative measurement. A third approach is to use aequorin, a naturally occurring protein found in jellyfish, which luminesces when bound to $Ca^{2+}$. Aequorin can be targeted to $Ca^{2+}$ stores in cells of other organisms by genetic techniques.

## 5.1.1  Flux Balance Equations with Rapid Buffering

We begin with a general description applicable to any cell with a cytosolic compartment (subscripted by i, originally for "intracellular") and an ER (subscripted by ER) and one species of $Ca^{2+}$ buffer in each compartment. That gives a total of four species, with bound and free $Ca^{2+}$ in each of the two compartments.

We define $N_i$ = total number of $Ca^{2+}$ ions, bound and free, in the cytoplasm and $N_{ER}$ = total number of $Ca^{2+}$ ions in the ER. Then

$$[Ca^{2+}]_i^{tot} = \frac{N_i}{\overline{V}_i}, \quad [Ca^{2+}]_{ER}^{tot} = \frac{N_{ER}}{\overline{V}_{ER}}, \tag{5.1}$$

where the $\overline{V}$'s representing the volumes of the cytosol and ER have overbars to distinguish them from membrane potentials.

The fundamental physical principle used in constructing equations for the four $Ca^{2+}$ concentrations is conservation of $Ca^{2+}$ ions. The number of ions in each compartment changes due to fluxes across each of the membranes separating the compartments:

$$\frac{dN_i}{dt} = J_{PM}^{in} - J_{PM}^{out} - J_{ER}^{in} + J_{ER}^{out} \tag{5.2}$$

$$\frac{dN_{ER}}{dt} = J_{ER}^{in} - J_{ER}^{out}. \tag{5.3}$$

The $J$'s have units of $\mu$moles/s and represent common fluxes found in cells (see Figure 5.1 and Table 5.1). In terms of concentrations we have

$$\frac{d[Ca^{2+}]_i^{tot}}{dt} = \frac{1}{\overline{V}_i}(J_{PM}^{in} - J_{PM}^{out} - J_{ER}^{in} + J_{ER}^{out}) = j_{PM}^{in} - j_{PM}^{out} - j_{ER}^{in} + j_{ER}^{out}, \tag{5.4}$$

$$\frac{d[Ca^{2+}]_{ER}^{tot}}{dt} = \frac{1}{\overline{V}_{ER}}(J_{ER}^{in} - J_{ER}^{out}) = \frac{\overline{V}_i}{\overline{V}_{ER}}(j_{ER}^{in} - j_{ER}^{out}), \tag{5.5}$$

**Table 5.1** Key to Ca$^{2+}$ Levels and Fluxes

| Ca$^{2+}$ levels | $[Ca^{2+}]_i^{tot}$ | bound (buffered) + free cytosolic Ca$^{2+}$ |
|---|---|---|
| | $[Ca^{2+}]_{ER}^{tot}$ | bound + free ER Ca$^{2+}$ |
| | $[Ca^{2+}]_i$ | free cytosolic Ca$^{2+}$ |
| | $[Ca^{2+} \cdot B]_i$ | bound cytosolic Ca$^{2+}$ |
| Fluxes | $J_{PM}^{in}$ | influx through PM Ca$^{2+}$ channel |
| | $j_{PM}^{in}$ | $J_{PM}^{in}$ divided by cytosolic volume $\overline{V}_i$ |
| | $J_{PM}^{out}$ | extrusion by plasma membrane Ca$^{2+}$ ATP-ase |
| | $j_{PM}^{out}$ | $J_{PM}^{out}$ divided by cytosolic volume $\overline{V}_i$ |
| | $J_{ER}^{in}$ | pumping into ER by SERCAs |
| | $J_{ER}^{out}$ | flux from ER to cytosol through IP3R or RYR |

where we have absorbed the volumes into the $j$'s, giving them units of concentration/time, typically $\mu$M/s.

Now we come to grips with the buffering equations. The term $[Ca^{2+}]^{tot}$ consists of free Ca$^{2+}$ plus Ca$^{2+}$ bound to buffer B. In the cytosol,

$$[Ca^{2+}]_i + [Ca^{2+} \cdot B]_i = [Ca^{2+}]_i^{tot}, \tag{5.6}$$

$$[B]_i + [Ca^{2+} \cdot B]_i = [B]_i^{tot}, \tag{5.7}$$

with similar expressions for the ER. We avoid adding a differential equation to describe these reactions by assuming rapid equilibrium (see Section 4.1) between Ca$^{2+}$ and buffer:

$$[Ca^{2+}]_i = \frac{K_i[Ca^{2+} \cdot B]_i}{[B]_i}. \tag{5.8}$$

Methods for treating Ca$^{2+}$ buffering are discussed in more detail in Chapter 8. Combining (5.6) and (5.8) gives

$$[Ca^{2+}]_i^{tot} = [Ca^{2+}]_i \left( 1 + \frac{[B]_i}{K_i} \right). \tag{5.9}$$

Combining (5.7) and (5.8) gives

$$[B]_i^{tot} = [B]_i \left( 1 + \frac{[Ca^{2+}]_i}{K_i} \right). \tag{5.10}$$

Finally, solving (5.10) for $[B]_i$ and substituting into (5.9) gives the desired algebraic relation between $[Ca^{2+}]_i^{tot}$ and $[Ca^{2+}]_i$:

$$[Ca^{2+}]_i^{tot} = [Ca^{2+}]_i \left( 1 + \frac{[B]_i^{tot}}{K_i + [Ca^{2+}]_i} \right). \tag{5.11}$$

To get the balance equation for $[Ca^{2+}]_i$, we apply the chain rule and (5.11) to obtain

$$\frac{d[Ca^{2+}]_i^{tot}}{dt} = \frac{d[Ca^{2+}]_i^{tot}}{d[Ca^{2+}]_i} \frac{d[Ca^{2+}]_i}{dt} = \frac{1}{f_i([Ca^{2+}]_i)} \frac{d[Ca^{2+}]_i}{dt}, \tag{5.12}$$

where

$$f_i([Ca^{2+}]_i) = \left(1 + \frac{K_i[B]_i^{tot}}{(K_i + [Ca^{2+}]_i)^2}\right)^{-1}. \qquad (5.13)$$

This is a form of the rapid buffering approximation (RBA) (Wagner and Keizer 1994), specialized to cells that may be considered spatially homogeneous. The extension to situations in which diffusion of $[Ca^{2+}]_i$ and buffer cannot be ignored is taken up in Chapter 8.

In general, $f_i$ is a function of $[Ca^{2+}]_i$, but for low–affinity buffers $K_i \gg [Ca^{2+}]_i$, and $f_i$ can be taken as constant:

$$f_i \approx \frac{1}{1 + ([B]_i^{tot}/K_i)}. \qquad (5.14)$$

In this case, $f_i$ can be interpreted as the fraction of $[Ca^{2+}]_i^{tot}$ that is free. Typical measured values for $f_i$ are 0.01–0.05.

Combining (5.12) and the corresponding equation for the ER with (5.4) and (5.5) gives

$$\frac{d[Ca^{2+}]_i}{dt} = f_i(j_{PM}^{in} - j_{PM}^{out} - j_{ER}^{in} + j_{ER}^{out}) \qquad (5.15)$$

$$\frac{d[Ca^{2+}]_{ER}}{dt} = \frac{\overline{V}_i f_{ER}}{\overline{V}_{ER}}(j_{ER}^{in} - j_{ER}^{out}). \qquad (5.16)$$

Note that as $[B]_i^{tot}$ increases, $f_i$ decreases and $[Ca^{2+}]_i$ becomes slower. This can be a problem when measuring $[Ca^{2+}]_i$ using fluorescent dyes, which are themselves $Ca^{2+}$ buffers. On the other hand, changing the cell's buffering power with dyes or other exogenous buffers can also be a useful experimental tool.

Equation (5.16) is often rewritten

$$\frac{d[Ca^{2+}]_{ER}}{dt} = \frac{f_i}{\sigma}(j_{ER}^{in} - j_{ER}^{out}), \qquad (5.17)$$

where

$$\sigma = \frac{\overline{V}_{ER} f_i}{\overline{V}_i f_{ER}}. \qquad (5.18)$$

We can interpret $\overline{V}_{ER}/f_{ER}$ and $\overline{V}_i/f_i$ as the "effective volumes" of the ER and cytosol, respectively, that is, the volumes taking into account the fraction of $Ca^{2+}$ that is free in each compartment. Then $\sigma$ is the ratio of effective volumes, an important dimensionless quantity that captures the combined effects of volume and buffering differences between the ER and the cytosol. Note that (5.15) and (5.17) have two fewer parameters than the original set (5.15) and (5.16). Thus, two of the parameters have been revealed to be redundant by this simple substitution. A more extensive example of the use of dimensionless parameters is given in Section 5.3 for the gonadotroph model.

If we multiply (5.17) by $\sigma$ and add it to (5.15), the ER flux terms cancel, and we obtain

$$\frac{d[\text{Ca}^{2+}]_{\text{T}}}{dt} = f_{\text{i}}(j_{\text{PM}}^{\text{in}} - j_{\text{PM}}^{\text{out}}), \tag{5.19}$$

where

$$[\text{Ca}^{2+}]_{\text{T}} = [\text{Ca}^{2+}]_{\text{i}} + \sigma[\text{Ca}^{2+}]_{\text{ER}}. \tag{5.20}$$

The reader may check that

$$[\text{Ca}^{2+}]_{\text{T}} = \frac{f_{\text{i}}(N_{\text{i}} + N_{\text{ER}})}{\bar{V}_{\text{i}}} \tag{5.21}$$

and $[\text{Ca}^{2+}]_{\text{T}}$ may thus be interpreted as the "total free $\text{Ca}^{2+}$" of the cell with the cytosolic volume as the reference volume.

In several examples to follow we will find it convenient to replace $[\text{Ca}^{2+}]_{\text{ER}}$ by $[\text{Ca}^{2+}]_{\text{T}}$. For example, if the plasma membrane is impermeable to $\text{Ca}^{2+}$ (the closed-cell model), then $j_{\text{PM}}^{\text{in}} = j_{\text{PM}}^{\text{out}} = 0$ and $[\text{Ca}^{2+}]_{\text{T}}$ is constant, even if $[\text{Ca}^{2+}]_{\text{i}}$ and $[\text{Ca}^{2+}]_{\text{ER}}$ fluctuate. This allows us to eliminate one differential equation. Because $\text{Ca}^{2+}$ is conserved, no information is lost, and we can recover $[\text{Ca}^{2+}]_{\text{ER}}$ from the algebraic relation

$$[\text{Ca}^{2+}]_{\text{ER}} = ([\text{Ca}^{2+}]_{\text{T}} - [\text{Ca}^{2+}]_{\text{i}})/\sigma. \tag{5.22}$$

Although the ER is necessarily smaller in volume than the cytosol, it can have greater surface area because it is convoluted. If pump density is the same, the ER fluxes will then be larger than the plasma membrane fluxes. One can study the influence of ER surface area by multiplying the ER fluxes by a common factor, but it may be more convenient to factor out the ratio and make it explicit in the equations. Equations (5.15) and (5.17) can be rearranged to give

$$\frac{d[\text{Ca}^{2+}]_{\text{i}}}{dt} = f_{\text{i}}\left(j_{\text{PM}}^{\text{in}} - j_{\text{PM}}^{\text{out}} - \frac{1}{\lambda_{\text{ER}}}\left(\bar{j}_{\text{ER}}^{\text{in}} - \bar{j}_{\text{ER}}^{\text{out}}\right)\right) \tag{5.23}$$

$$\frac{d[\text{Ca}^{2+}]_{\text{ER}}}{dt} = \frac{f_{\text{i}}}{\sigma\lambda_{\text{ER}}}\left(\bar{j}_{\text{ER}}^{\text{in}} - \bar{j}_{\text{ER}}^{\text{out}}\right), \tag{5.24}$$

where $\bar{j}_{\text{ER}}^{\text{in}} = \lambda_{\text{ER}} j_{\text{ER}}^{\text{in}}$, $\bar{j}_{\text{ER}}^{\text{out}} = \lambda_{\text{ER}} j_{\text{ER}}^{\text{out}}$, and $\lambda_{\text{ER}} = A_{\text{PM}}/A_{\text{ER}}$ is the ratio of plasma membrane to ER membrane surface area. This ratio will be important in understanding the $\beta$-cell and gonadotroph models below, and will be derived by nondimensionalizing the latter model (see Exercise 8).

## 5.1.2   Expressions for the Fluxes

The final step in specifying a particular model is to replace the general terms $j_{\text{PM}}^{\text{in}}, j_{\text{PM}}^{\text{out}}, j_{\text{ER}}^{\text{in}}$, and $j_{\text{ER}}^{\text{out}}$ by appropriate biophysical expressions. Efflux from the cell is generally mediated by a pump flux, $j_{\text{PMCA}}$, and influx into the ER from the cytosol by

$j_{SERCA}$. Similar pumps were discussed in detail in Chapter 3. In whole cell models we generally use empirical, Hill-type formulas for pump rate, of the form

$$j_{SERCA} = \frac{v_{SERCA} \cdot [Ca^{2+}]_i^2}{K_{SERCA}^2 + [Ca^{2+}]_i^2},$$    (5.25)

rather than detailed kinetic models. The maximal pump rate $v_{SERCA}$ is proportional to the total number of SERCA or PMCA pumps and to the rate of a single pump, and is typically expressed here in $\mu M/s$.

Efflux from the ER to the cytosol often includes a constant, unregulated leak conductance of the form

$$j_{LEAK} = v_{LEAK} \cdot ([Ca^{2+}]_{ER} - [Ca^{2+}]_i),$$    (5.26)

where $v_{LEAK}$ is the leak permeability (also expressed in $\mu M/s$) and $([Ca^{2+}]_{ER} - [Ca^{2+}]_i)$ is the thermodynamic driving force for a symmetric channel (see Exercise 2). Note that there is no contribution of membrane potential to the ER driving force. It is generally, though not universally, believed that there is no significant ER membrane potential because of counterions that balance the fluxes of $Ca^{2+}$.

Interesting dynamics such as oscillations generally require nonlinear feedback on the efflux rate by $Ca^{2+}$. This can be mediated by the RyR, for which a detailed model is presented in Section 5.2, or the IP$_3$R, which was discussed in Chapter 4 and is further developed in Section 5.3.

$Ca^{2+}$ generally enters cells through ion channels. Ionic currents are measured in pA ($10^{-12}$ C/s) or nA ($10^{-9}$ C/s), so an additional factor is needed to convert to $\mu$moles:

$$j_{PM}^{in} = -\alpha I_{Ca},$$    (5.27)

where $\alpha = 1/(2F\overline{V}_i)$, $F = 96480$ C/mol is Faraday's constant ($F = eA$ = the elementary charge × Avogadro's number), and the factor 2 accounts for the two positive charges of a $Ca^{2+}$ ion. As an example, if $I_{Ca}$ is measured in pA, and cytosolic volume is measured in $\mu m^3$, then $j_{PM}^{in}$ will have units of M/ms. Note that by the Hodgkin–Huxley convention, an inward cation current such as $I_{Ca}$ is negative, so the flux is positive.

# 5.2 Calcium Oscillations in the Bullfrog Sympathetic Ganglion Neuron

Bullfrog sympathetic ganglion neurons (BFSG) are excitable cells with a full complement of voltage-dependent ion channels (Yamada et al. 1998). However, the $Ca^{2+}$ oscillations modeled here are driven by nonlinearity in the ER, with the plasma membrane playing only a passive role. As early as 1976, Kuba and Nishi observed rhythmic hyperpolarizations of the resting membrane potential when the neurons were exposed to caffeine (Kuba and Nishi 1976). See Figure 5.2. The BFSG is a good model system for studying RyR-mediated $Ca^{2+}$ oscillations in neurons.

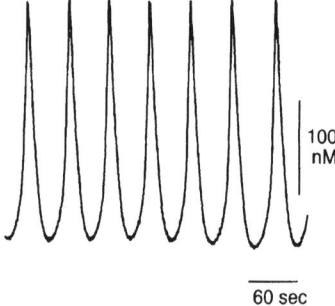

100 nM

60 sec

**Figure 5.2** Steady state $[Ca^{2+}]_i$ oscillations recorded from a fura-2 loaded bullfrog sympathetic ganglion neuron during continuous exposure to caffeine (5 mM) under mildly depolarizing conditions (30 mM $K^+$). Data provided by Dr. David Friel, Case Western Reserve University.

These caffeine-induced oscillations also occur when the membrane voltage is clamped at a fixed value, suggesting that voltage-gated ion channels on the plasma membrane are not involved in producing the oscillations. Caffeine has been found to activate the RyR by shifting the $Ca^{2+}$ dependence of channel opening to lower levels of $[Ca^{2+}]_i$. Ryanodine, a specific ligand of the RyR, blocks the response of these neurons to caffeine by locking them into a low–conductance state. The oscillations induced by caffeine have a very long period (several minutes), and the $[Ca^{2+}]_i$ spikes are characterized by a sharp upstroke followed by a plateau phase and a subsequent downstroke. The upstroke is associated with the dumping of $Ca^{2+}$ by the ER store; the plateau phase is largely due to $Ca^{2+}$ extrusion; and the downstroke is due to $Ca^{2+}$ uptake by the store. Sustained $Ca^{2+}$ entry is required during the interspike intervals, suggesting that store refilling driven by $Ca^{2+}$ influx is necessary after each spike for the generation of the subsequent spike. Therefore, increasing the rate of $Ca^{2+}$ entry by depolarizing the membrane leads to shortened interspike interval and increased amplitude (Friel and Tsien 1992; Friel 1995; Nohmi et al. 1992).

## 5.2.1  Ryanodine Receptor Kinetics: The Keizer–Levine Model

Experimental investigations of the $Ca^{2+}$-dependent gating properties of the RyR channels have provided data for the development of a mathematical model based on channel kinetics. We begin with equations for the RyR receptor itself as studied in conditions where $[Ca^{2+}]_i$ is a parameter imposed by the experimenter. For example, in the experiments of Györke and Fill, cardiac RyRs in lipid bilayers were exposed to controlled $Ca^{2+}$ elevations by flash photolysis of a caged $Ca^{2+}$ compound, DM-nitrophen (Györke and Fill 1993).

Here we present a minimal model designed to capture the key features observed by Györke and Fill. (For an alternative model see Tang and Othmer (1994).) Like the $IP_3$ receptor, the RyR exhibits fast activation and slow inactivation in response to a rise in $Ca^{2+}$(Figure 5.3). A particularly interesting feature is that a second $Ca^{2+}$ flash can "wake up" the channel and elicit more current. Another difference with the $IP_3$ receptor is that the steady-state open probability is monotonic with respect to $[Ca^{2+}]_i$, not bell-

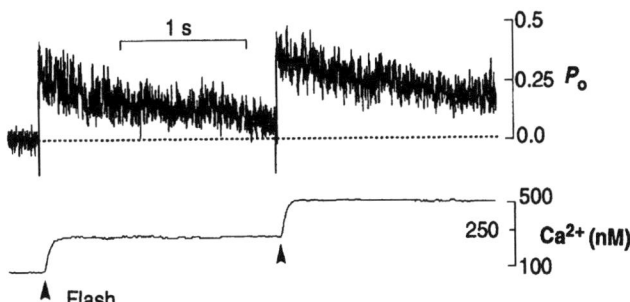

**Figure 5.3** Adaptation response of cardiac RyR in lipid bilayers to release of caged $Ca^{2+}$. Reprinted from Györke and Fill (1993).

shaped. Keizer and Levine developed a model to account for all three features observed in those experiments (Keizer and Levine 1996). The state diagram they devised (Figure 5.5) consists of two closed states, $C_1$ and $C_2$, and two open states, $O_1$ and $O_2$. At rest (corresponding to a background $[Ca^{2+}]_i$ level of 0.1 $\mu$M), most of the receptors are in state $C_1$.

Both open states have the same conductance, so the net open probability is $P_O = P_{O1} + P_{O2}$. The receptors respond to a step increase in $Ca^{2+}$ with a rapid rise in $P_O$ as

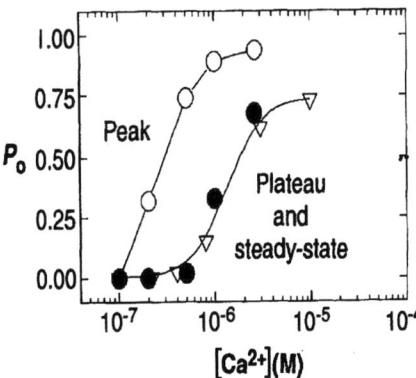

**Figure 5.4** Dependence of peak and steady-state RyR current on $Ca^{2+}$. Reprinted from Györke and Fill (1993).

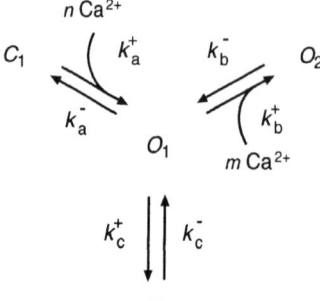

**Figure 5.5**  Kinetic states of the Keizer-Levine model for the RyR.

states $O_1$ and $O_2$ are populated, followed by a slow decline to a steady, plateau level as states flow into $C_2$ (Figure 5.6A, solid curve). Figure 5.6B shows the peak and plateau values attained after a step from $[Ca^{2+}]_i = 0.1$ $\mu M$ to the indicated value. (Note that the jump to the peak depends on the initial conditions, which determine the number of receptors in $C_1$ available to be recruited by the $Ca^{2+}$ step.) During the short time shown in Figure 5.6, there is little back flow out of $C_2$, but a further increase in $[Ca^{2+}]_i$ draws more receptors out of $C_1$ into the open states (Figure 5.6A, arrow and dashed curve).

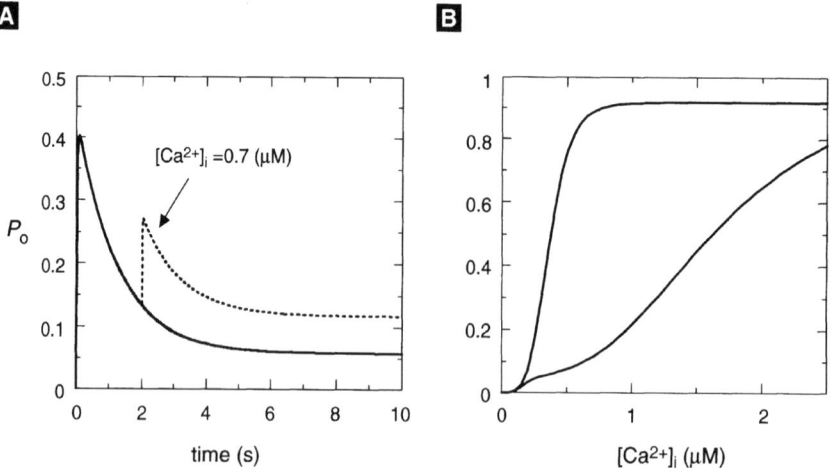

**Figure 5.6**  Keizer–Levine model: (A) Time course of response of $P_O$ to a step increase in $[Ca^{2+}]_i$ from 0.1 to 0.35 $\mu M$ (solid). Dotted curve shows "adaptation" when a second step to 0.7 $\mu M$ is applied at $t = 2$ s. (B) Peak and steady-state levels of $P_O$ for different values of $[Ca^{2+}]_i$. (see Keizer and Levine (1996), Figure 5.3, and Figure 5.4, this chapter.) Parameters: $k_a^+ = 1500 /(\mu M^4$ s), $k_a^- = 28.8/s$, $k_b^+ = 1500 /(\mu M^3$ s), $k_b^- = 385.9/s$, $k_c^+ = 1.75/s$, $k_c^- = 0.1/s$).

The full four–state model is described by three independent ODEs plus a conservation condition for the total number of channels: $P_{O1} + P_{O2} + P_{C1} + P_{C2} = 1$. We can apply the methods of Chapter 4 to derive a quasi-steady-state approximation (Exercise 3) assuming that $P_{C2}$ is slow, while the other states are in rapid equilibrium:

$$P_O \approx P_O^{slow} = \frac{w \cdot (1 + ([Ca^{2+}]_i/K_b)^3)}{1 + (K_a/[Ca^{2+}]_i)^4 + ([Ca^{2+}]_i/K_b)^3}, \tag{5.28}$$

where $P_O^{slow}$ is proportional to $w = 1 - P_{C2}$, which is the fraction of noninactivated receptors. In terms of probabilities,

$$Pr\{Open\} = Pr\{Not\ Inactivated\} \cdot Pr\{Open|Not\ Inactivated\}.$$

Given $[Ca^{2+}]_i$, $w$ is calculated from the differential equation

$$\frac{dw}{dt} = \frac{(w_\infty - w)}{\tau} \tag{5.29}$$

with

$$w_\infty = \frac{1 + (K_a/[Ca^{2+}]_i)^4 + ([Ca^{2+}]_i/K_b)^3}{1 + (1/K_c) + (K_a/[Ca^{2+}]_i)^4 + ([Ca^{2+}]_i/K_b)^3} \tag{5.30}$$

and

$$\tau = \frac{w_\infty}{k_c^-}. \tag{5.31}$$

As we did for the gating variables in Chapter 2, it is natural to write a pseudo-exponential rate equation for $w$ because changes in $w$ are exponential for fixed $[Ca^{2+}]_i$. Note that the rate constants of the fast processes no longer appear, just the dissociation constants, defined by $K_a^4 = k_a^-/k_a^+$, $K_b^3 = k_b^-/k_b^+$, and $K_c = k_c^-/k_c^+$. The slow time scale $\tau$ is inversely proportional to $k_c^-$, the rate of transition out of state $C_2$.

### 5.2.2 Bullfrog Sympathetic Ganglion Neuron Closed–Cell Model

We proceed to write down the equations for a closed cell using (5.15) (with $j_{PM}^{in} = j_{PM}^{out} = 0$) together with appropriate expressions for the fluxes due to the ryanodine receptors, a leak out of the ER, and a SERCA pump:

$$\frac{d[Ca^{2+}]_i}{dt} = f_i(j_{RyR} + j_{LEAK} - j_{SERCA}). \tag{5.32}$$

The RyR flux is

$$j_{RyR} = v_{RyR} P_O \cdot ([Ca^{2+}]_{ER} - [Ca^{2+}]_i), \tag{5.33}$$

where $v_{RyR}$ (also in $\mu M/s$) is proportional to the number of RyRs and $P_O$ is the probability that a receptor is open. (Note that here we are generalizing (5.26) for a symmetric leak channel, but symmetry is only an approximation for the RyR. See Exercise 2.) We use the quasi-steady-state approximation (5.28)–(5.31) of the previous section for $P_O$.

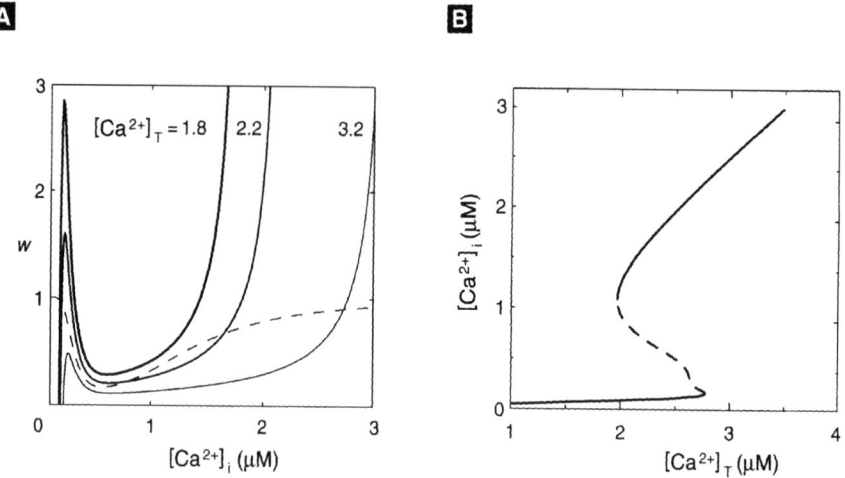

**Figure 5.7**  Closed-cell Keizer–Levine model: (A) Nullclines for $[Ca^{2+}]_i$ (solid) and $w$ (dashed). (B) Intersections of nullclines (steady–states) in (A) for various values of $[Ca^{2+}]_T$. Parameters: As in Figure 5.6 plus: $f_i = 0.01$, $v_{RyR} = 5/s$, $v_{LEAK} = 0.2/s$, $v_{SERCA} = 100\ \mu M/s$, $K_{SERCA} = 0.2\ \mu M$, $\sigma = 0.02$, $K_a = 0.4\ \mu M$, $K_b = 0.6\ \mu M$, $K_c = 0.1\ \mu M$, $k_c^- = 0.1/s$, and $[Ca^{2+}]_T$ is in $\mu M$ as indicated.

Because of conservation of ions, we do not need a differential equation for $[Ca^{2+}]_{ER}$, but just the algebraic relation (5.22).

The leak is given by

$$j_{LEAK} = v_{LEAK} \cdot ([Ca^{2+}]_{ER} - [Ca^{2+}]_i), \qquad (5.34)$$

and the pump flux is

$$j_{SERCA} = v_{SERCA} \cdot \frac{[Ca^{2+}]_i^2}{[Ca^{2+}]_i^2 + K_{SERCA}^2}. \qquad (5.35)$$

The closed cell model consists of the two differential equations for $[Ca^{2+}]_i$ (5.32) and $w$ (5.29) plus the associated algebraic equations.

With the parameters given in Figure 5.7, the model is not able to produce oscillations, but instead is bistable; it can assume either a high-$[Ca^{2+}]_i$ steady state or a low-$[Ca^{2+}]_i$ steady state. This behavior can be understood by examining the phase plane, which is possible because we reduced the kinetic equations for the RyR to one slow equation. In the closed-cell model total $Ca^{2+}$ ($[Ca^{2+}]_T$) is constant, and this is a natural parameter to use to characterize the system. For low values of $[Ca^{2+}]_T$, the steady-state $[Ca^{2+}]_i$ is low, and $w$ is high (few receptors are adapted). For high values of $[Ca^{2+}]_T$, steady-state $[Ca^{2+}]_i$ is high, and $w$ is low (many receptors are adapted). For intermediate values of $[Ca^{2+}]_T$ there are three steady states. Examination of the eigenvalues of the Jacobian in this case reveals that the low and high steady states are stable nodes (see Appendix A), while the middle steady state is a saddle. Sample nullclines are plotted in Figure 5.7A, and the summary of how the steady states vary with $[Ca^{2+}]_T$ (i.e., the bifurcation diagram) is plotted in Figure 5.7B. This global view reveals a large region of bistability or hysteresis bracketed by two saddle-node bifurcations (the turning points, where the determinant changes sign) that give birth to all of the steady states in the system.

There is also a value of $[Ca^{2+}]_T$ near the bottom turning point where there is a Hopf bifurcation (the trace of the Jacobian changes sign; see Appendix A). Geometrically, the $w$-nullcline intersects the $[Ca^{2+}]_i$-nullcline with a more negative slope. The oscillations born at this point, however, are small in amplitude and unstable. By making a small change in the $Ca^{2+}$-handling parameters it is possible to obtain stable oscillations with the closed-cell model (see Exercise 4). Nonetheless, Keizer and Levine concluded that that this is not a good model for the oscillations observed by Friel and Tsien; they are not robust (i.e., the oscillations only exist for a small range of $[Ca^{2+}]_T$), and it is difficult to get oscillations with period greater than one minute. This is because the time scale of the oscillations is determined by the reciproal of $k_c^-$, which is about 10 seconds. In order to obtain robust oscillations on the minute time scale, it is necessary to add another, slower process to the system, which we do in the next section.

### 5.2.3 Bullfrog Sympathetic Ganglion Neuron Open–Cell Model

In the previous section we saw that adaptation of the RyR is not likely the mechanism behind slow oscillations in the BFSG cell, and that another, slower, process is needed. A natural extension suggested by the analysis of the closed cell is to make $[Ca^{2+}]_T$ a slow variable rather than a constant parameter. Biophysically, this allows for larger variation in the filling state of the ER. We will see that in each cycle the stores almost completely empty and refill. In other words, the net flux in and out of the cell over a cycle is 0, but there is a large efflux at some points in the cycle that is balanced by a large influx at other points.

The open–cell model expands on the closed–cell model by adding an equation to describe changes in $[Ca^{2+}]_T$ due to fluxes into the cell and PMCA pumps that remove $Ca^{2+}$:

$$\frac{d[Ca^{2+}]_T}{dt} = f_i(j_{IN} - j_{PMCA}). \tag{5.36}$$

The PMCA flux is given by

$$j_{PMCA} = \frac{v_{PMCA}[Ca^{2+}]_i^2}{K_{PMCA}^2 + [Ca^{2+}]_i^2}. \tag{5.37}$$

Since oscillations occur under voltage clamp, we can represent influx of $Ca^{2+}$ as a constant $j_{IN}$ (understood as influx across the plasma membrane, in $\mu M/s$). The plasma membrane fluxes must also be included in the equation for $[Ca^{2+}]_i$:

$$\frac{d[Ca^{2+}]_i}{dt} = f_i(j_{RyR} + j_{LEAK} - j_{SERCA} + j_{IN} - j_{PMCA}). \tag{5.38}$$

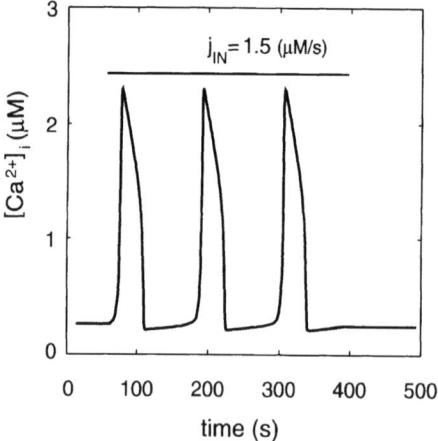

**Figure 5.8** Reduced open-cell Keizer–Levine model. At time = 50 seconds, oscillations are initiated by stepping $j_{IN}$ from 0 to 1.5. At time = 400 seconds, oscillations are terminated by reducing $j_{IN}$ back to 0. Parameters as in Figure 5.7 plus: $v_{PMCA} = 5.0$ $\mu$M/s, $K_{PMCA} = 0.6$ $\mu$M, $j_{IN} = 1.5$ $\mu$M/s (cf.Keizer and Levine (1996), Figure 5).

The open cell model thus consists of three differential equations (5.29), (5.36), and (5.38), associated algebraic expressions for the fluxes, and the conservation relationship given by (5.20). We can collapse this system back into just two equations for $[Ca^{2+}]_i$ and $[Ca^{2+}]_T$. This is justified because variation in $w$ is fast compared to the oscillation period (see Exercise 6). We eliminate the equation for $w$ by writing $w = w_\infty$. The RyR channel flux is then given by

$$j_{RyR} = v_{RyR} \cdot \frac{w_\infty(1 + ([Ca^{2+}]/K_b)^3)}{1 + (K_a/[Ca^{2+}]_i)^4 + ([Ca^{2+}]_i/K_b)^3} ([Ca^{2+}]_{ER} - [Ca^{2+}]_i). \qquad (5.39)$$

Solutions of the reduced open cell model ((5.36) and (5.38) with $j_{RyR}$ replaced by the expression in (5.39)) are shown in Figure 5.8. With adequate $Ca^{2+}$ influx, oscillations of the right shape and duration are seen. However, if $Ca^{2+}$ influx is suppressed, say by removal of external $Ca^{2+}$, the oscillations cease immediately, also in agreement with experiment (Friel and Tsien 1992).

The closed cell model can be viewed as the *fast subsystem* of the open cell. That is, the bifurcation diagram of the closed cell model with $[Ca^{2+}]_T$ as a parameter is a phase plane of the reduced open cell model with $[Ca^{2+}]_T$ as a slow variable (slow compared to $[Ca^{2+}]_i$). We illustrate this by projecting the trajectory of the open cell onto the diagram computed with the closed cell, supplemented with the nullcline for $[Ca^{2+}]_T$ (Figure 5.9). Moreover, we can predict from this that if $j_{IN}$ is increased so that the $[Ca^{2+}]_T$ nullcline intersects the S-shaped $[Ca^{2+}]_i$ nullcline above the upper limit point or below the lower limit point, there will be no oscillations (see Exercise 5). This is confirmed by the bifurcation diagram of the reduced open cell with respect to $j_{IN}$ (Figure 5.10). Within the oscillatory range, variation of $j_{IN}$ produces a broad range of frequencies, but almost no change in amplitude. This follows from the relaxation oscillator character of the model (compare with discussion of the Morris–Lecar model, Chapter 2).

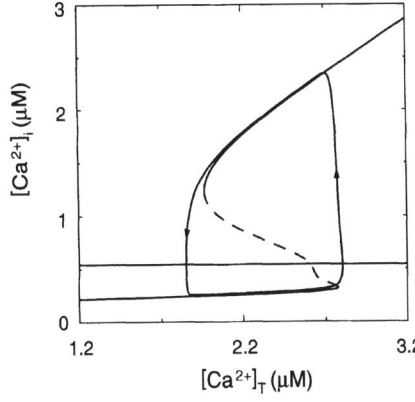

**Figure 5.9** The bifurcation diagram of the closed–cell Keizer–Levine model Figure 5.7 reinterpreted as a phase plane of the reduced open–cell model Figure 5.8, with superimposed trajectory and $[Ca^{2+}]_T$ nullcline (horizontal line). Direction of motion along trajectory indicated by arrows. Parameters as in Figure 5.7 but with $j_{IN} = 1.5\mu M/s$ (cf. Keizer and Levine (1996), Figure 3).

## 5.3  The Pituitary Gonadotroph

Our next example of a whole–cell model is the coupling of electrical signaling and $Ca^{2+}$ signaling in gonadotrophs. Gonadotrophs are hormone–secreting (endocrine) cells in the anterior part of the pituitary gland. Their physiological function is to release gonadotropins (gonad-stimulating hormones) when they themselves are stimulated by a hormone, gonadotropin-releasing hormone (GnRH), that is secreted by endocrine neurons in the hypothalamus of the brain. $Ca^{2+}$ signaling is crucial for these cells to fulfill their secretory functions, which are part of the complex neuro-endocrine control of reproduction in mammals. Physiological levels of GnRH have been shown to trigger large–amplitude $Ca^{2+}$ oscillations in cultured gonadotrophs. It has further been shown that these oscillatory $Ca^{2+}$ signals are directly coupled to the exocytosis of hormone-containing vesicles in these cells (Tse et al. 1993). These $Ca^{2+}$ signals can become more

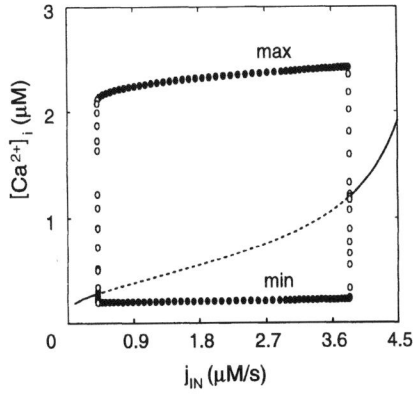

**Figure 5.10** Bifurcation diagram of the reduced open cell version of Keizer–Levine, showing the behavior of the model as in Figure 5.7 and Figure 5.8 for a range of values of $j_{IN}$ (cf. Keizer and Levine (1996), Figure 8).

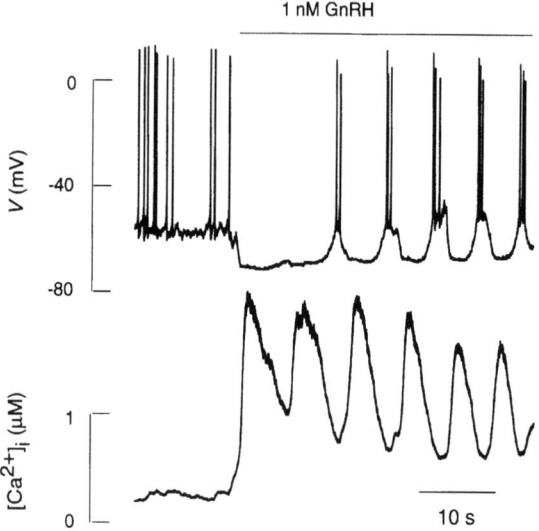

**Figure 5.11** Burst firing driven by repetitive releases of Ca$^{2+}$ from the ER in a pituitary gonadotroph stimulated by GnRH. Data provided by Drs. Stanko Stojilković and Fred van Goor, National Institute of Child Health and Development, N.I.H.

complex, because gonadotrophs are excitable cells capable of generating Ca$^{2+}$ action potentials that drive Ca$^{2+}$ from the extracellular medium into the cell in an oscillatory manner. Thus, gonadotrophs are equipped with two distinct mechanisms of generating dynamic Ca$^{2+}$ signals, one due to voltage gated Ca$^{2+}$ entry (PM oscillator) and the other due to the oscillatory release of Ca$^{2+}$ from the intracellular store (ER oscillator). These two dynamic signaling systems are coupled by the Ca$^{2+}$ and Ca$^{2+}$-activated K$^+$ ion channels in the plasma membrane. This makes the gonadotroph an ideal system to study complex Ca$^{2+}$ signaling mechanisms and to apply the modular approach of model development.

One dramatic manifestation of the ER–plasma membrane interaction is shown in Figure 5.11. The unstimulated gonadotroph exhibits low-frequency spiking and low [Ca$^{2+}$]$_i$. After GnRH is applied, Ca$^{2+}$ oscillations due to repetitive releases of Ca$^{2+}$ from the ER appear. Each Ca$^{2+}$ rise turns on Ca$^{2+}$-activated K$^+$ channels and transiently interrupts the firing, resulting in repetitive bursting.

We begin with the ER oscillator and then add in the plasma membrane oscillator, culminating in a comprehensive model that can reproduce the burst pattern like that in Figure 5.11.

### 5.3.1   The ER Oscillator in a Closed Cell

As for the BFSG, it is useful to study the gonadotroph first in the closed-cell case, when the ER oscillator is decoupled from the PM oscillator. This allows us to concentrate on the intracellular mechanisms that are independent of the influence of Ca$^{2+}$ exchange across the plasma membrane.

ER-mediated $Ca^{2+}$ oscillations in gonadotrophs are linked to the activation of $IP_3$ receptor channels. In contrast to the caffeine-induced $Ca^{2+}$ oscillations in the bullfrog sympathetic neurons, mediated by the RyR, the ER-dependent $Ca^{2+}$ oscillations in gonadotrophs are induced by the endogenous intracellular messenger $IP_3$ produced when the physiological ligand GnRH binds to its receptors on the surface of the cell. Agonist-induced $Ca^{2+}$ oscillations in a large number of cell types including gonadotrophs are generated by the dynamic interactions between $IP_3$ channels and SERCA pumps expressed in the membrane of the intracellular $Ca^{2+}$ store (ER or SR). Of crucial importance are the $Ca^{2+}$-dependent gating properties of the $IP_3$ channels. As discussed in Chapter 4, cytosolic free $Ca^{2+}$ can both activate and inactivate the opening of $IP_3$ channels. This leads to the characteristic bell-shaped $Ca^{2+}$-dependence of steady-state channel opening with the peak typically located at the mid-submicromolar level of $Ca^{2+}$ (see Figure 4.11 in Chapter 4; Iino (1990); Bezprozvanny et al. (1991)). It has also been shown that $Ca^{2+}$-dependent activation occurs on a faster time scale than $Ca^{2+}$-dependent inactivation (Parker and Ivorra 1990; Finch et al. 1991).

The De Young–Keizer model that was discussed in Chapter 4 was the first model to incorporate all these experimental data (DeYoung and Keizer 1992). Other models based on similar data were developed later, including important simplifications by Li and Rinzel (1994), Atri et al. (1993), and Othmer and Tang (1993). These models have in common that both positive and negative feedback reside in the $IP_3R$ itself, in contrast with two earlier classes of models. In one class of models, oscillations depend on the interaction of two separate internal pools of $Ca^{2+}$, one responsive to $IP_3$ and one to $Ca^{2+}$(Kuba and Takeshita 1981; Goldbeter et al. 1990). In another, $IP_3$ itself had to oscillate (Meyer and Stryer 1988; Cuthbertson and Chay 1991). It is possible that the latter class of models corresponds to a distinct class of oscillations in cells (Thomas et al. 1996). Yet another model that differs from all of the above in that it neither requires $IP_3$ nor involves calcium-induced calcium release is proposed in (Dolmetsch and Lewis 1994). Though we will not describe these other models in detail, they make good starting points for independent study.

Differences in the time scales of channel gating are attributed to differences in the binding constants of $IP_3$ and $Ca^{2+}$ to their binding sites on the $IP_3$ channels. This assumption, although not necessarily an accurate description of the real system, leads to a simple model that gives results that are identical to the observed ones. Even if the time scale differences are actually due to mechanisms other than differences in the binding constants, the resulting model would be qualtitatively the same mathematically. Here we use a simplified version of the model due to Li and Rinzel (1994). For a closed cell at a fixed level of $IP_3$, the Li–Rinzel simplification involves only two dynamic variables: $[Ca^{2+}]_i$ and $h$, the fraction of channels not inactivated by $Ca^{2+}$ and hence available to open. The equations are

$$\frac{d[Ca^{2+}]_i}{dt} = \frac{f_i}{V_i}\left(J_{ER}^{out} - J_{ER}^{in}\right), \tag{5.40}$$

$$\frac{dh}{dt} = A\left[K_d - ([Ca^{2+}]_i + K_d)h\right],\tag{5.41}$$

with $A$ a variable to control the relative time scale between the two differential equations. This simplified model gives as good a fit to the experimental data as the original model it was based on. Filling in expressions for $J_{ER}^{out}$, which is the sum of a small constitutive leak and the $[Ca^{2+}]_i$ and $IP_3$ regulated flux through the $IP_3$ receptors, and for $J_{ER}^{in}$ we have

$$J_{ER}^{out} = \left[L + \frac{P_{IP3R}[IP_3]^3[Ca^{2+}]_i^3 h^3}{([IP_3] + K_i)^3([Ca^{2+}]_i + K_a)^3}\right]([Ca^{2+}]_{ER} - [Ca^{2+}]_i),\tag{5.42}$$

$$J_{ER}^{in} = \frac{V_{SERCA}[Ca^{2+}]_i^2}{[Ca^{2+}]_i^2 + K_{SERCA}^2},\tag{5.43}$$

where $P_{IP3R}$ is the maximum total permeability of $IP_3$ channels (in pL/s for consistency) $L$ is the ER leak permeability (also in pL/s) and $V_{SERCA}$ is the nonscaled maximum SERCA pump rate (in aMol/s = $10^{-18}$mol/s). The constant $K_{SERCA}$ is the $Ca^{2+}$ level for half maximal activation of the SERCA pump and $K_i$, $K_a$, and $K_d$ are the dissociation constants of $IP_3$, $Ca^{2+}$-activation, and $Ca^{2+}$-inactivation sites on the $IP_3R$, respectively (all expressed in $\mu M$). As for the BSFG closed-cell model, $[Ca^{2+}]_T$ is a given constant, and we obtain $[Ca^{2+}]_{ER}$ from $([Ca^{2+}]_T - [Ca^{2+}]_i)/\sigma$.

For any model, it can greatly deepen our understanding to scale the variables and parameters to achieve a dimensionless form of the equations. This process often reveals that some of the parameters do not affect the dynamical behavior independently, and they can be combined. In the present case, the number of parameters can be reduced from 10 in the original equations (including $\sigma$) to 6 in the scaled form. (Compare (5.42) and (5.43) with (5.44) and (5.45).) There are often multiple ways of scaling the same system. One strategy is to scale the dependent variables so that their magnitudes are of order one, which facilitates comparison of magnitudes, and to achieve a maximum reduction in the number of independent parameters. The $h$ variable in the equations above is already dimensionless. We can scale $[Ca^{2+}]_i$ by either $K_a$ or $K_d$. We choose $K_d$, because this leaves only one independent parameter in the $h$ equation. Once we have made this choice, $[Ca^{2+}]_{ER}$, $[Ca^{2+}]_T$, $K_a$, and $K_{SERCA}$ should also be scaled by $K_d$:

$$\hat{c} = [Ca^{2+}]_i/K_d,$$
$$\hat{c}_E = [Ca^{2+}]_{ER}/K_d,$$
$$\hat{c}_T = [Ca^{2+}]_T/K_d,$$
$$\hat{k}_a = K_a/K_d,$$
$$\hat{k}_S = K_{SERCA}/K_d.$$

It is natural to scale the $IP_3$ concentration by $K_i$, and as a result, $L$ and $V_{SERCA}$ need to be scaled accordingly:

$$\hat{i} = I/K_i,$$
$$\hat{l} = L/P_{IP3R},$$

$$\hat{v}_S = \left(\frac{\sigma}{K_d P_{IP3R}}\right) V_{SERCA}.$$

Finally, scaling time as follows cleans up all the constants on the left–hand side of the $[Ca^{2+}]_i$ equation:

$$\hat{t} = \left(\frac{P_{IP3R} f_i}{\sigma \overline{V}_i}\right) t,$$

which leaves the following to clean up the $h$ equation:

$$\hat{a} = \left(\frac{\sigma \overline{V}_i K_d}{P_{IP3R} f_i}\right) A.$$

The final result is

$$\frac{d\hat{c}}{d\hat{t}} = \left[\hat{i} + \left(\frac{\hat{i}\hat{c}h}{(\hat{i}+1)(\hat{c}+\hat{k}_a)}\right)^3\right](\hat{c}_T - \sigma_1\hat{c}) - \frac{\hat{v}_S\hat{c}^2}{\hat{c}^2 + \hat{k}_S^2}, \tag{5.44}$$

$$\frac{dh}{d\hat{t}} = \hat{a}[1 - (\hat{c}+1)h], \tag{5.45}$$

where $\sigma_1 = 1 + \sigma$ and $\hat{c}_E$ can be recovered from $\hat{c}_E = (\hat{c}_T - \hat{c})/\sigma$. Note that, in order to have $Ca^{2+}$ and time come out as recognizable quantities, the plots to follow in Section 5.3 have all been created with the *dimensional* forms of the models discussed, and therefore are expressed in "real world" units. Several exercises consist of comparing the dimensional and dimensionless versions of the models.

Agonist-induced and $IP_3R$ $Ca^{2+}$ oscillations in different cell types can have very different frequency, amplitude, and spiking profiles. However, they all share the following features: (1) oscillations occur only at intermediate levels of $IP_3$ concentration; (2) the oscillation amplitude is almost constant for different levels of $IP_3$; (3) the oscillation period decreases as the $IP_3$ level increases; (4) the oscillations can occur without a constant $Ca^{2+}$ influx from the extracellular medium, even if the stores are depletable ($Ca^{2+}$ just shuttles back and forth between the ER and the cytosol). Because the above model was based on key experimental data and is a general model of $IP_3$-triggered $Ca^{2+}$ oscillations, it should reproduce all these common features. Feature (4) is automatically satisfied here, because the model is for a closed cell that is isolated from the extracellular medium. Figure 5.12, created with the dimensional version of the model, shows that the model indeed exhibits all the other features. Properties (2) and (3) suggest that the agonist concentration is transmitted to the cell in terms of frequency rather than amplitude, termed frequency encoding (Goldbeter et al. 1990).

This very simple model not only reproduces well-established experimental observations but also predicts other possible ways $Ca^{2+}$ oscillations can be induced. These can easily be shown by the phase plane analysis of (5.44)–(5.45). The equations for the

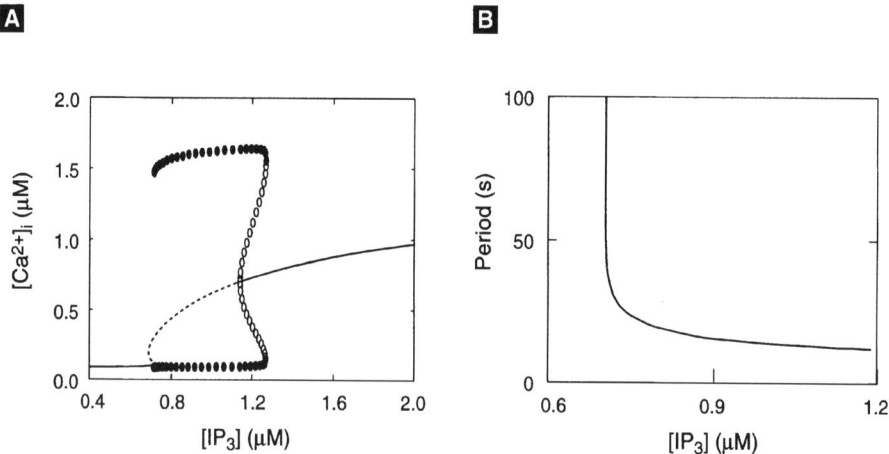

**Figure 5.12** Closed-cell gonadotroph model (dimensional equations (5.40)–(5.43)): (A) Steady-state (curves) and periodic (circles) solutions plotted against [IP$_3$]. Dotted curves and open circles denote unstable steady-state and periodic solutions. The amplitude remains almost unchanged within the range of [IP$_3$] values that produce oscillations. (B) Oscillation period is a decreasing function of [IP$_3$]. Parameters: $f_i = 0.01$, $\bar{V}_i = 4$ pL, $L = 0.37$ pL/s, $P_{IP3R} = 26640$ pL/s, $K_i = 1.0$ $\mu$M, $K_a = 0.4$ $\mu$M, $V_{SERCA} = 400$ aMol/s, $K_{SERCA} = 0.2$ $\mu$M, $A = 0.5$/s, $K_d = 0.4$ $\mu$M, $\sigma = 0.185$, $[Ca^{2+}]_T = 2$ $\mu$M, $[IP_3] = \mu$M as indicated.

two nondimensional nullclines can be solved explicitly:

$$\text{The } \hat{c}\text{-nullcline: } h = \frac{(1+\hat{\imath})(\hat{c}+\hat{k}_a)}{\hat{\imath}\hat{c}}\left[\frac{\hat{v}_S\hat{c}^2}{(\hat{c}^2+\hat{k}_S^2)(\hat{c}_T-\sigma_1\hat{c})}-\hat{\imath}\right]^{1/3}, \tag{5.46}$$

$$\text{The } h\text{-nullcline: } h = \frac{1}{1+\hat{c}}. \tag{5.47}$$

If we look at plots of the *dimensional* versions of the nullclines shown in Figure 5.13, the [Ca$^{2+}$]$_i$-nullcline is "N-shaped" between two asymptotes: One is the vertical axis (i.e., [Ca$^{2+}$]$_i$ = 0) where $h$ approaches negative infinity, and the other one is at [Ca$^{2+}$]$_i$ = [Ca$^{2+}$]$_T/\sigma_1$ where $h$ approaches positive infinity. This nullcline crosses the horizontal axis at [Ca$^{2+}$]$_i \approx$ [Ca$^{2+}$]$_i^* = K_{SERCA}\sqrt{L[Ca^{2+}]_T/V_{SERCA}}$. The value [Ca$^{2+}$]$_i^*$ is also the approximate value of the basal level of Ca$^{2+}$ at equilibrium. This expression thus shows the dependence of the basal Ca$^{2+}$ level on the key parameters $K_{SERCA}$, $L$, [Ca$^{2+}$]$_T$ and $V_{SERCA}$. The $h$-nullcline is a monotonically decreasing function of [Ca$^{2+}$]$_i$. These two nullclines can intersect each other at either one single point or at three points depending on the choice of parameter values. Since we are interested in the conditions for the occurrence of oscillations, we focus here on the case in which the two intersect at one single point (Figure 5.13). This point is the equilibrium, or steady-state, solution

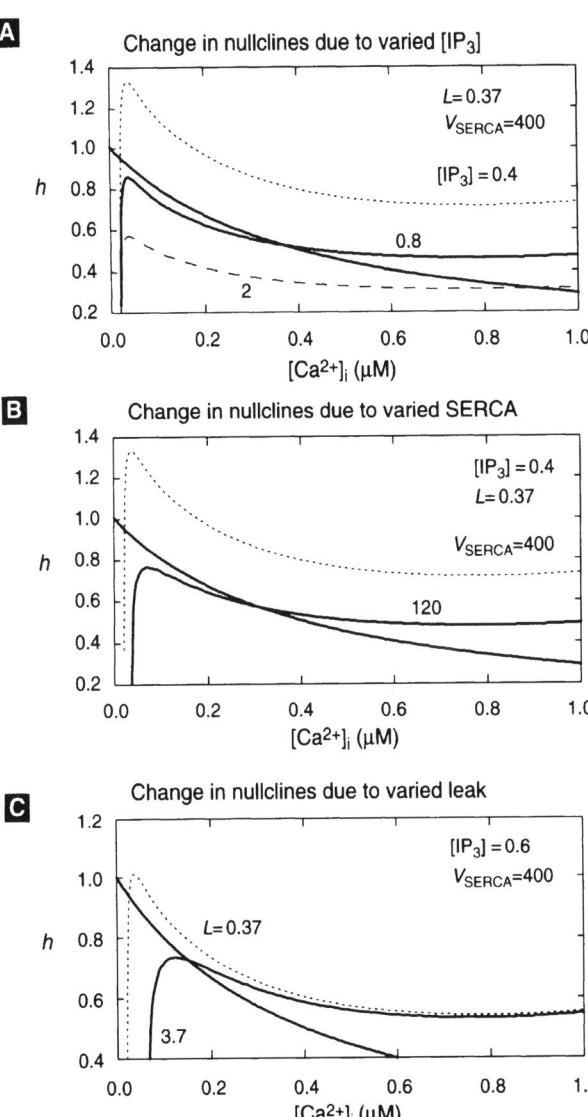

**Figure 5.13** Phase plane diagrams for the dimensional version of closed-cell gonadotroph model showing how the $[Ca^{2+}]_i$-nullcline moves in response to changes in (A) $[IP_3]$, (B) SERCA pump rate $V_{SERCA}$, and (C) the leak permeability $L$. In each panel, the monotonically decreasing $h$-nullcline is always plotted in thick, solid curve while the $[Ca^{2+}]_i$-nullclines are plotted in different line types, because they correspond to different parameter values. The thick, solid $[Ca^{2+}]_i$-nullcline in each panel corresponds to the parameter value that destabilizes the steady state and yields oscillations. Parameters as in Figure 5.12 except as indicated in the figure.

of the system at which the rate of $Ca^{2+}$ release from the store exactly matches the rate of $Ca^{2+}$ uptake. Oscillations occur when this steady state becomes unstable.

Phase plane analysis (Figure 5.13) indicates that instability of the steady state can be achieved in several possible ways. In all cases, the presence of a basal level of $IP_3$ is a prerequisite. Notice that the $h$-nullcline is independent of any parameter and remains unchanged in all panels (the solid, monotonically decreasing curve). In the top panel, the $[Ca^{2+}]_i$-nullcline is plotted for three different values of $[IP_3]$. The $[Ca^{2+}]_i$-nullcline moves downward as $[IP_3]$ increases such that the intersection between the two null-clines occurs at a larger value of $[Ca^{2+}]_i$ and a lower value of $h$. The following is an intuitive explanation of why this is the natural consequence when $[IP_3]$ is increased. In order to maintain balanced rates of $Ca^{2+}$ release and uptake at the equilibrium when $Ca^{2+}$ permeability of $IP_3$ channels is increased, more channels need to be inactivated at the same level of $[Ca^{2+}]_i$. Because increased inactivation happens at lower values of $h$, the nullcline moves downward when $[IP_3]$ increases. Instability occurs at intermediate values of $[IP_3]$, i.e., when the $[Ca^{2+}]_i$-nullcline is moved downward so that it crosses the $h$-nullcline at locations where its slope is more negative. But the downward move-ment of the $[Ca^{2+}]_i$-nullcline can also be achieved at fixed basal $[IP_3]$ levels that are not themselves high enough to destabilize the equilibrium. One alternative is to decrease the value of $V_{SERCA}$ (middle panel). This can be achieved experimentally by applying a drug called thapsigargin that specifically blocks the SERCA pump. Another alternative is to increase the leak permeability of the ER membrane (bottom panel). This can be realized by applying ionomycin, a drug that makes holes in the ER membrane. There are two other ways to cause instability at basal levels of $[IP_3]$ (Exercise 7). One is to lower the value of $K_a$, which means increased affinity of the activation site for $Ca^{2+}$. The other is to increase $[Ca^{2+}]_T$. This is equivalent to overloading the ER store, which is known experimentally to result in increased excitability of the system.

### 5.3.2  Open–Cell Model with Constant Calcium Influx

Now let us study the same model in the case where the cell is no longer closed. In an open cell, $Ca^{2+}$ can flow into the cell from the extracellular medium and vice versa. As in the BFSG, $\hat{c}_T$ is no longer a constant since unbalanced $Ca^{2+}$ exchange across the cell surface will change its value. Returning again to the nondimensional form, such an open cell model involves three dynamical variables, $\hat{c}$, $\hat{c}_E$, and $h$, satisfying

$$\frac{d\hat{c}}{d\hat{t}} = \hat{j}_{rel}(\hat{c}, h, \hat{c}_T) - \hat{j}_{fil}(\hat{c}) + \epsilon\left[\hat{j}_{in} - \hat{j}_{out}(\hat{c})\right], \tag{5.48}$$

$$\frac{d\hat{c}_E}{d\hat{t}} = -[\hat{j}_{rel}(\hat{c}, h, \hat{c}_T) - \hat{j}_{fil}(\hat{c})]/\sigma, \tag{5.49}$$

$$\frac{dh}{d\hat{t}} = \hat{a}\left[1 - (\hat{c} + 1)h\right], \tag{5.50}$$

where $\hat{\jmath}_{rel}$ and $\hat{\jmath}_{fil}$ are the nondimensional ER fluxes from (5.44)–(5.45):

$$\hat{\jmath}_{rel} = \left[1 + \left(\frac{\hat{\imath}\hat{c}h}{(\hat{\imath}+1)(\hat{c}+\hat{k}_a)}\right)^3\right](\hat{c}_T - \sigma_1 \hat{c}),$$

$$\hat{\jmath}_{fil} = \frac{\hat{v}_s \hat{c}^2}{\hat{c}^2 + \hat{k}_S^2},$$

and

$$\hat{\jmath}_{out} = \frac{\hat{v}_p \hat{c}^2}{\hat{c}^2 + \hat{k}_P^2} \tag{5.51}$$

is the nondimensional form (see Exercise 8) of the plasma membrane PMCA pump (5.37).

Note that the plasma membrane flux terms in (5.48) are multiplied by a small dimensionless parameter $\epsilon = \lambda_{ER}$, which is the ratio of plasma membrane to ER surface area (see (5.23) and (5.24)). The fluxes, $j_{PM}^{in}$, $j_{PM}^{out}$, $j_{ER}^{in}$, and $j_{ER}^{out}$ are implicitly proportional to the surface area of the respective membranes, whereas the introduction of $\epsilon$ allows explicit comparison between the magnitudes of the respective $Ca^{2+}$ fluxes.

Using $\hat{c}_T = \hat{c} + \sigma \hat{c}_E$, we can replace the $\hat{c}_E$ equation with an equation for $\hat{c}_T$:

$$\frac{d\hat{c}_T}{d\hat{t}} = \epsilon \left[\hat{\jmath}_{in} - \frac{\hat{v}_p \hat{c}^2}{\hat{c}^2 + \hat{k}_p^2}\right]. \tag{5.52}$$

The nondimensional formulation makes it clear that $\hat{c}_T$ is a slow variable because plasma membrane fluxes are much smaller than ER fluxes in the gonadotroph, and hence $\epsilon$ is very small.

For simplicity, we first study the case where $\hat{\jmath}_{in}$ is a constant. This corresponds to the experimental situation in which the cells are voltage clamped (Kukuljan et al. 1994). Figure 5.14 shows a case, for the *dimensional* version of the model, in which the cell is clamped first at a voltage that is not too negative and $J_{IN}$ is large enough to sustain large–amplitude $Ca^{2+}$ oscillations and a filled store. However, when the clamping voltage is switched to a more negative value (at $t = 40$), $J_{IN}$ is nearly 0, so that there is only $Ca^{2+}$ extrusion but no $Ca^{2+}$ influx at the cell surface. A similar situation can occur when the cells are placed in $Ca^{2+}$-deficient or zero $Ca^{2+}$ medium. Contrary to the caffeine-induced $Ca^{2+}$ oscillations in bullfrog sympathetic neurons of Figure 5.8, the absence of $Ca^{2+}$ influx does not kill the oscillation immediately. Instead, the amplitude gradually decreases as the store empties and stops only when the store is nearly depleted (Figure 5.14B).

This phenomenon can be understood by the bifurcation analysis of the "fast sub-system" of the open cell model. Recall that $\hat{c}_T$ is a "slow" variable and can be treated as a parameter of the fast subsystem. Assuming that $\epsilon = 0$, this fast subsystem is identical to the closed–cell model presented in the previous subsection. The bifurcation diagram for the *dimensional* version of this model with respect to $[Ca^{2+}]_T$ for a fixed value of

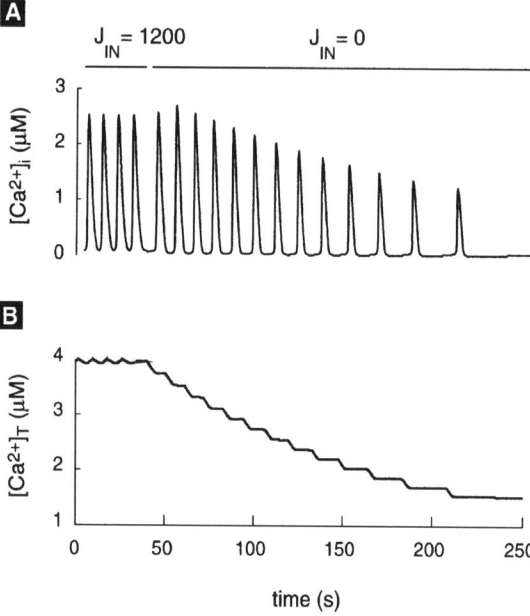

**Figure 5.14** Open-cell gonadotroph model (*dimensional* form of (5.48)–(5.52), see Exercise 8) with constant $Ca^{2+}$ influx $J_{IN}$. Time series for $[Ca^{2+}]_i$ (A) and $[Ca^{2+}]_T$ (B) are shown for two different values of $J_{IN}$. When $J_{IN} = 1200$ (time between 0 and 40), sustained oscillations in $[Ca^{2+}]_i$ are obtained with high $[Ca^{2+}]_T$. After $J_{IN}$ is switched from 1200 to 0 (time > 40), the oscillations continue, but the amplitude and the frequency decrease. Eventually, the store becomes depleted, as reflected by the low level of $[Ca^{2+}]_T$. The oscillations stop when the store becomes depleted. Parameters as in Figure 5.12 except: $[IP_3] = 0.9\ \mu M$, $\epsilon = 0.01$, $V_{PMCA} = 2000$ aMol/s, $K_{PMCA} = 0.3\ \mu M$, $J_{IN}$ aMol/s as indicated.

$[IP_3]$ is shown in Figure 5.15A. It shows that oscillations can occur only at intermediate values of $[Ca^{2+}]_T$ and that the oscillation amplitude decreases as $[Ca^{2+}]_T$ decreases. Figure 5.15B shows that the oscillation period increases (i.e., the frequency decreases) as $[Ca^{2+}]_T$ decreases. Superimposed on the bifurcation diagram in Figure 5.15A are the projected trajectories taken from Figure 5.14A.

## 5.3.3 The Plasma Membrane Oscillator

Extensive studies of the plasma membrane (PM) oscillator have revealed the major types of ion channels that are expressed in gonadotrophs as well as the detailed gating properties of these channels. The oscillations in the PM potential occur spontaneously in the absence of any hormonal signal. They are generated by several voltage-gated ion currents including L-type (noninactivating) and T-type (transient) $Ca^{2+}$ currents, a $Na^+$ current that is not essential, and a delayed rectifier $K^+$ current. There exists another $K^+$ channel that is of great importance to this cell: the K(Ca) ($Ca^{2+}$-activated $K^+$) channel that is sensitive to apamin (see, e.g., Hille (2001)). Although the K(Ca) channels are not involved in producing the voltage spikes, they provide the key link between the ER oscillator and the PM oscillator. This is because these channels are activated when the cytosolic $Ca^{2+}$ level reaches peak values, resulting in the disruption of tonic spiking in PM potential and the hyperpolarization of the cell. A realistic model of this oscillator based on experimental data can be found in Li et al. (1995a). Here, however, we use a

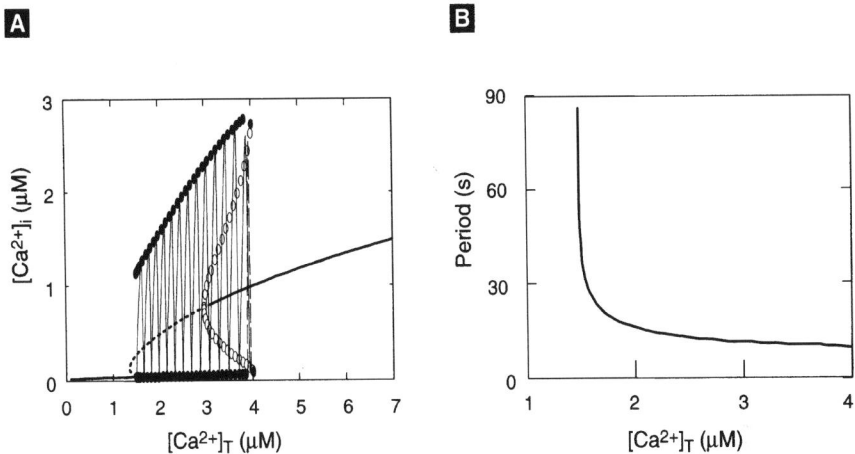

**Figure 5.15** Bifurcation in the *dimensional* version of the open-cell gonadotroph model: (A) Steady states (curves) and periodic solutions (circles) versus the total intracellular $Ca^{2+}$ level $[Ca^{2+}]_T$. The oscillation amplitude decreases significantly as $[Ca^{2+}]_T$ decreases. (B) The oscillation period increases as $[Ca^{2+}]_T$ decreases. Superimposed on the bifurcation diagram in (A) are the projected trajectories for $J_{IN} = 1200$ (dashed) and $J_{IN} = 0$ (solid) from Figure 5.14.

much simpler model of the PM oscillator based on a modified Morris–Lecar model to demonstrate how the system works.

All we need is a model that generates $Ca^{2+}$ action potentials spontaneously such that each AP spike is associated with the entry of a quantum of $Ca^{2+}$ into the cell. The Morris–Lecar model (see Section 2.4) is the simplest possible model of this kind, involving only an L-type $Ca^{2+}$ current and a delayed rectifier $K^+$ current. The main modification involves introducing the K(Ca) current, which is absent in the original Morris–Lecar model but is crucial for gonadotrophs. Note that $I_{app}$ and the leak currents are removed from the original model (2.30)–(2.34). The equations governing the gonadotroph plasma membrane oscillator are then

$$C_m \frac{dV}{dt} = -g_{Ca} m_\infty (V - V_{Ca}) - \left[ g_K w + \frac{g_{K(Ca)}[Ca^{2+}]_i^{\,4}}{[Ca^{2+}]_i^{\,4} + K_{K(Ca)}^4} \right](V - V_K), \qquad (5.53)$$

$$\frac{dw}{dt} = \phi(w_\infty - w)/\tau, \qquad (5.54)$$

where $[Ca^{2+}]_i$ is again the $Ca^{2+}$ level in the cytosol but is here a parameter for the plasma membrane model.

The voltage–dependent equilibrium open fractions for the channel states and the voltage–dependent activation time constant for the delayed rectifier are, as before:

$$m_\infty = 0.5[1 + \tanh((V - v_1)/v_2)],$$

$$w_\infty = 0.5[1 + \tanh((V - v_3)/v_4)],$$
$$\tau = 1/\cosh((V - v_3)/(2 \cdot v_4)).$$

At low levels of $[Ca^{2+}]_i$, the system generates continuous spiking. When $[Ca^{2+}]_i$ is larger than a certain threshold value, it hyperpolarizes the membrane potential. Also, when the plasma membrane voltage undergoes tonic spiking, $Ca^{2+}$ enters the cell and increases $[Ca^{2+}]_i$. We take this into account in the next section.

### 5.3.4  Bursting Driven by the ER in the Full Model

A full model of $Ca^{2+}$ signaling in agonist-stimulated gonadotrophs involves both the ER oscillator and the PM oscillator. Having studied all the ingredients of this model, we now combine the modules into a complete, realistic model described by the following five differential equations:

$$\frac{dV}{d\hat{t}} = -\hat{i}_{Ca} - \hat{i}_K - \hat{i}_{K(Ca)}, \tag{5.55}$$

$$\frac{dw}{d\hat{t}} = \phi(w_\infty - w)/\hat{\tau}, \tag{5.56}$$

$$\frac{d\hat{c}}{d\hat{t}} = \hat{j}_{rel}(\hat{c}, h, \hat{c}_T) - \hat{j}_{fil}(\hat{c}) + \epsilon\left[-\hat{\alpha}\,\hat{i}_{Ca} - \frac{\hat{v}_p \hat{c}^2}{\hat{c}^2 + \hat{k}_p^2}\right], \tag{5.57}$$

$$\frac{dh}{d\hat{t}} = \hat{a}[1 - (\hat{c} + 1)h], \tag{5.58}$$

$$\frac{d\hat{c}_T}{d\hat{t}} = \epsilon\left[-\hat{\alpha}\,\hat{i}_{Ca} - \frac{\hat{v}_p \hat{c}^2}{\hat{c}^2 + \hat{k}_p^2}\right], \tag{5.59}$$

where

$$\hat{i}_{Ca} = \hat{g}_{Ca} m_\infty (V - V_{Ca}),$$
$$\hat{i}_K = \hat{g}_K w (V - V_K),$$
$$\hat{i}_{K(Ca)} = \hat{g}_{K(Ca)} \frac{\hat{c}^4}{\hat{c}^4 + \hat{k}_{K(Ca)}^4}(V - V_K),$$

and $\hat{\alpha}$ is the non-dimensionalized factor that converts the current flux into the ion flux of $Ca^{2+}$ (5.27). Note that $I_{K(Ca)}$ is now expressed in terms of the nondimensional $\hat{c}$ (with $\hat{k}_{K(Ca)} = K_{K(Ca)}/K_d$), and the conductances have been scaled, eliminating $C_m$ as a parameter. For example,

$$\hat{g}_{Ca} = g_{Ca}\left(\frac{\sigma \overline{V}_i}{C_m P_{IP3R} f_i}\right).$$

The nondimensional voltage–dependent time constant $\tau(V)$ is given by

$$\hat{\tau}(V) = \tau(V)\left(\frac{P_{IP3R} f_i}{\sigma \overline{V}_i}\right).$$

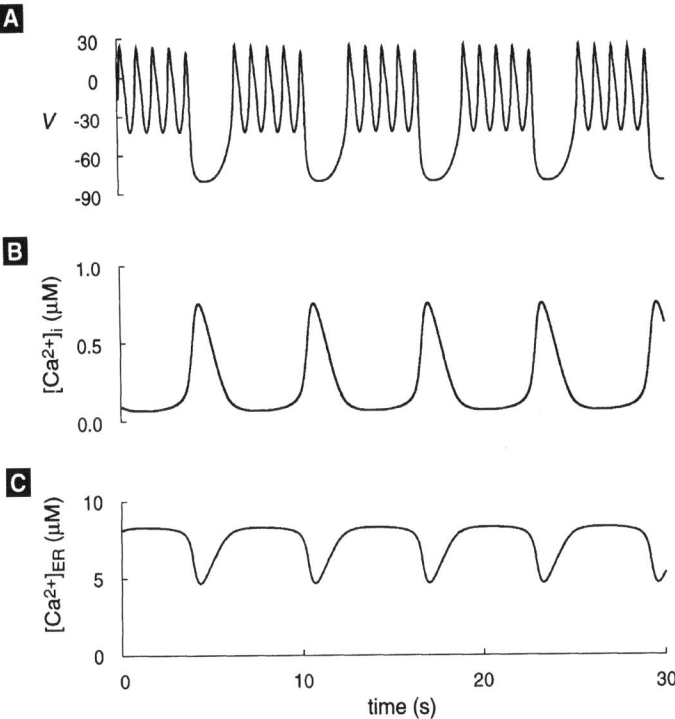

**Figure 5.16** Full gonadotroph model (Equations (5.55)–(5.59) in *dimensional* form): bursting as a result of interaction between the plasma membrane oscillator and the ER oscillator. Plasma membrane voltage $V$ (A), cytosolic calcium level $[Ca^{2+}]_i$ (B), and the store calcium content $[Ca^{2+}]_{ER}$ (C) are shown. Each spike in $[Ca^{2+}]_i$ hyperpolarizes the plasma membrane by activating calcium–sensitive potassium channels. ER parameters as in Figure 5.14 except: $L = 1.48$ pL/s, $P = 2960$ pL/s, $K_i = 0$ $\mu$M, $A = 2$/s, $V_{SERCA} = 480$ aMol/s, $V_p = 400$ aMol/s. Plasma membrane (Morris–Lecar) parameters: $V_K = -85$ mV, $V_{Ca} = 120$ mV, $K_{K(Ca)} = 0.5$ $\mu$M, $C_m = 1$ $\mu$F/cm$^2$, $g_K = 20$ $\mu$S/cm$^2$, $g_{Ca} = 20$ $\mu$S/cm$^2$, $g_{K(Ca)} = 8$ $\mu$S/cm$^2$, $\phi = 12$/s, $v_1 = -3$ mV, $v_2 = 30$ mV, $v_3 = -20$ mV, $v_4 = 30$ mV, $\alpha = 0.2$ (aMol·cm$^2$)/nC.

The effects of the ER oscillator on the PM oscillator are determined by $I_{K(Ca)}$. Whenever $\hat{c}$ is high, it activates $I_{K(Ca)}$, which hyperpolarizes the PM potential and inhibits the PM oscillator. The effects of the PM oscillator on the ER oscillator are mediated by the term $-\alpha I_{Ca}$, which describes the Ca$^{2+}$ influx through voltage gated Ca$^{2+}$ channels. As shown in the previous section, this influx leads to a fuller ER store and makes the ER more excitable. Thus, the voltage spikes activate the ER oscillator. The interaction of the two oscillators gives rise to bursting of the PM potential (Figure 5.16). Compare with Figure 5.11. Such bursting is mainly driven by the ER oscillator, which periodically hyperpolarizes the plasma membrane. However, the PM oscillations are indispensable for maintaining the bursting, because the Ca$^{2+}$ entry that accompanies

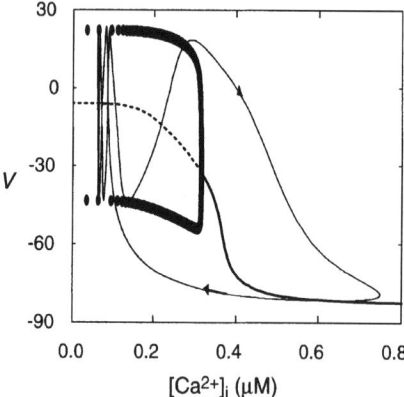

**Figure 5.17** Full gonadotroph model: bifurcation diagram for the *dimensional* form of the fast subsystem (plasma membrane oscillator) with superimposed trajectory from Figure 5.16.

each voltage spike compensates the loss of $Ca^{2+}$ to the extracellular medium and is thus key in preventing the store from depleting (see Li et al. (1997) for details of such interactions).

In Figure 5.17 the *dimensional* bifurcation diagram of the fast subsystem (the PM oscillator) with respect to $[Ca^{2+}]_i$ is shown with the trajectory of the bursting oscillations in the top panel of Figure 5.16 superimposed. Notice that the PM voltage $V$ follows the diagram well during the spiking phase, when $[Ca^{2+}]_i$ changes slowly. But during the $[Ca^{2+}]_i$ spike, which is fast, the trajectory does not follow the diagram closely.

## 5.4  The Pancreatic Beta Cell

Pancreatic $\beta$-cells secrete insulin, the hormone that maintains blood plasma glucose within narrow limits in the face of variable food ingestion and physical activity patterns. Insulin allows muscle to take up glucose for immediate energy, causes the liver to store glucose as glycogen for medium-term energy storage, and signals fat cells to use glucose for fat deposition for long-term energy storage. Diabetes is a disease in which glucose is chronically elevated, leading to blindness, kidney failure, limb amputation, cardiovascular disease, and death. Type I ("juvenile") diabetes is the result of an absolute lack of insulin following auto-immune destruction of the $\beta$-cells. Type II ("adult onset") diabetes, the more common variety, involves a relative lack of insulin, usually as a result of two defects: insulin resistance (higher than normal concentrations of insulin are required for glucose processing) and failure of the $\beta$-cells to produce enough insulin to compensate. The rising tide of diabetes in the industrialized nations suggests that Type II diabetes is a maladaptive response to a toxically energy-rich food environment, with genes governing the susceptibility of individuals. For an entree into the vast literature on glucose homeostasis and diabetes see Porte, Jr. (1990) and Taylor (1999).

A number of aspects of this system have been modeled, including the kinetics of insulin secretion (Grodsky 1972), the etiology of diabetes as $\beta$-cells fail to compensate (Topp et al. 2000), insulin action (Quon and Campfield 1991a; Quon and Campfield 1991b), and whole-body plasma insulin oscillations (see Keener and Sneyd (1998) Chapter 19).

Here we limit our attention to the electrical activity and $Ca^{2+}$ oscillations used by the $\beta$-cell to regulate insulin secretion. In terms of cell physiology, the $\beta$-cell closely resembles the pituitary gonadotroph. Both are endocrine cells in which secretion is controlled by $[Ca^{2+}]_i$, which is in turn regulated to a large degree by bursting, and they share many mechanistic elements such as K(Ca) channels and $IP_3$ receptors. However, there are important and interesting differences from that point on. Bursting in $\beta$-cells is primarily driven by the plasma membrane oscillator, though the ER probably plays a significant role at least in modulating the burst mechanism.

These characteristics are shared with many neurons, and $\beta$-cells provided an early paradigm for modeling of bursting. See Rinzel and Ermentrout (1998) or Keener and Sneyd (1998).

The models discussed in this section are complementary to the Keizer–Maki model in Chapter 4. The latter considers insulin oscillations but does not explicitly treat the $Ca^{2+}$ influx that leads to insulin secretion. Here we treat $Ca^{2+}$ influx in detail, but assume that secretion increases with mean $[Ca^{2+}]_i$.

We will focus on membrane-potential-driven $[Ca^{2+}]_i$ oscillations, though $\beta$-cells have occasionally been observed to exhibit oscillations driven by the ER, similar to those in gonadotrophs. Ironically, the Keizer–DeYoung model for the $IP_3$ receptor was originally developed to explain this marginal phenomenon in $\beta$-cells, and was then ready at hand for application to gonadotrophs.

One indication of the differences between $\beta$-cell and gonadotroph bursting is apparent in Figure 5.18. This simultaneous recording of membrane potential and $[Ca^{2+}]_i$ oscillations shows that $[Ca^{2+}]_i$ is high during the depolarized spiking phase of the bursts, rather than during the hyperpolarized silent phase. Compare with Figure 5.11.

An important aspect of $\beta$-cell function that we will also have to neglect in this chapter is the organization of the cells into electrically coupled populations, called the islets of Langerhans. Here we take advantage of the observation that the cells in an islet are synchronized. That is why the voltage of the single peripheral cell shown in Figure 5.18B is in register with the $[Ca^{2+}]_i$ signal (Figure 5.18A), which comes from many cells in the islet. This allows us to study a simple single-cell model, which can be taken as representative of the whole islet. More subtle effects of electrical coupling will be discussed in Chapter 6. See also Sherman (1997).

## 5.4.1 Chay–Keizer Model

The model we use to illustrate $\beta$-cell bursting is based loosely on the model of Sherman et al. (1988), recast in Morris–Lecar form. Bursting occurs in response to glucose, so no applied current is needed. Morris–Lecar by itself can account for the spiking during the

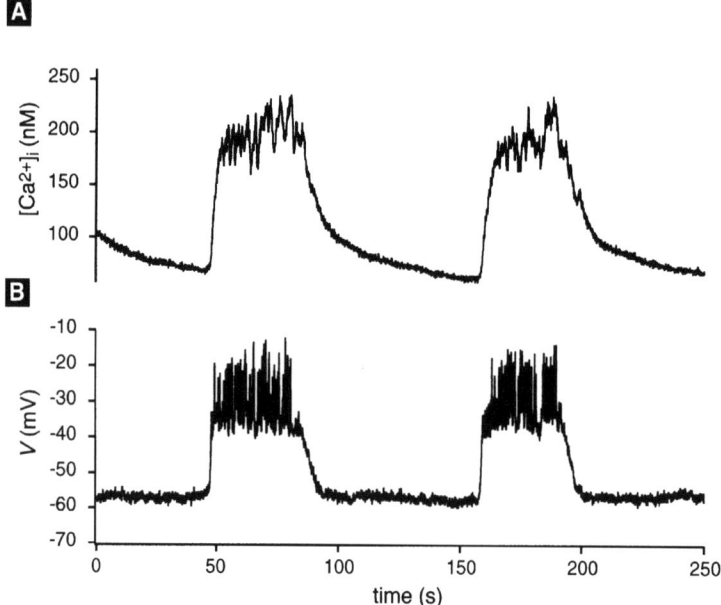

**Figure 5.18** Simultaneous recording of membrane potential bursts and $Ca^{2+}$ oscillations from a pancreatic islet of Langerhans. $V$ recorded by perforated patch on a peripheral $\beta$-cell. $[Ca^{2+}]_i$ recorded from whole islet using fura-2. Data provided by Drs. Les Satin and Min Zhang, Virginia Commonwealth University.

active phase of a burst. In order to obtain cycling between bursting and silent phases, we need to add a slow negative–feedback current. The first hypothesis, proposed by Atwater, Rojas, and colleagues (Atwater et al. 1980) and made into a mathematical model by Chay and Keizer (1983), was that bursting was mediated by a K(Ca) current. The idea was that $[Ca^{2+}]_i$ would build up slowly during the spiking phase until the inhibitory effect of the increased $K^+$ current reached a sufficiently high level to terminate the spiking. The K(Ca) current is represented as

$$I_{K(Ca)} = g_{K(Ca)} \frac{[Ca^{2+}]_i}{K_{K(Ca)} + [Ca^{2+}]_i}(V - V_K), \qquad (5.60)$$

and the modified Morris–Lecar equations, supplemented by a slow equation for $[Ca^{2+}]_i$, are

$$C_m \frac{dV}{dt} = -I_{Ca} - I_K - I_L - I_{K(Ca)}, \qquad (5.61)$$

$$\frac{dw}{dt} = \phi \frac{w_\infty - w}{\tau}, \qquad (5.62)$$

$$\frac{d[Ca^{2+}]_i}{dt} = f_i(-\alpha I_{Ca} - v_{LPM}[Ca^{2+}]_i). \qquad (5.63)$$

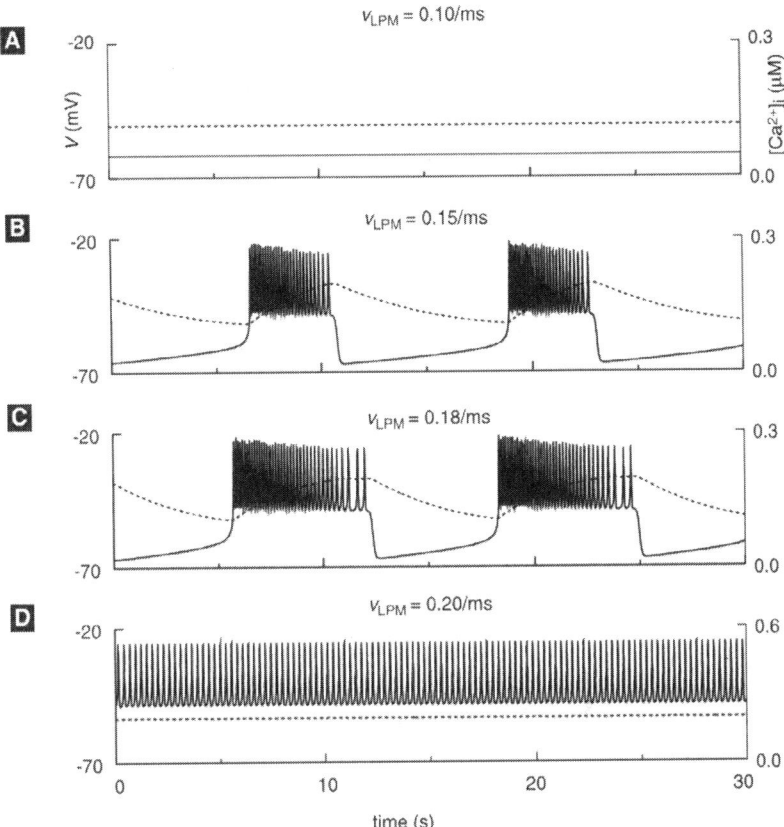

**Figure 5.19** Bursting and glucose sensing of the Chay–Keizer model, (5.61)–(5.63). Parameters: $C_m = 5300$ fF, $g_{Ca} = 1000$ pS, $V_{Ca} = 25$ mV, $g_K = 2700$ pS, $V_K = -75$ mV, $I_{app} = 0$, $v_1 = -20$ mV, $v_2 = 24$, $v_3 = -16$ mV, $v_4 = 11.2$ mV, $\phi = 0.035$/ms, $g_L = 150$ pS, $V_L = -75$ mV, $g_{K(Ca)} = 2000$ pS, $K_{K(Ca)} = 5$ $\mu$M, $f = 0.001$, $\alpha = 4.5 \times 10^{-6}$ $\mu$M/(fA· ms), and $v_{LPM}$ as indicated.

Here $f_i$ is the fraction of free $[Ca^{2+}]_i$, and $\alpha = 10^3/(2F\overline{V}_i)$ converts current in fA to $\mu$M/ms (cf. (5.27)). The term $v_{LPM}[Ca^{2+}]_i$ is a linearized representation of the PMCA. We will refer to this model as "Chay–Keizer" for brevity because it is based on the same mechanisms, but see (Chay and Keizer 1983) for details of the original. For nondimensionalization and applications, see (Pernarowski et al. 1992).

Numerical results are shown in Figure 5.19. Focusing first on panels B and C, we see that bursting can occur without participation of the ER; all that is needed is a slow negative–feedback process, here supplied by $[Ca^{2+}]_i$.

Because $[Ca^{2+}]_i$ varies very slowly, we can analyze the dynamics by decomposing the system into simpler components, one fast and one slow. The first step is to view

$[Ca^{2+}]_i$ as a parameter that controls the behavior of the fast subsystem, consisting of $V$ and $n$. In analogy to Figure 5.9, we construct the bifurcation diagram of the fast subsystem with respect to $[Ca^{2+}]_i$. Figure 5.20A shows the resulting three-branched Z-shaped curve (Z-curve) of steady states. The Z-curve is actually the projection into the $V$-$[Ca^{2+}]_i$ plane of the three-dimensional curve on which $V$ and $n$ are at steady state. When $[Ca^{2+}]_i$ is high, $g_{K(Ca)}$ is large, and the system goes to the lower branch of low-voltage steady states. When $[Ca^{2+}]_i$ is low, $g_{K(Ca)}$ is small, and $V$ rises above the threshold for activation of the voltage-dependent $Ca^{2+}$ and $K^+$ currents, generating the depolarized upper branch. For intermediate values of $[Ca^{2+}]_i$ there are three steady states. The new steady states are born via saddle-node bifurcations, with the saddle points making up the middle branch of the Z-curve.

The fast subsystem consists of two variables, not just one as was the case in Figure 5.9, so it can itself undergo oscillations, even with $[Ca^{2+}]_i$ fixed. These oscillations arise via a Hopf bifurcation on the upper branch and correspond to the fast spikes during a burst. For intermediate values of $[Ca^{2+}]_i$, the system is bistable: It can either be at rest on the lower branch of the Z-curve or oscillate on the upper branch. Bursting is a repetitive alternation between these two states. In order for this to occur, $[Ca^{2+}]_i$ must have appropriate slow dynamics, such that it increases when $V$ is high and decreases when $V$ is low. We can interpret this geometrically by ignoring $n$ and viewing the $V$-$[Ca^{2+}]_i$ bifurcation diagram as a $V$-$[Ca^{2+}]_i$ phase plane (Figure 5.20B). The Z-curve serves as a nullcline for $V$, and we add a nullcline for $[Ca^{2+}]_i$, which increases when the phase point is above the $[Ca^{2+}]_i$ nullcline and decreases when the point is below the nullcline. Thus, in order for $[Ca^{2+}]_i$ to rise during the active phase of each burst and fall during the silent phase, the $[Ca^{2+}]_i$-nullcline must intersect the middle branch of the Z-curve. If instead the $[Ca^{2+}]_i$-nullcline intersects the lower branch, the cell will remain silent. If the $[Ca^{2+}]_i$-nullcline intersects the spiking branch, the cell can fire continuously; the influx of $[Ca^{2+}]_i$ during each spike exactly balances the removal.

To understand this analysis in depth, it is helpful to construct the phase planes corresponding to different values of $[Ca^{2+}]_i$ (Exercise 13). Of particular interest is the key role of the homoclinic orbit (labeled HC in Figure 5.20A) in terminating the active phase of each burst. See also Sherman (1997) or Rinzel and Ermentrout (1998).

Returning to Figure 5.19, the sequence of panels illustrates how $\beta$-cells modify their electrical activity according to the level of glucose. As glucose concentration increases, the rate of the PMCA is here hypothesized to increase as well. For low glucose (low pump rate), the cell is electrically silent (top panel). For glucose above a threshold concentration, bursting appears (second panel). Further increases in glucose result in longer active spiking phases, or plateaus, and shorter silent phases (third panel). Finally, for very high glucose, the cell remains permanently in the active phase, spiking continuously. Comparing these statements to the bifurcation analysis (Figure 5.20), we find that those about glucose and pump rate are equivalent to statements about how the $[Ca^{2+}]_i$-nullcline intersects the Z-curve. An alternative scenario for glucose sensing is described in Exercise 10.

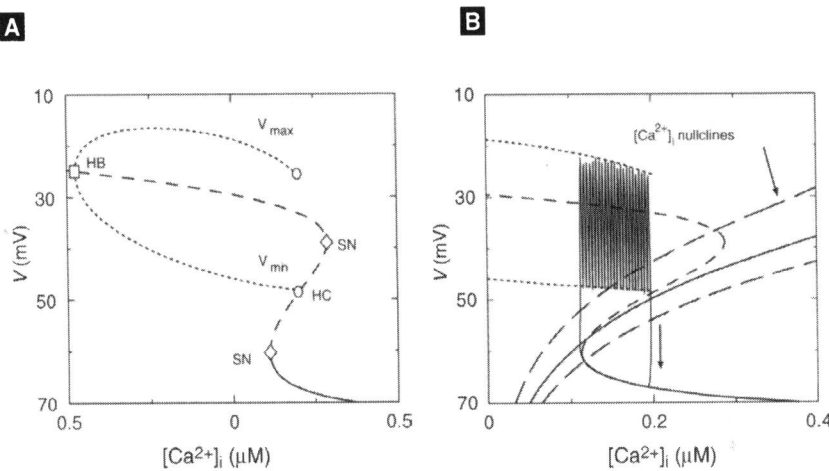

**Figure 5.20** (A) Bifurcation diagram of Chay–Keizer fast equations (5.61)–(5.62) using $[Ca^{2+}]_i$ as a parameter. Hopf bifurcation (HB – square); saddle-nodes (SN – diamonds); homoclinic orbit (HC – circles). (B) Projection of $V$-$[Ca^{2+}]_i$ trajectory for $v_{LPM} = 0.13$ ms$^{-1}$ onto the bifurcation diagram. Vertical arrow next to trajectory indicates direction of increasing time. Nullclines for $[Ca^{2+}]_i$ are shown for three cases: $v_{LPM} = 0.10$ (bottom), $v_{LPM} = 0.13$ (middle), and $v_{LPM} = 0.20$ (top). The corresponding solutions are silent, bursting, and continuously spiking, respectively. Compare with Figure 5.19.

### 5.4.2 Chay–Keizer with an ER

The ER is significant in $\beta$-cells for two reasons. One is that an important potentiator of insulin secretion, acetylcholine (ACh), works by promoting production of IP$_3$ and dumping the stores, similar to GnRH in gonadotrophs. This has been modeled in Bertram et al. (1995) and Chay (1997).

The second reason is that the ER strongly influences the kinetics of $[Ca^{2+}]_i$, and hence all $[Ca^{2+}]_i$-dependent processes. Here we consider only the latter aspect, showing that even a passive ER, which does not actively dump Ca$^{2+}$, can have profound effects on bursting. To do this we append an equation for $[Ca^{2+}]_{ER}$ to the Chay–Keizer model and add appropriate flux terms, a passive conductance and a linearized version of the SERCA pump (with leading constant $v_{LSP}$), to the $[Ca^{2+}]_i$ equation:

$$C_m \frac{dV}{dt} = -I_{Ca} - I_K - I_{K(ATP)} - I_{K(Ca)}, \tag{5.64}$$

$$\frac{dw}{dt} = \phi \frac{w_\infty - w}{\tau}, \tag{5.65}$$

$$\frac{d[Ca^{2+}]_i}{dt} = f_i(-\alpha I_{Ca} - v_{LPM}[Ca^{2+}]_i)$$

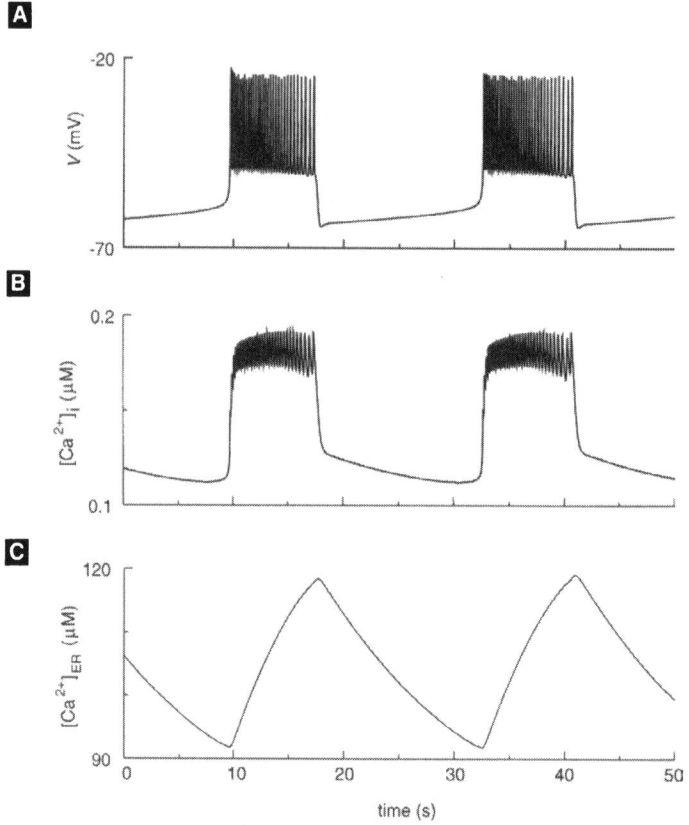

**Figure 5.21** Bursting in the Chay–Keizer model with ER, (5.64)–(5.67). Parameters as in Figure 5.19, except $f_i = 0.01$, with the following parameters added for the ER: $P_{IP3R} = 0.0008$/ms, $\lambda = 2$, $\sigma = 0.032$, $v_{LSP} = 0.6$/ms.

$$+ \frac{f_i}{\lambda_{ER}} \left( P_{IP3R}([Ca^{2+}]_{ER} - [Ca^{2+}]_i) - v_{LSP}[Ca^{2+}]_i \right), \qquad (5.66)$$

$$\frac{d[Ca^{2+}]_{ER}}{dt} = \frac{f_i}{\sigma\lambda_{ER}} \left( -P_{IP3R}([Ca^{2+}]_{ER} - [Ca^{2+}]_i) + v_{LSP}[Ca^{2+}]_i \right). \qquad (5.67)$$

Compare these equations with (5.23) and (5.24). Although $[Ca^{2+}]_{ER}$ does not directly affect the plasma membrane, it has indirect effects through $I_{K(Ca)}$. This model is essentially equivalent to Theresa Chay's last $\beta$-cell model (Chay 1997).

Bursting with this model is shown in Figure 5.21. Here $f_i$ has been increased to 0.01, which means that in the absence of the ER only very fast bursting can occur (see Exercise 12). In this case, bursting with periods of 10–60 seconds, as typically observed in islets, depends on slow kinetics supplied by the ER. Note that $[Ca^{2+}]_{ER}$ rises and falls slowly the way $[Ca^{2+}]_i$ does in the original Chay–Keizer model (cf. Figure 5.19). In contrast, $[Ca^{2+}]_i$ shows two distinct time scales, fast jumps at the beginning and end of each burst, a slowly rising plateau during the active phase, and a slow tail during the silent phase. The fast jumps reflect the intrinsic kinetics of $[Ca^{2+}]_i$, while the slow portions reflect slow uptake and release of $Ca^{2+}$ by the ER. This complex time course matches the characteristics of $[Ca^{2+}]_i$ seen in experiments (see Figure 5.18 and also Valdeolmillos et al. (1989)) better than Chay–Keizer (Figure 5.19). It also demonstrates that it is possible for slow negative feedback to operate through the K(Ca) channel, or some other $Ca^{2+}$-sensitive channel, even though $[Ca^{2+}]_i$ does not itself appear to be slow.

The examples discussed here represent only the tip of the iceberg of the $\beta$-cell field. The precise contributions of the mechanisms we have treated, K(Ca) channels, $g_{K(ATP)}$ channels, and the ER, as well as some that we have not treated, such as inactivation of $Ca^{2+}$ channels, are not settled. However, it seems likely that complex interactions of all of these will be necessary to explain the diverse phenomena observed. The mechanisms of other important regulators of cell electrical activity and $[Ca^{2+}]_i$, such as cAMP and epinephrine, remain to be elucidated. Nonetheless, the basic mechanisms and concepts presented here should prepare the reader sufficiently to explore the exercises and the literature on his or her own.

# Suggestions for Further Reading

- *Ryanodine receptor adaptation and $Ca^{2+}$-induced $Ca^{2+}$ release-dependent $Ca^{2+}$ oscillations*, Joel Keizer and Leslie Levine. This paper is the original source for the Keizer–Levine model (Keizer and Levine 1996).
- *$Ca^{2+}$ excitability of the ER membrane: an explanation for $IP_3$-induced $Ca^{2+}$ oscillations*, Yue Xian Li, Joel Keizer, Stanko S. Stojilković, and John Rinzel. A review of how the gonadotroph model described here was developed, with some equations and discussion of scaling and references to the physiolgical literature (Li et al. 1995b).
- *$InsP_3$-induced $Ca^{2+}$ excitability of the endoplasmic reticulum*, Joel Keizer, Yue Xian Li, Stanko Stojilković, and John Rinzel. Another review of the gonadotroph, but in words and pictures. This review contains many references to the physiological literature (Keizer et al. 1995).
- *Contributions of modeling to understanding stimulus-secretion coupling in pancreatic $\beta$-cells*, Arthur Sherman. A review of $\beta$-cell modeling oriented toward biologists (Sherman 1996).
- *Calcium and membrane potential oscillations in pancreatic $\beta$-cells*, Arthur Sherman. A mathematical tutorial centered on $\beta$-cell models with some connections to general modeling of bursting. Covers phase plane and bifurcation analysis (Sherman 1997).

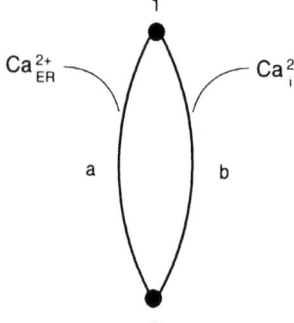

**Figure 5.22**   Kinetic diagram for ER leak.

## 5.5  EXERCISES

1. Generalize (5.13) to the case of (a) two and (b) arbitrarily many buffers.

2. The expression for ER leak (5.26) can be derived from a symmetric cycle model (Fig. 5.22) representing a pore that can exist in either an unbound state (1) or a bound state (2) and bind $Ca^{2+}$ on either the ER or the cytosolic side. A cycle from 1 to 2, binding $Ca^{2+}$ on the ER side, and back to 1, releasing $Ca^{2+}$ on the cytosolic side transports 1 ion from the ER to the cytosol. Assume symmetry, that is, the rates of binding $[Ca^{2+}]_i$ and $[Ca^{2+}]_{ER}$ are equal, $k_{12}^a = k_{12}^b = k_{12}^*$, and the rates of releasing $[Ca^{2+}]_i$ and $[Ca^{2+}]_{ER}$ are equal, $k_{21}^a = k_{21}^b = k_{21}$, and use the diagrammatic method of Chapter 3 to show that

$$J^{ss} = J_{21}^{ss_b} = \frac{k_{12}^*}{2}([Ca^{2+}]_{ER} - [Ca^{2+}]_i) \qquad (5.68)$$

in an appropriate limit. (The quantity $[Ca^{2+}]_{ER} - [Ca^{2+}]_i$ is called the thermodynamic driving force.)

. 3. (a) Write down the mass action equations corresponding to the Keizer–Levine kinetic diagram in Figure 5.5 and calculate the steady-state fraction of open channels (those in states O1 or O2) as a function of $Ca^{2+}$. Compare to the plateau curve in Figure 5.6.

   (b) Approximate the peak open fraction following a step of $Ca^{2+}$ from rest by assuming that C2 does not change over short times. Compare to the peak curve in Figure 5.6.

   (c) Derive the quasi-steady-state approximation (5.28)–(5.31). Hint: Combine the result in (b) with a differential equation for $w = 1 - P_{C2}$ assuming that transitions among O1, O2, and C1 are in rapid equilibrium. Simulate the two-pulse experiment of Figure 5.6 and verify that the quasi-steady-state approximation retains the feature of adaptation.

   (d) Draw the reduced diagram. What are the expressions for the reduced rate constants?

4. (a) By calculating the Jacobian of the closed-cell Keizer–Levine model (Section 5.2.2), show that for oscillations to arise via a Hopf bifurcation, it is necessary for the $w$ and $[Ca^{2+}]_i$ nullclines to intersect with negative slope and with the $w$ nullcline steeper.

   (b) Using the result of (a), find parameter values for which there are stable oscillations graphically, by plotting the nullclines, or analytically, by solving the $w$ and $[Ca^{2+}]_i$ equations in terms of $w$. Hint: One way is to play with the affinity of the SERCA pump.

(c) (optional) Construct the bifurcation diagram for the modified system and plot the period vs. $[Ca^{2+}]_T$. Correlate changes in the period with changes in the phase plane, particularly the invariant sets of the saddle point.

(d) Find the s-curve analytically by isolating $[Ca^{2+}]_T$.

5. (a) Plot period vs. $j_{IN}$ for the reduced open-cell Keizer–Levine model (Section 5.2.3).

(b) Compare the range of $j_{IN}$ for which there are oscillations predicted by the phase plane to those calculated in the bifurcation diagram. Using the phase plane in Figure 5.9, explain the variation in period.

(c) (optional) Compare the variation in period with that in the Chay–Keizer $\beta$-cell model when $g_{K(Ca)}$ is varied (Figure 5.19, Figure 5.20).

6. (a) Solve the Keizer–Levine open-cell system and find values of $j_{IN}$ that support oscillations. Compare the extent of store dumping with the closed–cell oscillations (Exercise 4). Plot the fluxes across the plasma membrane and ER through the cycle.

(b) Investigate the effect of increasing $k_c^-$ and explain what this implies about the reduced open–cell model with $w = w_\infty$.

(c) (optional) Construct the bifurcation diagram with $j_{IN}$ as a parameter and compare to Figure 5.10.

7. (a) Create a table of nondimensional parameter values from the dimensional parameters corresponding to Figure 5.12 for the closed-cell gonadotroph model (Section 5.3.1).

(b) Implement and solve both the dimensional and nondimensional forms of closed-cell gonadotroph model (Section 5.3.1) at $[IP_3] = 1.0$ and compare the results.

(c) For the same parameter values as in Figure 5.13B (dotted curve), show that oscillations can occur by either decreasing the value of $K_a$ ($Ca^{2+}$ sensitivity of the activation gate) or increasing the value of $[Ca^{2+}]_T$.

8. (a) Write out the dimensional version of the open–cell gonadotroph model (5.48)–(5.50). Start by adding $j_{IN}$ and $j_{PMCA}$ (cf. (5.37)) to (5.40) and writing an equation for $[Ca^{2+}]_{ER}$. Then nondimensionalize to obtain (5.48)–(5.50). Hint: Express the fluxes in terms of flux per unit area. For example, define $V_{PMCA} = V_{PMCA}/A_{PM}$, where $A_{PM}$ is the plasma membrane area.

(b) Simulate the open-cell gonadotroph model (5.48)–(5.52) with the same parameter values as in Figure 5.14.

(c) Show that at small, positive, and constant values of $\hat{j}_{in}$, intermittent intermittent or waxing-and-waning type of "bursting" can ooccur in the open-cell gonadotroph model.

(d) Explain this phenomenon by using a bifurcation diagram similar to that in Figure 5.15.

9. Slow, influx-driven $Ca^{2+}$ oscillations observed in bullfrog sympathetic neurons (see Figure 5.8) can also occur in cells that express only $IP_3R$ channels. This can be demonstrated by the open-cell gonadotroph model (Section 5.3.2) described by (5.48)–(5.52).

(a) Let $\epsilon = 0$. Plot the bifurcation diagram of the resulting 2-variable closed-cell model as a function of $\hat{c}_T$ at decreasing values of $\hat{k}_S$ while keeping other parameter values identical to those given in Figure 5.15. Determine a value of $\hat{k}_S$ that yields an "S"-shaped, bistable diagram similar to that in Figure 5.7B. Then, simulate the whole model to generate slow oscillations similar to those in Figure 5.8.

(b) For the same parameter values used in (a), show that you get almost identical slow oscillations by eliminating the $h$ equation and replacing $h$ with $h_\infty = 1/(1 + \hat{c})$ in the other two equations. Explain why.

10. One criticism of the glucose-sensing in the Chay–Keizer model (Figure 5.19) is that even though the cell visually appears to be more active electrically in higher glucose, mean $[Ca^{2+}]_i$ does not increase much. An alternative model that addresses this problem uses the K(ATP) channel, a $K^+$ channel that is inhibited by ATP and activated by ADP. It is thus a natural link between glucose metabolism and membrane potential.

(a) Starting with Chay–Keizer, set $g_L = 0$ and add the current

$$I_{K(ATP)} = g_{K(ATP)}(V - V_K) \tag{5.69}$$

to the $V$-equation. Calculate the solution for $g_{K(ATP)} = 160, 150, 135,$ and $130$ and describe the changes in plateau fraction and $[Ca^{2+}]_i$ compared to Figure 5.19.

(b) Explain your results in terms of the biophysics of $g_{K(ATP)}$.

(c) Derive a formula for the Z-curve by solving for $[Ca^{2+}]_i$. Describe algebraically and geometrically the effect of varying $g_{K(ATP)}$.

(d) Using the result of (c) or just constructing bifurcation diagrams for different values of $g_{K(ATP)}$, explain geometrically the effects of $g_{K(ATP)}$.

11. Keizer and Magnus (1989) proposed that $g_{K(ATP)}$ played not simply a modulatory role in bursting, but could itself provide slow negative feedback. They hypothesized that a rise in $[Ca^{2+}]_i$ would lead to mitochondrial uptake of $Ca^{2+}$ and dissipate the mitochondrial membrane potential that provides the energy for ATP synthesis. Thus, as $[Ca^{2+}]_i$ rises, ATP falls and ADP rises.

(a) Show, assuming that ATP and ADP compete for the same binding site, that

$$g_{K(ATP)} = \bar{g}_{K(ATP)} \frac{1 + [ADP]/K_1}{1 + [ADP]/K_1 + [ATP]/K_2}. \tag{5.70}$$

(b) Add the following simplified equation for mitochondrial dynamics to Chay–Keizer (with $I_{K(ATP)}$ as in (5.69) included in the $V$ equation):

$$\frac{d[ADP]}{dt} = v_{MITO}\left([ATP] - [ADP]\exp\left(R\left(1 - \frac{[Ca^{2+}]_i}{R_1}\right)\right)\right). \tag{5.71}$$

Assume that

$$[ADP] + [ATP] = 1 \text{ mM} \tag{5.72}$$

and simulate the solution to get bursting using the parameter values from Figure 5.19 except that $K_c = 0.15/ms$, $g_{K(Ca)} = 0$ pS, $g_L = 0$ pS, $f_i = 0.01$, and add the following new parameters for metabolic feedback: $\bar{g}_{K(ATP)} = 6000$ pS, $K_1 = 0.45$ mM, $K_2 = 0.012$ mM, $R = 0.9$, $R_1 = 0.35$ mM, $v_{MITO} = 5.0 \times 10^{-5}/ms$.

(c) Show that the parameter $R$ can function as a glucose sensor. Should $R$ increase or decrease with glucose?

12. (a) Remove the ER from the Chay–Keizer model with ER and solve, retaining all other parameters from Figure 5.21. What is the burst period?

(b)  Add the ER back, and construct the bifurcation diagram of the model (5.64)–(5.67) using $[Ca^{2+}]_{ER}$ as the bifurcation parameter. Explain how $[Ca^{2+}]_{ER}$ can act as a slow variable even though it does not appear in the $V$ equation. Use the diagram to explain how varying $\sigma$ (5.67) changes the burst period.

(c)  Make another bifurcation diagram for this model using $[Ca^{2+}]_i$ as the bifurcation parameter, with $[Ca^{2+}]_{ER}$ held constant. Overlay the $[Ca^{2+}]_i$ nullcline. How does the diagram change as $[Ca^{2+}]_{ER}$ is varied?

(d)  Increase $P_{IP3R}$ (5.66) to 0.008/ ms. Use the two bifurcation diagrams from (b) and (c) to analyze the effect of $P_{IP3R}$ on burst frequency. Compare with (Chay 1997).

(e)  (optional) Restore $P_{IP3R}$ to 0.0008/ms and increase $\lambda$ (5.66) to 800. Use the bifurcation diagram from (c) to explain why the burst frequency increases.

13.  Construct phase planes of the Chay–Keizer fast subsystem (5.61), (5.62) with $[Ca^{2+}]_i$ as a parameter and correlate the various patterns with the bifurcation diagram Figure 5.20. Look for stable and unstable nodes, saddle points, and periodic orbits. Choose values in each of the regimes defined by the bifurcation diagram:

   - just to the left and just to the right of the Hopf bifurcation

   - just to the left and just to the right of the point labeled SN

   - just to the left and just to the right of the point labeled HC (for homoclinic orbit).

   Explain why the period increases sharply as the HC is approached from the left. Explain why the period solution disappears at HC. It will help to draw the stable and unstable manifolds of the saddle point.

# CHAPTER 6

# Intercellular Communication

## John Rinzel

Orchestrating the activity of cell populations for physiological functioning of the brain, organs, and musculature depends on transmission of signals, learning and memory devices, and feedback control systems. By what biophysical mechanisms do cells communicate in order to coordinate their activity as local ensembles, as multimodal circuits, and across system levels? Here we only scratch the surface of this fascinating topic. We will focus on electrically active cells; for this, you can have in mind, for example, cardiac cells, many types of secretory cells, and neurons.

We know that ionic currents underlie cellular electrical activity. Hence, one way that cells can interact is by directly passing ionic current between each other. Perhaps the simplest mechanism for such communication is the analogue of resistive coupling between units, i.e., with the intercellular current being proportional to the voltage difference between cells. Heart cells, and many other types of cells, communicate in this manner, with the ions flowing directly between two coupled cells. In this case, referred to as electrical coupling (and sometimes, in the neural context, electrotonic coupling), the current flows through channel proteins that span the plasma membranes of both cells as shown in Figure 6.1. The clusters of such channel proteins that are found at cell-to-cell contacts are called gap junctions.

While gap junctions are occasionally found in neural circuits, they seem to be more common during development or they may not constitute an exclusive means of interaction. Neurons have a rich repertoire of other ways for exciting or inhibiting other target neurons, via the indirect means of chemical synaptic transmission depicted in Figure 6.5. At a terminal of a neuron's axon the neurotransmitter of one type or another is bundled in vesicles that are released with increased probability when the

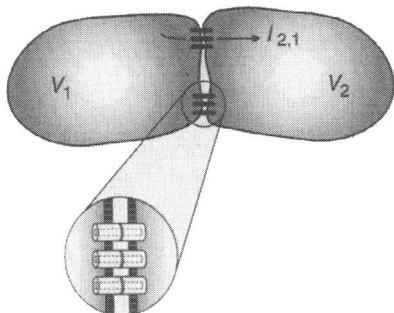

**Figure 6.1** Schematic of gap junctions between two cells. Coupling current from cell #1 to #2 is $I_{2,1} = g_c(V_1 - V_2)$, the exact negative of $I_{1,2}$. Redrawn from Hille (2001).

terminal is active (depolarized). The transmitter, released into the extracellular space in the terminal region, diffuses and binds to receptors on the postsynaptic cell's nearby membrane. These receptors may be part of a receptor–channel complex or be linked indirectly through second messengers to a nearby channel. The activated channels then lead to the postsynaptic current and action. The variety of transmitters and receptors enable many different time scales and "sign" of the input that is being delivered to the target cell. We find synapses that are excitatory or inhibitory; they can be fast or slow; they can be shunting; their synaptic parameters can change with usage, either, depressing or facilitating; they can be voltage gated or not on the postsynaptic side.

Gap-junctional coupling is typically localized, certainly for cells without spatially extended processes, to nearby neighbors. Prime examples include the heart and islets of Langerhans in the pancreas. In contrast, neurons can interact across distances that are many times greater than a cell body diameter, by means of their potentially far-reaching axonal and dendritic arbors. The synaptic interactions enable them to participate in local calculations and with distant assemblies, on selective time scales, fast or slow and to various degrees. For long–distance communication between cells propagated action potentials typically mediate signal transfer to the synapses via axons.

Given the variety of coupling mechanisms and connectivity patterns there are many possible behavioral modes, spontaneous and/or stimulus-driven, that such circuits may exhibit. Obviously we cannot elaborate on many of these behaviors; we will restrict attention to a few examples and focus on pairwise interactions. Also we choose not to present here some other mechanisms for intercellular interaction such as coupling by cell-generated electrical fields, diffusion and exchange of ions (that effect Nernst equilibrium potentials) and second messengers, mechanical and hormonal effects (Keener and Sneyd 1998).

# 6.1 Electrical Coupling and Gap Junctions

The proteins that constitute the channels at gap junctions are of the connexin family. Connexin molecules in the plasma membrane of one cell link up those in an adjacent cell to form the channels. These channels pass most ions as well as various molecules up to molecular weights of tens of kilodaltons, including those involved in second messenger systems such as $IP_3$. The single channel conductance can be in the range of 75–150 pS, and in most known instances they are relatively voltage-independent. The variety of connexins provides a rich repertoire for diverse modulatory possibilities by changes in intracellular pH, second messengers, neurotransmitters, or voltage (Perez Velasquez and Carlen 2000). For example, the reduction of gap–junctional coupling due to increasing levels of $Ca^{2+}$ is described as a protection mechanism.

For cell-level modeling we typically represent the net conductance of a gap junction as a constant, the product of the mean number of open channels at the junction and the single channel conductance. Simultaneous pairwise recordings, at least for geometrically simple neighboring cells, can be used to estimate this conductance. For spatially extended cells, like neurons with branching dendrites, such measurements would be confounded especially if the gap junctions are at electrically remote sites. Gap junctions may also be detected by dye coupling, although this method is problematic without good controls to test how readily the dye passes.

## 6.1.1  Synchronization of Two Oscillators

Consider the case of two cells idealized as isopotential compartments as shown in Figure 6.1. The gap–junctional current $I_{2,1}$ that flows from cell 1 into cell 2 is written as $g_c(V_1 - V_2)$, where $g_c$ is the net coupling conductance of all the gap junctions formed between the two cells. This current appears as a source term in the current balance equation for cell 2. Defined as such, with the current as leaving from cell 1, it appears as a sink term for cell 1 (or equivalently, $g_c(V_2 - V_1)$ is a source term into cell 1). Thus we have

$$C_m \frac{dV_1}{dt} = -I_{ion,1} + I_{app,1} + g_c(V_2 - V_1), \tag{6.1}$$

$$C_m \frac{dV_2}{dt} = -I_{ion,2} + I_{app,2} + g_c(V_1 - V_2), \tag{6.2}$$

where $I_{ion,1}$ is a function of $V_1$ and the set of gating variables $w_1$ in cell 1, and $I_{ion,2}$ is similarly dependent on $V_2$ and $w_2$. Let us first predict the behavior, say in the case where the gap junction conductance is large. This means that the cells are very tightly electrically coupled. We would then expect them to have approximately the same voltage. Indeed if the cells are identical, then $V_1 = V_2$ is always a solution: The coupling current would be zero in this case. Of course, a perturbation (e.g., brief current pulse) to just one cell would make their $V$'s differ, transiently. By subtracting the two equations and then dividing by $g_c$ we see, after dropping the term of order $1/g_c$, that

$$\tau_c \frac{d(V_1 - V_2)}{dt} \approx -(V_1 - V_2), \tag{6.3}$$

where $\tau_c = C_m/g_c$. This shows that the cells will re-establish uniformity with an effective time constant of $\tau_c$, that is very short when $g_c$ is large.

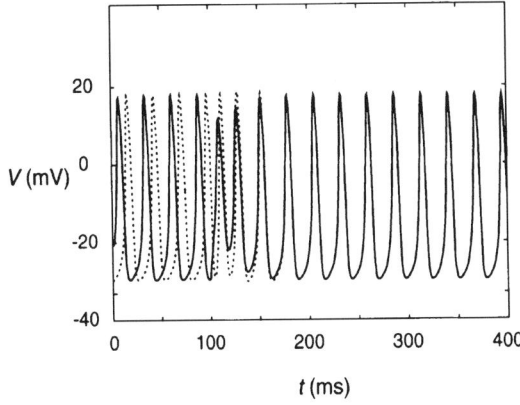

**Figure 6.2** Two identical Morris–Lecar neuron models, auto-rhythmic, synchronize with electrical coupling (turned on at $t = 100$ ms). Parameter values for the model are those of Chapter 2 with the following exceptions: $v3 = -5$ mV, $v4 = 10$ mV, $\phi = 0.5$, $g_{Ca} = 8$ mS/cm$^2$, $I_{app} = -10\mu$A/cm$^2$. Net gap junction coupling conductance is $g_c = 2$ mS/cm$^2$. Initial conditions: $V_1 = -20$ mV, $w_1 = 0.2$, $V_2 = -30$ mV, $w_2 = 0$.

This is illustrated in Figure 6.2 for two Morris–Lecar cells that are tuned into an oscillatory regime (as discussed in Chapter 2). In this example the cells are initially out of phase and uncoupled. Synchronization occurs promptly after the coupling is introduced at $t = 100$ ms.

## 6.1.2 Asynchrony Between Oscillators

While electrical coupling is typically considered as a mechanism for uniformizing cells, one should be aware that the outcome depends considerably on coupling strength and dynamic features of the individual cells. For example, the same cells used above when coupled weakly with gap junctions do not synchronize. Instead, they establish a rhythm with the two cells in a stable antiphase–locked pattern: just the opposite of togetherness (Figure 6.3). Also, for this situation the period of the network oscillation depends on the value of $g_c$. This is not true for the in-phase pattern. When the coupling strength is

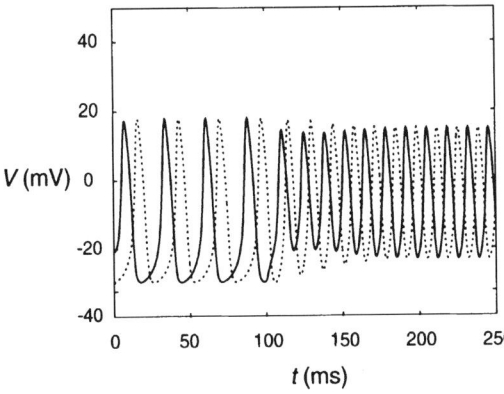

**Figure 6.3** Time courses of $V_1$, $V_2$ showing antiphase behavior when weak gap–junctional coupling is turned on at $t = 100$ ms. Same initial conditions and parameters as in Figure 6.2, except that here $g_c = 1$ mS/cm$^2$.

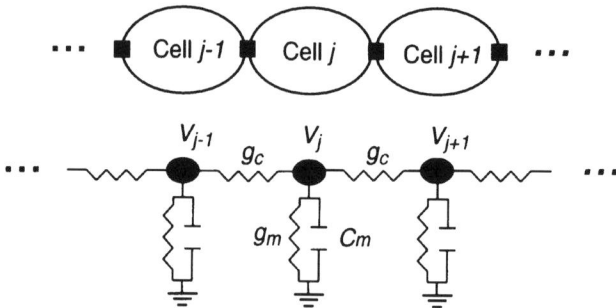

**Figure 6.4** Schematic of linear array of passive cells coupled (nearest neighbor) by gap junctions.

large enough, the period (for a pair of identical cells) is that of the isolated cell and is therefore independent of $g_c$.

## 6.1.3   Cell Ensembles, Electrical Coupling Length Scale

Such synchronization also occurs for spatially distributed multicell systems; even though electrical coupling is localized it can, when large enough, synchronize a population that might be spread over a sizable spatial region. In order to formalize this, one can introduce the concept of an electrical length scale that involves coupling conductance as well as leakage conductance. This emphasizes the relativeness of intercellular current to the current flowing across a cell's plasma membrane. By using this notion we can refer to the "electrical size" of a cellular array; it could be quite compact even though the spatial extent might not be.

In order to formalize this notion we consider the simple case of Figure 6.4, where we have a chain of gap-junction-coupled cells that have passive membrane properties, i.e., $I_{ion} = g_m V$. Thus we have the equation for a typical cell, cell $j$:

$$\frac{dV_j}{dt} = \frac{g_c(V_{j-1} - V_j) - g_m V_j + g_c(V_{j+1} - V_j)}{C_m}. \tag{6.4}$$

Suppose that we have a very long chain of cells (and for now, ignore end effects) and suppose we voltage clamp the cell in the middle for which $j = 0$ ($j > 0$ refers to cells to the right and $j < 0$ is for cells to the left). In this passive system the voltage distribution will go to a steady state after some transient, and the voltage will decrease from $V_0$ as $|j|$ increases. This attenuation occurs as current spreads from one cell to the next away from $j = 0$ and as current is lost through the "leaky" membrane of each cell.

The steady voltage decays in an exponential fashion with distance. The decay "rate" $\gamma$ can be found by considering the steady–state case of (6.4) (i.e., setting $dV_j/dt = 0$) and seeking a solution of the form $V_j = V_0 \gamma^j$. This leads to a quadratic equation for $\gamma$:

$$\gamma^2 - (2 + g_m/g_c)\gamma + 1 = 0. \tag{6.5}$$

Hence, there are two values of $\gamma$:

$$\gamma_{\pm} = 1 + g_m/(2g_c) \pm \sqrt{[1 + g_m/(2g_c)]^2 - 1}. \tag{6.6}$$

Note that both roots are positive with $\gamma_+ > 1$ and $\gamma_- < 1$.

For a voltage profile that decays with distance to the right of $j = 0$ we must disallow $\gamma_+$ (such a term would grow with $j$). Thus, if we ignore end effects due to termination of the chain, the steady state profile for $j \geq 0$ has the form

$$V_j = V_0 \gamma_-^j. \tag{6.7}$$

The solution for $j \leq 0$ is $V_j = V_0 \gamma_+^j = V_0 \gamma_-^{-j}$, since $\gamma_+ \gamma_-^{-1} = 1$.

We can test our intuition by considering some limiting cases. When $g_m/g_c$ is large, the leakage conductance is dominant and the voltage should attenuate rapidly with distance from $j = 0$. Indeed, $\gamma_- \approx g_c/g_m$, so that $V_j$ decays steeply. On the other hand, when $g_m/g_c$ is small, then $\gamma_- \approx 1 - \sqrt{g_m/g_c}$ and the spatial decay is gradual.

Pursuing this latter case a bit more, if cell-to-cell attenuation is small and the spatial profile changes smoothly, we might treat the one-dimensional cellular array approximately as a continuum with position $x \approx j\Delta x$, where $\Delta x$ is a cell–size parameter, considered small. The leakage conductance $g_m$ is proportional to a cell's surface area, which is the perimeter of a cross section times $\Delta x$. We will take some care with the representation of $g_c$. Let's think about the reciprocal, $r_c = 1/g_c$, the resistance to longitudinal current flow between cells, say from one cell center to the next. Then, $r_c$ involves resistance due to the gap junctions as well as cytoplasmic resistance along the cell's axial direction. If there are many gap junctions between adjacent cells, then the major contribution to $r_c$ will be the cytoplasmic term, which is proportional to the area of a cross section times $\Delta x$. Consequently, the ratio $g_m/g_c$ ($= g_m r_c$) is proportional to $\Delta x^2$ and may be written as $(\Delta x/\lambda)^2$. Finally, combining this with our expression for $\gamma_-$ (from (6.6) for small $g_m/g_c$) and noting that $\log(\gamma_-) \approx -\sqrt{g_m/g_c}$, we see that the solution form for the continuum approximation is

$$V(x) = V_j = V_0 \exp(j \log \gamma_-) \approx V_0 \exp(-j\Delta x/\lambda) = V_0 \exp(-x/\lambda). \tag{6.8}$$

Thus, $\lambda$ is the electrical length constant for the continuum approximation.

This treatment relates nicely to continuum models for current spread in individual cells that have extended processes, like neurons with their dendrites or axons. These processes behave like electrical cables; current flows along the axis and leaks through the membrane. In this case of a neuronal process, we could think of it for simulation purposes as a chain of short compartments (length $\Delta x$) each with a membrane surface area and connected by axial resistances; there are no gap junctions between the fictitious adjacent compartments. If $d$ is the diameter of the cross section, then we would have $g_m = \Delta x \pi d G_m$, where $G_m$ is the leak conductance density (S/cm$^2$) and $r_c = \Delta x 4 R_i/(\pi d^2)$ where $R_i$ is the cytoplasmic–specific resistivity (ohm·cm). Putting this together we obtain for the electrical length constant of a passive neuronal dendrite or axon $\lambda = \sqrt{d/(4G_m R_i)}$. As an illustration, for some neurons estimates are in the range

$G_m = 0.25$ mS/cm$^2$, $R_i = 100$ $\Omega$·cm, so that with $d = 2.5$ $\mu$m yields $\lambda = 500$ $\mu$m. This means that for a dendritic branch of physical length 250 $\mu$m, the corresponding electrotonic length is about 0.5. Therefore, we do not expect excessive attenuation for steady or slow voltage changes. Transients decay more abruptly. Also, the attenuations are more severe from dendritic locations to the cell body than vice versa. In electrically coupled excitable systems with weaker gap junctions or extended electrical size (physical length larger than $\lambda$) synchrony is not quickly established, and localized perturbations may lead to waves or other spatiotemporal patterning.

The treatment of dendrites as cables and the compartmental method is discussed in Chapter 7, and in more depth in Johnston and Wu (1995), Koch and Segev (1998), Koch (1999), and Keener and Sneyd (1998). We note that this discretization of neuronal cable-like processes is called the compartmental method and was first developed by Rall for treating the effect of spatiotemporally distributed synaptic inputs over a neuron's somatic-dendritic area (Segev et al. 1995).

## 6.2  Synaptic Transmission Between Neurons

In the preceding section we saw that strong gap junctions can be used for coordinating cellular electrical activity, subserving functions in which cells might work together in approximate synchrony. The bidirectionality and instantaneous nature of gap–junction coupling are well suited for achieving these goals. In the nervous system there are demands for more complex patterning. Individual cells generally are likely involved in numerous different computations, with possibly different time scales, sometimes being called into action with a brief wake-up call but in other cases only after a long barrage of inputs. In some cases the precise timing of action potentials might be important, and we would expect fast coupling mechanisms. But when firing rate, rather than spike timing, is more important for signaling, perhaps the synchronization properties of fast coupling may be less critical and slower coupling mechanisms might dominate. Some flexibility for multiplexing of these signaling modes is attainable by exploiting the potentially large variety of intrinsic mechanisms, i.e., the many different ionic channels and modulators for them. Even greater computational power can be achieved, however, by selective adjustment of the many synapses on a cell, maybe in an activity–dependent fashion. One can imagine a great many neural ensembles with possibly shared units that can be dynamically constructed, recruited, and dismissed as needed for memory storage, sensory processing and perception, and executive command functions.

It is hard to imagine how instantaneous bidirectional coupling could allow in an efficient way such a rich set of alternatives. However, the one-way signaling via chemical synapses enables the system to employ a few transmitters but yet enrich the possibilities for postsynaptic response by having many choices for the postsynaptic receptors and channels. Sites for modulatory action can be implemented on either the postsynaptic or presynaptic side. Thus while vesicle-packaged transmitter is released in punctate fashion, like the action potential, the time scale and even the "sign" of the response are determined by the machinery on the postsynaptic cell.

## 6.2.1  Kinetics of Postsynaptic Current

We will formulate an idealized model for the current generated in a postsynaptic cell due to neurotransmitter release by a presynaptic cell in which an action potential has occurred. The sender's action potential opens voltage gated calcium channels at the axon terminal where vesicles are poised ready to fuse with the membrane and release transmitter.

Whether or not release occurs is a probabilistic event. The failure rate can be high at some, e.g., cortical, synapses. There is also a small probability of spontaneous release. Experiments at the neuromuscular junction and central synapses use the quantal release hypothesis to analyze the statistics of postsynaptic responses for spontaneous and evoked release to estimate the number of active sites, quantal content, and release probability (Johnston and Wu 1995). We will not consider these issues here, nor the details of transmitter diffusion in and removal from the synaptic cleft, the small extracellular space between the pre- and postsynaptic sites.

We suppose that transmitter is available briefly, and that during this time it can bind to receptors, actually receptor–channel complexes, on the postsynaptic membrane. For this simple model we imagine a two-state channel. Binding of transmitter to a receptor molecule favors opening of the channel and unbinding leads to closing. A kinetic

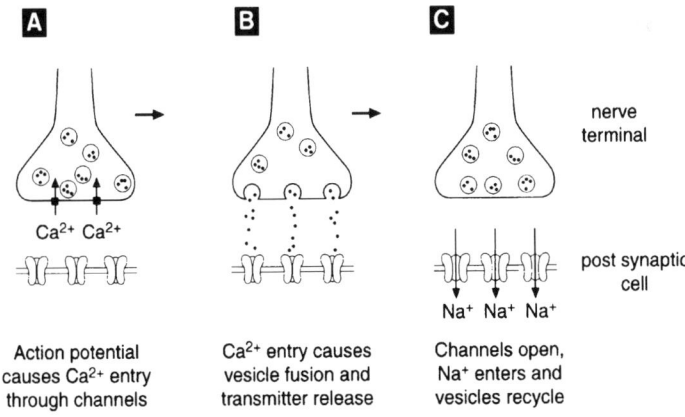

**Figure 6.5** Steps in chemical synaptic transmission. Action potential arrives at terminal of presynaptic axon; depolarization opens voltage gated calcium channels. Resulting calcium influx causes vesicle fusion and transmitter release. Transmitter diffuses, reaches postsynaptic membrane, and binds to receptors. The activated receptors cause postsynaptic channels to open. Depending on which type of ions are allowed to pass (sodium shown here) the postsynaptic response can be depolarization or hyperpolarization. Adapted from Koch (1999).

scheme for this is the following:

$$C + T \underset{\beta'}{\overset{\alpha'}{\rightleftharpoons}} O.$$

If $s$ is the fraction of complexes in the open state $O$, then $1 - s$ is the fraction closed, and we have, given a transmitter concentration [T],

$$\frac{ds}{dt} = \alpha'[T](1 - s) - \beta's. \tag{6.9}$$

This representation assumes unlimited transmitter and receptor availability. Also, it requires us to specify a time course for [T]; for example, we might assume a square pulse over some time duration (perhaps with a modest delay, on the order of a millisecond or less) when the presynaptic cell fires. We will adopt a simpler, and easy to implement, scheme, supposing that maximal transmitter is available when the presynaptic membrane potential $V_{pre}$ is above some level, say $\theta_{syn}$. This would mean $[T] = T_{max}s_\infty$ with $s_\infty = H(V_{pre} - \theta_{syn})$, where $H(x)$ is the Heaviside step function. We will smooth this out using a sigmoidal function for $s_\infty$. Redefining the rate constants, which we view as adjustable depending on the desired time course of postsynaptic conductance, we get

$$\frac{ds}{dt} = \alpha s_\infty(1 - s) - \beta s, \tag{6.10}$$

with $s_\infty = 1/[1 + \exp(-(V - \theta_{syn})/k_{syn})]$, with $k_{syn}$ positive. The current through the post-synaptic membrane is

$$I_{syn} = g_{syn}s(V - V_{syn}), \tag{6.11}$$

where the reversal potential $V_{syn}$ depends on the concentration differences for the ion species that flow through the open receptor channels in the postsynaptic membrane; $g_{syn}$ is the maximal conductance at the synapse if all the available channels are open.

## 6.2.2  Synapses: Excitatory and Inhibitory; Fast and Slow

This type of model has been used to describe some common types of excitatory and inhibitory synapses that have relatively simple kinetics. The different transmitters and different receptors and different ions that pass through the synaptic channels lead to a variety of synaptic types. The most commonly known transmitter that is used for excitatory synapses is glutamate. It can activate AMPA receptor channels that typically pass inward current (carried mostly by sodium ions, but also some potassium and other ions). The postsynaptic conductance is relatively fast with a rise time ($\approx \alpha^{-1}$) of order 1 millisecond and decay times ($\approx \beta^{-1}$) of a few to tens of milliseconds. Expressions such as (6.10) have been used for these synapses. The value of $V_{syn}$ in this case is typically about 100 mV above rest. Glutamate can also evoke slower postsynaptic responses by activating NMDA receptors (fast rise, of order milliseconds, and slow decay, of order many tens to hundreds of ms) or a number of different types of metabotropic gluta-mate receptors with very slow rise and slow decay of order seconds. The conductance associated with NMDA receptors is, curiously, also gated by the postsynaptic voltage $V_{post}$. If $V_{post}$ is not adequately above $V_{rest}$, the NMDA-associated channels are blocked by magnesium ions from the outside; sufficient depolarization relieves this block, and the channels can open. The dependence of synaptic current on pre- and postsynaptic activity implicate the NMDA conductance in various models for associative learning and synaptic plasticity generally. The very slow metabotropic glutamate–mediated ef-fects are believed to be primarily modulatory, acting, for example, to effectively change the firing threshold of a cell. In the examples below of excitatory synaptic coupling we will be thinking primarily of AMPA-mediated excitatory synapses, for which (6.10) is a good first approximation. More complete descriptions, even for this simple receptor-channel complex, might involve kinetic models with additional states (Destexhe et al. 1998).

Fast inhibitory synapses are activated by the transmitter GABA when they bind to $GABA_A$ receptor channels. The postsynaptic conductance has a fast rise time, one or a few milliseconds, and somewhat slower decay than the AMPA-excitatory conductance, on the order of tens of milliseconds. The current is carried largely by chloride ions. Interestingly, the value for $V_{syn}$ can be quite variable, typically −60 mV to −80 mV but in some cases as depolarized as −35 mV; this variability likely reflects different types or activities of chloride pumps that regulate intracellular $Cl^-$ concentration. If a cell is sitting near its resting potential of say −65 mV, the synaptic current that is generated by activating these channels would be outward (hyperpolarizing) if $V_{syn}$ is below −65 mV or inward (depolarizing) if $V_{syn}$ is above −65 mV. It is not uncommon for $GABA_A$- mediated currents to be depolarizing in developing neural tissue. Thus it is not strictly correct, although frequently done, to refer to GABA as an inhibitory trans-mitter. One should take care in describing a $GABA_A$- mediated synapse as inhibitory; the current's sign depends on where $V_{syn}$ is relative to $V$. Note that if $V$ is close to $V_{syn}$, little synaptic current will be generated, even if a large conductance $g_{syn}s$ is activated. On the other hand, the membrane potential would be effectively clamped to $V_{syn}$ until the conductance deactivates. In this case the $GABA_A$ synapse acts as a strong shunt in the membrane; other modest-sized synaptic inputs would be ignored during this time.

The transmitter GABA acting through $GABA_B$ receptors can lead to a very slow inhibition with a conductance that rises and decays on the order of 100 or more mil-liseconds. The current is carried primarily by $K^+$ ions and so the reversal potential may be −70 to −90 mV. The simple model in (6.10) cannot account for this slow current. One shortcoming is that the model predicts a peak shortly after the depolarization of $V_{pre}$, while $GABA_B$–mediated inhibition peaks only much later. A minimal model for this current would involve at least two dynamic variables.

Typical sizes of $I_{syn}$ at a single synapse are a few picoamperes and may evoke responses (e.g., in cortical neurons) of 0.1 to 1 mV. A number of inputs must be summed in order to bring the neuron to firing threshold. Generally, the postsynaptic response to multiple inputs, even in the subthreshold regime, is not, however, a linear summation of individual inputs. This is because the synaptic current depends on $V_{post}$, in particular through the driving force $V_{post} - V_{syn}$. If $V_{post}$ is far from $V_{syn}$, then the dependence on $V_{post}$ is very weak, as, say, for AMPA-mediated excitation if $V_{post}$ is near rest. But in other cases, say for a GABA$_A$-mediated synapse, we would not get linear summation: Doubling the number of inputs does not double the synaptic current. This is easy to see for a passive membrane system, as we considered above, with steady synaptic conductance input $g_s = constant = g_{syn}s$. The steady response $\overline{V}$ would be

$$\overline{V} = V_{syn}g_s/(g_m + g_s),\tag{6.12}$$

saturating for large $g_s$ at $V_{syn}$, as we expect. While increasing $g_s$ drives $V_{post}$ closer to $V_{syn}$, the increment in synaptic current diminishes for large $g_s$: $dI_{syn}/dg_s$ behaves like $g_s^{-2}$ for large $g_s$; this is the sublinear summation effect.

When synaptic inputs are activated repeatedly the response may not remain uniform from one stimulus to the next in a train. This dynamic aspect may involve facilitation or depression of successive responses on a range of time scales. For example, the response might be depressed by 50% or more to a steady level after less than ten stimulus repeats (say at 50 Hz), and then recovery may take a second or so (O'Donovan et al. 1998). A favored mechanism for depression is presynaptic depletion of ready-to-release transmitter. Synaptic depression is analogous to receptor desensitization in other contexts, such as hormonal communication (Goldbeter 1996).

Various possible functional roles for such synaptic dynamics in network activity have been suggested. One hypothesis is that synaptic depression could mediate phasic behavior of a feedforward network. That is, if the steady response saturates beyond some input rate (so the tonic output conveys no information about the input rate), the network could still be responsive to transient changes in input rate; just after a sudden increase in input rate the synapses would not yet be adapted to their new target level of depression, so the network's output would briefly rise and then fall back to the saturation level. In the context of recurrently connected networks synaptic depression provides a mechanism of slow negative feedback for the intercellular coupling. This can underlie rhythmic population activity even if isolated units are capable only of tonic firing for steady inputs; here the autocatalysis could be provided by recurrent excitatory connections (Tabak et al. 2000).

## 6.3   When Synapses Might (or Might Not) Synchronize Active Cells

In order to illustrate some effects of mutual synaptic coupling between cells we will consider the simple case of two identical cells, each of which is autorhythmic, and examine conditions for which the cells tend to fire together or apart. We are asking about the synchronization patterns in this simple two-cell network. Classical notions are that mutual excitation tends to make cells synchronous (in phase), while mutual inhibition pushes them apart, leading to antiphase behavior. That is, if two excitatory cells are somewhat out of phase, the firing of the leading cell encourages the follower to fire sooner, thus bringing the cells more nearly in phase, while for inhibitory cells that are somewhat out of phase the leading cell's firing will delay the follower, increasing their phase difference. Successive cycles would increase their phase difference to 180° but not beyond, since then the follower would act as the leader, pushing the phase difference back toward 180°. These expectations and some surprises will be illustrated below when we consider the effects of synaptic time scales.

Here we will use the Morris–Lecar model in Type I mode (see Chapter 2), so that near the threshold for repetitive firing the steady firing frequency can be made arbitrarily low. We dictate here that the gating variable's kinetics are relatively slow and that the stimulating current is adjusted so that the cell model is firing slowly, about 15 Hz.

First we consider the case of mutual excitation. In panels A and B of Figure 6.6 we confirm the expectation that the cells will fire together. The cells are substantially out of phase before we actually implement the coupling, but then afterwards they converge to a pattern of near synchrony, with one cell preceding the other by just a slight bit. When the coupling is first turned on the follower cell immediately advances, and throughout the transient phase of synchronization the cells are causing each other to fire faster. Interestingly, as synchrony is established the cells slow down to nearly their intrinsic frequencies, as if the coupling is only to synchronize them. When comparing the time courses we see that during the transient phase the leader cell's voltage is more affected than the follower's. The synaptic current from the follower at first reduces the leader's postspike hyperpolarization. Then, as the follower catches up, this synaptic current is delivered when the leader is still strongly depolarized, and its intrinsic conductances swamp the perturbing effect of the synaptic input.

For the preceding example the synaptic conductance time course has fast rise and fast decay phases; the conductance is essentially activated only during the presynaptic depolarization. If we allow for the synaptic decay to be much slower, with all other parameters and initial conditions unchanged, we find that these two cells now fire in antiphase, as shown in Figure 6.6C and D. The frequency of each cell is nearly twice that of an isolated cell. This is understandable. Since the synaptic current decays more slowly, it provides a longer-lived depolarizing influence, and it is strongest during a cell's trajectory as it rises toward threshold, when it is most responsive to depolarizing influences. Note that if the output of this two-cell network converges onto a common target, the effective delivery rate, because of the antiphase pattern, is twice that of each, four times that of an individual cell.

For the examples in this section we have chosen parameter values in order to emphasize the importance of synaptic time scales in determining firing patterns in networks. This is an active area of research; for example, see Wang and Rinzel (1992) and Van Vreeswijk et al. (1994). Deeper understanding will be achieved, and for now, we caution that it should not be taken as universal that fast (slow) excitatory synaptic coupling leads to in-phase (antiphase) firing. In fact, for the example in

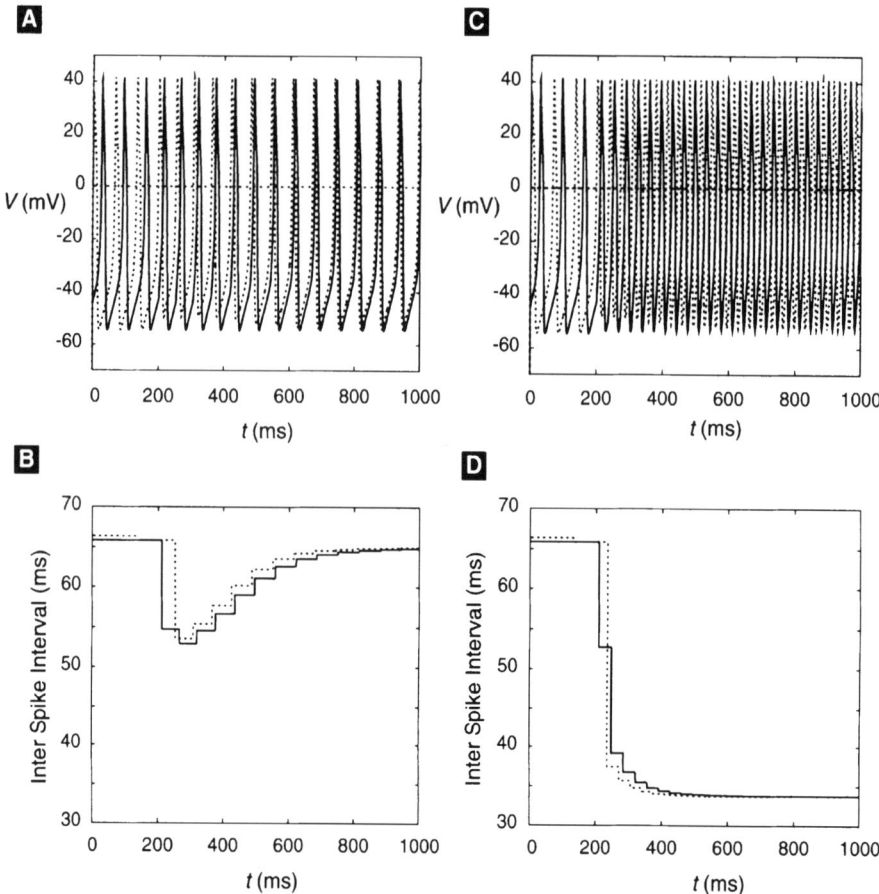

**Figure 6.6** The kinetics of synaptic excitation affect synchronization properties of cells, as illustrated with two identical coupled cells. Neuron models are uncoupled until $t = 200$ ms. (A and B) fast excitatory synapses lead to synchrony; time courses of $V_1$, $V_2$ in (A) and of instantaneous interspike interval in (B). (C and D) antiphase pattern evolves with slow excitatory synapses; time courses of $V_1$, $V_2$ in (C) and of instantaneous interspike interval in (D). Parameter values for $v3$, $v4$, $g_{Ca}$ are as in Figure 6.2–Figure 6.3, except that here, $\phi = 0.1$ and $I_{app} = 15 \ \mu A/cm^2$; for the synaptic kinetics, $\theta_{syn} = 20$ mV, $k_{syn} = 2$ mV with $\alpha = 3$/ms, $\beta = 1$/ms in (A and B) and $\alpha = 3$/ms, $\beta = 0.1$/ms in (C and D). Initial conditions: $V_1 = -44$ mV, $w_1 = s_1 = 0$, $V_2 = 0$ mV, $w_2 = 0.3$, $s_2 = 0.3$.

Figure 6.6A and B, different initial conditions to such coupled cells can lead to antiphase locking even for these fast synapses.

Next we consider the case of two cells (the same cells as above) coupled with inhibitory synapses. We start the cells with different initial voltages and watch them settle

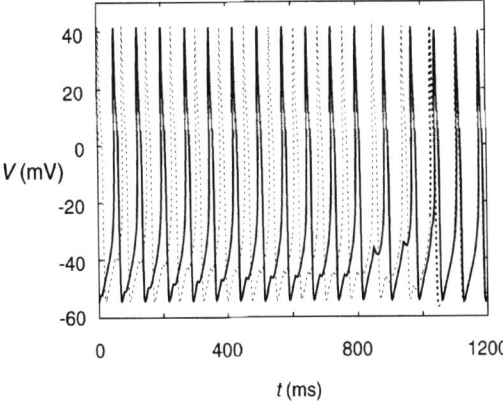

**Figure 6.7** Kinetics of synaptic inhibition can affect synchronization. Fast inhibition leads to antiphase behavior for $t < 800$ ms; switching to slowly decaying inhibition for $t > 800$ ms leads to in-phase behavior. Parameters are the same as in Figure 6.6 except that here $\beta = 1$/ms for $t < 800$ ms and $\beta = 0.1$/ms for $t > 800$ ms. Initial conditions: $V_1 = -55$ mV, $w_1 = s_1 = 0$, $V_2 = -20$ mV, $w_2 = s_2 = 0$.

slowly into an antiphase rhythm (Figure 6.7) during the first 800 ms of the simulation. During this portion the inhibitory synapses have fast kinetics, and we are confirming the classical expectation that inhibition leads to antiphase firing between a cell pair. In these voltage time courses one can clearly see the hyperpolarizing effect of the synaptic inputs. As a cell's voltage rises from its minimum the partner fires and sends a brief pulse of outward current, which halts transiently the rising voltage. At $t = 800$ ms the decay rate of inhibition is slowed from $\beta = 1$/ms to $\beta = 0.1$/ms. Within a few cycles the cells lock into perfect synchrony. Slowly decaying mutual inhibition can lead to in-phase locking among neural oscillators. This behavior has been proposed as the mechanism for gamma rhythms that are seen in various brain regions, and are believed to have a functional role in some cognitive processes (Wang and Buzsaki 1996; Traub et al. 1996).

As a secondary note, although we do not show it here, this particular network with the fast inhibitory synapses is bistable. In addition to the antiphase behavior (for $t < 800$ ms) it also has a stable in-phase behavior for some set of initial states.

The examples above are highly idealized and primarily directed toward rhythmicity. The parameters have been chosen in order for us to illustrate several features about temporal patterning in mutually coupled pairs with just minimal adjustments from one case to the next. Important questions arise about how these features might carry over to larger networks. In a case of mixed fast and slow synaptic coupling are the synchronizing or desynchronizing effects more important? Are the patterns robust to effects of noise and heterogeneity? What if there is a mixture of gap–junctional and synaptic coupling, in regimes where their effects counteract each other?

# 6.4 Neural Circuits as Computational Devices

Up to this point we have examined how various coupling mechanisms can coordinate cellular electrical activity in the context of stereotypical steady firing patterns:

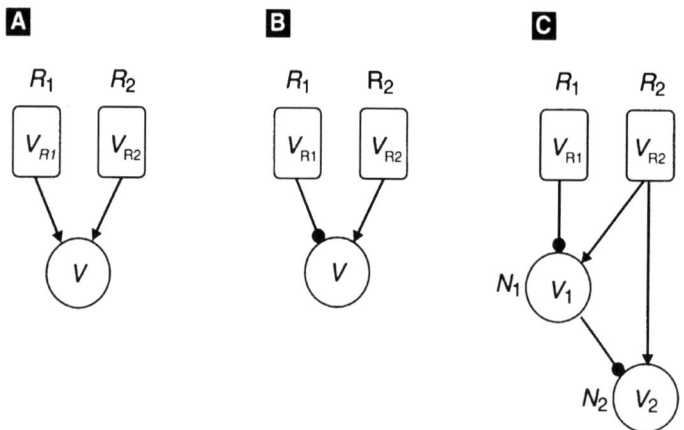

**Figure 6.8** Schematic of receptors ($R_1$, $R_2$), neurons and connections for implementing neural computations. (A) $R_1$ and $R_2$ activate excitatory inputs on target neuron. (B) $R_1$ sends inhibition; $R_2$ sends excitation. (C) First–layer neuron $N_1$ (same configuration as in (B) sends inhibition to second–layer neuron $N_2$, which also receives direct excitation from $R_2$.

synchronous or antisynchronous oscillations. Beyond these examples one seeks to understand, in the context of neural systems, how circuits (cells and their connectivity) might be implemented to carry out various computations or computational strategies.

We will develop examples to illustrate idealized mechanisms for carrying out a few computational tasks: temporally specific addition or coincidence detection, temporally specific subtraction or vetoing, and direction selectivity. The setup will be similar in each case. There will be two input units, $R_1$ and $R_2$, that we might think of as sensory neurons, receptors to the outside world. They feed into a target neuron that does the computation. (Figure 6.8A). This is a feedforward circuit. The information will be passed by means of chemical synapses. Depending on the combined synaptic input (triggered by the firings of $R_1$ and $R_2$) the target neuron may or may not elicit a spike. The "answer" from the computation will simply be yes or no, whether the target neuron fires or does not fire an action potential for our specified input pattern. This will depend on whether $R_1$ and $R_2$ are both activating excitatory synapses (Figure 6.8A) or one of them is sending inhibition and the other excitation (Figure 6.8B), and of course the relative timing will matter.

For a neuron in which inputs are distributed over the dendritic tree, the spatial as well the temporal distribution of incoming signals determines the net response. For example, suppose there is simultaneously timed excitation and inhibition delivered at different locations along one dendritic branch. An inhibitory synapse located closer to the soma will be more effective at reducing the response to the excitatory input than if it is located more distally on the branch.

Let $s_1$ and $s_2$ be the (fractional) postsynaptic conductances that arise in the target cell from activity of $R_1$ and $R_2$, respectively. We will consider the results of only a single firing of each of the input neurons. The conductances $s_1$ and $s_2$ will be determined from the kinetics as in (6.10). The kinetic parameters are chosen so that an excitatory conductance time course is brief and an inhibitory conductance is slower, on time scales comparable to spike dynamics. For simplicity we do not describe the action potentials of $R_1$ and $R_2$ in detail. Instead, we represent the receptor voltage responses, $V_{R1}$ and $V_{R2}$, as brief (2 ms), square-pulse, depolarizing events. These will yield appropriate time courses for $s_1$ and $s_2$.

The parameters for our Morris–Lecar target neurons are the standard ones from Table 2.4 for Type II behavior; here, we set $I_{app} = 0$. The equations that govern the response of the target neuron then are as follows:

$$C\frac{dV}{dt} = -I_{ion} - g_1 s_1 (V - V_{syn,1}) - g_2 s_2 (V - V_{syn,2}), \qquad (6.13)$$

$$\frac{dw}{dt} = \frac{\phi[w_\infty - w]}{\tau_w}, \qquad (6.14)$$

$$\frac{ds_1}{dt} = \alpha_1 s_{\infty,1}(V_{R1})(1 - s_1) - \beta_1 s_1, \qquad (6.15)$$

$$\frac{ds_2}{dt} = \alpha_2 s_{\infty,2}(V_{R2})(1 - s_2) - \beta_2 s_2. \qquad (6.16)$$

The parameters $g_1$ and $g_2$ represent the synaptic weights, the maximum postsynaptic conductances that can be evoked by activity coming from $R_1$ and $R_2$. The values of $V_{syn,1}$ and $V_{syn,2}$, the synaptic reversal potentials, will determine whether the synapse is excitatory (with $V_{syn}$ far above $V_{rest}$, so that the postsynaptic current is inward) or inhibitory (with $V_{syn}$ below $V_{rest}$, say equal to $V_K$).

We begin with a simple case in which $R_1$ and $R_2$ both deliver brief excitatory input to the target neuron. The initiation times of $V_{R1}$ and $V_{R2}$ are respectively $t_1$ and $t_2$. In our examples, we will fix $t_2$ and vary $t_1$, where $t_1 = t_2 + \Delta t$ so that the input from $R_1$ precedes by $|\Delta t|$ the input from $R_2$ (if $\Delta t$ is negative; a positive value of $\Delta t$ means that $R_2$ fired before $R_1$). Suppose that the synapses are excitatory and identical, having the same synaptic weights and the same time courses except for initiation times. Assume that each input is subthreshold, but when delivered simultaneously they can fire the target cell. We see in Figure 6.9A that when $R_1$ fires far earlier than $R_2$ then two nearly identical input current events are seen in the target cell, and neither leads to a target cell firing. No nonlinearities are evident in this case. However, when the inputs are delivered within a restricted time window they can cause the target cell to fire (Figure 6.9B,C). Thus the neuron's integration and threshold properties make it act like a coincidence detector for excitatory input. The regenerative spike–generating mechanism is quite nonlinear; its output does not reflect linear summation of inputs: one plus one is not necessarily equal to two.

The response plot (Figure 6.9D) summarizes the sensitivity to $\Delta t$; it is the tuning curve for the target cell as a coincidence detector. Notice how sharp the sensitivity is. If

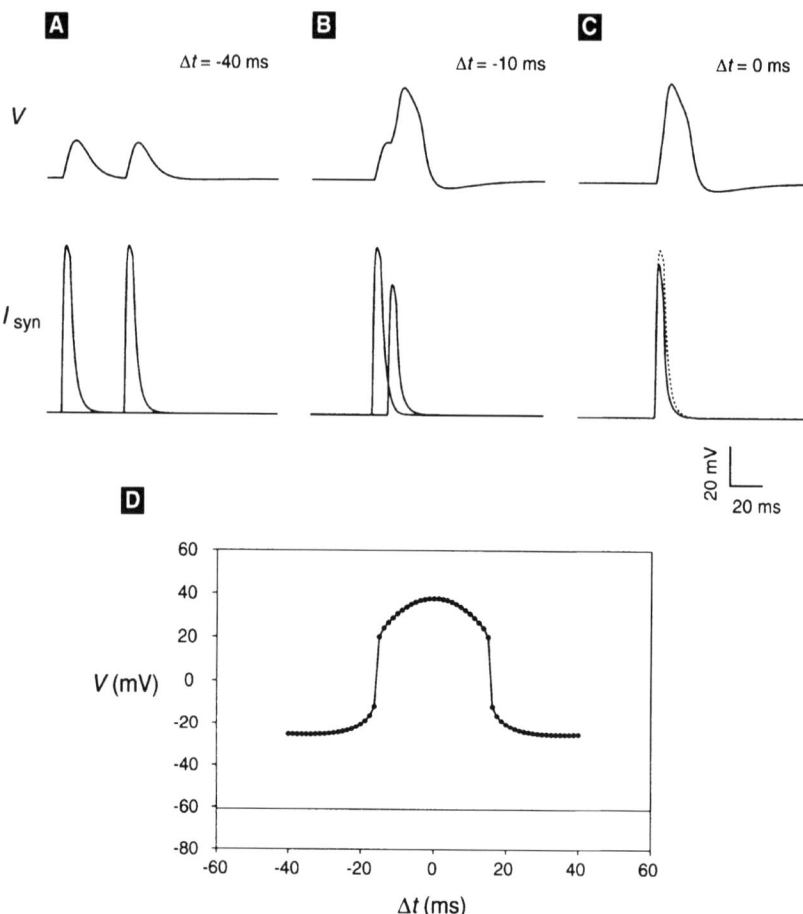

**Figure 6.9**  Coincidence detection for excitatory inputs. $V$ time courses of target neuron (upper records) for excitatory inputs (lower records are synaptic currents) from receptor cells $R_1, R_2$ that are offset in time by $\Delta t$. Results for three different values of $\Delta t$ (in ms): −40 in (A), −10 in (B), 0 in (C). Note: The current record shown dashed in (C) is reproduced from (A) to show the effect of reduced driving force when inputs occur simultaneously. (D) Response plot of peak of $V$ (following second input) versus $\Delta t$. Parameter values for excitatory synaptic inputs: $\alpha_1 = \alpha_2 = 1.0$/ms, $\beta_1 = \beta_2 = 0.3$/ms, $V_{syn,1} = V_{syn,2} = 100$ mV, $\theta_{syn} = 20$ mV, $k_{syn} = 2$ mV; $g_1 = g_2 = 1.0$ mS/cm². At $t = 0$ neurons are at rest, $V = −60.9$ mV, $w = 0.0149$ and synaptic gating variables are zero.

this target neuron activates synapses that feed higher order neurons and if the synapses require pre-synaptic depolarization above 0 mV, then the next level would effectively see the yes/no answer. Emphasizing the computational analogy one might say that the target neuron is performing the logical AND operation. Notice further that the tuning curve is symmetric about $\Delta t = 0$, as expected. Since the inputs are identical, the target cell's response is the same for $R_1$ preceding or following $R_2$ by $\Delta t$.

In Section 6.2 we demonstrated the feature of sublinear summation of synaptic inputs that are generated by conductance changes. The basis for this is that the synaptic current depends on the membrane potential through the driving potential $V - V_{syn}$. Thus the current for a given postsynaptic conductance is reduced when $V$ operates closer to $V_{syn}$. This effect is seen in Figure 6.9B for the case of nearly (but not exactly) simultaneous transient inputs. When $|\Delta t|$ is small enough the target neuron is still depolarized when the second input arrives. Hence current injected by this second input is smaller than that resulting from the earlier input. Similarly, when the two inputs occur simultaneously there is a reduction in the unitary input (compare solid with dashed records in the lower part of Figure 6.9C), but the effect is less pronounced, since the input current is complete before significant depolarization occurs.

Next we illustrate the vetoing power that inhibition can exercise over a suprathreshold excitatory input. For this case $R_2$ yields an excitatory conductance as in Figure 6.9 except doubled in size, adequate to evoke by itself a spike in the target neuron. Also, $R_1$ activates an inhibitory synaptic conductance that has a slower time course than an excitatory one. If inhibition is delivered sufficiently in advance of excitation, then it has no deleterious effect on the response to excitation (Figure 6.10A1). However, if it occurs in a narrow temporal window that starts just before excitation, it can overrule the excitation, vetoing an action potential in the target cell (Figure 6.10A2). Even if it occurs slightly after the excitatory input, it can still prevent spiking or at least shorten the target neuron's action potential (Figure 6.10A3). The tuning curve for this vetoing or subtraction effect is not symmetric (Figure 6.10A4). The inhibition with its slowly decaying tail is more effective if delivered with or reasonably before the excitation. But it is ineffective if delivered afterward even by a modest amount. Once the intrinsic spike-generating conductances are activated only a very strong and not too delayed inhibition can cancel the spike. Such an input–output relation might be likened to the logical AND-NOT operation: An output spike occurs only if $R_2$ fires and $R_1$ does not.

If one imagines that the receptor cells $R_1$ and $R_2$ are activated in temporal sequence (with delay $\Delta t$) by a moving stimulus in the environment, our target neuron shows sensitivity to the direction of this movement. Suppose that a stimulus moving from left to right activates $R_1$ and then $R_2$. In this case our tuning curve shows that vetoing occurs far more effectively for movement from left to right ($\Delta t$ negative) than from right to left ($\Delta t$ positive). Now suppose that we want to go the next step by having an output neuron that acts as a directionally selective motion detector. Can we create a unit that fires only when the stimulus moves across the receptor pair from left to right and in a confined range of $\Delta t$ (this means either that the stimulus is moving fast enough or it is assumed of moderate speed and in the spatially localized receptive fields of these receptors)? We want a tuning curve that looks like Figure 6.10B4.

A simple way to solve conceptually this design task is to imagine subtracting the tuning curve of Figure 6.10A4 from the response curve of $R_2$ delivered alone to the target

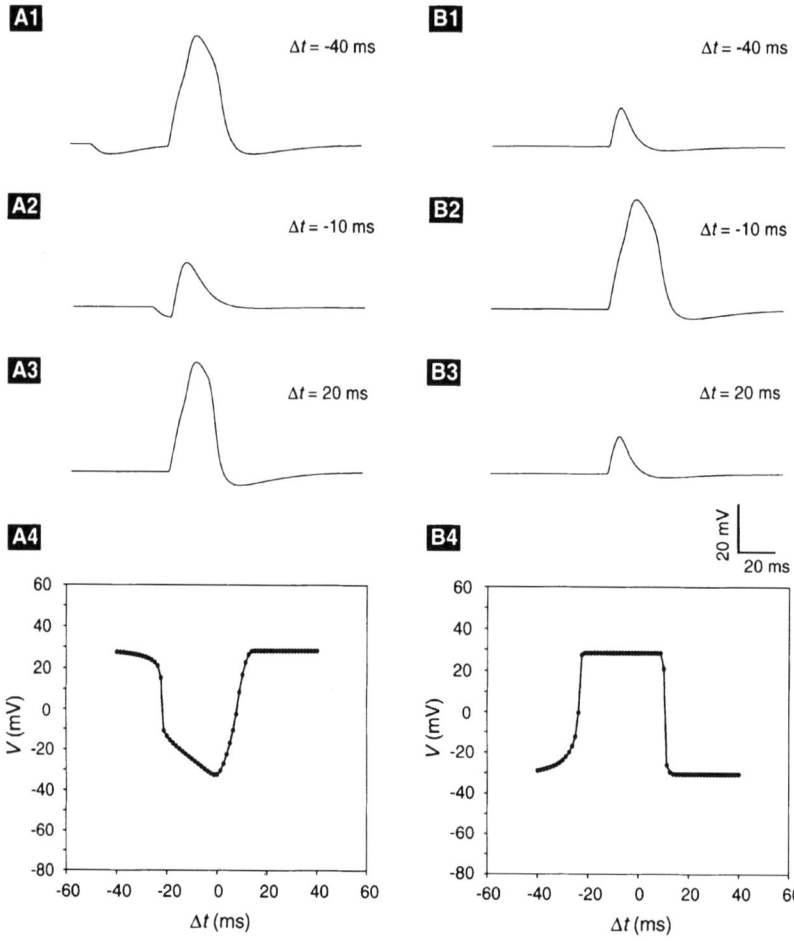

**Figure 6.10** (A1–A4) Implementing subtraction with excitation and inhibition. (A1–A3) membrane potential time courses $V$ of target neuron for excitatory input from receptor cell $R_2$ and inhibitory input from $R_1$, offset in time by $\Delta t$ (see schematic in Figure 6.8B). Results for different values of $\Delta t$, −40, −10, 20 ms, are shown in (A1–A3), respectively. (A4) Response plot of peak of $V_1$ (associated with input from $R_1$) versus $\Delta t$. Parameter values for excitatory inputs as in Figure 6.9; for inhibitory input kinetics: $\alpha_1 = 0.5$/ms, $\beta_1 = 0.1$/ms, $V_{syn,1} = -84$ mV, $\theta_{syn} = 20$ mV, $k_{syn} = 2$ mV, $g_1 = 2.5$ mS/cm$^2$, $g_2 = 2.0$ mS/cm$^2$. (B1–B4) Directionally selective, motion–detecting neuron. (B1–B3) Membrane potential time courses $V_2$ of second layer neuron $N_2$ (Figure 6.8C) that receives excitatory input from receptor cell $R_2$ and inhibitory input from $N_1$ (A1–A4). Results for different values of $\Delta t$, −40, −10, 20 ms, are shown in panels (B1–B3), respectively. (B4) Response plot of peak of $V_2$ versus $\Delta t$. Parameter values for $N_1$ and the inputs to $N_1$ are same as in the left panels; for $N_2$ and inputs, the parameters are identical as onto $N_1$ (with $g_{N2,N1} = 2.5$ mS/cm$^2$, $g_{N2,R2} = 2.0$ mS/cm$^2$). Input from $R_2$ to $N_2$ is delayed in each case by 10 ms. At $t = 0$ neurons are at rest, $V = -60.9$ mV, $w = 0.0149$ and synaptic gating variables are zero.

neuron (the latter being just flat at 32 mV, independent of $\Delta t$). One way to implement this in neuralware is by introducing a downstream neuron $N_2$ (with voltage $V_2$) whose tuning curve will have this property. Let us refer to the target neuron in earlier examples as $N_1$ (with voltage $V_1$). The two Morris–Lecar units $N_1, N_2$ have identical intrinsic properties. To carry out the subtraction we now feed excitatory input from $R_2$ into $N_2$, with some propagation delay, and have $N_1$ inhibit $N_2$ (Figure 6.8C). Thus for large $|\Delta t|$ the inhibition from $N_1$ vetoes the excitation from $R_2$ with the result that $N_2$ does not fire (Figure 6.10B1,B3). For modest $|\Delta t|$ the inhibition is preferentially absent for $\Delta t$ negative and $N_2$ can fire (Figure 6.10B2). Figure 6.10B4 shows the tuning curve of the output neuron $N_2$. With this configuration we identify $N_1$ and $N_2$ as neurons in the first and second processing layers after the network's input layer of receptor units. Other types of feedforward connectivities can lead to different feature–detecting properties; e.g., in the primary visual cortex there are neurons ("simple cells") that respond to bars of light with specific orientation, and the sensory signals coming to these neurons are mapped with specific alignments from the thalamus (which shows, to first approximation, no orientation specificity).

The examples above have exploited the sensitivity of the model neuron's responsiveness to timing of its inputs. There is good evidence that sound localization as carried out by the auditory system (in some frequency ranges) can involve neurons capable of coincidence discriminationsat submillisecond time scales. These neurons have been shown in some species to have potassium currents that are believed to render them unresponsive to tonic inputs but highly sensitive to transient inputs (Trussell 1999). Whether or not neural computations generally are performed based on precise spike timing raises interesting questions about the nature of the neural code and mechanisms (Shadlen and Newsome 1998; Rieke et al. 1997; Dayan and Abbott 2001). How could spike timing be preserved in the face of neural variability, including the possible sources of heterogeneity of neuronal and coupling properties, sparseness of coupling, and the stochastic nature of synaptic transmission? Can some type of population averaging or multi–layer processing recover timing information, or must it necessarily be degraded as activity passes through successive layers? In some contexts firing rate (short time averaged) is likely the code carrier.

## 6.5   Large–Scale Networks

The previous sections have dealt with simple 2 or 3-cell networks, but how does one develop and implement formulations to treat large cellular ensembles? One approach, of course, is just to extend to $N$ units the cell-based descriptions that we have been using above. This means specifying the biophysical details of each unit, the coupling dynamics (e.g., synaptic conductances or diffusion of intercellular messengers), and the coupling topology (e.g., nearest–neighbor or all-to-all coupling or something intermediate). In the case of spiking units this would mean that we are attempting to keep track of each spike in the network and time course of intrinsic variables (membrane potential and gating variable for ionic conductances at all times) and coupling variables associated with the intercellular communication. Such approaches have been used to model electrical activity in the heart (Keener and Sneyd 1998), the dynamics of hormonal secretory systems like the pancreas (Sherman 1996), and behaviors of neural systems like epileptic waves in brain slices (Traub et al. 1999) or receptive field properties of visual cortex (Troyer et al. 1998; McLaughlin et al. 2000).

An alternative approach to such detailed cell-based models is to use mean field descriptions. For this one assumes that the space and time scales of interest do not demand knowing what each individual unit is doing at every moment in time; that the amount of heterogeneity and fluctuation in the system is enough but not too much so that averaging over cellular properties and small time and space scales is allowable to yield a description for the dynamics of the ensemble averages. For example, instead of instantaneous membrane potential (for each cell, on the time scale of milliseconds) being the primary observable, we might use the firing rate of a cell on the time scale of tens of milliseconds and a spatial scale that includes several cells in a unit length. If you think about it, the Hodgkin-Huxley model of electrical excitability is a mean field description at the cellular level with respect to the population of ionic channels. The random openings and closings of individual channels are not described; only the probability or fraction of gating variables being in a particular state is represented (see Chapter 11, as well as Johnston and Wu (1995) and Koch (1999)). An implicit assumption in using the mean field is that the individuals are behaving asynchronously. We would not use the HH model if we thought that the channels were all opening and closing precisely together in lock step. Similarly, using a mean field model for a cellular network means that we assume that the cells are not synchronized on a fine temporal scale, say spike for spike. Consequently, we would not use the mean field models to answer questions about precise spike timing, as we have been doing in the previous sections.

So how does one obtain the mean field model for an ensemble? This is a difficult and significant question, or several questions, actually: How do you retain some information about biophysical properties of cells and synapses when you go from a "micro" to a "macro" description? Can you be sure if you do a cell-based network simulation that your units will actually behave asynchronously? How do you decide whether your mean field model should retain explicitly some form of cell heterogeneity? How does one average over the different sources of stochasticity? These issues will keep researchers in computational cell biology quite busy over the next several years.

In practice, mean field or population-averaged models in cell/neuro biology are typically not derived from first principles. They are more or less developed from a set of assumptions that involve disregarding some of the difficult issues. One may lose some connection with the biophysical details in this way.

Here's one example along these lines of a derivable mean field description. Suppose we treat a spike as an event without considering the ionic currents that generate the spike. The leaky integrate-and-fire model of neuronal excitability does this. Here one assumes that below "firing threshold" the neuron has a linear I-V relation and the membrane potential is described by an RC circuit:

$$C_\mathrm{m} \frac{dV}{dt} = -\frac{V}{R_\mathrm{m}} - I_\mathrm{syn} + I_\mathrm{app}, \qquad (6.17)$$

where $I_\mathrm{syn}$ is the summed synaptic input from the other cells making synaptic contact with this one.

When $V$ reaches threshold $V_\mathrm{th}$ then we say that a spike occurred, and $V$ is reset to some level, typically to $V_\mathrm{rest}$. In a case like this the model is linear except for the resetting events, and on the face of it each cell has one variable, $V$. So, what about the synaptic variables? One could assume a very simple model for a synaptic input that given a presynaptic spike the postsynaptic $V$ jumps instantaneously up or down by a specified amount. This assumption throws away the time course of the postsynaptic conductance. However, on the plus side one may now derive legitimately a mean field population model, getting a (partial) differential equation for the probability density function $p(v, t)$ of membrane potential (say for $V_\mathrm{rest} \leq V \leq V_\mathrm{th}$) as a function of time. This description does indeed well describe a network of $n$ such integrate-and-fire cells, sparsely coupled, in the limit of large $n$. It has been used to describe a variety of behaviors (Nykamp and Tranchina 2000; Omurtag et al. 2000). However, as one seeks to restore some of the biophysical variables, the description can become impractical. Each new dynamic variable, synaptic or intrinsic, that one includes becomes another independent variable in $p(v, \ldots, t)$ leading to a high-dimensional system growing geometrically in size. In contrast, the problem size grows linearly if one adds additional variables in the case of an $n$-cell integrate-and-fire system. One of the insights that emerge from approaches like these is that if cells are really asynchronous, then at each instant in time they will have $V$-values spread over the interval, and one should expect that some fraction (perhaps small) of cells are near to $V_\mathrm{th}$. Consequently, a transient excitatory input will lead to almost instantaneous firing of some cells. This means that the network can exhibit very fast response times for some inputs.

A less rigorously supported but attractive mean field model was popularized by Wilson and Cowan (Wilson and Cowan 1972). This formulation jumps right to mean field quantities, which in the simplest interpretation are the probabilities of firing (per unit time) of the typical, or "mean," cell in the given populations: $E$ and $I$ for, say, excitatory and inhibitory subpopulations of interacting neurons. Note that this Wilson–Cowan formulation implicitly assumes random, sparse coupling without spatial structure. One assumes that the neurons have input/output relations represented by $f_E$ and $f_I$ so that under steady input to, say, the "$E$" cells, they would fire with steady rate $f_E$ (and analogously for "$I$" cells). The shape of $f_E$ is usually taken to be sigmoidal, the rate increasing to a saturation level as input$_E$ grows large. The network is considered not able to equilibrate instantaneously to dynamic inputs but that it "integrates" with a time constant $\tau_E$.

For now, consider an isolated population of mutually excitatory cells. We would
have

$$\tau_E \frac{dE}{dt} = -E + f_E, \tag{6.18}$$

where $f_E$ is a function (such as that explored in Figure 6.12) of input to the $E$ population
from both extrinsic sources $P_E$ and from the other cells in the network. In the mean field
approach the other cells are also firing at rate $E$, so the synaptic field is proportional
to $E$. Thus,

$$input_E = a \cdot E + P_E. \tag{6.19}$$

Note that the parameter $a$ incorporates the transformation from presynaptic firing
rate to postsynaptic conductance and to postsynaptic current, etc. One should also
appreciate that in using the sigmoidal-like input/output relation there is not a strict
threshold for cell firing. The firing probability decreases toward zero for decreasing input strength, but it is never really zero even for negative inputs. An implicit assumption
here is that noise in the system smoothes the abrupt thresholds as found in frequency–
current relations of, say, cellular models like Morris–Lecar without noise. Also, in any
real network cells are not prefectly identical, so this heterogeneity also contributes to
such smoothing in the mean field models.

Now let us warm up by considering just the isolated $E$ network and ask what steady
states of activity are possible. At steady state we must have

$$\frac{dE}{dt} = 0, \tag{6.20}$$

so that from (6.18),

$$E = f_E(a \cdot E + P_E). \tag{6.21}$$

The solutions to this nonlinear transcendental equation are most easily analyzed graphically. That is, plot the left-hand side and the right-hand side on the same axes as
functions of $E$. Their intersections correspond to the possible solutions. The parameter $a$ acts primarily on the steepness of the sigmoidal curve, while $P_E$ is more related
to left/right translation. In Figure 6.11 we can see that for each value of $a$ there is an
intersection near $E = 0$. That is, the network has a stable rest state for these parameter
values. On the other hand, we see that if $a$ is large enough, then there are multiple steady
states. For this one-dimensional dynamical system the middle state of moderate activity
is unstable, while the upper state of high activity is stable. In this parameter regime
the network is bistable. If the cells are at rest (very low firing rate) and no transient
input is given, then they will remain at rest. However, if a sufficiently strong brief input
is delivered, then the cells will become active and because of the recurrent excitation
in the network the cells will continue to drive each other, and the network reaches a
steady state of high activity. This recurrent excitation is the autocatalytic mechanism
that underlies the bistability. We may view $a$ as the relative weight of the autocatalytic

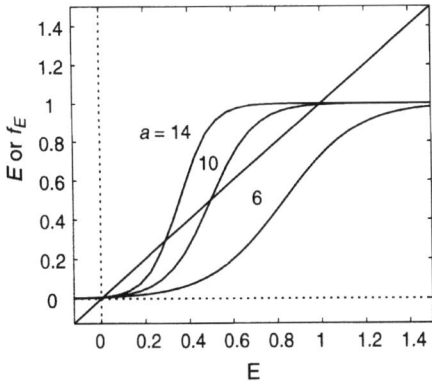

**Figure 6.11** Graphical depiction for determination of steady states to single population of recurrently connected excitatory neurons. Left– and right–hand sides of (6.21), plotted as functions of $E$, firing rate. Definitions: $f_E = 1/[1 + \exp(-input_E)]$ and $p_E = -5$; right–hand side plotted for three different values $a = 6, 10, 14$.

feedback, the strength of excitatory interactions. Note that because the input/output relation is saturating the activity level does not grow without bound.

Now we are ready to consider the two populations of excitatory and inhibitory cells. Extending the development above, we write down the model equations as

$$\tau_E \frac{dE}{dt} = -E + f_E, \tag{6.22}$$

$$\tau_I \frac{dI}{dt} = -I + f_I, \tag{6.23}$$

where $f_E$ is a function of $input_E = a \cdot E - b \cdot I + P_E$ and $f_I$ is a function of $input_I = c \cdot E - d \cdot I + P_I$. This pair of ordinary differential equations governs the behavior, in the mean field sense, of the interacting $E$ and $I$ populations. Note the signs of the interaction terms inside of $input_E$ and $input_I$. Positive $a$ means recurrent excitation, autocatalysis; positive $c$ means that $I$ is driven by $E$; positive $c \cdot b$ (with the minus sign on $b$) means that negative feedback through $I$ onto $E$; and $-d \cdot I$ means $I$ cells inhibit each other.

The model equations can be analyzed with phase plane methods. For the case shown in Figure 6.12 the network has a stable rest state (the only intersection of the $E$– and $I$–nullclines) with both populations at very low firing rate. This state is globally attracting. For any initial conditions the network has a transient response and then returns to rest. The three different trajectories correspond to responses of $E$ and $I$ for brief stimuli to the $E$ population. For a small stimulus only a few $E$ cells fire, not enough to amplify the excitatory activity. However, with a larger initial level of $E$ we see some regenerative effects. First, $E$ grows beyond its starting value: Some nonlinear amplification occurs. Then, $E$ peaks as the $I$ population is recruited, driven by the firing of the excitatory cells. Eventually, all cells return to the rest state. This illustrates the excitability of the network.

The phase plane portrait of the network is qualitatively similar to that of the basic Morris–Lecar model of cellular electrical excitability. The activator variable (in this

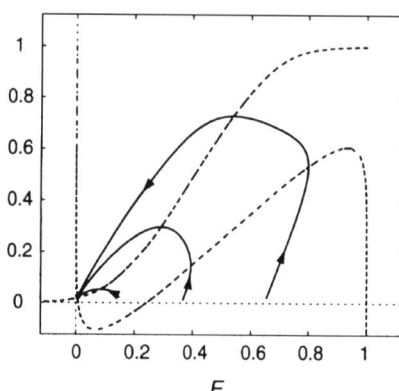

**Figure 6.12** Network excitability. Phase plane portrait for firing-rate model of neuronal network with interacting excitatory and inhibitory subpopulations, based on (6.23). The cellular input/output relations (rate vs. input) are $f_E = 1/[1 + \exp(-input_E)]$ as in Figure 6.11 and $f_I = 1/[1 + \exp(-input_I)]$. The $E$–nullcline is the cubic-shaped dashed line; $I$–nullcline is the sigmoidal dashed line. Trajectories (solid), for three different initial conditions that represent perturbations from rest to the $E$ cells, all tend back to the rest state for large time. Parameter values: $a = 16, b = 12, c = 16, d = 5, P_E = -5, P_I = -4, \tau_E = \tau_I = 1$. Note that the model equations have been expressed in dimensionless form.

case $E$) has a cubic-shaped nullcline; the left, middle, and right branches correspond to the three roots of (6.21) for a fixed level of $P_E - b \cdot I$. The variable $I$ here corresponds to a recovery process like $w$ in the Morris–Lecar model. Its nullcline is monotonic.

If we vary the parameter $P_E$, the $E$–nullcline changes, while the $I$–nullcline remains fixed. For example, increasing a bit the extrinsic input to the $E$ population, say imagining more feedforward excitation onto the $E$ cells from some other network, will cause them to fire faster, and thereby recruit more inhibition. The steady state will migrate to the right and upward along the $I$–nullcline. If one increases $P_E$ enough, then the steady state will move onto the middle branch and perhaps destabilize. This is seen in Figure 6.13A, where the network is now in an oscillatory mode, having gone through a Hopf bifurcation. The negative–feedback inhibition lags somewhat behind the excitatory activity, seen in this phase plane portrait as well as in the time courses shown in Figure 6.13B. If the $E$-population is driven hard enough, the system stops oscillating. The steady state, for large enough $P_E$, restabilizes and is on the right branch of the $E$-nullcline. This corresponds to a balanced state with strong steady firing in each population.

These examples are only illustrative of this class of mean field firing–rate models. Some specific neural systems where such formulations have been applied include Pinto et al. (1996), Ermentrout (1998), and Hansel and Sompolinsky (1998); in the paper by Pinto et al. you will find a pseudo-derivation of the model based on cellular properties.

# Suggestions for Further Reading

- *Spikes*, Fred Rieke, David Warland, Rob de Ruyter van Steveninck, and William Bialek. This influential book treats neuronal communication from a highly information-theoretic point of view (Rieke et al. 1997).

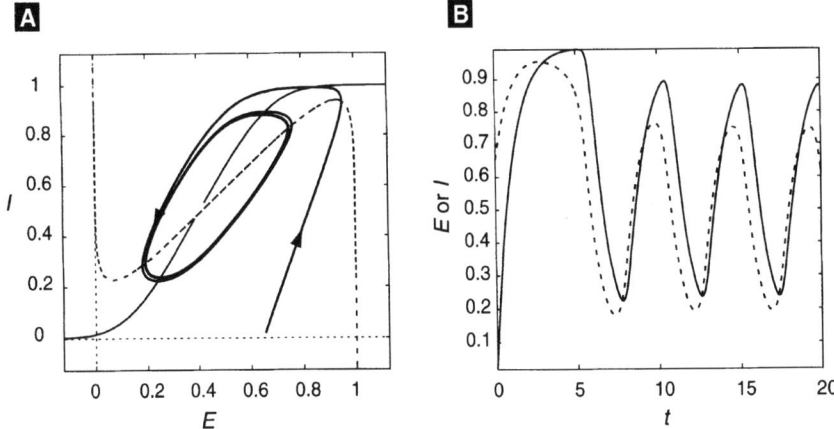

**Figure 6.13** Network oscillations. (A) Phase plane and (B) time courses for the firing rate model, with increased input to the excitatory population: $P_E = -1$. Other parameters are as in Figure 6.12. Dashed line is for $E$-nullcline in (A) and $E$ vs. $t$ in (B); solid line is for $I$-nullcline in (A) and for $I$ vs. $t$ in (B). Darker solid curve in (A) is the phase plot.

- *Neural networks as spatio-temporal pattern-forming systems*, Bard Ermentrout. This paper reviews pattern formation in neural networks, including the continuous networks discussed in this chapter (Ermentrout 1998).
- *Mathematical Physiology*, James Keener and James Sneyd. Keener and Sneyd treat some of the cellular level topics presented in this chapter, as well as many other topics in systems level physiology, from a more analytic perspective as opposed to the computational focus presented here (Keener and Sneyd 1998).
- *Methods in Neuronal Modeling*, Christof Koch and Idan Segev, editors. This is a compilation of chapters from various authors on a wide variety of topics related to neuronal modeling, both at the cell and at the network level (Koch and Segev 1998).
- *Foundations of Cellular Neurophysiology*, Daniel Johnston and Samuel Wu. Covers in more depth topics such as transmitter release, plasticity, elementary networks and extracellular electrical behavior (Johnston and Wu 1995).
- *Theoretical Neuroscience: Computational and Mathematical Modeling of Neural Systems*, Peter Dayan and Larry Abbott. This book addresses neural coding issues with several different types of models (Dayan and Abbott 2001).
- *Biophysics of Computation*, Christof Koch. This book covers many aspects of cell-based neurophysiology from a modeling perspective (Koch 1999).

## 6.6 EXERCISES

1. Consider the dynamics of two identical electrically passive cells with gap–junctional coupling:

$$\frac{dV_1}{dt} = (-g_m V_1 + g_c(V_2 - V_1) + I_1)/C_m,$$

$$\frac{dV_2}{dt} = (-g_m V_2 + g_c(V_1 - V_2) + I_2)/C_m.$$

Let $g_m = 0.05$ mS/cm$^2$, $C_m = 1\ \mu$ F/cm$^2$. Consider $g_c = 0.1$ mS/cm$^2$ (Note: We have divided the equations by cell area). For now, let $I_1 = I_2 = 0$.

(a)  Solve the equations with $V_1(0) = 1$, $V_2(0) = 0$ and observe, over 60 ms, the convergence and eventual decay back to rest of both cells.

(b)  This is a linear system. Its solution is a sum of two exponentials. Find the characteristic decay rates (eigenvalues) and the corresponding time constants. (Hint: Find singular points.)

(c)  Evaluate and plot $V_s = (V_1 + V_2)/2$ and $V_d = (V_1 - V_2)$. Notice that they appear to be single exponentials. Confirm this by plotting $V_s$ and $V_d$ on a semilog scale and identify the slopes with the eigenvalues.

(d)  Derive differential equations for $V_s$ and $V_d$ by adding and subtracting the equations for $V_1$ and $V_2$. See that these equations are uncoupled, and so each of $V_s$ and $V_d$ decays as a single exponential. Identify the eigenvalue for $V_d$ as the reciprocal of the system's equalization time constant.

(e)  Do the eigenvalues change if $I_1$ and/or $I_2$ are nonzero?

2.  Determine the two time constants of the following system: See that compartment 1 acts as a driver for compartment 2:

$$\frac{dV_1}{dt} = (-g_m V_1 + \alpha g_c(V_2 - V_1) + I_1)/C_m,$$

$$\frac{dV_2}{dt} = (-g_m V_2 + g_c(V_1 - V_2) + I_2)/C_m.$$

Let $g_m = 0.05$ mS/cm$^2$, $C_m = 1\mu$F/cm$^2$, $g_c = 0.1$ mS/cm$^2$. Noting that we have divided the equations by cell area, $\alpha$ is the ratio of areas: Cell 2's area divided by cell 1's. Consider $\alpha = 1$ or 0.1. For now, let $I_1 = I_2 = 0$. Note: In this formulation $g_c$ is the total gap junctional conductance divided by the area of cell 2.

(a)  Compute and plot the time courses for $V_1$, $V_2$ (on the same axes) for the two values of $\alpha$. Interpret the results. Why does $V_1$ decay more slowly for $\alpha = 0.1$? What happened to the early decaying component?

(b)  Estimate (analytically) the eigenvalues in the limit of $\alpha$ very small. Explain the differences between these eigenvalues for the two cases of $\alpha$, 1 or small.

(c)  How well is the behavior of $V_1$ and $V_2$ approximated for small $\alpha$ if you just set $\alpha = 0$ in the equations? Discuss the behavior in terms of cell 1 being a driver of cell 2 (think of the larger membrane area in the dendrites). Now do the simulation with initial conditions $V_1(0) = 0$, $V_2(0) = 0$. How effectively does compartment 2 drive compartment 1?

(d)  Suppose the steady input $I_1$ is delivered to compartment 1. Find the attenuation factor for the steady response from the dendrite to the soma; $V_1/V_2$. Do the same for $I_2$

given to compartment 2. Compare and discuss the two attenuation factors and their differences.

3. Consider two Morris–Lecar cells with gap–junction coupling; use the same intrinsic parameters as in Figure 6.2 and Figure 6.3 (except $I_1 = I_2 = -5$, unless stated otherwise).

   (a) Formulate the equations, as in Exercise 2, allowing for possibly different areas.

   (b) Consider first an isolated cell. Show that it is bistable, with an attracting oscillation mode in the depolarized regime and a stable steady state at low membrane potential (look at and describe the phase plane: nullclines, steady states, and their stability, etc.). Demonstrate this bistability over a range of $I$ values.

   (c) Show that for weak gap–junction conductance $g_c$, the pair can exist with one cell, say cell 1, in the "upper" (oscillating) state and cell 2 in the lower (steady) state. But as $g_c$ increases to intermediate values (say beyond $g_c = 0.3$), the cells tend to be in the same state. If $g_c$ is not too large the cells can oscillate in antiphase but for for large $g_c$ the cells oscillate in-phase. Show that there is some range of $g_c$ where both oscillating states coexist.

   (d) Consider the effect of unequal areas. Suppose cell 1 (as in Exercise 2) is larger and $\alpha$ is the ratio of their areas (as defined in Exercise 2): $\alpha = $ area2/area1. Let $g_c = 0.3$. Verify that the state of cell 1 up (and oscillating) with cell 2 down is still valid. Compare its frequency to that of the control case ($\alpha = 1$) and explain the difference. Of course one can one also get the state with both cells up and oscillating, but now with a different phase relationship than control. Explain the difference. Is there an analog here for the two states of in-phase and anti-phase as in (c)? Give an example if you say Yes.

4. We have discussed the Wilson–Cowan paradigm in excitable and in oscillatory modes.

   (a) Using a numerical package, compute the bifurcation diagram with stimulation to the "E" cells as a parameter.

   (b) Plot the frequency vs stimulus strength and discuss the low frequency behavior.

   (c) Can you find other parameters that would give a bistable network?

5. Consider just an excitatory population. As a function of input, this population shows a range of bistability.

   (a) Introduce a slow negative feedback, Q, as an intrinsic property. Q should grow with the firing rate of the excitatory cells, and result in spike frequency adaptation. The Q dynamics could be something like

   $$\frac{dQ}{dt} = (Q_\infty - Q)/\tau_q$$

   The sigmoid $Q_\infty$ should be positioned in the E-Q plane to create oscillations or excitability. This is one way to give the single population some interesting dynamics in the context of a rate model like Wilson-Cowan.

   (b) Try using $\theta_Q$ (the location of $Q_\infty = 0.5$) as the slow variable.

6. Intercellular communication and the interesting dynamics that result is not limited to coupling through voltage. Coupling can occur via extracellular messengers or by the exchange of $Ca^{2+}$ or other species via gap junctions.

   (a) Create a model of two cells coupled via $Ca^{2+}$ using the bullfrog sympathetic ganglion neuron open cell model discussed in Section 5.2. Couple the cells using a simple linear "diffusion" coefficient mediating transfer between $Ca^{2+}$ concentrations in the two cells, analagous to the conductance that couples the voltage differences seen already in this chapter.

(b) Vary the "diffusion" coefficient for $Ca^{2+}$ between the two cells when they are both oscillating due to $Ca^{2+}$ influx across the plasma membrane. Over what range do the cells synchronize?

(c) Can $Ca^{2+}$ influx in only one of the cells (resulting in oscillations) lead to oscillations in both of the coupled cells?

# PART II

# ADVANCED MATERIAL

# CHAPTER 7

# Spatial Modeling

## James P. Keener

All of the models considered in previous chapters have relied on the implicit assumption that chemical concentrations are uniform in space. This assumption is reasonable when the region of space in which the reaction takes place is confined and quite small. However, there are many situations in which chemical concentrations are not uniform in space. A well-known example in which nonuniform distributions are crucial is the propagation of an action potential along the axon of a nerve fiber (Figure 1.1). When a nerve cell "fires," a wave of membrane depolarization is initiated at the base of the axon (where it connects to the cell body; see Figure 2.1) and propagates along the axon out to its terminus. During propagation, large spatial gradients in membrane potential and local currents are created. The interaction between these spatial gradients and voltage-sensitive ion channels in the axonal membrane drives the wave along the axon. In order to understand the propagation of a nerve impulse, we must first master the basic principles of molecular diffusion and the interactions between chemical reaction and diffusion.

Many other questions arise in molecular cell biology that demand at least an elementary understanding of molecular diffusion. For instance, how long does it take for a chemical signal generated at the cell membrane to diffuse to the nucleus? Why are expensive transport systems required to move some materials in cells, for example between a nerve cell body and synapses in axons and dendrites? How fast can molecules or ions pass through protein channels in membranes?

In Chapter 4 and Chapter 5 we faced the problem of nonuniform $Ca^{2+}$ concentration in the vicinity of $Ca^{2+}$ channels. There we made a simplifying assumption that the $Ca^{2+}$ concentration is high in a small region adjacent to the channel (domain $Ca^{2+}$). To

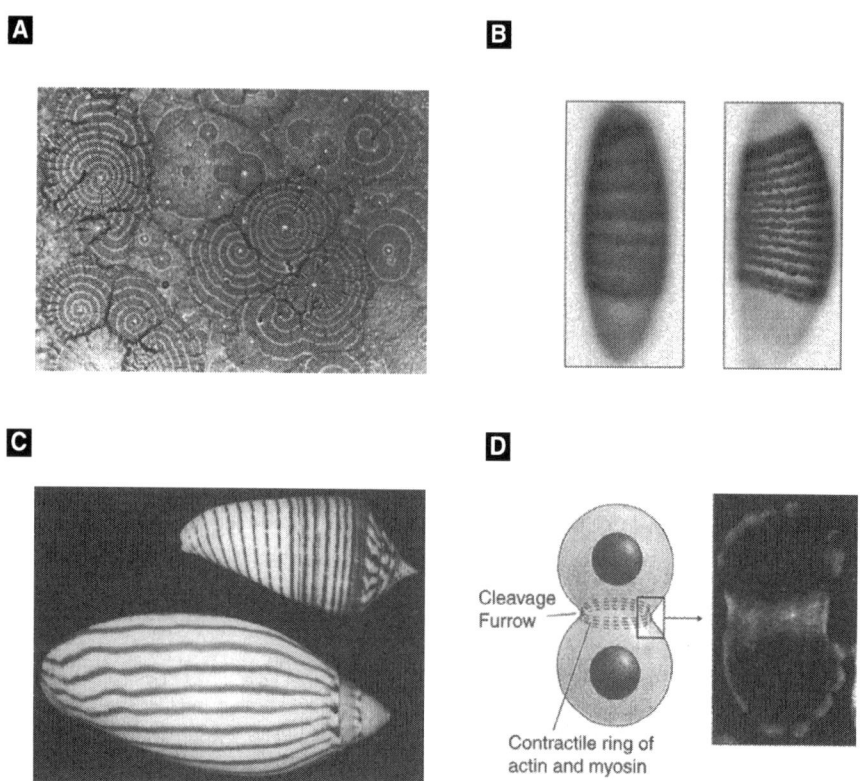

**Figure 7.1** Spatial phenomena in cell biology. (A) Waves of aggregation in fields of slime mold amoebae. The light and dark bands correspond to regions where the amoebae are actively moving or not. The amoebae collect at the center of each pattern, where they form a multicellular slug. The collective motion of the cells is organized by waves of cyclic AMP that propagate throught the extracellular medium. Courtesy of Peter Newell. (B) Stripes of gene expression (*ftz* and *eve*) in a fruit fly embryo at (left) 3 hr. after fertilization and (right) 3 1/2 hr. after fertilization. Reprinted from Lawrence (1992). For an introductory discussion of the segmentation genes in Drosophila, see Alberts et al. (1994). (C) Pigmentation patterns on sea shells. From Meinhardt (1998). (D) The cleavage furrow in a dividing cell. Reprinted from Alberts et al. (1994); original by Yoshio Fukui. A dividing slime mold amoeba is stained for actin and myosin. The actomyosin ring in the center of the cell contracts like a purse-string to divide the cell in half.

improve on the domain $Ca^{2+}$ approximation and other simplified approaches to spatial nonuniformity, we will need the spatial modeling principles described here.

A more sophisticated example of spatiotemporal organization in living cells is the phenomenon of $Ca^{2+}$ waves that propagate through eggs after fertilization. These waves will be modeled in great detail in Chapter 8, after we have studied reaction–diffusion equations in this chapter. Similar to $Ca^{2+}$ waves in eggs are waves of cyclic AMP that propagate through fields of slime mold amoebae shown in Figure 7.1A. By directing the motion of the amoebae, these chemical waves organize the complex behaviors of this primitive multicellular organism: aggregation of simple amoebae into a multicellular slug, motility of the slug, and formation of the fruiting body. Other interesting examples of spatial organization include gap-gene expression in early fruit fly embryos (Figure 7.1B), seashell patterns (Figure 7.1C), and medial ring placement at cell division (Figure 7.1D). Although we will not attempt to model any of these phenomena in this book, a starting point for such investigations is this chapter.

The chapter is organized along the following lines: First, we consider diffusion in one dimension, such as we might find in a long thin tube like a nerve axon. We distinguish between a *conservation law* (how the law of conservation of matter relates molecular flux to local changes in concentration) and a *constitutive relation* (how molecular flux is determined by concentration gradients, fluid transport, and electrophoresis). These principles are expressed in the precise mathematical terms of *partial differential equations* (PDEs). We show the exact solution to these equations for a number of important illustrative cases. Because PDEs cannot be solved exactly in most realistic situations, we next describe a numerical procedure, called the method of lines, that is easily implemented. Also, because very few spatial nonuniformities are effectively one-dimensional, we show how to formulate the conservation law and constitutive relations in two and three dimensions. We then couple molecular diffusion to nonlinear chemical reactions in order to study wave propagation in one spatial dimension. The theory is applied to the FitzHugh–Nagumo equations of nerve impulse propagation introduced in Chapter 2.

## 7.1  One-Dimensional Formulation

### 7.1.1  Conservation in One Dimension

Many equations in biology are consequences of *conservation laws*. A conservation law is simply a mathematical statement describing how some quantity is created or destroyed or moves about.

Consider a chemical species C whose concentration $c(x,t)$ varies in time and space, where the spatial variation is restricted to one spatial variable $x$. This situation is illustrated in Figure 7.2, where the chemical species C is contained in a long, thin tube with

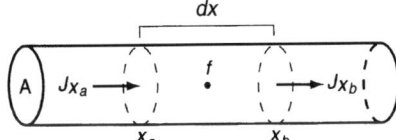

**Figure 7.2**  Conservation in one dimension.

constant cross-sectional area A. In any fixed region R along the tube, the conservation of C can be expressed in words as

> time rate of change of the total amount of C within R =
>
> rate at which C flows in to R
>
> − rate at which C flows out of R
>
> + rate at which C is produced within R
>
> − rate at which C is destroyed within R.                                                    (7.1)

The total amount of chemical C contained in a small slice of tube between $x$ and $x + dx$ is $c(x,t)A\,dx$. At any time $t$, the total amount of C in some arbitrary interval $x_a < x < x_b$ can be computed by integrating $c(x,t)A$ over that interval:

$$\text{total amount of C in the interval } [x_a, x_b] = \int_{x_a}^{x_b} c(x,t)\,A\,dx. \tag{7.2}$$

It is important to distinguish between concentration (amount/ volume) and the "total amount." If $c$ has units of micromolar (micromol/liter), then the total amount has units of micromoles.

Now suppose that C is free to move about inside the tube, so that C moves in and out of the interval by crossing the boundaries of the interval at $x = x_a$ and $x = x_b$. If we denote by $J(x,t)$ the rate at which C moves across the boundary at position $x$ from left to right at time $t$, then the net movement, or flux, of C into the interval is

$$\text{net rate of entry of C} = AJ(x_a, t) - AJ(x_b, t). \tag{7.3}$$

Since the net rate of entry has units of amount/time and $A$ has units of area, the flux rate $J(x,t)$ has units of amount/area/time. It is also important to remember that $J(x,t)$ is positive when C moves to the right, and negative when C moves to the left.

The total amount of C in the interval can also change because of the production or destruction of C within the interval. If we let $f(x,t,c)$ denote the net rate of increase of C (production − destruction) per unit volume at location $x$ and time $t$, then the total amount of C produced in the interval at time $t$ is

$$\text{net rate of production of C} = \int_{x_a}^{x_b} f(x,t,c(x,t))\,A\,dx. \tag{7.4}$$

Note that the presence of $c$ in the definition of $f$ allows for the possibility that the rate of production of C depends on $c$ itself. Since the units of the net rate of production of C are amount/time, the units of $f$ must be amount/time/volume. When $f$ is positive, the region is a source (leading to an increase in the total amount of C), and when $f$ is negative, it is a sink. The function $f$ is often called a source function.

The conservation law (7.1) can now be written in mathematical symbols as

$$\frac{d}{dt}\int_{a}^{b} c(x,t)\,dx = J(x_a, t) - J(x_b, t) + \int_{x_a}^{x_b} f(x,t,c(x,t))\,dx, \tag{7.5}$$

where the constant $A$ has been factored out. The flux terms can be replaced by

$$J(x_b, t) - J(x_a, t) = \int_{x_a}^{x_b} \frac{\partial}{\partial x} J(x, t) \, dx, \qquad (7.6)$$

allowing all the terms in (7.5) to be written as integrals:

$$\frac{d}{dt} \int_{x_a}^{x_b} c(x, t) \, dx = \int_{x_a}^{x_b} \frac{\partial}{\partial x} J(x, t) \, dx + \int_{x_a}^{x_b} f(x, t, c(x, t)) \, dx. \qquad (7.7)$$

If the function $c(x, t)$ is smooth enough, then the differentiation and integration can be interchanged, and (7.7) can be rewritten as

$$\int_{x_a}^{x_b} \left[ \frac{\partial}{\partial t} c(x, t) \, dx - \frac{\partial}{\partial x} J(x, t) - f(x, t, c(x, t)) \right] dx = 0. \qquad (7.8)$$

Since the interval is arbitrary, the only way this equality can hold is if the integrand is zero. Therefore, we replace (7.8) by the equivalent conservation law in differential form:

$$\frac{\partial c}{\partial t} - \frac{\partial J}{\partial x} = f(x, t, c). \qquad (7.9)$$

Notice that in this equation there are two independent variables ($x$ and $t$), and that the equation contains partial derivatives with respect to both of these. Such equations are called partial differential equations. Since time is one of the independent variables, and this equation describes the evolution of $c(x, t)$ in time, (7.9) is called an evolution equation because it describes how the concentration of C evolves (changes) as time proceeds.

## 7.1.2 Fick's Law of Diffusion

Equation (7.9) is underdetermined because it is a single equation relating two unknowns: the concentration $c$ and the flux $J$. To resolve this problem, an additional equation relating $c$ and $J$ is needed.

In contrast to the conservation law (7.9), which follows indubitably from the general principle of material conservation, the relation between $c$ and $J$ must be determined empirically and is not universally valid. To make this distinction, the secondary relation between $c$ and $J$ is usually called a constitutive equation.

One such constitutive relation is called Fick's law, and states that C moves from regions of high concentration to regions of low concentration, at a rate proportional to the concentration gradient. In mathematical symbols, this diffusive flux is

$$J(x, t) = -D \frac{\partial}{\partial x} c(x, t), \qquad (7.10)$$

where the proportionality constant $D$ is called the diffusion constant. The negative sign signifies that C moves spontaneously from regions of high concentrations to regions of low concentrations. The value of $D$ depends on the size of C, as well as properties of the

**Table 7.1** Molecular weight and diffusion coefficients of some biochemical substances in dilute aqueous solution.

| Substance | Molecular Weight | $D/10^7 \text{cm}^2/\text{s}$ |
|---|---|---|
| glucose | 192 | 660 |
| insulin | 5734 | 210 |
| cytochrome c | 13,370 | 11.4 |
| myoglobin | 16,900 | 11.3 |
| $\beta$-lacroglobulin | 37,100 | 7.5 |
| serum albumin | 68,500 | 6.1 |
| hemoglobin | 64,500 | 6.9 |
| catalase | 247,500 | 4.1 |
| urease | 482,700 | 3.46 |
| fibrinogen | 339,700 | 1.98 |
| myosin | 524,800 | 1.10 |
| tobacco mosaic virus | 40,590,000 | 0.46 |

medium in which it is diffusing. The constant $D$ has units of length$^2$/time. Diffusion coefficients of some typical biochemicals are given in Table 7.1.

Using Fick's law, (7.9) becomes the reaction–diffusion equation

$$\frac{\partial c}{\partial t} - \frac{\partial}{\partial x}\left(D\frac{\partial c}{\partial x}\right) = f(x,t,c). \tag{7.11}$$

In this equation, the term $\frac{\partial}{\partial x}\left(D\frac{\partial c}{\partial x}\right)$ is the diffusion term, and $f$ is the reaction term. When $f$ is zero, that is, when there are no sources or sinks, (7.11) becomes the diffusion equation

$$\frac{\partial c}{\partial t} = \frac{\partial}{\partial x}\left(D\frac{\partial c}{\partial x}\right). \tag{7.12}$$

## 7.1.3 Advection

Suppose that there is a uniform macroscopic flow of the solvent, with speed $v$ along the $x$-axis, which carries solutes along with it. Then, during a small time $\Delta t$, all of the C between $x = x_a$ and $x = x_a - v\Delta t$ will cross the point $x = x_a$. The total amount of C crossing $x_a$ during this time is found by multiplying the concentration $c(x,t)$ by the fluid volume $Av\Delta t$. The corresponding flux is therefore (after dividing by $\Delta t$ to get amount per unit time)

$$J(x,t) = vc(x,t). \tag{7.13}$$

This flux is called the advective flux. Note that whereas the diffusive flux was proportional to the concentration gradient, the advective flux is proportional to the concentration itself.

If there is both diffusive flux and advective flux, then the total flux is the sum of the two:

$$J(x,t) = vc(x,t) - D\frac{\partial}{\partial x}c(x,t). \tag{7.14}$$

Using this constitutive relation, (7.9) becomes a reaction–advection–diffusion equation,

$$\frac{\partial c}{\partial t} + \frac{\partial}{\partial x}\left(vc - D\frac{\partial c}{\partial x}\right) = f(x,t,c). \tag{7.15}$$

### 7.1.4 Flux of Ions in a Field

If the substance C is an ion and there is an electrical potential gradient, then there will also be a flux of C because of the influence of the potential on the ion. In this case the flux of ions is given by the Nernst–Planck equation

$$J = -D\left(\frac{\partial c}{\partial x} + \frac{zF}{RT}c\frac{\partial \phi}{\partial x}\right), \tag{7.16}$$

where $\phi$ is the electric potential, $z$ is the number of positive charges on the ion (a negative integer if the ion is negatively charged), $F$ is Faraday's constant, $R$ is the universal gas constant, and $T$ is absolute temperature. Notice that according to this equation, there is movement because of both the concentration gradient and the potential gradient.

### 7.1.5 The Cable Equation

Suppose that our long one-dimensional tube is bounded by a membrane, as in a nerve axon. In this case, we wish to keep track of the electrical potential across the membrane, rather than some chemical species within the tube. Nonetheless, the rules of conservation are the same, so the derivation of the governing equation is similar.

Suppose the total current along the interior of the axon is $I$, positive from left to right, and the transmembrane current per unit membrane area is $I_T$, positive outward. Then, conservation of current implies that

$$I(x_a,t) - I(x_b,t) = \int_{x_a}^{x_b} SI_T dx, \tag{7.17}$$

where $S$ is the circumference of the tube. This conservation law can be expressed using integrals as

$$-\int_{x_a}^{x_b} \frac{\partial I}{\partial x}dx = \int_{x_a}^{x_b} SI_T dx, \tag{7.18}$$

and since the interval is arbitrary, the integrands must be equal, so that

$$-\frac{\partial I}{\partial x} = SI_T. \tag{7.19}$$

Recall from Chapter 1 that the total transmembrane current consists of two components, a capacitive current and the ionic currents

$$-\frac{\partial I}{\partial x} = S\left(C_{\mathrm{m}}\frac{\partial V}{\partial t} + I_{\mathrm{ion}}\right), \tag{7.20}$$

where V is the transmembrane potential. Finally, the relationship between current and potential is given by the constitutive relationship known as Ohm's law (also called the core conductor assumption),

$$I = -\frac{A}{R_{\mathrm{c}}}\frac{\partial \phi_i}{\partial x}, \tag{7.21}$$

where $R_{\mathrm{c}}$ is the cytoplasmic resistance (with units Ohms length), and $\phi_i$ is the intracellular potential. With this constitutive relationship, our equation becomes

$$\frac{\partial}{\partial x}\left(\frac{A}{R_{\mathrm{c}}}\frac{\partial \phi_i}{\partial x}\right) = S\left(C_{\mathrm{m}}\frac{\partial V}{\partial t} + I_{\mathrm{ion}}\right). \tag{7.22}$$

Finally, we close the model by assuming that the membrane is in a highly conductive bath, so that the extracellular potential $\phi_e$ is a constant. Since $V = \phi_i - \phi_e$, we arrive at the cable equation

$$\frac{\partial}{\partial x}\left(\frac{A}{R_{\mathrm{c}}}\frac{\partial V}{\partial x}\right) = S\left(C_{\mathrm{m}}\frac{\partial V}{\partial t} + I_{\mathrm{ion}}\right). \tag{7.23}$$

For a tube of uniform circular cross section and diameter $d$, $A/s = d/4$. Typical parameter values for a variety of cells are shown in Table 7.2.

## 7.1.6   Boundary and Initial Conditions

In the study of ordinary differential equations, it is necessary to specify initial data before one can find a solution trajectory. With partial differential equations, one must specify both initial data and boundary data before a solution can be found. Roughly

Table 7.2 Typical cable parameter values for a variety of excitable cells. From Keener and Sneyd (1998).

| parameter | $d$ | $R_{\mathrm{c}}$ | $R_{\mathrm{m}}$ | $C_{\mathrm{m}}$ | $\lambda_{\mathrm{m}}$ |
|---|---|---|---|---|---|
| units | $10^{-4}$ cm | $\Omega$ cm | $10^3\ \Omega$ cm$^2$ | $\mu$F/cm$^2$ | cm |
| squid giant axon | 500 | 30 | 1 | 1 | 0.65 |
| lobster giant axon | 75 | 60 | 2 | 1 | 0.25 |
| crab giant axon | 30 | 90 | 7 | 1 | 0.24 |
| earthworm giant axon | 105 | 200 | 12 | 0.3 | 0.4 |
| marine worm giant axon | 560 | 57 | 1.2 | 0.75 | 0.54 |
| mammalian cardiac cell | 20 | 150 | 7 | 1.2 | 0.15 |
| barnacle muscle fiber | 400 | 30 | .23 | 20 | 0.28 |

speaking, there must be one condition for each degree of freedom. Thus, since reaction–diffusion equations are of first order in time, there must be one initial condition for each unknown function. Since they are of second order in space, there must be two boundary conditions (conditions at some points in space) for each unknown function.

Initial conditions usually specify the values of the dependent variables at some initial time (usually $t = 0$) at which the solution is known or specified by experimental conditions. Boundary conditions reflect certain physical conditions of the experiment. For example, if the concentration $c$ is specified to be some function $f(t)$ at some boundary point, say $x = x_a$, then the condition $c(x_a, t) = f(t)$ is applied, called a Dirichlet boundary condition. If, on the other hand, the flux at a point is specified, then the condition $-D\frac{\partial c}{\partial x}(x_a, t) = g(t)$, called a Neumann boundary condition, is applied. If the flux is related to the value of $c$ at the boundary, then the Robin condition, $-D\frac{\partial c}{\partial x}(x_a, t) = h(t) - \alpha c(x_a, t)$, is applied.

It is often convenient to assume that a domain is infinite, even though there is no such thing as an infinitely long tube. Even with infinite domains, however, boundary conditions must be specified as constraints on the behavior of the dependent variable in the limit that $x \to \pm\infty$.

## 7.2 Important Examples with Analytic Solutions

### 7.2.1 Diffusion Through a Membrane

Consider a membrane separating two large regions of space that contain some chemical C. The concentration on the left is $c_1$, and the concentration on the right is $c_2$ (Figure 7.3). There is a small pore in the membrane (a one-dimensional channel of length $L$) through which the chemical C can freely pass. Suppose that the two regions of space are so large that their concentrations are not changing, even if chemical is flowing from one region to the other.

Let us assume that the transport of C across the membrane has been going on for some time, so that the process is at steady state, i.e., the concentration $c(x,t)$ is independent of time ($\frac{\partial c}{\partial t} = 0$). In this case, $c(x)$ must satisfy the "boundary value problem"

$$\frac{\partial^2 c}{\partial x^2} = 0, \qquad c(0) = c_1, \; c(L) = c_2. \tag{7.24}$$

cell membrane

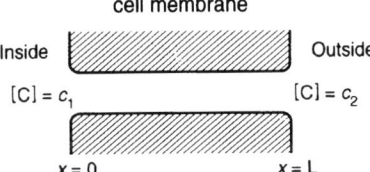

Inside        Outside

$[C] = c_1$           $[C] = c_2$

$x = 0$       $x = L$

**Figure 7.3** Simple diagram of a pore through a membrane.

The solution of this problem is quite easy to find, being

$$c(x) = c_1 \left(1 - \frac{x}{L}\right) + c_2 \frac{x}{L}. \tag{7.25}$$

Consequently, the steady flux through the channel is proportional to the concentration difference across the membrane,

$$J = -Dc_x = \frac{D}{L}(c_1 - c_2). \tag{7.26}$$

## 7.2.2 Ion Flux Through a Channel

Suppose that the chemical moving through the channel is an ion, and that there is a potential difference across the channel, with $\phi(0) = \phi_1$ and $\phi(L) = \phi_2$. We make the simplifying approximation that the potential gradient through the channel is constant:

$$\frac{d\phi}{dx} = \frac{\Delta\phi}{L} = \frac{V}{L}, \text{ where } V = \phi_1 - \phi_2. \tag{7.27}$$

If the process is in steady state so that the ion flux everywhere in the channel is the same constant, then, from (7.16),

$$J = -D\left(c_x + \alpha c \frac{V}{L}\right), \tag{7.28}$$

where $\alpha = zF/RT$. We solve this differential equation for $c(x)$ subject to the boundary condition that $c(0) = c_1$ and find that

$$c(x) = c_1 e^{-\alpha V \frac{x}{L}} - \frac{JL}{D\alpha V}\left(1 - e^{-\alpha \Delta\phi x/L}\right). \tag{7.29}$$

Now we can determine the flux $J$ by requiring that $c(L) = c_2$, so that

$$J = -\frac{D}{L}\alpha V \left(\frac{c_2 - c_1 e^{-\alpha V}}{1 - e^{-\alpha V}}\right). \tag{7.30}$$

This expression for flux can be converted to an ionic current $I_c$ by multiplying by $zF$, in which case we obtain

$$I_c = -\frac{D}{L}zF\alpha V \left(\frac{c_2 - c_1 e^{-\alpha V}}{1 - e^{-\alpha V}}\right). \tag{7.31}$$

This expression is the famous Goldman–Hodgkin–Katz current equation, and it has the important property that $I_c = 0$ when

$$\Delta V = \frac{RT}{zF} \ln \frac{c_1}{c_2}, \tag{7.32}$$

which is called the reversal potential or Nernst potential for the channel. The Nernst potential is the transmembrane potential when the ion is at equilibrium across the membrane for a given transmembrane concentration ratio $c_1/c_2$.

### 7.2.3  Voltage Clamping

A typical experiment in electrophysiology is to hold fixed the transmembrane potential at some place on the membrane. For example, with a long axon, one might clamp the voltage at one end and determine the resulting voltage profile along the axon.

For a passive membrane (e.g., a dendritic membrane), the voltage profile should satisfy (7.23) with $I_{ion} = V/R_m$, where $R_m$ is the passive membrane resistance (ohm $\cdot$ cm$^2$). The steady–state voltage profile $V(x)$ must satisfy

$$\frac{A}{R_c}\frac{\partial^2 V}{\partial^2 x} = \frac{SV}{R_m},\tag{7.33}$$

subject to the boundary conditions $V(0) = V_{fixed}$ and $\frac{\partial V}{\partial x}(L) = 0$ if the far end is sealed. The solution of this problem is

$$V(x) = V_{fixed}\frac{e^{(L-x)/\lambda_m} - e^{(x-L)\lambda_m}}{e^{L/\lambda_m} - e^{-L/\lambda_m}},\tag{7.34}$$

where $\lambda_m = \sqrt{\frac{AR_m}{SR_c}}$ is the length constant for the axon. For a long axon ($L$ is many length constants), this solution reduces to

$$V(x) = V_{fixed}e^{-\frac{x}{\lambda_m}},\tag{7.35}$$

a simple exponential decay away from the voltage–clamped end. Some examples of space constants for a variety of excitable tissues are included in Table 7.2.

### 7.2.4  Diffusion in a Long Dendrite

All of the above examples examined steady behavior, after initial transients have decayed. However, reaction–diffusion equations also contain information about the temporal evolution of the process to steady state.

Consider calcium diffusing in a long dendrite. Suppose caged calcium is photoreleased from a small region around $x = 0$. If we denote by $c(x,t)$ the concentration of calcium along the length of the dendrite at each time $t$, then the model becomes

$$\frac{\partial c}{\partial t} = D\frac{\partial^2 c}{\partial x^2}, \quad -\infty < x < \infty, \quad t > 0,\tag{7.36}$$

$$c(x,0) = C_0\delta(x),\tag{7.37}$$

where $C_0$ is the total amount of released calcium, and $\delta(x)$ is the Dirac delta function. Because the dendrite is long, we view the domain as infinite. Since we do not expect the concentration of calcium to become appreciable at $x = \pm\infty$ in any finite time, we require $\lim_{x\to\pm\infty} c(x,t) = 0$.

It can be shown (Exercise 2) that the solution of this model is

$$c(x,t) = \frac{C_0}{\sqrt{4\pi Dt}}\exp\left(-\frac{x^2}{4Dt}\right),\tag{7.38}$$

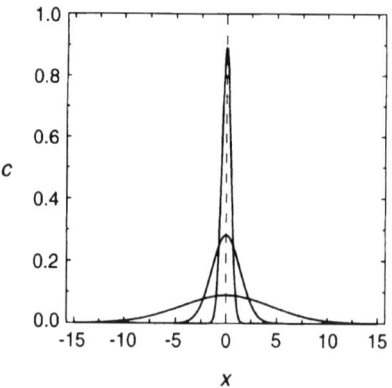

**Figure 7.4** Gaussian at 0.1, 1, and 10 s

which is illustrated in Figure 7.4. For each fixed $t$, this solution is a Gaussian function, and over time, the function becomes wider and the maximal value (at $x = 0$) declines,

$$c(0,t) = \frac{C_0}{\sqrt{4\pi Dt}}. \tag{7.39}$$

At any other point $x \neq 0$, the solution is biphasic, initially increasing to a maximum value and then decreasing back to zero. The maximum is attained when $Dt/x^2 = \frac{1}{2}$. This time behavior is illustrated in Figure 7.5.

We can readily calculate that

$$\langle x^2 \rangle = \int_{-\infty}^{\infty} x^2 c(x,t)\, dx = 2Dt, \tag{7.40}$$

so that the "root mean square" (rms) distance moved in time $t$ is

$$\sqrt{\langle x^2 \rangle} = x_{\text{rms}} = \sqrt{2Dt}. \tag{7.41}$$

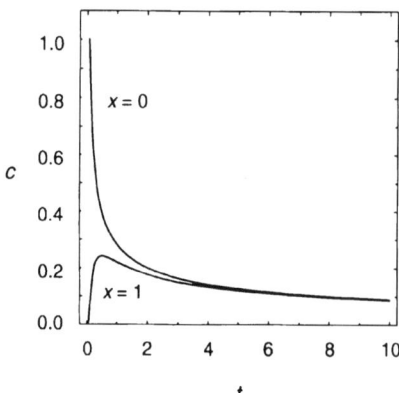

**Figure 7.5** Plot of $c(0,t)$ and $c(x \neq 0, t)$ from (7.38).

## 7.2.5 Diffusion into a Capillary

Suppose that a long capillary (open at one end) filled with water is inserted into a solution of known chemical concentration $C_0$, and the chemical species diffuses into the capillary through the open end. The concentration of the chemical species should depend only on the distance down the tube and so is governed by the diffusion equation

$$\frac{\partial c}{\partial t} = D\frac{\partial^2 c}{\partial x^2}, \qquad 0 < x < \infty, \qquad t > 0, \tag{7.42}$$

where for convenience we assume that the capillary is infinitely long. Because the solute bath in which the capillary sits is large, it is reasonable to assume that the chemical concentration at the tip is fixed at $c(0,t) = C_0$, and because the tube is initially filled with pure water, $c(x,0) = 0$.

The solution of this problem is given by

$$c(x,t) = 2C_0\left(1 - \frac{1}{\sqrt{2\pi}}\int_{-\infty}^{z}\exp\left(-\frac{s^2}{2}ds\right)\right), \qquad z = \frac{x}{\sqrt{2Dt}}. \tag{7.43}$$

If the cross-sectional area of the capillary is $A$, then the total number of molecules that enter the capillary in a fixed time $T$ is

$$N = A\int_{0}^{\infty}c(x,T)dx = 2C_0 A\sqrt{\frac{TD}{\pi}}. \tag{7.44}$$

From this equation it is possible to determine the diffusion coefficient by solving (7.44) for $D$, yielding

$$D = \frac{\pi N^2}{4C_0^2 A^2 T}. \tag{7.45}$$

Segel, Chet, and Henis used this formula to estimate the diffusion coefficient for bacteria (Segel et al. 1977). With $C_0$ at $7 \times 10^7$/ml, and times $T = 2, 5, 10, 12.5, 15$, and 20 minutes, they counted $N$ of 1800, 3700, 4800, 5500, 6700, and 8000 bacteria in a capillary of length 32 mm with 1 $\mu$l total capacity. In addition, with concentrations $C_0$ of 2.5, 4.6, 5.0, and $12.0 \times 10^7$ bacteria per milliliter, counts of 1350, 2300, 3400, and 6200 were found at $T = 10$ minutes. Using (7.45) a value of $D$ in the range of 0.1–0.3 cm$^2$/hour was found.

A second useful piece of information is found from (7.43) by observing that $c(x,t)/C_0$ is constant on any curve for which $z$ is constant. Thus, the curve $t = x^2/D$ is a level curve for the concentration, and gives a measure of how fast the substance is moving into the capillary. The time $t = x^2/D$ is called the diffusion time for the process. To give some idea of the effectiveness of diffusion in various cellular contexts, in Table 7.3 are shown typical diffusion times for a variety of cellular structures. Clearly, diffusion is quite effective when distances are short, but totally inadequate for longer distances, such as along a nerve axon. Obviously, biological systems must employ other transport mechanisms in these situations in order to survive.

**Table 7.3** Estimates of diffusion times for cellular structures of typical dimensions, computed from the relation $t = x^2/D$ using $D = 10^{-5} \text{cm}^2/\text{s}$.

| $x$ | $t$ | Example |
|---|---|---|
| 10 nm | 100 ns | thickness of cell membrane |
| $1\mu$m | 1 ms | size of mitochondrion |
| $10\mu$m | 100 ms | radius of small mammalian cell |
| $100\mu$m | 10s | diameter of a large muscle fiber |
| $250\mu$m | 60 s | radius of squid giant axon |
| 1 mm | 16.7 min | half-thickness of frog sartorius muscle |
| 2 mm | 1.1 h | half-thickness of lens in the eye |
| 5 mm | 6.9 h | radius of mature ovarian follicle |
| 2 cm | 2.6 d | thickness of ventricular myocardium |
| 1 m | 31.7 yrs | length of a nerve axon |

# 7.3   Numerical Solution of the Diffusion Equation

In order to attack more complex situations of reaction and diffusion, it is usually necessary to resort to numerical solutions of the partial differential equation. With the advent of cheap powerful computers, this approach has become increasingly useful. Here we describe the simplest numerical method to solve reaction–diffusion equations. While other more sophisticated numerical methods are available, this method is adequate for our purposes, and can be readily implemented using your favorite numerical integrator.

Consider the problem of determining calcium concentration following the photorelease of caged calcium in a sealed dendrite 40 microns long. We define the spatial variable $x$ to extend from 0 to 40 microns, and the starting time $t = 0$ to be the time at which the caged calcium is released. The equations we wish to solve are

$$\frac{\partial c}{\partial t} = D\frac{\partial^2 c}{\partial x^2},$$ 

(7.46)

where

$$c(x,0) = \begin{cases} C_0 & 20\ \mu\text{m} < x < 30\ \mu\text{m} \\ 0 & \text{elsewhere} \end{cases}$$ 

(7.47)

(7.48)

and

$$c_x(0,t) = c_x(40,t) = 0,$$ 

(7.49)

and where $C_0$ is the concentration of calcium released. Because the dendrite is closed to calcium flux at its ends, no-flux boundary conditions are specified at both ends. The caged calcium is initially confined to the region between 20 and 30 microns.

To solve this problem numerically, we subdivide the spatial domain $(0 < x < 40)$ into $N$ equal intervals, with $\Delta x = 40/N$ denoting the length of each interval. If the $N+1$ endpoints of these intervals are denoted by $x_n$, where $n = 0, 1, 2, \ldots, N$, then we define an approximation to $c(x, t)$ at these points by $c(x_n, t) = c_n(t)$.

Recall from calculus that the definition of the partial derivative of $c$ is

$$\frac{\partial c(x, t)}{\partial x} = \lim_{\Delta x \to 0} \frac{c(x + \Delta x, t) - c(x, t)}{\Delta x}. \tag{7.50}$$

It follows that if $\Delta x$ is small, but not zero, we have an approximation to the partial derivative:

$$\frac{\partial c(x, t)}{\partial x} \approx \frac{c(x + \Delta x, t) - c(x, t)}{\Delta x}. \tag{7.51}$$

In a similar way an approximation to the second partial derivative is found with $\Delta x$ small, but not zero:

$$\frac{\partial^2 c(x, t)}{\partial x^2} \approx \frac{c(x + \Delta x, t) - 2c(x, t) + c(x - \Delta x, t)}{\Delta x^2}. \tag{7.52}$$

Using this approximation at each spatial grid point $x = x_n$, we derive a set of ordinary differential equations

$$\frac{\partial c_n(t)}{\partial t} = \frac{D}{\Delta x^2} (c_{n+1}(t) - 2c_n(t) + c_{n-1}(t)). \tag{7.53}$$

Notice that this approximation is valid only at interior grid points with $n = 1, 2, \ldots, N-1$, since for $n = 0$ or $n = N$ equation (7.53) references points $c_{-1}$ and $c_{N+1}$ that are outside the domain, and therefore are not known. However, if we invoke the no-flux boundary conditions, and use the approximation (7.51), we learn that

$$c_{-1}(t) = c_0(t), \qquad c_{N+1}(t) = c_N(t). \tag{7.54}$$

These we apply to (7.53) for $n = 0$ and $n = N$ and obtain

$$\frac{\partial c_0(t)}{\partial t} = \frac{D}{\Delta x^2} (c_1(t) - c_0(t)), \tag{7.55}$$

and

$$\frac{\partial c_N(t)}{\partial t} = \frac{D}{\Delta x^2} (c_{N-1}(t) - c_N(t)). \tag{7.56}$$

The system of ordinary differential equations (7.53), (7.55), and (7.56) is a closed system of $N+1$ equations in $N+1$ unknowns that can be simulated with any standard differential equation solver. This conversion of a partial differential equation to a system of ordinary differential equations using difference approximations for the spatial derivatives is called the method of lines.

The initial conditions are found directly from the initial condition for the partial differential equation, with one minor adjustment. Since $c_n(t) = c(x_n, t)$, we set $c_n(0) = c(x_n, 0)$, wherever that is well-defined. However, the initial profile has a jump

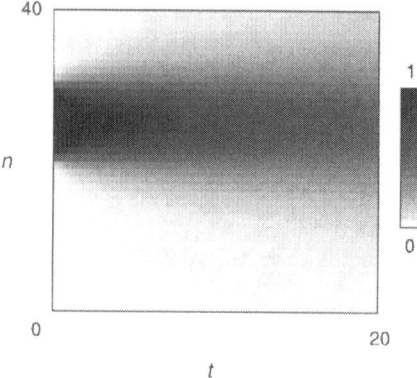

40

1

$n$

0

0                                                      20

$t$

**Figure 7.6** Numerical solution of equations (7.46)–(7.49).

discontinuity at $x = 20$ and $x = 30$, so it is preferable to define the value of $c$ at these points to be the average of the limiting values from the left and right.

The solution of this problem is shown in Figure 7.6.

# 7.4   Multidimensional Problems

The multidimensional formulation of a reaction–diffusion equation is an easy generalization from one dimension. The primary difference is that in multiple dimensions, flux is a vector rather than a scalar. As a vector, flux indicates not only the rate, but also the direction, of transport, and the derivation of the conservation law is an exercise in multidimensional calculus.

## 7.4.1   Conservation Law in Multiple Dimensions

Consider a chemical species C whose concentration $c(x, y, z, t)$ varies in both time and in some three-dimensional region with volume $V$. The verbal expression of conservation (7.1) remains valid. At any time $t$, the total amount of C in the volume can be computed by integrating $c(x, y, z, t)$ over the volume:

$$\text{total amount of C} = \int_V c(x, y, z, t)\, dV. \tag{7.57}$$

Now suppose that C is free to move about randomly, so that C moves in and out of the volume by passing through the volume's surface S. The flux $J(x, y, z, t)$ is a vector, since C can move in any direction. If we denote by $\mathbf{n}(x, y, z)$ the outward unit normal vector on S (see Figure 7.7), then the net flux of C into the region is given by

$$\text{net rate of entry of C} = -\int_S J(x, y, z, t) \cdot \mathbf{n}(x, y, z)\, dA, \tag{7.58}$$

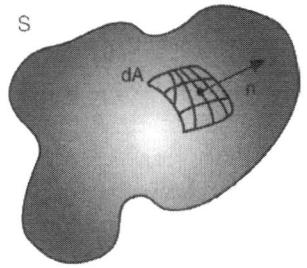

**Figure 7.7** Schematic diagram of a multi–dimensional region.

where $dA$ is the surface integration element. Because $\mathbf{n}$ is the outward normal, $J \cdot \mathbf{n}$ is positive when the motion is from inside to outside, which accounts for the negative sign in this equation. The rate of production of C in the volume can be written as $f(x,y,z,t,c)$, where as before, this rate is allowed to depend on $c$ itself. Thus, the total rate of production of C in the region is given by

$$\text{net rate of production of C} = \int_V f(x,y,z,t,c(x,y,z,t))\,dV. \tag{7.59}$$

The conservation equation can now be expressed mathematically as

$$\frac{d}{dt}\int_V c\,dV = -\int_S J \cdot \mathbf{n}\,dA + \int_V f\,dV. \tag{7.60}$$

The surface integral can be replaced by a volume integral using the divergence theorem, which yields the multidimensional integral form

$$\frac{d}{dt}\int_V c\,dV = -\int_V \nabla \cdot J\,dV + \int_V f\,dV, \tag{7.61}$$

where $\nabla\cdot$ is the divergence operator. As before, if the function $c(x,y,z,t)$ is smooth enough, and since the volume $V$ is arbitrary, we can rewrite (7.8) in differential form:

$$\frac{\partial c}{\partial t} + \nabla \cdot J = f. \tag{7.62}$$

Note that there are four independent variables ($x$, $y$, $z$, and $t$) and that the equation contains partial derivatives with respect to all four variables.

## 7.4.2   Fick's Law in Multiple Dimensions

Fick's law states that C moves from regions of high concentration to regions of low concentration, at a rate proportional to the concentration gradient. Thus, in multiple dimensions, Fick's law takes the form

$$J(x,y,z,t) = -D\nabla c(x,y,z,t), \tag{7.63}$$

where the diffusion constant $D$ is the proportionality constant, and the negative sign ensures that C moves down the concentration gradient. Even in multiple dimensions, the units of $D$ are length$^2$/time.

Using Fick's law, (7.62) can be rewritten as a reaction–diffusion equation:

$$\frac{\partial c}{\partial t} - \nabla \cdot (D\nabla c) = f. \tag{7.64}$$

### 7.4.3   Advection in Multiple Dimensions

Multidimensional advective flux has the same appearance as in the one-dimensional case:

$$J(x, y, z, t) = vc(x, y, z, t). \tag{7.65}$$

Notice, however, that the velocity $v$ is a vector, so the flux vector points in the direction of the velocity vector.

If the random and biased directional motions coexist, the total flux is the vector sum of the diffusive and drift fluxes:

$$J(x, y, z, t) = vc(x, y, z, t) - D\nabla c(x, y, z, t). \tag{7.66}$$

Using this constitutive relation in (7.62), the multidimensional reaction–advection–diffusion equation is

$$\frac{\partial c}{\partial t} + v\nabla c - \nabla \cdot (D\nabla c) = f. \tag{7.67}$$

### 7.4.4   Boundary and Initial Conditions for Multiple Dimensions

As in one dimension, we must specify both initial and boundary conditions to pose the problem completely. The only difference here is that the functions involved are multidimensional, and so, when the spatial domain is complex, can be quite complicated.

The Dirichlet boundary condition in multiple dimensions specifies the values of the dependent variable $c$ on the boundary, via $c(x, y, z, t) = f(x, y, z, t)$ with $x, y, z$ restricted to the boundary. Similarly, the Neumann boundary condition specifies the flux of $c$ on the boundary via $\mathbf{n} \cdot \nabla c = g$. Finally, the Robin condition specifies some relationship between the flux of $c$ and the value of $c$ on the boundary via $-\mathbf{n} \cdot D\nabla c = h + \alpha c$.

### 7.4.5   Diffusion in Multiple Dimensions: Symmetry

If the diffusion constant $D$ does not vary in space or time, then the diffusive term can be written

$$\nabla \cdot (D\nabla c) = D\nabla \cdot (\nabla c) = D\nabla^2 c. \tag{7.68}$$

In this expression, $\nabla^2$ is the "Laplacian operator," which in Cartesian coordinates is

$$\nabla^2 c = \frac{\partial^2 c}{\partial x^2} + \frac{\partial^2 c}{\partial y^2} + \frac{\partial^2 c}{\partial z^2}. \tag{7.69}$$

If the spatial domain is more naturally described by other coordinate systems, then the representation of the Laplacian changes accordingly. For example, if the domain is a long cylindrical tube, and the concentration is not expected to be uniform in tubular cross-sections, then cylindrical coordinates $(r, \theta, z)$, where $x = r \cos \theta, y = r \sin \theta$, are most appropriate. In these coordinates

$$\nabla^2 c = \frac{1}{r} \frac{\partial}{\partial r} \left( r \frac{\partial c}{\partial r} \right) + \frac{1}{r^2} \frac{\partial^2 c}{\partial \theta^2} + \frac{\partial c^2}{\partial z^2}. \tag{7.70}$$

If the domain is a sphere, then spherical coordinates $(r, \theta, \phi)$, where $x = r \sin \phi \cos \theta, y = r \sin \phi \sin \theta, z = r \cos \phi$, are most appropriate, in which case the Laplacian operator is

$$\nabla^2 c = \frac{1}{r^2} \frac{\partial}{\partial r} \left( r^2 \frac{\partial c}{\partial r} \right) + \frac{1}{r^2 \sin \theta} \frac{\partial}{\partial \theta} \left( \sin \theta \frac{\partial c}{\partial \theta} \right) + \frac{1}{r^2 \sin^2 \theta} \frac{\partial^2 c}{\partial \phi^2}. \tag{7.71}$$

An important reason for using other coordinate systems is that there may be symmetries that allow the problem to be reduced. For example, suppose that a spherical cell of radius $R$ is suddenly immersed into a large bath containing a high concentration of glucose, and that the glucose can move across the membrane and then diffuse throughout the cell. If the concentration of glucose in the cell is initially uniform ($c = c_0$), then the solution should be independent of $\phi$ and $\theta$ for all time. This implies that

$$\frac{\partial c}{\partial \theta} = \frac{\partial c}{\partial \phi} = \frac{\partial^2 c}{\partial \phi^2} = 0. \tag{7.72}$$

Thus, a reasonable model for this problem is

$$\frac{\partial c}{\partial t} = \frac{D}{r^2} \frac{\partial}{\partial r} \left( r^2 \frac{\partial c}{\partial r} \right), \tag{7.73}$$

$$c(r, 0) = c_0, \tag{7.74}$$

$$D \frac{dc}{dr} = j \text{ at } r = R, \tag{7.75}$$

where $c_0$ is the initial cytosolic glucose concentration, and $j$ is the rate of entry of glucose through the plasma membrane.

## 7.5  Traveling Waves in Nonlinear Reaction–Diffusion Equations

Consider a reaction–diffusion equation with a nonlinear source term:

$$\frac{\partial c}{\partial t} = D \frac{\partial^2 c}{\partial x^2} + f(c), \tag{7.76}$$

where $f(c)$ is the cubic polynomial $f(c) = Ac(1 - c)(c - \alpha)$, with $0 < \alpha < \frac{1}{2}$. While real chemical reactions are not modeled exactly by a cubic polynomial, the reaction term has features that resemble those of several more realistic reactions, and so is worthy of our attention. This equation can be (and has been) used to understand features of action potential propagation in nerve axons, calcium fertilization waves in frog eggs, and cyclic AMP waves in slime molds.

A key feature of this reaction term is that it has three zeros ($0$, $\alpha$, and $1$), two of which ($0$ and $1$) are stable. Linear stability is determined by the sign of $f'(c)$ at the rest point, and if $f'(c_0) < 0$, the rest point $c_0$ is stable. In this problem, however, there is a stronger type of stability in that the solution of the ordinary differential equation $\frac{dc}{dt} = f(c)$ approaches either $c = 0$ or $c = 1$ starting from any initial position except $c = \alpha$.

The function $f(c)$ can be thought of as a switch. If $c$ is somehow pushed slightly away from $0$, it returns quickly to $0$. However, if $c$ is pushed away from $0$ and exceeds $\alpha$, then it goes to $1$. Thus, the level $\alpha$ is a threshold for $c$. Because it has two stable rest points, equation (7.76) is often called the bistable equation.

## 7.5.1 Traveling Wave Solutions

An interesting and important problem is to determine the behavior of the bistable equation when a portion of the region is initially above the threshold $\alpha$ and the remainder is initially at zero. To get some idea of what to expect it is useful to perform a numerical simulation. For this numerical simulation we use the method of lines to solve the differential equations

$$\frac{dc_0}{dt} = \frac{D}{\Delta x^2}(c_1(t) - c_0(t)) + f(c_0), \tag{7.77}$$

$$\frac{dc_n}{dt} = \frac{D}{\Delta x^2}(c_{n+1}(t) - 2c_n(t) + c_{n-1}(t)) + f(c_n), \qquad n = 1, 2, \ldots, N-1, \tag{7.78}$$

$$\frac{dc_N}{dt} = \frac{D}{\Delta x^2}(c_{N-1}(t) - c_N(t)) + f(c_N). \tag{7.79}$$

The simulation shows that the variable $c$ quickly changes into a profile that is a transition between $c = 0$ on the bottom and $c = 1$ on the top (Figure 7.8). After this transitional profile is formed, it moves without change of shape from top to bottom at (what appears to be) a constant velocity.

This numerical solution suggests that we should try to find a translationally invariant solution. A translationally invariant solution is one that does not change its value along any straight line $x + st = x_0$, for an appropriately chosen value of $s$, the wave speed. Thus, we look for special solutions of the bistable equation of the form

$$c(x,t) = U(x + st), \tag{7.80}$$

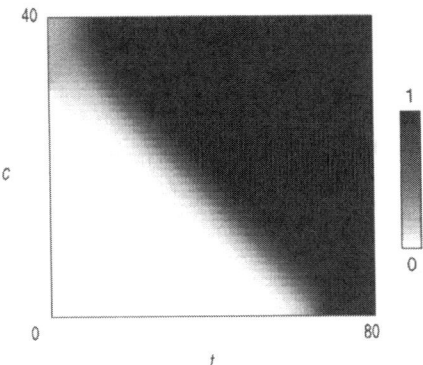

40

$c$

0                                         80

$t$

1

0

**Figure 7.8** Numerically computed solution of the bistable equation, with $A = 1$, $\alpha=0.1$ and $D = 1$.

with the additional property that $\lim_{\xi \to -\infty} U(\xi) = 0$, $\lim_{\xi \to \infty} U(\xi) = 1$. Notice that since

$$\frac{\partial c(x,t)}{\partial x} = \frac{d}{d\xi}U(\xi)\frac{\partial \xi}{\partial x} = \frac{d}{d\xi}U(\xi) \text{ and } \frac{\partial c(x,t)}{\partial t} = \frac{d}{d\xi}U(\xi)\frac{\partial \xi}{\partial t} = s\frac{d}{d\xi}U(\xi), \quad (7.81)$$

in this translating coordinate system, the bistable equation becomes the ordinary differential equation

$$s\frac{dU}{d\xi} = D\frac{d^2U}{d\xi^2} + f(U). \quad (7.82)$$

There are two ways to try to solve (7.82). An exact solution can be found in the special case that $f$ is a cubic polynomial. There are several other examples of functions $f$ for which exact solutions can be found, but this method does not work in most cases. A more general method is to examine (7.82) in the phase plane, which we will do below.

The exact solution can be found for the cubic polynomial $f$ as follows. Since we want $\lim_{\xi \to -\infty} U(\xi) = 0$, $\lim_{\xi \to \infty} U(\xi) = 1$, we guess a relationship between $dU/d\xi$ and $U$ of the form

$$\frac{dU}{d\xi} = aU(1 - U), \quad (7.83)$$

for some positive number $a$. It follows that

$$\frac{d^2U}{d\xi^2} = a(1 - 2U)\frac{dU}{d\xi}. \quad (7.84)$$

Substituting this into (7.82) and factoring out $U(1 - U)$, we find that

$$as = a^2D(1 - 2U) + A(U - \alpha). \quad (7.85)$$

This identity holds for all $U$ only if

$$a^2 = \frac{A}{2D}, \quad s = \sqrt{AD/2} \cdot (1 - 2\alpha). \quad (7.86)$$

The solution is found by quadrature from (7.83) to be

$$U(\xi) = \frac{1}{2} + \frac{1}{2}\tanh\left(\frac{1}{2}\sqrt{\frac{A}{2D}}\xi\right). \tag{7.87}$$

The analysis used for finding traveling waves using phase portraits works for any bistable function $f$. We begin by writing the traveling wave (7.82) as the two-dimensional system

$$\frac{dU}{d\xi} = W, \tag{7.88}$$

$$\frac{dW}{d\xi} = sW - f(U). \tag{7.89}$$

This system has three critical points, at $(U, W) = (0,0), (\alpha, 0)$, and $(1,0)$. The linearized stability of these critical points is determined by the roots of the characteristic equation

$$\lambda^2 - s\lambda + f'(U_0) = 0, \tag{7.90}$$

where $U_0$ is any one of the three steady rest values of $U$.

If $f'(U_0)$ is negative, then the critical point $(U_0, 0)$ is a saddle point. To find a traveling wave solution, we seek a trajectory that leaves the saddle point at $(U, W) = (0,0)$ and ends up at the saddle point at $(U, W) = (1,0)$. We can implement this (almost) numerically. If we start with an initial point close to the origin along the straight line $W = \lambda U$ in the positive quadrant, with $\lambda$ the positive root of the characteristic equation $\lambda^2 - s\lambda + f'(0) = 0$, and integrate for a while, one of two things will occur. If $s$ is relatively small, the trajectory will cross the $U$–axis before reaching $U = 1$, while if $s$ is relatively large, the trajectory will increase beyond $U = 1$ and become quite large. By adjusting the parameter $s$ one can find trajectories that barely miss hitting the point $(U, W) = (1,0)$ by crossing the $U$–axis or by exceeding $U = 1$ and becoming large (Figure 7.9). A trajectory that comes close to the saddle point at $(U, W) = (1,0)$ is a numerical approximation to the traveling wave solution, and the value of $s$ for which this nearly connecting trajectory is attained is a good approximation for the wave speed.

## 7.5.2 Traveling Wave in the FitzHugh–Nagumo Equations

As we have seen in earlier chapters, chemical reaction schemes in cell biology can be quite complicated, involving many species. Furthermore, some species may be free to move, while others are not. Models of nerve axons, for example, include both diffusing species (transmembrane potential) and nondiffusing variables (the ion-gating variables, because ion channels are embedded in the membrane and do not move on the millisecond time scale of an action potential). Similarly, the $Ca^{2+}$ wave induced by fertilization of a frog egg involves both cytosolic calcium, which is a diffusing variable, and ER calcium which, (to a first approximation) is not.

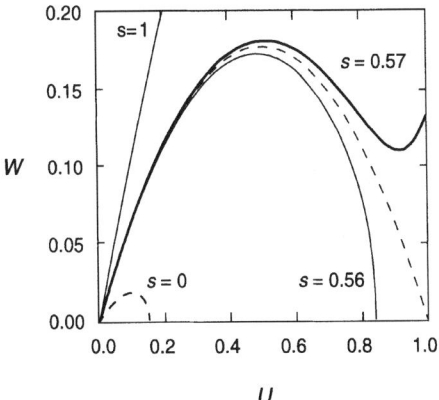

**Figure 7.9** Phase plane portrait of possible traveling wave trajectories.

Perhaps the best known example of a reaction–diffusion system is the Hodgkin–Huxley equations discussed in Section 2.5, which describe action potential propagation in a nerve axon. In this model there are four dependent variables: transmembrane potential $V$ and three gating variables, $m$, $n$, and $h$. The equation for transmembrane potential $V(x,t)$ is the cable equation

$$C_m \frac{\partial V}{\partial t} = \frac{R}{2\rho} \frac{\partial^2 V}{\partial x^2} + I_{\text{ion}}(V, m, n, h), \tag{7.91}$$

where $C_m$ is the membrane capacitance, $R$ is the axonal radius, $\rho$ is the axoplasmic resistivity, and $I_{\text{ion}}$ is the current carried into the axon by ions crossing through voltage-sensitive channels, and $\frac{R}{2\rho} \frac{\partial^2 V}{\partial x^2}$ is the net current along the axon carried by ions in response to spatial gradients of intracellular potential. For the giant axon of the squid, Hodgkin and Huxley report that $R = 240\ \mu\text{m}$, $\rho = 0.35\ \Omega\cdot\text{m}$, and $C_m = 0.01\ \text{F/m}^2$.

In the Hodgkin–Huxley equations given in Section 2.5, $I_{\text{ion}}$ is a complicated function of transmembrane potential and the gating variables. To simplify the function, FitzHugh lumped the three gating variables into one (called $w$). The resulting equations in spatial form are

$$\frac{\partial V}{\partial t} = D \frac{\partial^2 V}{\partial x^2} + \frac{B}{V_1 V_2} V(V - V_1)(V_2 - V) - C\sqrt{V_1 V_2}\, w, \tag{7.92}$$

$$\frac{\partial w}{\partial t} = \frac{\epsilon}{\sqrt{V_1 V_2}} (V - V_3 w), \tag{7.93}$$

where $D = R/(2\rho C_m) \approx 0.03\ \text{m}^2/\text{s}$ (for squid giant axon), $V_1, V_2$, and $V_3$ are positive "voltage" constants, and $B$, $C$, and $\epsilon$ are rate constants with units 1/s. It is also assumed that $\epsilon \ll B, C$.

By defining $v = V/\sqrt{V_1 V_2}$, we transform (7.92) and (7.93) into

$$\frac{\partial v}{\partial t} = D \frac{\partial^2 v}{\partial x^2} + Bv(v - \beta)(\delta - v) - Cw, \tag{7.94}$$

$$\frac{\partial w}{\partial t} = \epsilon(v - \gamma w), \qquad (7.95)$$

where $\beta = V_1/\sqrt{V_1 V_2}$, $\delta = V_2/\sqrt{V_1 V_2}$, and $\gamma = V_3/\sqrt{V_1 V_2}$. Because $\epsilon \ll B, C$, we can use reduction of scale arguments (see Chapter 4) to justify the assumption that $w(x,t) = w_0 = $ constant. In this case, (7.94) and (7.95) reduce to a single reaction–diffusion equation:

$$\frac{\partial v}{\partial t} = D\frac{\partial^2 v}{\partial x^2} + Bv(v - \beta)(\delta - v) - Cw_0. \qquad (7.96)$$

Let us assume that the "reaction" part of (7.96), $G(v) = Bv(v - \beta)(\delta - v) - Cw_0$, has three real steady states

$$G(v_i) \equiv Bv_i(v_i - \beta)(\delta - v_i) - Cw_0 = 0, \text{ for } i = 1, 2, 3; \ v_1 < v_2 < v_3.$$

By defining $c = (v - v_1)/(v_3 - v_1)$, (7.96) becomes identical to (7.76), with $A$ now some nonlinear function of $B$, $\beta$, $\delta$, and $Cw_0$. Hence, from the results leading to (7.86), we know that for appropriate choices of $\beta$, $\delta$, and $Cw_0$, (7.96) supports traveling wave solutions of velocity

$$s = \sqrt{AD/2} \cdot (1 - 2\alpha).$$

To estimate the velocity of propagation of an action potential wave front, we must have, in addition to $D \approx 0.03$ m$^2$/s, estimates of the rate constant $A$ and the threshold $\alpha$ in the $f(c)$ term of (7.76). Given that the amplitude of an action potential is $\approx 100$ mV, and that the threshold for initiation is $\approx 20$ mV (from rest), we set $\alpha \approx 0.2$. During the rise of an action potential, $V$ increases with a doubling time of a fraction of a millisecond (say, 0.2 ms). Trajectories of the reaction equation $dc/dt = f(c)$ depart from the unstable steady state according to

$$c(t) - \alpha = (c_0 - \alpha)\exp(A\alpha(1 - \alpha)t). \qquad (7.97)$$

To verify this, set $c = y + \alpha$ and linearize $dc/dt = f(c)$ to get $dy/dt = f'(\alpha)y$ and then solve to obtain (7.97). The doubling time for departure from the unstable steady state is

$$\frac{\ln 2}{A\alpha(1 - \alpha)} \approx 0.2 \text{ ms}.$$

or $A \approx 2 \cdot 10^4$/s. Hence, if the upstroke of the action potential can be approximated by the FitzHugh–Nagumo equations, it should propagate at velocity

$$s \approx \sqrt{\frac{(2 \cdot 10^4/\text{s})(3 \cdot 10^2 \text{m}^2/\text{s})}{2}} \cdot (1 - 0.4),$$

or approximately 10 m/s. This compares favorably with the observed velocity of 20 m/s.

# Suggestions for Further Reading

- *Random Walks in Biology*, Howard Berg. This is a lovely introductory book on diffusion processes in biology (Berg 1993).
- *Mathematical Problems in the Biological Sciences*, S. Rubinow. Chapter 5 gives a nice introduction to diffusion processes (Rubinow 1973).
- *Diffusional mobility of golgi proteins in membranes of living cells*, N.B. Cole, C.L. Smith, N. Sciaky, M. Terasaki, and M. Edidin. This paper gives an example of how diffusion coefficients are measured in a specific biological context (Cole et al. 1996).
- *Complex patterns in a simple system*, John Pearson. Reaction diffusion equations are used to model many interesting phenomena. A sampler of the kinds of patterns that are seen in reaction diffusion systems is given in this paper (Pearson 1993).
- *The theoretical foundation of dendritic function*, Idan Segev, John Rinzel, and Gordon Shepard. This book contains the collected papers of Wilfrid Rall, a pioneer in the application of cable theory and compartment modeling to neuronal dendrites (Segev et al. 1995).
- *Mathematical Physiology*, James Keener and James Sneyd. Several of the topics presented in this chapter are covered here in more depth (Keener and Sneyd 1998).

## 7.6 EXERCISES

1. A rule of thumb (derived by Einstein) is that the diffusion coefficient for a globular molecule satisfies $D \approx M^{-1/3}$ where $M$ is the molecular weight. Determine how well this relationship holds for the substances listed in Table 7.1 by plotting $D$ and $M$ on a log-log plot.

2. (a) Verify that the solution of (7.36)–(7.37) is given by (7.38). Verify (7.41).

   (b) Show that the total amount of C, given by $\int_{-\infty}^{\infty} c(x,t)\,dx$, is constant for all time. What is the constant?

3. Verify that (7.43) satisfies the diffusion equation with boundary data $c(0,t) = C_0$ and initial data $c(x,0) = 0$.

4. Using the data given in the text and equation (7.45), estimate the diffusion coefficient for bacteria.

5. Numerically simulate the differential equations (7.53), (7.55), and (7.56) with initial data corresponding to $c(x,0) = 1$ for $20 < x < 30$ and $c(x,0) = 0$ elsewhere, using $N = 40$ discrete intervals and $D = 2.25 \times 10^{-6}$ cm$^2$/s as a typical diffusion coefficient for calcium. What is the final steady–state distribution of calcium and what is the approximate time constant of decay to this steady solution?

6. Numerically simulate a voltage–clamp experiment on a spatial domain that is 4 space constants long with $V(0,t) = 1$, and $V(x,0) = 0$, using constants appropriate for barnacle fiber and squid giant axon. What are the observable differences between these two simulations?

7. (a) Show that the function

$$c(r,t) = \frac{1}{4\pi Dt} e^{\left(-\frac{r^2}{4Dt}\right)}$$

satisfies the diffusion equation in two spatial dimensions,

$$c_t = \frac{D}{r} \frac{\partial}{\partial r} \left( r \frac{\partial c}{\partial r} \right),$$

where $r^2 = x^2 + y^2$.

(b)  Show that the total amount of C, given by $2\pi \int_0^\infty c(r, t) r \, dr$, is constant for all time. What is the constant?

(c)  Evaluate $r_{rms}$ where $r_{rms}^2 = 2\pi \int_0^\infty r^2 c(r, t) r \, dr$.

(d)  When is the maximal value of $c(r, t)$ achieved and what is the maximal value?

8.  (a)  Show that the function

$$c(r, t) = \frac{1}{(4\pi Dt)^{3/2}} e^{\left( -\frac{r^2}{4Dt} \right)}$$

satisfies the diffusion equation in three spatial dimensions,

$$c_t = \frac{D}{r^2} \frac{\partial}{\partial r} \left( r^2 \frac{\partial c}{\partial r} \right),$$

where $r^2 = x^2 + y^2 + z^2$.

(b)  Show that the total amount of C, given by $4\pi \int_0^\infty c(r, t) r^2 \, dr$, is constant for all time. What is the constant?

(c)  Evaluate $r_{rms}$ where $r_{rms}^2 = 4\pi \int_0^\infty r^2 c(r, t) r^2 \, dr$.

(d)  When is the maximal value of $c(r, t)$ achieved and what is the maximal value?

9.  A quantitative estimate of the way proteins diffuse on membranes is provided by *fluorescence recovery after photobleaching* (FRAP) studies, wherein cells are treated with a fluorescent reagent that binds to a specific surface protein, which is uniformly distributed on the surface. A laser light is then focused onto a small area of the surface, irreversibly bleaching the bound reagent and thus reducing the fluorescence in the illuminated area. In time, the fluorescence of the bleached area increases because the unbleached fluorescent surface molecules diffuse into the bleached area while the bleached molecules diffuse out. Model and simulate this experiment in two ways:

(a)  Make a one–dimensional model for a domain 10 microns long, with no-flux boundary conditions at both ends. Assume that the first micron is initially bleached and the remaining space is initially unbleached. Assume that the diffusion coefficient of the molecules is $10^{-7}$ cm$^2$/s. Determine the spatial profile as a function of time, and the final uniform distribution of unbleached protein.

(b)  Make a two–dimensional model for a perfectly circular domain of radius 10 microns with a one–micron circular region at the center that is initially bleached. Assume that $\partial c / \partial r = 0$ at both $r = 0$ and $r = 10$ $\mu$m. Use the discretization of the diffusion operator given by

$$\frac{1}{r} \frac{\partial}{\partial r} \left( r \frac{\partial c}{\partial r} \right) \approx \frac{1}{2r_n} \left( (r_{n+1} + r_n)(c_{n+1} - c_n) - (r_{n-1} + r_n)(c_n - c_{n-1}) \right).$$

Determine the spatial profile as a function of time and the final uniform distribution. What differences are there between the two-dimensional and the one-dimensional models?

10. Simulate the bistable equation starting from initial data having $V(x, 0) > \alpha$ for a small region on the left end of the domain, and $V(x, 0) = 0$ elsewhere. What is the speed of the traveling wave that forms?

11. Numerically simulate an experiment on an idealized nerve axon that is stimulated at one end with a time–dependent current input. The equations are

$$\frac{\partial \phi}{\partial t} = \frac{\partial^2 \phi}{\partial x^2} + f(\phi) - w, \qquad f(\phi) = \phi(\phi - 1)(0.1 - \phi),$$

$$\frac{\partial w}{\partial t} = 0.01(\phi - 0.5w),$$

subject to boundary conditions $\partial \phi(0, t)/\partial x = I(t)$, $\partial \phi(10, t)/\partial x = 0$. Pick $I(t)$ to be a square pulse. Vary the height and length of the pulse in order to initiate a traveling wave. Describe the response when the amplitude and/or duration of the stimulating pulse is too small to initiate a traveling wave.

12. Using the method described in Section 7.5.1, compute a traveling wavefront solution to the Morris–Lecar equations described in Section 2.4:

$$\frac{\partial V}{\partial t} = D\frac{\partial^2 V}{\partial x^2} - g_{Ca}m_\infty(V)(V - V_{Ca}) - g_k w(V - V_K) - g_L(V - V_L) + I_{app}$$

where $m_\infty(V) = 0.5[1 + \tanh((V - v_1)/v_2)]$. Use $D = 300$ cm$^2$/s, $I_{app} = 60$ pA, and all other parameter values as in Table 2.4. In this approximation, assume that $w(x, t) = w_0 =$ constant; try $w_0 = 0.1$. Plot $V(x, t)$ as in Figure 7.8, and estimate the speed of propagation of the wavefront in cm/s.

# CHAPTER 8

# Modeling Intracellular Calcium Waves and Sparks

Gregory D. Smith, John E. Pearson, and Joel E. Keizer

In this chapter we shall discuss a variety of intracellular $Ca^{2+}$ wave phenomena, but always from the perspective that the distance scales of interest are large enough that $Ca^{2+}$ transport is well-modeled by conservation equations based on a continuum description of matter (recall Chapter 7). Although recent experimental and theoretical work suggests that the macroscopic behavior of propagating $Ca^{2+}$ waves (e.g., wave speed) may depend in subtle ways on the density and distribution of intracellular $Ca^{2+}$ release channels, we postpone consideration of intracellular heterogeneities such as clusters of $Ca^{2+}$ release channels until later in the chapter. This makes sense because both the mathematics and simulation methods used to study nonlinear wave propagation in *homogeneous* media are simpler than the *heterogeneous* case. This simplicity should facilitate the development of intuition regarding nonlinear wave propagation. Throughout the chapter a recurring theme will be the manner in which $Ca^{2+}$ buffers, through their important association with free $Ca^{2+}$, can influence wave phenomena dependent on diffusion. The chapter concludes with calculations of localized $Ca^{2+}$ elevations due to intracellular $Ca^{2+}$ release, i.e., $Ca^{2+}$ "puffs" or "sparks," elementary events that sum to produce $Ca^{2+}$ waves.

## 8.1 Microfluorometric Measurements

Chapter 5 described several examples of global or cell-wide $Ca^{2+}$ excitability and oscillations. Experimental observations of such intracellular $Ca^{2+}$ dynamics are often made using microfluorimetry, an experimental technique that involves loading $Ca^{2+}$

indicator dyes into cells and instrumentation that optically excites these indicators and measures emission. These $Ca^{2+}$ indicator dyes are themselves $Ca^{2+}$ buffers (often highly mobile) that can potentially affect intracellular $Ca^{2+}$ signaling. For example, the differential fraction of free to bound cytosolic $Ca^{2+}$, denoted by $f_i$ $([Ca^{2+}]_i)$ in (5.13), will be determined by both exogenous as well as endogenous $Ca^{2+}$ buffers.

Although a measured fluorescence signal is only indirectly related to the dynamics of intracellular $Ca^{2+}$, it is relatively straightforward to determine the free $Ca^{2+}$ concentration during a cell-wide $Ca^{2+}$ response using the time course of measured fluorescence (Grynkiewicz et al. 1985). If the $Ca^{2+}$ and indicator dye concentrations are homogeneous throughout the cell, the equilibrium relation

$$[CaB] = \frac{[Ca^{2+}]_i[B]_T}{K + [Ca^{2+}]_i} \tag{8.1}$$

is valid as long as $[Ca^{2+}]_i$ changes slowly compared to the equilibration time of the buffers. Because this equilibration time is on the order of milliseconds, this condition is usually satisfied for global $Ca^{2+}$ responses, which occur with a time scale of seconds or tens of seconds.

If the equilibrium relation (8.1) accurately describes the relationship between the concentration of bound indicator dye ($[CaB]$) and free $Ca^{2+}$ ($[Ca^{2+}]_i$), then it is a simple matter to "backcalculate" the free $Ca^{2+}$ concentration as a function of time. For a single excitation wavelength measurement (e.g., using a nonratiometric dye such as fluo-3 at low concentration), we can idealize the indicator fluorescence as the sum of two components,

$$F = \eta_B[B] + \eta_{CaB}[CaB], \tag{8.2}$$

where $\eta_B$ and $\eta_{CaB}$ are proportionality constants for free and bound dye, respectively. When $\eta_B < \eta_{CaB}$, the maximum and minimum observable fluorescences are given by $F_{min} = \lim_{[Ca^{2+}]_i \to 0} F = \eta_B[B]_T$ and $F_{max} = \lim_{[Ca^{2+}]_i \to \infty} F = \eta_{CaB}[B]_T$. Using the equilibrium relation it can be shown (see Exercise 1) that

$$[Ca^{2+}]_i = K\frac{[CaB]}{[B]} = K\frac{F - F_{min}}{F_{max} - F}. \tag{8.3}$$

If only the $Ca^{2+}$-bound indicator fluoresces strongly, then $F \approx \eta_{CaB}[CaB]$, and a slightly simpler expression results from substituting $F_{min} = 0$ in (8.3).

It should be noted that the validity of (8.3) relies on the stability of instrument sensitivity, optical path length, and dye concentration between measurements of $F$, $F_{min}$, and $F_{max}$. Because determining $F_{min}$ and $F_{max}$ usually involves titrating the indicator released from lysed cells, this is difficult to achieve in practice (Grynkiewicz et al. 1985).

In whole-cell $Ca^{2+}$ measurements, fluorescence intensities can be measured at two excitation wavelengths ($\lambda$ and $\lambda'$) using indicator dyes such as fura-2. Such *ratiometric* measurements can be related to the underlying free $Ca^{2+}$ signal by supplementing (8.2) with

$$F' = \eta'_B[B] + \eta'_{CaB}[CaB], \tag{8.4}$$

where the primes indicate the second excitation wavelength. Using the first equality of (8.3), the fluorescence ratio $R = F/F'$ can be inverted to give (see Exercise 2)

$$[Ca^{2+}]_i = K \frac{\eta_B - \eta'_B R}{\eta'_{CaB} R - \eta_{CaB}} = K \left( \frac{R - R_{min}}{R_{max} - R} \right) \left( \frac{\eta'_B}{\eta'_{CaB}} \right), \qquad (8.5)$$

where for the second equality we use $R_{min} = \lim_{[Ca^{2+}]_i \to 0} F/F' = \eta_B/\eta'_B$ and $R_{max} = \lim_{[Ca^{2+}]_i \to \infty} F/F' = \eta_{CaB}/\eta'_{CaB}$. If $\lambda'$ is chosen to be a wavelength at which the calibration spectra at different $Ca^{2+}$ concentrations cross one another, then $\eta'_B \approx \eta'_{CaB}$, and the last factor in (8.5) is eliminated. An advantage of the ratiometric method is its insensitivity to changes in dye concentration and instrument sensitivity between measurements.

## 8.2   A Model of the Fertilization Calcium Wave

When mature *Xenopus laevis* oocytes (eggs) are loaded with an indicator dye (e.g., $Ca^{2+}$-green dextran) and stimulated by the fusion of sperm, a propagating wave of intracellular $Ca^{2+}$ release can be observed by backcalculating the free $Ca^{2+}$ concentration ($[Ca^{2+}]_i(x,t)$) from a time-dependent fluorescence signal ($F(x,t)$) according to (8.3) or (8.5). This *fertilization* $Ca^{2+}$ *wave* is an important step in early development. It triggers the fusion of cortical granules (vesicles) with the plasma membrane, a process that initiates the raising of the viteline envelope and a long-lasting block to polyspermy. The cell divisions that initiate development of *Xenopus* begin only after the fertilization $Ca^{2+}$ wave has propagated throughout the entire cell. The eggs of many species, from starfish to mammals, exhibit propagating $Ca^{2+}$ waves upon fertilization.

Because fertilization $Ca^{2+}$ waves such as those shown in Figure 8.1 still occur when the extracellular medium is nominally $Ca^{2+}$ free, the phenomenon appears to be largely independent of $Ca^{2+}$ influx. On the other hand, the fertilization $Ca^{2+}$ wave absolutely requires functional IP₃ receptors. When IP₃-mediated $Ca^{2+}$ release is blocked by any one of several experimental manipulations (e.g., upon introduction of heparin into the cytosol), the fertilization $Ca^{2+}$ wave is not observed (Nuccitelli et al. 1993).

We expect $Ca^{2+}$ diffusion to play an important role in the fertilization $Ca^{2+}$ wave because *Xenopus laevis* eggs are large (approximately 1.2 mm in diameter). Indeed, Figure 8.1 indicates that the $Ca^{2+}$ concentration within the egg depends very much on both spatial position and time. For this reason a whole cell model of this phenomenon would be deficient, and instead researchers mathematically describe the fertilization $Ca^{2+}$ wave in the *Xenopus* egg and other cell types using a combination of spatial modeling and whole–cell modeling approaches (Atri et al. 1993; Girard et al. 1992; Dupont and Goldbeter 1994; Jafri and Keizer 1994; Wagner et al. 1998).

Here we introduce such a spatial whole–cell model of the fertilization $Ca^{2+}$ wave by recalling that the Li–Rinzel reduction of the DeYoung–Keizer model of the IP₃R can be combined with sigmoidal SERCA pump kinetics and a passive ER leak to create a whole–cell model of $Ca^{2+}$ handling in pituitary gonadotrophs (Section 5.3). Using a variation of this model to represent $Ca^{2+}$-induced $Ca^{2+}$ release (CICR) and reuptake by

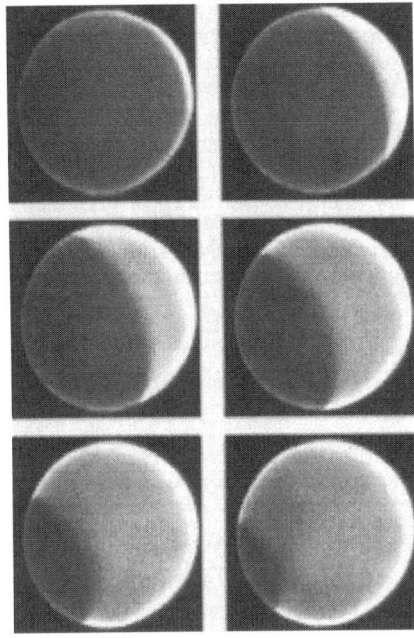

**Figure 8.1** Microfluorimetric images of sperm-induced fertilization calcium waves in mature *Xenopus laevis* eggs loaded with the indicator dye $Ca^{2+}$-green dextran. The wave takes approximately 5 minutes to cross the length of the egg (1.2 mm). The wave begins in the top left frame and is almost complete in the bottom right frame. Image provided by Richard Nuccitelli, University of California at Davis.

the endoplasmic reticulum (ER), we follow Section 7.5.2 and account for $Ca^{2+}$ diffusion in both the cytosol and ER by writing the following reaction–diffusion system:

$$\frac{\partial [Ca^{2+}]_i}{\partial t} = f_i \left[ D_i \nabla^2 [Ca^{2+}]_i + j_{IP_3R} + j_{LEAK} - j_{SERCA} \right], \tag{8.6}$$

$$\frac{\partial w}{\partial t} = \left[ w_\infty \left( [Ca^{2+}]_i, [IP_3] \right) - w \right] / \tau, \tag{8.7}$$

$$\frac{\partial [Ca^{2+}]_{ER}}{\partial t} = f_{ER} \left[ D_{ER} \nabla^2 [Ca^{2+}]_{ER} - (\bar{V}_i / \bar{V}_{ER}) (j_{IP_3R} - j_{LEAK} + j_{SERCA}) \right], \tag{8.8}$$

where $j_{LEAK} = v_{LEAK} ([Ca^{2+}]_{ER} - [Ca^{2+}]_i)$, $j_{IP_3R} = v_{IP_3R} m_\infty^3 w^3 ([Ca^{2+}]_{ER} - [Ca^{2+}]_i)$, $j_{SERCA} = v_{SERCA} [Ca^{2+}]_i^2 / ([Ca^{2+}]_i^2 + K_{SERCA}^2)$, and $\bar{V}$'s are the volumes as before. In these equations, $w$ is the fraction of IP$_3$Rs *not* inactivated and the open probability of the IP$_3$Rs is given by $P_O = m_\infty^3 w^3$ where the fraction of activated IP$_3$Rs $(m_\infty^3)$ is assumed to be an instantaneous function of $Ca^{2+}$ and IP$_3$ concentrations,

$$m_\infty^3 = \left( \frac{[IP_3]}{[IP_3] + d_1} \right)^3 \left( \frac{[Ca^{2+}]_i}{[Ca^{2+}]_i + d_5} \right)^3. \tag{8.9}$$

Although the Li–Rinzel reduction of the DeYoung–Kiezer model gives a time constant of IP$_3$R inactivation that is dependent on $Ca^{2+}$ and IP$_3$ concentration, $\tau ([Ca^{2+}]_i, [IP_3])$, for simplicity we will assume that $\tau$ is constant (2 sec). A diagram of the fertilization $Ca^{2+}$ wave model components and fluxes is presented in Figure 8.2.

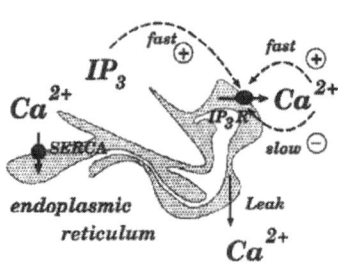

**Figure 8.2** Schematic diagram of the *Xenupus* egg fertilization Ca²⁺ wave model. Ca²⁺ enters the cytosol from the ER via a passive leak and the IP₃R, which is activated by both Ca²⁺ and IP₃ on a fast time scale and inhibited by Ca²⁺ on a slow time scale, all at the cytoplasmic face. The ER is refilled by a SERCA-type Ca²⁺-ATPase pump. Diffusion of Ca²⁺ in both the cytosol and ER is accounted for using effective diffusion coefficients (constant) that account for interactions with Ca²⁺ buffers (not shown) and the volume fractions of both compartments. Reprinted from Jafri and Keizer (1994).

Note that in (8.6)–(8.8) the maximum conductance of IP₃-mediated Ca²⁺ release ($v_{IP_3R}$), passive leak rate ($v_{LEAK}$), and maximum rate of ATP-dependent reuptake ($v_{SERCA}$) are constants. Thus, the model assumes homogeneous Ca²⁺ release and reuptake dynamics, that is, a uniform and high density of inositol 1,4,5-trisphosphate (IP₃) receptors and sarco-endoplasmic reticulum Ca²⁺-ATPases. Also note that (8.7) includes no diffusion term for the simple reason that IP₃Rs (and thus the gating variable representing their inactivation) has a fixed spatial location and does not diffuse. Throughout this chapter we will assume that [IP₃] is uniform and constant (but see Wagner et al. (1998)).

## 8.3   Including Calcium Buffers in Spatial Models

In this spatial whole–cell model given by (8.6)–(8.8), the quantities $f_i$ and $f_{ER}$ multiply both reaction and diffusion terms. As discussed in Chapter 5, these factors scale the Ca²⁺ release and reuptake rates to account for the proportion of Ca²⁺ bound to buffer in both cytosolic and ER compartments.

Note that the coefficients of the Laplacians in (8.6) and (8.8) imply constant *effective diffusion coefficients* for the cytosol and ER given by $D_i^{eff} = f_i D_i$ and $D_{ER}^{eff} = f_{ER} D_{ER}$, both of which are expected to be much smaller than the free Ca²⁺ diffusion coefficient. This assumption is here made for convenience and is strictly true only for low–affinity Ca²⁺ buffers.

To see that this is the case, consider how one Ca²⁺ buffer (perhaps representing a Ca²⁺ indicator dye) would be added to the fertilization Ca²⁺ wave model. The association and dissociation of Ca²⁺ with the indicator dye would contribute bimolecular reaction terms in the cytosolic compartment, giving

$$\frac{\partial [Ca^{2+}]_i}{\partial t} = D_C \nabla^2 [Ca^{2+}]_i + R + j_T, \tag{8.10}$$

$$\frac{\partial [B]}{\partial t} = D_B \nabla^2 [B] + R, \tag{8.11}$$

$$\frac{\partial[\text{CaB}]}{\partial t} = D_{\text{CB}}\nabla^2[\text{CaB}] - R, \tag{8.12}$$

where $D_C$, $D_B$, and $D_{CB}$ are diffusion constants for free $Ca^{2+}$, free buffer, and bound buffer, respectively,

$$R = -k^+[\text{B}][\text{Ca}^{2+}]_i + k^-[\text{CaB}],$$

and $j_T = j_{\text{IP}_3\text{R}} + j_{\text{LEAK}} - j_{\text{SERCA}}$ is the sum of all $Ca^{2+}$ fluxes into and out of the ER.

Notice that if the diffusion of buffer doesn't depend on whether or not $Ca^{2+}$ is bound ($D_B \approx D_{CB}$), then (8.11) and (8.12) can be summed to give

$$\frac{\partial[\text{B}]_T}{\partial t} = D_B\nabla^2[\text{B}]_T,$$

where $[\text{B}]_T = [\text{B}] + [\text{CaB}]$ is the total buffer concentration profile (free plus bound). This equation implies that if $[\text{B}]_T$ is initially uniform (not a function of position), it will remain uniform for all time. We can thus eliminate (8.11) and obtain a reduced system given by (8.10) and (8.12) with

$$R = -k^+ ([\text{B}]_T - [\text{CaB}])[\text{Ca}^{2+}]_i + k^-[\text{CaB}].$$

## 8.4  The Effective Diffusion Coefficient

If the buffer reactions are rapid with respect to the diffusion, it is possible to simplify our model further using the rapid buffer approximation (recall Chapter 5) (Wagner and Keizer 1994). Assuming local equilibrium (i.e., chemical equilibrium at every point in space), we can write an equilibrium expression for total cell $Ca^{2+}$ concentration:

$$[\text{Ca}^{2+}]_T = [\text{Ca}^{2+}]_i + [\text{CaB}] = [\text{Ca}^{2+}]_i + \frac{[\text{Ca}^{2+}]_i[\text{B}]_T}{[\text{Ca}^{2+}]_i + K}.$$

Next, we eliminate reactions terms (and the fast time scale) by adding (8.10) and (8.12), to give

$$\frac{\partial[\text{Ca}^{2+}]_T}{\partial t} = \frac{\partial[\text{Ca}^{2+}]_i}{\partial t} + \frac{\partial[\text{CaB}]}{\partial t} = D_C\nabla^2[\text{Ca}^{2+}]_i + D_B\nabla^2[\text{CaB}] + j_T. \tag{8.13}$$

In Exercise 3 the reader can confirm that substituting $[\text{Ca}^{2+}]_T$ and $[\text{CaB}]$ in terms of $[\text{Ca}^{2+}]_i$ and taking derivatives gives the so-called rapid buffer approximation,

$$\frac{\partial[\text{Ca}^{2+}]_i}{\partial t} = \beta\left((D_C + D_B\gamma)\nabla^2[\text{Ca}^{2+}]_i - \frac{2\gamma D_B}{K + [\text{Ca}^{2+}]_i}\left(\nabla^2[\text{Ca}^{2+}]_i\right)^2 + j_T\right), \tag{8.14}$$

where $\gamma = K[\text{B}]_T/(K + [\text{Ca}^{2+}]_i)^2$ and $\beta = 1/(1 + \gamma)$. Although this equation may appear ominous, note that using the rapid buffer approximation we have been able to eliminate the two extra equations (for [B] and [CaB]) needed to explicitly account for an indicator dye in the cytosolic compartment.

Another valuable feature of (8.14) is that it provides some insight into the effect of $Ca^{2+}$ buffers on $Ca^{2+}$ transport in cells by allowing us to identify the $Ca^{2+}$-dependent effective diffusion coefficient,

$$D^{\text{eff}} = \beta \left( D_C + \gamma D_B \right).$$

Furthermore, if this buffer has low affinity so that $[Ca^{2+}]_i \ll K$ and $\gamma \approx [B]_T/K$, then as promised at the beginning of this section this diffusion coefficient is approximately constant:

$$D^{\text{eff}} \approx \frac{K}{K + [B]_T} \left( D_C + \frac{[B]_T}{K} D_B \right).$$

For simplicity the remainder of this chapter assumes a rapid and low–affinity buffer.

## 8.5   Simulation of a Fertilization Calcium Wave

Now that we have justified the reduced effective diffusion coefficients used in the fertilization $Ca^{2+}$ wave model, let us further simplify (8.6)–(8.8) by assuming that ER depletion is minimal, i.e., $[Ca^{2+}]_{ER} = c_{ER}$ is approximately constant. We will also assume that the dynamics of $[Ca^{2+}]_i$ are much slower than the gating variable for $Ca^{2+}$ inactivation of the $IP_3R$, so that $w$ is well approximated by $w_\infty([Ca^{2+}]_i)$. With these assumptions, (8.6) and (8.7) reduce to the single-variable reaction–diffusion equation

$$\frac{\partial c}{\partial t} = D\nabla^2 c + f(c), \tag{8.15}$$

where we have written $c = [Ca^{2+}]_i$, $D = f_i D_i$, and

$$f(c) = f_i \left( \left[ v_{\text{LEAK}} + v_{IP_3R}(c) \right] \left[ c_{ER} - c \right] - \frac{v_{\text{SERCA}} c^2}{c^2 + K_{\text{SERCA}}^2} \right), \tag{8.16}$$

and where

$$v_{IP_3R}(c) = \left[ m_\infty(c) w_\infty(c) \right]^3.$$

Using parameters given in Exercise 4, Figure 8.3 shows that $f(c)$ is a cubic function with three zeros. Equilibrium point 1 represents a stable resting state at basal $[Ca^{2+}]_i$ (ER replete), while equilibrium 3 is a stable resting point at high $[Ca^{2+}]_i$ (ER empty). Equilibrium 2 lies between points 1 and 3, is unstable, and corresponds to a threshold for $Ca^{2+}$-induced $Ca^{2+}$ release.

## 8.6   Simulation of a Traveling Front

We follow Section 7.3 to simulate our minimal model of a fertilization $Ca^{2+}$ wave given by (8.15). Assuming a one-dimensional geometry ($0 \leq x \leq L$), we discretize space so

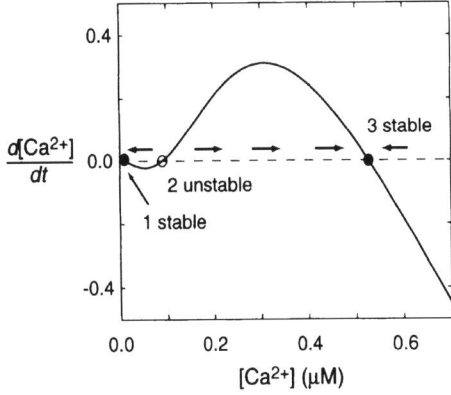

**Figure 8.3** Using parameters as in Exercise 4, the cubic rate function $dc/dt = f(c)$ of the fertilization $Ca^{2+}$ wave model, (8.15), leads to a bistable phase portrait (x-axis). Equilibrium point 1 represents a stable resting state at basal $[Ca^{2+}]_i$ (ER replete) while equilibrium 3 is a stable resting point at high $[Ca^{2+}]_i$ (ER empty). Equilibrium 2 lies between points 1 and 3, is unstable, and corresponds to a threshold for $Ca^{2+}$-induced $Ca^{2+}$ release.

that $c_j(t)$ is an approximation to $c(x_j, t)$, where $x_j = j\Delta x$, $0 \leq j \leq J$, and $\Delta x = L/J$. In this way we arrive at the system of ODEs

$$\frac{dc_0}{dt} = D\frac{c_1 - c_0}{(\Delta x)^2} + f(c_0),$$

$$\frac{dc_j}{dt} = D\frac{c_{j+1} - 2c_j + c_{j-1}}{(\Delta x)^2} + f(c_j), \tag{8.17}$$

$$\frac{dc_J}{dt} = D\frac{c_{J-1} - c_J}{(\Delta x)^2} + f(c_J),$$

where the equations for $c_0$ and $c_J$ are found using a discretized version of no–flux boundary conditions $\partial c/\partial x|_{x=0,L} = 0$ to specify values at the "ghost points" $c_{-1} = c_0$ and $c_{J+1} = c_J$.

Figure 8.4 shows a simulation of a fertilization $Ca^{2+}$ wave calculated using (8.17). Our assumed one dimensional geometry suggests that we interpret the calculation as a propagating planar signal ($[Ca^{2+}]_i$ not a function of $y$ or $z$, but only $x$). Initial conditions are chosen so that elevated $[Ca^{2+}]_i$ in a small region triggers a wave traveling leftwards at a velocity of 3.45 $\mu$m/s. The basal $[Ca^{2+}]_i$ before the front passes is given by equilibrium point 1 (0.01 $\mu$M), and after the front passes elevated $[Ca^{2+}]_i$ is given by equilibrium point 3 (0.53 $\mu$M). Because $[Ca^{2+}]_i$ doesn't return to basal values after the wave passes, this type of traveling wave is referred to as a *propagating front*.

In the fertilization $Ca^{2+}$ wave model, we could say that the physiological state of the cell cytoplasm is bistable, that is, if the spatial component of the model is neglected (for example, by setting $D = 0$ in (8.15)), the scalar ODE ($dc/dt = f(c)$) that remains

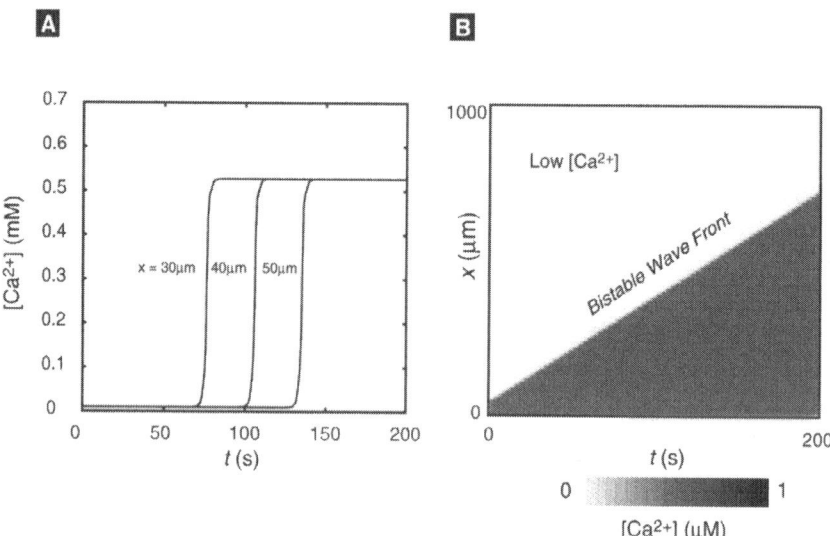

**Figure 8.4** Simulation of a fertilization $Ca^{2+}$ wave following (8.15) results in a traveling front. (A) Time courses $c(x_i, t)$ (the $Ca^{2+}$ profile) at three spatial positions $x_i = 30, 40, 50 \, \mu m$. (B) An array plot of $c(x, t)$. The slope of the front gives a velocity of 3.45 $\mu m/s$. Parameters as in Exercise 4 and Exercise 6.

exhibits bistability. Alternatively, imagine walling off (or lifting out) a small region of the spatial model and investigating its dynamics. If the region is small enough, the time scale for diffusion in the isolated region ($T = L^2/D$) will be much faster than the dynamics associated with the reaction terms $f(c)$. Thus, the $Ca^{2+}$ profile will be approximately uniform ($\partial c/\partial x = 0$), and the region will behave as a *compartmental model* in which diffusion no longer plays a role. The propagating front observed in the spatial model, Figure 8.4, is possible precisely because any such isolated compartmental model would exhibit bistability. Indeed, if the physiological state of the cytosol were not bistable, the basal and elevated $[Ca^{2+}]_i$ at the wave front and back could not persist. We will see below that spatial phenomena observed in reaction–diffusion models can be categorized by the qualitative dynamics of the ODEs obtained by assuming that all profiles are uniform.

Following Section 7.5.1 we can analyze the fertilization $Ca^{2+}$ wave model (8.15) by looking for a (rightward) traveling wave solution $c(x, t) = C(x - vt) = C(z)$. Making this substitution as well as $\partial C/\partial x = (dC/dz)(\partial z/\partial x) = dC/dz$ and $\partial C/\partial t = (dC/dz)(\partial z/\partial t) = -v \, dC/dz$ into this reaction–diffusion equation gives

$$-v\frac{dC}{dz} = D\frac{d^2C}{dz^2} + f(C), \tag{8.18}$$

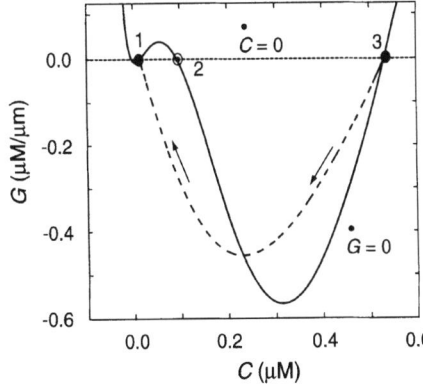

**Figure 8.5** The reaction–diffusion equation model of the fertilization Ca$^{2+}$ wave, (8.18), can be transformed into traveling wave coordinates ($z = x - vt$) resulting in a first–order system of ODEs, (8.19)–(8.20). $C(z)$ is Ca$^{2+}$ concentration and $G(z) = dC/dz$ is the rate function. Nullclines for $C$ and $G$ are shown (dotted and solid lines, respectively). When wave speed is $v = 4.96\ \mu$m/s a heteroclinic orbit (dashed line) connects the two stable equilibria (points 3 and 1), representing the Ca$^{2+}$ concentration before and after wave passage. See Exercise 6.

which can be written as the first–order system

$$\frac{dC}{dz} = G, \tag{8.19}$$

$$\frac{dG}{dz} = -\frac{1}{D}\left[vG + f(C)\right]. \tag{8.20}$$

The phase portrait in Figure 8.5 is a plot of the rate function $C' = G$ versus Ca$^{2+}$ concentration $C$. In Exercise 6 the reader can numerically demonstrate the existence of a well-behaved heteroclinic orbit when $v = 4.96\ \mu$m/s. This heteroclinic orbit does not exist for smaller or large values of $v$ and corresponds to the traveling front presented in Figure 8.4.

Intuitively, we expect the speed of traveling front solutions such as those presented in Figure 8.4 and Figure 8.5 to depend on the diffusion coefficient $D$. Indeed, when the rate function $f(c)$ is a cubic polynomial as in Chapter 7, the wave speed is proportional to the square root of the diffusion coefficient. This is clear from the discussion in Section 7.5.1. Although in the fertilization Ca$^{2+}$ wave model $f(c)$ is not a polynomial, we can demonstrate the same principle by noting that the substitution $D^* = \alpha^2 D$ in (8.18) can be offset by scaling the wave speed ($v^* = \alpha v$). To see this, we define a scaled wave coordinate $z^* = x - v^* t$ that implies $-v^*\, dC/dz^* = -v\, dC/dz$ or $dC/dz^* = \alpha^{-1} dC/dz$. Thus, any solution of (8.18) is also a solution of

$$v^* \frac{dC}{dz^*} = D^* \frac{d^2 C}{dz^{*2}} + f(C).$$

In Exercise 8 the reader can repeat the simulations of Figure 8.4 and/or Figure 8.5 to numerically confirm that the traveling front velocity is proportional to the square root of the diffusion coefficient. Interestingly, exogenous buffers have been shown to alter the speed of Ca$^{2+}$ waves in mature *Xenopus* oocytes (Nuccitelli et al. 1993), an effect that may be due to changes in the effective diffusion coefficient for Ca$^{2+}$.

The astute reader may have noticed that we have reported two different wave velocities for two different calculations of the same traveling front. Recall that Figure 8.4

used the method of lines to simulate the fertilization $Ca^{2+}$ wave model (8.15) and that we observed a speed of 3.45 $\mu$m/s. On the other hand, Figure 8.5 shows that the first–order system of ODEs resulting from a transformation into traveling wave coordinates, (8.19)–(8.20), has a heteroclinic orbit that connects the two stable equilibria for wave velocity of $v = 4.96$ $\mu$m/s. In Exercise 9 the reader can show that the later calculation is more accurate. The suppression of the wave speed in the calculation using the method of lines is due to *discretization error*. Figure 8.4 corresponds to Figure 8.5 as the mesh is refined ($\Delta x \to 0$).

## 8.7   Calcium Waves in the Immature Xenopus Oocycte

Although we have begun our discussion of models of intracellular $Ca^{2+}$ waves by focusing on traveling fronts in *mature Xenopus laevis* eggs, $IP_3$-dependent $Ca^{2+}$ responses occur in many cell types and take various forms. It is instructive to note here that $Ca^{2+}$ wave phenomena in the *immature X. laevis* oocyte (such as the spiral $Ca^{2+}$ wave shown in Figure 1.4) are qualitatively distinct from the traveling fronts discussed above. Although there are many differences between immature and mature *Xenopus* oocytes/eggs, from a dynamical point of view these different $Ca^{2+}$ wave phenomena are the result of different underlying dynamics of $Ca^{2+}$ handling that could be distinguished even in the absence of diffusion. While our model of the fertilization $Ca^{2+}$ wave demonstrated how bistable cellular dynamics can lead to a traveling front, the ER of the immature *Xenopus* oocyte is excitable (and may even be oscillatory) as opposed to bistable.

Figure 8.6 presents phase planes for the reaction terms of the fertilization $Ca^{2+}$ wave model (8.6)–(8.7), which illustrate the qualitative difference between excitable and bistable  dynamics. To make this comparison, we relax our assumption that the dynamics of $[Ca^{2+}]_i$ are much slower than the gating variable for $Ca^{2+}$ inactivation of the $IP_3R$ (so that $w$ is once again given by (8.7) rather than $w_\infty([Ca^{2+}]_i)$). The model now has two dynamic variables ($[Ca^{2+}]_i$ and $w$), the nullclines of which are shown in panel A of Figure 8.6. The intersections of these nullclines imply three equilibria (as before, points 1 and 3 are stable, while 2 is unstable). Conversely, panel B of Figure 8.6 presents the ($w$, $[Ca^{2+}]_i$) phase portrait with adjusted $IP_3R$ parameter values ($Ca^{2+}$-dependence of inactivation). The reaction terms of the model now represent excitable cytoplasm. A small perturbation in $Ca^{2+}$ concentration results in a long excursion of the trajectory representing $Ca^{2+}$ release from the ER, $IP_3R$ inactivation, and ATP-dependent $Ca^{2+}$ reuptake.

## 8.8   Simulation of a Traveling Pulse

A simulation of $Ca^{2+}$ wave phenomena mediated by ER $Ca^{2+}$ excitability can be implemented in a manner similar to (8.18) (see Exercise 7). To do this we discretize the

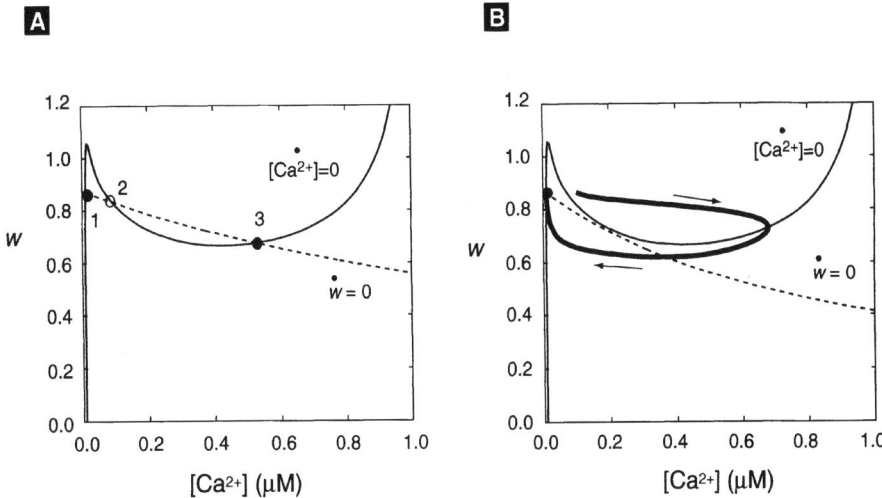

**Figure 8.6** The $(w, [Ca^{2+}]_i)$ phase plane for the reaction terms of the fertilization $Ca^{2+}$ wave model (8.6)–(8.7) can exhibit bistability (A) or excitability (B) with a change of IP$_3$R parameters. Parameters as in Exercise 4 except $d_2 = 2\,\mu M$ in the bistable case, while $d_2 = 1\,\mu M$ in the excitable case. In the Li–Rinzel reduction of the DeYoung–Keizer IP$_3$R model, this change in $d_2$ increases equilibrium IP$_3$R inactivation (note the downward/leftward shift in $w$ nullcline) and eliminates the high $[Ca^{2+}]_i$ equilibrium present in (A).

gating variable $w_j \approx w(x_j, t)$ at $J + 1$ mesh points $(0 \le j \le J)$ representing the fraction on noninactivated IP$_3$Rs at each location in space $(x_j = j\Delta x)$. Of course, there will be no Laplacian term in the equations for $w_j$ because the IP$_3$Rs (and thus their state) do not diffuse. The model now takes the form

$$\frac{dc_0}{dt} = D\frac{c_1 - c_0}{(\Delta x)^2} + g(c_0, w_0), \qquad \frac{dw_0}{dt} = [w_\infty(c_0) - w_0]/\tau,$$

$$\frac{dc_j}{dt} = D\frac{c_{j+1} - 2c_j + c_{j-1}}{(\Delta x)^2} + g(c_j, w_j), \qquad \frac{dw_j}{dt} = [w_\infty(c_j) - w_j]/\tau, \qquad (8.21)$$

$$\frac{dc_J}{dt} = D\frac{c_{J-1} - c_J}{(\Delta x)^2} + g(c_J, w_J), \qquad \frac{dw_J}{dt} = [w_\infty(c_J) - w_J]/\tau.$$

Figure 8.7 shows a propagating $Ca^{2+}$ wave that results when the cytosol, modeled using (8.21), is excited with the same initial conditions as in Figure 8.4. In this case, $[Ca^{2+}]_i$ returns to basal as the $Ca^{2+}$ transient propagates past a given point, and the cellular response is referred to as a *traveling pulse*. Traveling pulses produced by excitable dynamics are often called "trigger waves," because a perturbation at one end of the excitable medium triggers a signal that may propagate with undiminished amplitude.

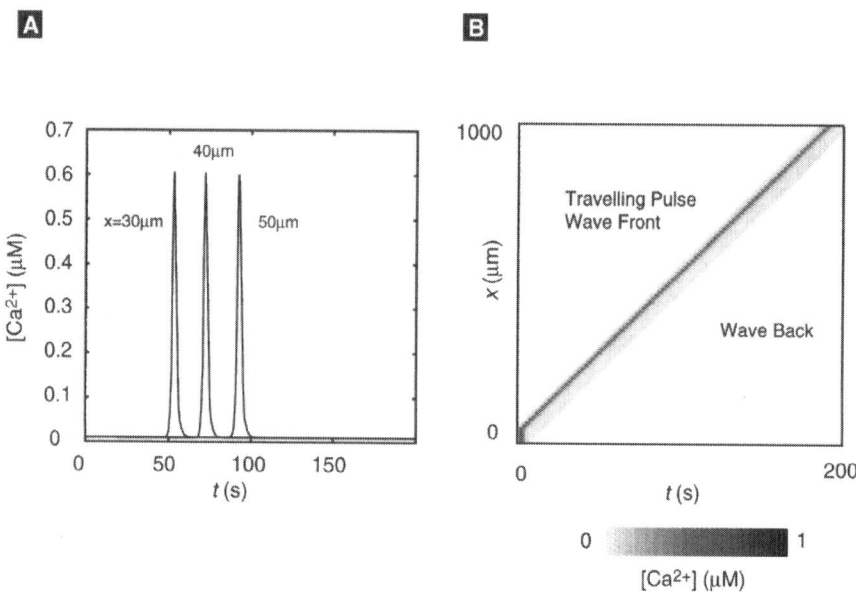

**Figure 8.7** $[Ca^{2+}]_i$ time courses and array plot similar to Figure 8.4 except that parameters are changed so that the $(w, [Ca^{2+}]_i)$ phase plane is excitable as in (B) of Figure 8.6. The simulation is implemented using (8.21) and results in a traveling pulse similar to $Ca^{2+}$ responses in the immature *Xenopus* oocyte. The slope of the front gives a velocity of 5 $\mu$m/s. Parameters given in Exercise 7.

# 8.9    Simulation of a Kinematic Wave

In previous sections we have discussed how the cytoplasm's physiological state influences $Ca^{2+}$ wave propagation. When the cytoplasm is bistable, a likely wave phenomenon is a traveling front; conversely, an excitable cytoplasm will likely support trigger waves, i.e., traveling pulses. These observations suggest that we consider the case of an oscillatory cytoplasm. As discussed in Chapter 5, $IP_3$-mediated whole–cell $Ca^{2+}$ responses are oscillatory in many cell types. Figure 8.8 demonstrates that it is an easy matter to adjust $IP_3R$ parameters ($Ca^{2+}$-dependence of activation and inactivation) in our model so that the $(w, [Ca^{2+}]_i)$ phase plane is oscillatory and a large–amplitude (0.01 $\mu$M < $[Ca^{2+}]_i$ < 0.65 $\mu$M) stable limit cycle exists.

Even in the absence of diffusion an oscillatory medium will support *kinematic waves*. Figure 8.9 illustrates this possibility using an array of clocks. Depending on initial conditions (i.e., the time to which the clocks are initially set), the absence of physical coupling between the clocks does not preclude the appearance of a traveling wave. Regardless of initial condition, all the clocks will eventually return to their initial

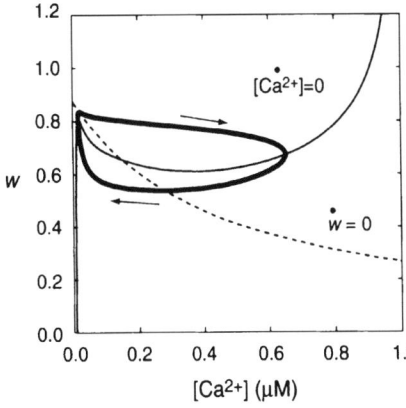

**Figure 8.8** $(w, [Ca^{2+}]_i)$ phase plane when the physiological state of the cytoplasm is oscillatory. Parameters as in Figure 8.6 except $d_2 = 0.5\,\mu M$ and $d_5 = 0.15\,\mu M$. In the Li–Rinzel reduction of the DeYoung–Keizer IP$_3$R model, $d_5$ and $d_2$ affect IP$_3$R activation and inactivation, respectively.

state. Thus, kinematic waves are distinct from traveling fronts and pulses in that they can occur with arbitrary shape.

Although a kinematic wave can be of arbitrary shape, deceptively shaped waves may look like traveling waves when there is a certain relationship between wavelength and the oscillatory frequency of the medium. To see this, let $c(x,0) = c_0(x)$ and $w(x,0) = w_0(x)$ be the initial Ca$^{2+}$ profile for our model tuned to exhibit limit cycle oscillations with period $\tau$. If we choose $(c_0(x), w_0(x))$ "on" the limit cycle (for all $x$), then in the absence of Ca$^{2+}$ diffusion the [Ca$^{2+}$]$_i$ will be temporally periodic at every spatial location, i.e., $c(x, t + \tau) = c(x, t)$. If the initial condition also happens to be spatially periodic, so that $c_0(x + \lambda) = c_0(x)$ (and similarly for $w_0$), then [Ca$^{2+}$]$_i$ will remain spatially periodic, i.e., $c(x + \lambda, t) = c(x, t)$, where $\lambda$ is the spatial period (the wavelength). As illustrated using the clock analogy in Figure 8.9, this simultaneous spatial and temporal periodicity implies a solution of the form $c(z) = c(x + vt)$ if $v = \lambda/\tau$.

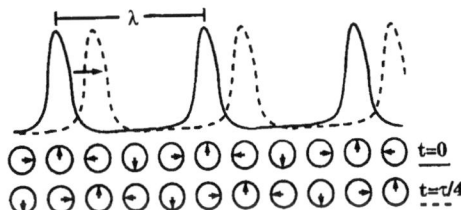

**Figure 8.9** An array of clocks illustrates a kinematic wave. The peak of the wave coincides with the hand of the clock pointing to 12 o'clock. The temporal period $(\tau)$ is the time it takes for the hand at a given location to make one complete clockwise revolution. The minimum distance between clocks with the same phase is the spatial period, or wavelength $(\lambda)$. A wave takes $\tau$ seconds to travel one wavelength, so that the wave speed is $v = \lambda/\tau$. From Jafri and Keizer (1994).

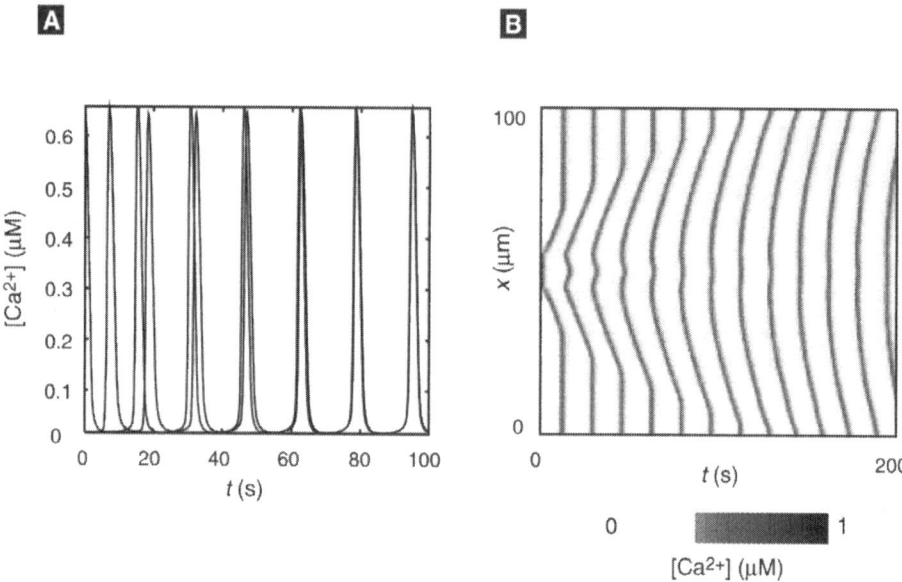

**Figure 8.10** $[Ca^{2+}]_i$ time courses and array plot similar to Figure 8.7 except that the $(w, [Ca^{2+}]_i)$ phase plane is oscillatory as in Figure 8.8. The simulation is implemented using (8.21) and results in kinematic wave (also called a phase wave). (A) Time courses $c(x_i, t)$ at $x_i = 50$ and $63\,\mu$m show that diffusion causes oscillations at different regions to synchronize. (B) An array plot illustrates the smoothing of the relative phases of the oscillations.

Using the parameters that generated the limit cycle shown in Figure 8.8, Figure 8.10 demonstrates that oscillatory $IP_3$-mediated $Ca^{2+}$ release may lead to kinematic waves. Initial conditions are chosen so that the phase of the oscillation is uniform throughout the medium, except for a central region that is phase advanced. Because $Ca^{2+}$ diffusion is included in these simulations following (8.21), the distribution of phases evolves as a function of time (note the changing spatial profile of the wave peak with each oscillation in Figure 8.10B). Because diffusion is now coupling cytoplasmic oscillations, pure kinematic waves no longer exist. Nevertheless, at each spatial location the solution remains "near" the limit cycle of Figure 8.8.

Although beyond the scope of this chapter, it can be shown that in the limit of slow diffusion (as expected here due to $Ca^{2+}$ buffers), the relative phases of the oscillations, referred to as the *phase gradient*, evolve according to Burger's equation (Murray 1989; Jafri and Keizer 1997). The important conceptual point is that buffered $Ca^{2+}$ diffusion causes the phase of the oscillations to synchronize over time. Figure 8.10A illustrates this with time courses $[Ca^{2+}]_i(x_i, t)$ from two spatial positions. Notice that the oscillations at these two locations synchronize; the (apparent) velocity of the repetitive wave between these two points increases. Indeed, the velocity of a kinematic wave can be arbitrarily large because it is not a propagating signal. In the array plot of Figure 8.10B, the diminution of the phase gradient with each cycle is striking.

As discussed at length in previous sections, the presence of endogenous and exogenous $Ca^{2+}$ buffers slows $Ca^{2+}$ diffusion. If the physiological state of a cell's cytoplasm is oscillatory, observed wave phenomena may very well be kinematic in nature, and the influence of $Ca^{2+}$ diffusion on propagation of the wave may be primarily through the evolution of the phase gradient. Conversely, an excitable cytosol absolutely requires diffusion to support a propagating fronts (i.e., trigger waves). While kinematic waves can exist even in the absence of diffusion, the wave speed of a propagating front scales with the effective diffusion coefficient in a predictable manner (at least in the limit of low affinity $Ca^{2+}$ buffers) (Jafri and Keizer 1994; Jafri and Keizer 1997; Sneyd et al. 1998). These distinctions are important and one may easily be deceived. For example, the observation of repetitive waves (i.e., a *wave train*) may suggest an oscillatory medium. On the other hand, such a wave train may be due to periodic forcing (or recurrent entry) of an excitable medium, as in target patterns and spiral waves imaged in the immature *Xenopus* oocyte. Is it obvious from Figure 1.4 whether the physiological state of the immature *Xenopus* oocyte cytoplasm is excitable or oscillatory?

## 8.10  Spark-Mediated Calcium Waves

Throughout this discussion we have, without exception, made the so-called *continuum approximation*. That is, we have made no attempt to explicitly model the stochastic dynamics of individual $Ca^{2+}$ ions undergoing Brownian motion. Indeed, an averaging process is implicit in any continuum description of matter. Consider that small molecules (such as free $Ca^{2+}$ ions) in micromolar solution are essentially point particles with mean intermolecular spacing of approximately 100 nm; that is, in a volume of $(100\,nm)^3$ a single solute molecule will be found on average. On a distance scale of 10 nm (a factor of ten smaller) the medium is highly heterogeneous. If thermal motion of both solvent and solute could be stopped, only one out of a thousand volume elements with dimension $(10\,nm)^3$ would be found to contain a solute molecule. Clearly, at distance scales of 10 nm (or less) a continuum description is invalid.

On the other hand, if a fluorescent molecule at micromolar concentrations is optically detected using a microscope with micron resolution, the relevant volume $(1\,\mu m^3 = 10^9\,nm^3)$ is likely to contain a thousand molecules. Though the actual number of molecules in the volume will vary as a function of time due to diffusion, these density fluctuations will be small. Thus, on a distance scale of a micron the medium appears spatially homogeneous, and the continuum approximation is legitimate.

The phrase "continuum approximation" can also refer to the fact that in our spatial whole–cell models we are assuming that the ER can be represented as a compartment continuously distributed throughout the  cytosol, albeit with prescribed volume fraction given by $f_{ER}$ in (8.8). Although it can sometimes be rigorously justified, anyone familiar with the mathematics of diffusion in heterogeneous media will recognize that this assumption is made largely for convenience (Crank 1975). Intracellular heterogeneities abound, and only sometimes can these heterogeneities be "averaged out" in a way that justifies a bidomain description (i.e., a mathematical technique known as homogenization).

One important cellular inhomogeneity known to have a profound impact on propagating $IP_3$-mediated $Ca^{2+}$ waves is the distribution of $IP_3Rs$. In the immature *Xenopus* oocyte, for example, $IP_3Rs$ occur in clusters of 10–100 with intercluster spacing on the order of a few microns (Berridge 1997). Under some conditions this organization of $Ca^{2+}$ release sites reveals itself in cauliflower-like wave fronts. Localized $Ca^{2+}$ elevations due to the activation of a single $Ca^{2+}$ release site (a $Ca^{2+}$ puff) and even a single channel (a $Ca^{2+}$ blip) are now regularly observed (Yao et al. 1995; Parker et al. 1996).

Localized $Ca^{2+}$ elevations due to intracellular $Ca^{2+}$ release have been observed in other cell types as well. When a $Ca^{2+}$ release event is mediated by ryanodine receptors (RyRs), as in cardiac myocytes, the phenomenon is referred to as a $Ca^{2+}$ spark (Cheng et al. 1993). Spark-mediated waves in cardiac myocytes are amenable to modeling because $Ca^{2+}$ waves often propagate along the longitudinal axis of a myocyte, i.e., perpendicular to the orientation of the sarcomeric Z-lines. The Z-lines, transverse tubules, and $Ca^{2+}$ release sites are oriented in the transverse direction and are regularly spaced with a separation of 2 $\mu$m. The ultrastructure of cardiac myocytes thus lends itself to one-dimensional modeling in which the cellular heterogeneity along the longitudinal axis is represented. Figure 8.11 shows some examples of propagating spark-mediated waves and isolated sparks in cardiac myocytes.

One such modeling approach (Keizer et al. 1998) is to augment the continuum bidomain description (8.8) with a spatially periodic maximum conductance for $Ca^{2+}$ release, say, $v_{RyR}(x)$ in analogy to $v_{IP_3R}$ in the $IP_3$-mediated wave model. Numerical studies show that spark-mediated $Ca^{2+}$ waves propagate in a saltatory manner, similar to experiment. In the following section we will consider a related and even more idealized model of spark-mediated $Ca^{2+}$ waves, the fire–diffuse–fire model.

# 8.11   The Fire–Diffuse–Fire Model

The essence of minimal models of spark-mediated $Ca^{2+}$ waves is to model $Ca^{2+}$ release sites as idealized point sources, that is, an array of Dirac delta functions, denoted by $\delta(x - x_i)$, where $x_i$ is the spatial position of the $i$th release site. While the Dirac delta function is actually not a function at all (rather, it is a distribution, the limit of a sequence of functions), it is often thought of as a sharply peaked function that is zero everywhere except $x = x_i$. In addition, the delta function is normalized so that

$$\int_{-\infty}^{\infty} \delta(x - x_i)dx = 1,  \tag{8.22}$$

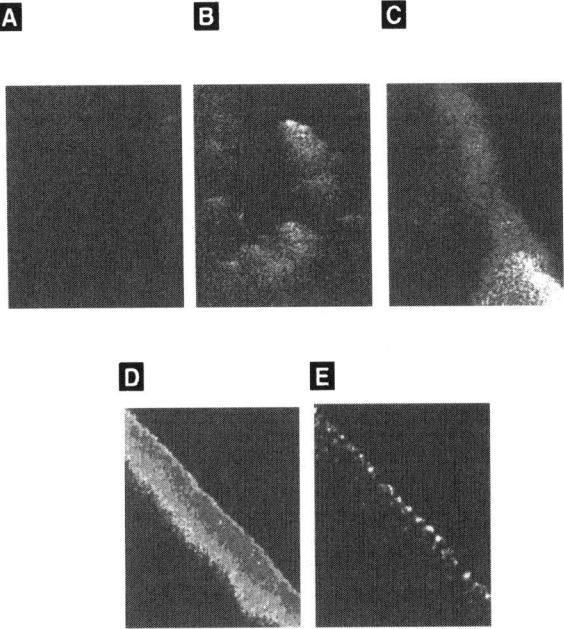

**Figure 8.11** Confocal line-scan images of isolated $Ca^{2+}$ sparks, wave initiation, and a spark-mediated propagating $Ca^{2+}$ wave in cardiac myocytes. Time is vertical (bar is 100 ms except for second panel which is 200 ms) and space is horizontal (bar is 5 $\mu$m). Image subtraction shows spatially localized $Ca^{2+}$ release originating from sites separated by $\approx 2\,\mu$m. From Cheng et al. (1996).

and it has the so-called sifting property

$$\int_{-\infty}^{\infty} f(x)\delta(x - x_i)dx = f(x_i). \qquad (8.23)$$

Armed with delta functions, we can model spark-mediated $Ca^{2+}$ waves by modifying the propagating front model (8.15) so that $Ca^{2+}$ release occurs only at regularly spaced release sites, that is,

$$\frac{\partial c}{\partial t} = D\frac{\partial^2 c}{\partial x^2} + f(c)\sum_i \delta(x - x_i), \qquad (8.24)$$

where $x_i$ is the location of the $i$th release site with kinetics given by $f(c)$ (Keener 2000).

We can further idealize this model for spark-mediated waves by removing the $Ca^{2+}$-dependence of the release rate $f(c)$. Instead, assume that the source strength $f_i(t)$ of each release site is a square pulse function of time,

$$f_i(t) = \frac{\sigma}{\tau_R}H(t - t_i)H(t_i + \tau_R - t).$$

In this expression $H(t)$ is the Heaviside step function, so that $H(t) = 0$ for $t < 0, H(t) = 1$ for $t \geq 0$. The product of these Heaviside functions represents the source turning "on" at time $t = t_i$ and remaining on for duration $\tau_R$, i.e., it turns "off" at time $t = t_i + \tau_R$. The constant $\sigma$ represents the source amplitude and has units of $\mu M \cdot \mu m$ because from (8.22) we see that the delta function has units of $1/\mu m$. The normalization factor $(1/\tau_R)$ is chosen so that

$$\int_{-\infty}^{\infty} f_i(t)\, dt = \sigma. \tag{8.25}$$

Substituting this form for the $Ca^{2+}$ release rate into (8.24) gives the "fire–diffuse–fire" model of spark-mediated $Ca^{2+}$ wave propagation,

$$\frac{\partial c}{\partial t} = D\frac{\partial^2 c}{\partial x^2} + \frac{\sigma}{\tau_R}\sum_i \delta(x - x_i)H(t - t_i)H(t_i + \tau_R - t). \tag{8.26}$$

To finish our presentation of the fire–diffuse–fire model, we must specify the location of each release site $(x_i)$ and the times $(t_i)$ at which $Ca^{2+}$ release begins. If we further assume a regular array of release sites, then $x_i = id$ where $d$ is the site spacing. Because we have $Ca^{2+}$-induced $Ca^{2+}$ release in mind, we assume that $Ca^{2+}$ release at site $i$ begins when the local $Ca^{2+}$ concentration $c(x_i, t)$ achieves a fixed threshold $Ca^{2+}$ concentration $c_\theta$ (here $0.1\ \mu M$). In Exercise 12 the reader can implement such a fire–diffuse–fire simulation.

Interestingly, Figure 8.12 shows that the fire–diffuse–fire model supports continuous as well as propagating $Ca^{2+}$ signals. In Figure 8.12A, the time constant for $Ca^{2+}$ release, $\tau_R$, is 1 s and the propagating signal is similar to the traveling front solutions presented earlier in this chapter. Conversely, Figure 8.12B presents a simulation using $\tau_R = 10$ ms. Here spark–like $Ca^{2+}$ releases lead to a propagating signal that is distinctly saltatory. Note that the continuous wave is traveling at 11.3 $\mu m/s$. while the saltatory wave is traveling at 67 $\mu m/s$. The long duration of $Ca^{2+}$ release in the continuous case appears to slow the velocity of the propagating signal.

The fire–diffuse–fire model can be analyzed to give insight into the continuous and saltatory limits of $Ca^{2+}$ wave propagation. As the reader may expect, it is not $\tau_R$ alone but rather a dimensionless parameter that determines the existence and form of propagating signals. Indeed, it can be shown that the relevant dimensionless quantity is $D\tau_R/d^2$. The continuous limit corresponds to $D\tau_R/d^2 \gg 1$ and the saltatory limit to $D\tau_R/d^2 \ll 1$. Below we study these two limits separately to determine (you guessed it) how diffusion influences wave velocity in both limits.

### Analysis of the Continuous Limit

The continuous limit pertains when $D\tau_R/d^2 \gg 1$, that is, when diffusion is fast and the release time long compared to the intersite separation. This limit could be achieved in a simulation by increasing the density of release sites $(d \to 0)$ while simultaneously decreasing the release rate $(\sigma)$ so that the release per unit length $(\sigma/d)$ is constant.

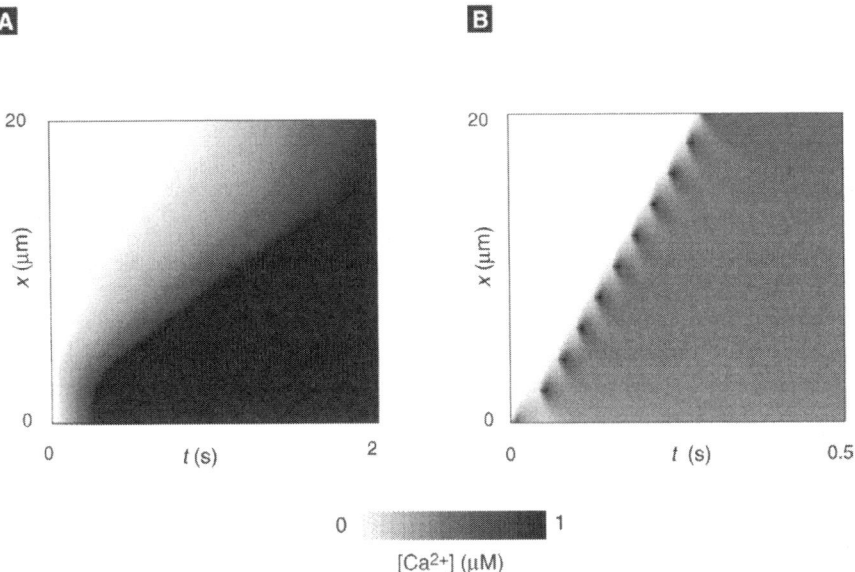

**Figure 8.12** Simulation of Fire–Diffuse–Fire model in both continuous (A) and saltatory (B) regimes using (8.26). The separation between release sites is $d = 2\,\mu$m. The continuous wave ($\tau_R = 1$ s) is traveling 11.3 $\mu$m/s, while the saltatory wave ($\tau_R = 10$ ms) is traveling 67 $\mu$m/s. Parameters as in Exercise 12.

Because the thresholding that determines Ca$^{2+}$ release times ($t_i$) is the only nonlinear part of our model (8.26), we can convolve the release rates $f_i(t)$ with the diffusion kernel to write an explicit expression for $c(x,t)$:

$$c(x,t) = \frac{\sigma}{\tau_R} \sum_{i=-(N-1)}^{N-1} \int_{t_i}^{\min(t,t_i+\tau_R)} \frac{dt'}{\sqrt{4\pi D(t-t')}} \exp\left[-\frac{(x-x_i)^2}{4D(t-t')}\right], \qquad \text{for} \quad t > t_{N-1},$$

$$(8.27)$$

where we have assumed a symmetric profile $c(x,t)$ and the firing of $N-1$ sites on either side of the origin. Equation (8.27) is not yet a solution because we have the unknown parameters $t_i$ to determine. To find the time at which site $N$ fires we evaluate (8.27) at $(t = t_N, x = x_N)$ and set $c(x_N, t_N) = c_\theta$. This yields

$$c_\theta = \frac{\sigma}{\tau_R} \sum_{i=-(N-1)}^{N-1} \int_{t_i}^{\min(t_N,t_i+\tau_R)} \frac{dt'}{\sqrt{4\pi D(t_N-t')}} \exp\left[-\frac{(x_N-x_i)^2}{4D(t_N-t')}\right],$$

an expression that can be solved for $t_N$. At this point we seek traveling–wave like solutions with regular firing times. This implies that $t_i = i\Delta$, where $\Delta$ is the time interval between adjacent site firings and a velocity of propagation given by $v = d/\Delta$. By substituting $\Delta = t_N/N$ and solving for $\Delta$ it can be shown (Keizer et al. 1998) that the velocity

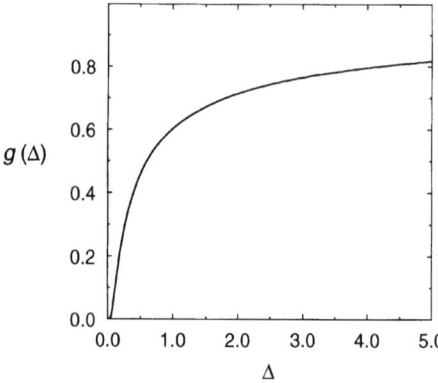

**Figure 8.13** Relationship between $\alpha = c_\theta d/\sigma$ and the dimensionless firing interval ($\Delta = 1/v$) for the fire–diffuse–fire model in the saltatory limit. Since $g(\Delta) < 1$, waves do not propagate for $\alpha > 1$. From Keizer et al. (1998).

of such a propagating wave has the following dependence on model parameters:

$$v \approx \sqrt{\left(\frac{\sigma}{dc_\theta}\right)\left(\frac{D}{\tau_R}\right)}. \tag{8.28}$$

Notice that the factor of $\sqrt{D/\tau_R}$ is similar to what we have seen before for traveling fronts and pulses. The existence of traveling wave solutions depends on $\sqrt{\sigma/dc_\theta}$.

### Analysis of the Saltatory Limit

The saltatory limit of the fire–diffuse–fire model corresponds to $D\tau_R/d^2 \ll 1$ in (8.26), that is, when diffusion is slow and release time short compared to the intersite separation. This limit could be achieved in a simulation by decreasing the time constant for Ca$^{2+}$ release ($\tau_R$) while maintaining a fixed density of release sites ($d$ constant). Because the normalization factor ($1/\tau_R$) is chosen to satisfy (8.25) regardless of $\tau_R$, the release rate per unit length ($\sigma/d$) remains constant as $\tau_R \to 0$, and (8.26) becomes

$$\frac{\partial c}{\partial t} = D\frac{\partial^2 c}{\partial x^2} + \sigma \sum_i \delta(x - x_i)\,\delta(t - t_i). \tag{8.29}$$

We now nondimensionalize space ($x^* = x/d, x_i^* = x_i/d = i$), time ($t^* = tD/d^2, t_i^* = t_i D/d^2$), and concentration ($c^* = c/c_\theta$) and drop asterisks to write

$$\frac{\partial c}{\partial t} = \frac{\partial^2 c}{\partial x^2} + \frac{1}{\alpha}\sum_i \delta(x - x_i)\,\delta(t - t_i), \tag{8.30}$$

where $\alpha = c_\theta d/\sigma$. Using the diffusion kernel we obtain an implicit expression for $c(x, t)$,

$$c(x, t) = \sum_i H(t - t_i)\sqrt{\frac{1}{4\pi\alpha^2(t - t_i)}}\,\exp\left[-\frac{(x - i)^2}{4(t - t_i)}\right], \tag{8.31}$$

where the presence of the Heaviside function indicates no contribution from sites that have not yet fired.

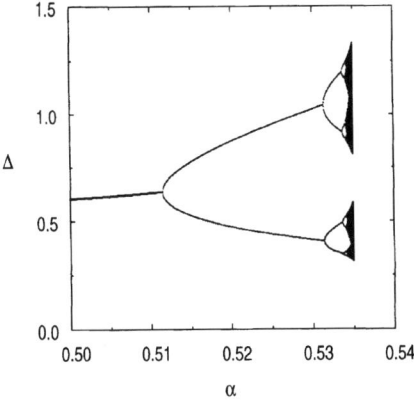

**Figure 8.14** For a range of values for $\alpha$, the saltatory limit of the fire–diffuse–fire model is simulated by successively calculating the dimensionless firing interval ($\Delta$) for $n = 1, 2, \cdots$ using (8.32) and the criterion $c(n, t_i) = 1$. The wave was initiated by simultaneously firing all sites for $-15 < n < 15$. A period doubling cascade begins at $\alpha \approx 0.512$ and terminates near $\alpha \approx 0.535$, beyond which waves do not propagate. From Keizer et al. (1998).

Assuming that the sites for which $-N < i < N$ have fired at times $t_i$ through $t_N$ (left and right pairs simultaneously) we want to determine $t_{N+1}$. The dimensionless threshold for firing is now $c = 1$, so from (8.31) we have

$$c(x_{N+1}, t_{N+1}) = 1 = \sum_{i=-N}^{N} \sqrt{\frac{1}{4\pi\alpha^2(t_{N+1} - t_i)}} \, \exp\left[-\frac{(N+1-i)^2}{4(t_{N+1} - t_i)}\right]. \qquad (8.32)$$

Because we are interested only in long time solutions, we consider the large $N$ limit and neglect all the terms in the above sum with $i \leq 0$, that is, we are following a rightward traveling wave that eventually is not influenced by the sites to the left of the origin. Under this approximation (8.32) simplifies to

$$\alpha = \sum_{i=1}^{N} \sqrt{\frac{1}{4\pi(t_{N+1} - t_i)}} \, \exp\left[-\frac{(N+1-i)^2}{4(t_{N+1} - t_i)}\right]. \qquad (8.33)$$

This expression is an implicit map for $t_{N+1}$ as a function of all the previous firing times, $(t_N, t_{N-1}, \ldots, t_1) \to t_{N+1}$. traveling-wave-like solutions correspond to fixed points of this map with regular firings, that is, $t_{N+1} = t_N + \Delta$ with constant $\Delta$ (giving a dimensionless velocity of $v = 1/\Delta$). Substituting $\Delta = t_{N+1} - t_N$ in (8.33) we obtain

$$\alpha = \sum_{i=1}^{N} \sqrt{\frac{1}{4\pi\Delta(N+1-i)}} \, \exp\left[-\frac{(N+1-i)}{4\Delta}\right]. \qquad (8.34)$$

Defining $n = N + 1 - i$ and taking the limit $N \to \infty$ we obtain

$$\alpha = \sum_{n=1}^{\infty} \sqrt{\frac{1}{4\pi n\Delta}} \, \exp\left(-n/4\Delta\right) \equiv g(\Delta). \qquad (8.35)$$

It can be shown that $0 \leq g(\Delta) \leq 1$. The first equality holds in the high–velocity limit ($\Delta \to 0$) and the second for low velocity ($\Delta \to \infty$). Since $g(\Delta)$ is monotonic, we can numerically calculate a unique solution $\Delta = g^{-1}(\alpha)$ when $0 < \alpha < 1$, i.e., the range of $g(\Delta)$. At this point it is worth remembering that $\alpha = c_\theta d/\sigma$. Thus, a necessary condition for the existence of a velocity is that $\alpha < 1$. If the sites are too far apart or too weak or the threshold too high, there can be no propagating waves. Interestingly, it can be shown that propagation failure occurs through a sequence of instabilities as $\alpha$ is increased well before the condition $\alpha = 1$ is reached (see further reading). When $\alpha$ is small ($\sigma/dc_\theta \gg 1$), propagating wave solutions exist and are stable. It can be shown that the dimensional velocity $v$ in this saltatory propagation limit is given by

$$v \approx \frac{4D}{d} \ln \left( \frac{\sigma}{dc_\theta} \right); \tag{8.36}$$

that is, it scales linearly with $D$, quite unlike propagating waves in the continuous limit, which have velocity that scales with $\sqrt{D}$ as in (8.28). If fact, whenever $\tau_R$ is sufficiently small (even if $\alpha$ is not particularly small), the saltatory wave velocity predicted by the fire–diffuse–fire model scales linearly with $D$.

# 8.12  Modeling Localized Calcium Elevations

Our discussion of spark-mediated $Ca^{2+}$ wave propagation began with a discussion of cellular heterogeneity. During this discussion of $Ca^{2+}$ wave phenomena it is important to remember that many cellular processes (including synaptic transmission, activity-dependent synaptic plasticity, and regulation of neuronal excitability) can be initiated by changes in intracellular $Ca^{2+}$ concentration in the absence of a global response such as a $Ca^{2+}$ oscillation or wave. For this reason, $Ca^{2+}$ sparks, puffs, and localized $Ca^{2+}$ elevations near voltage gated plasma membrane $Ca^{2+}$ channels (sometimes called $Ca^{2+}$ microdomains) are cellular signals of great interest. Localized $Ca^{2+}$ elevations are not only the "building blocks" of global $Ca^{2+}$ release events, but also highly specific regulators of cellular function (Berridge 1997; Berridge 1998).

As discussed above in the context of experimental observations of $Ca^{2+}$ waves, the interpretation of microfluorometric measurements of $Ca^{2+}$ puffs and sparks is complicated by the diffusion of $Ca^{2+}$, endogenous buffers, and indicator, all of which contribute to the dynamics of a fluorescence signal during and after $Ca^{2+}$ release. While in the case of global $Ca^{2+}$ responses and $Ca^{2+}$ waves the equilibrium relation (8.1) will likely hold between $Ca^{2+}$ and indicator, this is not so easily demonstrated in the case of localized $Ca^{2+}$ elevations (though it remains true in some cases (Smith et al. 1996)). A further complicating factor in interpreting *confocal* microfluorometric measurements is the optical blurring that occurs due to limited spatial resolution (Smith et al. 1998; Pratusevich and Balke 1996).

The reaction–diffusion equations for the buffered diffusion of intracellular $Ca^{2+}$, (8.10)–(8.12), are the starting point of a theoretical understanding the dynamics of localized $Ca^{2+}$ elevations. In the simplest scenario, a $Ca^{2+}$ puff or spark is due to $Ca^{2+}$ release through one channel or a tight cluster of channels. If $Ca^{2+}$ is released from intracellular $Ca^{2+}$ stores deep within a large cell (so that the plasma membrane is far

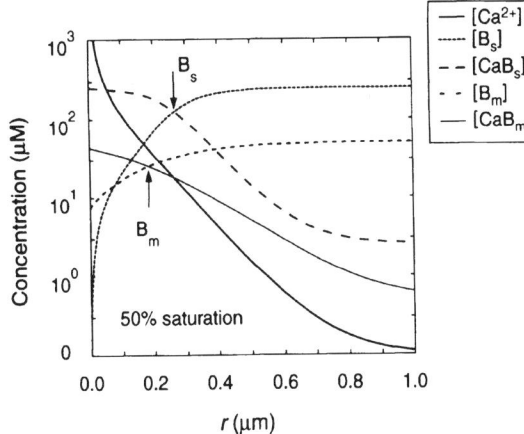

**Figure 8.15** Representative calculation of $Ca^{2+}$ profile near a point source for free $Ca^{2+}$. Source amplitude ($\sigma$) and elapsed time are 5 pA and 10 ms, respectively. From Smith et al. (1998).

away and doesn't influence the time–course of the event), and the intracellular milieu is homogeneous and isotropic, then we have spherical symmetry. In this case, the evolving profiles of $Ca^{2+}$ and buffer (though a function of time and distance from the source) will not be functions of the polar ($\phi$) or azimuthal ($\theta$) angle. In the case of such spherical or radial symmetry the Laplacian reduces to

$$\nabla^2 = \frac{1}{r^2}\frac{\partial}{\partial r}\left[r^2\frac{\partial}{\partial r}\right] = \frac{\partial^2}{\partial r^2} + \frac{2}{r}\frac{\partial}{\partial r}. \tag{8.37}$$

Figure 8.15 shows a spherically symmetric calculation of a localized $Ca^{2+}$ elevation using the full equations for the buffered diffusion of $Ca^{2+}$, (8.10)–(8.12), with parameters consistent with measurements of the effective diffusion coefficient in *Xenopus* oocyte cytoplasm (Allbritton, Meyer and Stryer 1992). Figure 8.15 is a numerically calculated snapshot of the concentration profiles for each species after an elapsed time of 10 ms. The concentration $[Ca^{2+}]_i$ is elevated near the source (*thick solid line*). Because released free $Ca^{2+}$ reacts with buffer, the concentration of bound buffer (*thin solid and thick dashed lines*) is elevated near the source. Conversely, the concentration of free buffer (*thin dotted and dashed lines*) decreases near the source. In this simulation, 250 $\mu$M stationary buffer was included in addition to 50 $\mu$M mobile buffer (both with $K$ of 10 $\mu$M). A source amplitude of 5 pA was used, corresponding to a tight cluster of IP$_3$Rs. Interestingly, Figure 8.15 shows that the mobile buffer is less easily saturated than stationary buffer, in spite of the fact that the stationary buffer is at fivefold higher concentration (note arrows).

Simulations such as these have played a role in understanding the dynamics of puffs and sparks. Figure 8.16 shows a $Ca^{2+}$ spark simulated using parameters consistent with experimental observation in cardiac myocytes (Smith et al. 1998). Such simulations confirm that the timecourse of observed fluorescence can be explained by a 2 pA, 15 ms

$Ca^{2+}$-release event from a tight cluster of RyRs located on the sarcoplasmic reticulum membrane.

Parameter studies using this model indicate that $Ca^{2+}$ spark properties (such as brightness, full width at half maximum, and decay time constant) are very dependent on indicator dye parameters (such as association rate constant, concentration, and diffusion coefficient). These relationships are not always intuitive. For example, increasing indicator dye concentration decreases the brightness of the simulated $Ca^{2+}$ spark in Figure 8.16. This is partly due to the fact that spark brightness is a normalized measure (peak/basal fluorescence), and partly due to the fact that high concentrations of indicator perturb the underlying free $Ca^{2+}$ signal.

## 8.13 Steady-State Localized Calcium Elevations

Numerical simulations like those in Figure 8.15 confirm that localized $Ca^{2+}$ elevations achieve steady–state values very rapidly (within microseconds) near point sources. Steady–state solutions to the full equations are thus of interest because they allow estimates of "domain" $Ca^{2+}$ concentration near open $Ca^{2+}$ channels. These steady-state solutions lend themselves to analysis, giving insight into the limiting (long time) "shape" of localized $Ca^{2+}$ elevations. In the case of one mobile buffer, steady-state solutions to (8.10)–(8.12) will satisfy the following boundary value problem (Roberts 1994):

$$0 = D_C \nabla^2 [Ca^{2+}]_i - k^+[B][Ca^{2+}]_i + k^- ([B]_T - [B]), \tag{8.38}$$

$$0 = D_B \nabla^2 [B] - k^+[B][Ca^{2+}]_i + k^- ([B]_T - [B]), \tag{8.39}$$

with associated boundary conditions

$$\lim_{r \to 0} \left\{ -4\pi r^2 D_C \frac{d[Ca^{2+}]_i}{dr} \right\} = \sigma, \qquad \lim_{r \to \infty} [Ca^{2+}]_i = [Ca^{2+}]_\infty,$$

$$\lim_{r \to 0} \left\{ -4\pi r^2 D_B \frac{d[B]}{dr} \right\} = 0, \qquad \lim_{r \to \infty} [B] = [B]_\infty = \frac{K[B]_T}{K + [Ca^{2+}]_\infty}. \tag{8.40}$$

Here we have eliminated the equation for [CaB] (rather than [B] as before) and written $D_C$ and $D_B$ for the diffusion coefficients of free $Ca^{2+}$ and free buffer, respectively.

Fixed buffers, while important for the time-dependent evolution of localized $Ca^{2+}$ elevations, have no influence on steady states. This can be seen by inspecting (8.10)–(8.12), where $D_{CB} = 0$ implies $R = -k_+[B][Ca^{2+}]_i + k_-[CaB] = 0$ at steady state.

As discussed in Chapter 4 and Chapter 5, there are many advantages to nondimensionalizing equations before preceding to analyze them. A convenient nondimensionalization of (8.38)–(8.40) begins by scaling $[Ca^{2+}]_i$ and [B] by representative concentrations, the dissociation constant ($K$) and total concentration of buffer ($[B]_T$) (Smith et al. 2001). This gives two dimensionless dependent variables, $\hat{c}$ and $\hat{b}$, given by $\hat{c} = [Ca^{2+}]_i/K$ and $\hat{b} = [B]/[B]_T$.

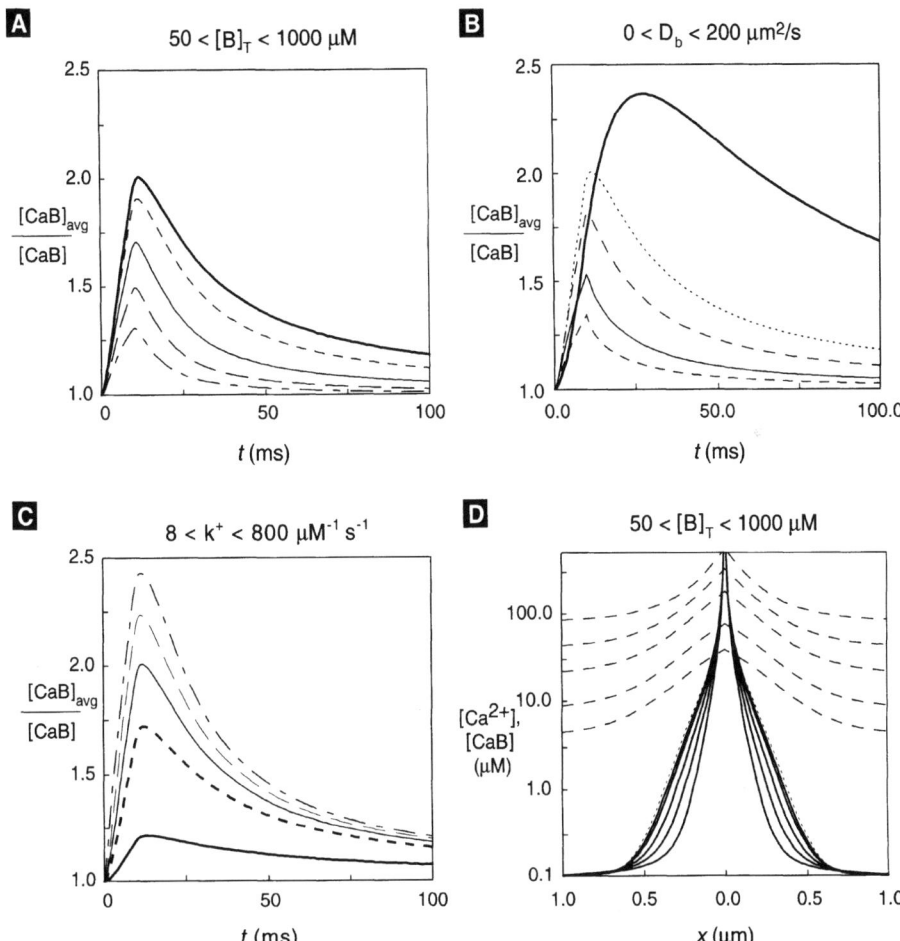

**Figure 8.16** Effects of indicator dye parameters on $Ca^{2+}$ spark properties. Source amplitude is 2 pA for 15 ms and simulated fluo-3 has $K$ of 1.13 $\mu M$. (A,B,C) Time course of normalized blurred fluorescence signal estimated according to (8.2) with $\eta_B = 0$. (D) Snapshot of $[Ca^{2+}]_i$ and $[CaB]$ profiles before termination of $Ca^{2+}$ release. Dotted line shows the $Ca^{2+}$ profile with no fluo-3. For details of the simulated confocal point spread function and buffer parameters. See Smith et al. (1998).

In Exercise 13 the reader can confirm that nondimensionalizing the independent variable ($\hat{\rho} = r/L$ with $L = \sigma/4\pi D_C K$) simplifies (8.38) and (8.39) to

$$\hat{\varepsilon}_c \nabla_{\hat{\rho}}^2 \hat{c} - \left(\hat{c}\hat{b} + \hat{b} - 1\right) = 0, \tag{8.41}$$

$$\hat{\varepsilon}_b \nabla_{\hat{\rho}}^2 \hat{b} - \left(\hat{c}\hat{b} + \hat{b} - 1\right) = 0, \tag{8.42}$$

where the subscript on the Laplacian indicates that differentiation is with respect to $\hat{\rho}$.

Two dimensionless diffusion coefficients ($\hat{\varepsilon}_c$ and $\hat{\varepsilon}_b$) appear in these equations. In terms of the original dimensional parameters of the problem, they are given by $\hat{\varepsilon}_c = \epsilon\alpha$ and $\hat{\varepsilon}_b = \epsilon D$, where $\alpha = K/[B]_T$ is a buffering factor (small when buffer is at high concentration compared to the dissociation constant), $D = D_B/D_C$ is a relative diffusion coefficient between buffer and $Ca^{2+}$, and the common factor $\epsilon$ is given by

$$\epsilon = (4\pi)^2 D_C^3 K/\sigma^2 k^+, \tag{8.43}$$

a quantity that is small for strong sources and/or fast buffers.

## 8.13.1  The Steady–State Excess Buffer Approximation (EBA)

The dimensionless steady-state equations for the buffered diffusion of $Ca^{2+}$ near a point source, (8.41) and (8.42), are nonlinear, and no general analytical solution is known for these equations. However, we can begin to understand the behavior of solutions (and the effect of $Ca^{2+}$ buffers on $Ca^{2+}$ domains) by considering (8.41) and (8.42) in limiting parameter regimes. The first such limit we will consider is called the "excess buffer approximation." If buffer is in excess, then the parameter $\alpha = K/[B]_T$ will be very small, and $\hat{\varepsilon}_c = \epsilon\alpha$ will also be small when $\epsilon = O(1)$. Therefore, we consider in detail (8.41) and (8.42) when $\hat{\varepsilon}_c \approx 0$, which in physical terms implies that the diffusion coefficient of $c$ is small compared to the size of the reaction terms in (8.41).

The mathematically inclined reader will notice that this is a singular perturbation problem. Because this technique goes beyond the scope of this chapter, we present only a heuristic analysis here. The interested reader is invited to consult Smith et al. (2001) for a more rigorous treatment.

With this caveat, we formally set $\hat{\varepsilon}_c = 0$ in (8.41), giving

$$\hat{c}\hat{b} + \hat{b} - 1 = 0, \tag{8.44}$$

which implies that (8.42) simplifies to

$$\nabla^2\hat{b} = 0.$$

When combined with the boundary conditions for $b$ (see Exercise 13), this equation implies

$$\hat{b} = \hat{b}_\infty, \tag{8.45}$$

where $\hat{b}_\infty = 1/(1 + \hat{c}_\infty)$ is the fraction of free buffer far from the source. Thus, our assumption that buffer is in excess ($\hat{\varepsilon}_c = 0$) implies that the buffer is not perturbed from its equilibrium value $\hat{b}_\infty$. Substituting (8.45) into (8.41) gives

$$\hat{\varepsilon}_c\nabla^2\hat{c} - \hat{b}_\infty(\hat{c} - \hat{c}_\infty) = 0.$$

This is a linear equation satisfied by

$$\hat{c} = \frac{1}{\hat{\rho}}e^{-\hat{\rho}/\Lambda} + \hat{c}_\infty, \tag{8.46}$$

where the dimensionless space constant $\Lambda = \sqrt{\hat{\varepsilon}_c/\hat{b}_\infty}$ and $\hat{c}_\infty = [Ca^{2+}]_i/K$ are chosen to satisfy the boundary conditions on $\hat{c}$ that can be derived from (8.40) (Carslaw and Jaeger 1959; Crank 1975).

When this result is expressed in dimensional form, we have

$$[Ca^{2+}]_i = \frac{\sigma}{4\pi D_C r}e^{-r/\lambda} + [Ca^{2+}]_\infty, \tag{8.47}$$

where $\lambda$ is the characteristic length constant for the mobile $Ca^{2+}$ buffer given by $\lambda = \sqrt{D_C/k^+[B]_\infty}$. This *excess buffer approximation*, first derived by Neher (Neher 1986), is valid when mobile buffer is in high concentration and/or when the source amplitude is small, that is, $\lim_{r\to 0}[B] \approx [B]_\infty$ (Neher 1986; Smith et al. 1996). Note that $\lambda$ decreases with increasing association rate constant ($k^+$) and free buffer concentration far from the source ($[B]_\infty$). When a buffer is in excess, we thus expect further increases in concentration to restrict localized $Ca^{2+}$ elevations. In addition, buffers with fast–reaction kinetics are expected to restrict localized $Ca^{2+}$ elevations more than slow buffers.

## 8.13.2 The Steady–State Rapid Buffer Approximation (RBA)

The *steady-state rapid buffer approximation* near a point source for $Ca^{2+}$ can be derived by noticing in (8.43) that rapid buffer (large $k^+$) leads to small values of $\epsilon$. This results in small values for both $\hat{\varepsilon}_c$ and $\hat{\varepsilon}_b$, which in physical terms implies that the diffusion coefficient of both $\hat{c}$ and $\hat{b}$ are small compared to the size of the reaction terms in (8.41) and (8.42).

If we formally set $\hat{\varepsilon}_c = \hat{\varepsilon}_b = 0$ in these equations, we find that, as before, (8.44) holds. Solving for $\hat{b}$, we find that at every spatial location $\hat{b}$ is given by

$$\hat{b} = \frac{1}{1+\hat{c}}, \tag{8.48}$$

or in dimensional terms,

$$[B] = \frac{K[B]_T}{K+[Ca^{2+}]_i}. \tag{8.49}$$

These equations are statements of *local equilibrium*, the fundamental assumption used in deriving the rapid buffer approximation in the context of traveling waves (recall (8.57)).

We proceed with this derivation of the steady-state RBA by subtracting (8.42) from (8.41) to give

$$\nabla_\rho^2\left(\hat{\varepsilon}_c\hat{c} - \hat{\varepsilon}_b\hat{b}\right) = 0. \tag{8.50}$$

In physical terms this expression is equivalent to the statement that at steady state the flux of total $Ca^{2+}$, diffusing in both free and bound forms across any spherical surface centered on the source is equal to the flux entering through the source (see Exercise 14). Integrating twice with respect to $\hat{\rho}$ and using the boundary conditions to determine the integration constants gives

$$\hat{\varepsilon}_c \hat{c} - \hat{\varepsilon}_b \hat{b} = \frac{\hat{\varepsilon}_c}{\hat{\rho}} + \hat{\varepsilon}_c \hat{c}_\infty - \hat{\varepsilon}_b \hat{b}_\infty. \tag{8.51}$$

Substituting (8.48) into this equation gives

$$\hat{\varepsilon}_c \hat{c} - \hat{\varepsilon}_b \left( \frac{1}{1+\hat{c}} \right) = \frac{\hat{\varepsilon}_c}{\hat{\rho}} + \hat{\varepsilon}_c \hat{c}_\infty - \hat{\varepsilon}_b \hat{b}_\infty, \tag{8.52}$$

which upon solving for $\hat{c}$ and converting back into dimensional form gives the steady-state RBA (Smith 1996)

$$[Ca^{2+}]_i = \frac{1}{2D_C} \left( -D_C K + \frac{\sigma}{4\pi r} + D_C[Ca^{2+}]_\infty - D_B[B]_\infty \right.$$

$$\left. + \sqrt{ \left( D_C K + \frac{\sigma}{4\pi r} + D_C[Ca^{2+}]_\infty - D_B[B]_\infty \right)^2 + 4D_C D_B[B]_T K } \right). \tag{8.53}$$

The steady–state RBA tends to be valid when $\epsilon$ is small, e.g., when buffers have large association and dissociation rate constants ($k^+$ and $k^-$). Interestingly, a sufficiently large source amplitude ($\sigma$) can compensate for modest binding rates, also causing $\epsilon$ to be small and the steady-state RBA to be valid (Smith et al. 1996).

## 8.13.3  Complementarity of the Steady-State EBA and RBA

The fundamental assumptions used in deriving the steady-state excess and rapid buffer approximations are significantly different. In the case of the RBA (8.53) we assumed that buffer and $Ca^{2+}$ were in local equilibrium, (8.49). Because $[Ca^{2+}]_i \to \infty$ as $r \to 0$ in (8.53), we see that according to the steady-state RBA

$$\lim_{r \to 0} [B] \approx 0 \quad (RBA). \tag{8.54}$$

Thus, the steady-state RBA cannot be valid unless the source is strong enough to saturate the buffer. On the other hand, in our derivation of the EBA we assumed that the buffer is not perturbed from its equilibrium value, (8.45). If this is true even near the source, then

$$\lim_{r \to 0} [B] \approx [B]_\infty \quad (EBA). \tag{8.55}$$

Thus, we expect the EBA and RBA approximations to be complementary, in the sense that the steady-state solution to the full equations for the buffered diffusion of $Ca^{2+}$ near a point source (the correct answer) cannot simultaneously be EBA-like, as in (8.55), and RBA-like, as in (8.54). In the process of extending both the EBA and RBA to higher order, this expectation has been confirmed (Smith et al. 2001).

# Suggestions for Further Reading

- *Simulation of the fertilization calcium wave in Xenopus laevis eggs*, John Wagner, Yue Xian Li, John Pearson, Joel Keizer. Modeling study of the fertilization $Ca^{2+}$ wave that suggests inhomogeneities in the $Ca^{2+}$ release properties near the plasma membrane are required to explain the shape and speed of these waves (Wagner et al. 1998).
- *Diffusion of inositol 1,4,5-trisphosphate but not $Ca^{2+}$ is necessary for a class of inositol 1,4,5-trisphosphate-induced $Ca^{2+}$ waves*, Saleet Jafri and Joel Keizer. A modeling study of kinematic $IP_3$-mediated $Ca^{2+}$ waves (Jafri and Keizer 1994).
- *Mathematical Biology*, James D. Murray. Advanced mathematical treatment of biological wave phenomena (Murray 1989).
- *Propagation of waves in an excitable medium with discrete release sites*, James Keener. Mathematical analysis of wave propagation in inhomogeneous bistable media (Keener 2000).
- *Fire–diffuse–fire model of dynamics of intracellular calcium waves*, Silvia Ponce-Dawson, Joel Keizer, John Pearson. The fire–diffuse–fire model is analyzed to illuminate the differences between continuous and saltatory $Ca^{2+}$ wave propagation (Ponce-Dawson et al. 1999).
- *Asymptotic analysis buffered $Ca^{2+}$ diffusion near a point source*, Greg Smith, Longxiang Dai, Robert Miura, and Arthur Sherman. Details the rapid and excess buffer approximations appropriate for modeling the steady–state $Ca^{2+}$ and buffer profiles of localized $Ca^{2+}$ elevations (Smith et al. 2001).
- *Modeling local and global $Ca^{2+}$ signals using reaction–diffusion equations*, Greg Smith. This book chapter provides more discussion on the effect of buffers on propagating $Ca^{2+}$ waves as well as simulations utilizing the rapid buffering approximation outside the low–affinity limit (Smith 2001).

## 8.14  Exercises

1. Show that the assumed form of the indicator dye fluorescence (8.2) and the equilibrium relation (8.1) imply the "backcalculation" formula (8.3).

2. Assuming (8.2) and (8.4), derive the backcalculation formula for a ratiometric indicator, (8.5).

3. Beginning with (8.13), finish the derivation of the rapid–buffer approximation by substituting equilibrium values for $[Ca^{2+}]_T$ and $[CaB]$,

$$\frac{\partial}{\partial t}\left([Ca^{2+}]_i + \frac{[Ca^{2+}]_i[B]_T}{[Ca^{2+}]_i + K}\right) = \nabla^2 \left(D_C[Ca^{2+}]_i + D_B\frac{[Ca^{2+}]_i[B]_T}{[Ca^{2+}]_i + K}\right) + j_T,$$

and taking spatial and temporal derivatives to obtain

$$\frac{1}{\beta}\frac{\partial[Ca^{2+}]_i}{\partial t} = (D_C + D_B\gamma)\nabla^2[Ca^{2+}]_i - \frac{2\gamma D_B}{K + [Ca^{2+}]_i}\left(\nabla^2[Ca^{2+}]_i\right)^2 + j_T$$

where $\gamma = K[B]_T/(K + [Ca^{2+}]_i)^2$ and $\beta = 1/(1 + \gamma)$.

4. Beginning with (8.16), numerically calculate $f(c)$ and confirm that it is a cubic function with three zeros as shown in Figure 8.3. Parameters are $[IP_3] = 0.7\,\mu M$, $a_2 = 0.2\,\mu M/s$, $[Ca^{2+}]_{ER} = 1\,\mu M$, $d_1 = 0.1\,\mu M$, $d_2 = 2\,\mu M$, $d_3 = 0.2\,\mu M$, $d_5 = 0.2\,\mu M$, $f_i v_{IP_3R} = 20\,\mu M/s$, $f_i v_{LEAK} = 0.004\,\mu M/s$, $f_i v_{SERCA} = 1.2\,\mu M/s$, $K_{SERCA} = 0.15\,\mu M$.

5. Using (8.18) reproduce the traveling front simulation shown in Figure 8.4. Parameters: $D = 16\,\mu m^2/s$, $\Delta x = 10\,\mu m$.

6. Analytically confirm that (8.15) can be transformed into traveling wave coordinates resulting in a first–order system of ODEs, (8.19)–(8.20). Follow Figure 8.5 and numerically check that a heteroclinic orbit connects the two stable equilibria (points 1 and 3) for $v = 4.96\,\mu m/s$. Parameters as in Exercise 4 and Exercise 5 with $C(0) = 0.52687\,\mu M$, $G(0) = 0\,\mu M/\mu m$ (i.e., units of $dC/dz$).

7. Implement a traveling pulse of $Ca^{2+}$-induced $Ca^{2+}$ release using (8.21). Parameters as in Exercise 4 and Exercise 6 except $d_2 = 1\,\mu M$ and $\tau = 2\,s$.

8. Repeat the simulations of Figure 8.5 using a range of values of $D$ and numerically confirm that the traveling front velocity is proportional to the square root of the diffusion coefficient. Plot each observed traveling front velocity against $\sqrt{D}$ to see a linear relation.

9. When Figure 8.4 uses the method of lines to simulate the fertilization $Ca^{2+}$ wave model (8.15), the traveling front is observed to have a speed of 3.45 $\mu m/s$. On the other hand, Figure 8.5 shows that the first–order system of ODEs (8.19)–(8.20) has a heteroclinic orbit that connects the two stable equilibria for a wave velocity of $v = 4.96\,\mu m/s$. This discrepancy is due to discretization error, i.e., the choice of $\Delta x = 10\,\mu m$ in Figure 8.4. Repeat Exercise 8 with different values of $\Delta x$ and show that $v \to 4.96$ as $\Delta x \to 0$ ($\Delta x = 1\,\mu m$ works reasonably well). As you refine the mesh, remember to simultaneously let $\Delta t \to 0$, so that $(\Delta x)^2/\Delta t$ remains constant (to maintain numerical stability).

10. Implement a kinematic wave following (8.21) with parameters such that the $(w, [Ca^{2+}]_i)$ phase plane is oscillatory as in Figure 8.8. Parameters as in Exercise 7 except $d_2 = 0.5\,\mu M$, $d_5 = 0.15\,\mu M$.

11. The delta function is the limit of a sequence of tall narrow functions of unit area. For example, show that the function

$$d_T(x - x') = \begin{cases} \dfrac{1}{T} & \text{for } x \in (x' - T/2, x' + T/2), \\ 0 & \text{otherwise,} \end{cases}$$

approaches $\delta(x - x')$ as $T \to 0$ and that the sifting property (8.23) holds in this limit.

12. Implement a fire–diffuse–fire simulation following (8.26) and reproduce Figure 8.12 in both the continuous and saltatory regimes. Parameters for panel A: $D = 30\,\mu m^2/s$, $\Delta x = 0.2\,\mu m$, $\sigma = 5\,\mu M \cdot \mu m$, $\tau_R = 0.01\,s$, $c_\theta = 0.1\,\mu M$. Release site separation is $d = 2\,\mu m$ (i.e., every tenth mesh point). Parameters for panel B are identical except $\tau_R = 1\,s$. The initital conditions are $c(x, 0) = 0$ with the one (three) leftmost release site(s) active in the saltatory (continuous) case.

13. Confirm that nondimensionalizing (8.38)–(8.39) as described in text results in (8.41)–(8.42). Show that the boundary conditions (8.40) become

$$\lim_{\rho \to 0} \left\{ -\hat{\rho}^2 \frac{d\hat{c}}{d\hat{\rho}} \right\} = 1, \quad \lim_{\hat{\rho} \to \infty} \hat{c} = \hat{c}_\infty, \tag{8.56}$$

$$\lim_{\hat{\rho} \to 0} \left\{ -\hat{\rho}^2 \frac{d\hat{b}}{d\hat{\rho}} \right\} = 0, \quad \lim_{\hat{\rho} \to \infty} \hat{b} = \hat{b}_\infty = \frac{1}{1 + \hat{c}_\infty}. \tag{8.57}$$

where $\hat{c}_\infty = [Ca^{2+}]_\infty / K$ and and $\hat{b}_\infty = [B]_\infty / [B]_T$.

14. Show that (8.50) implies that the steady–state flux of total $Ca^{2+}$ (diffusing in both free and bound forms) across any spherical surface centered on the source is equal to the source amplitude. First, convert (8.50) into dimensional form. Then substitute $[B] = [B]_T - [CaB]$ and $[B]_\infty = [B]_T - [CaB]_\infty$ and simplify. Finally, use this expression to calculate $J_{total} = J_{free} + J_{bound}$, where

$$J_{free} = -4\pi r^2 D_C \frac{d[Ca^{2+}]_i}{dr} \qquad J_{bound} = -4\pi r^2 D_B \frac{d[CaB]}{dr},$$

and confirm that $J_{total} = \sigma$.

15. To rigorously derive the steady-state EBA and RBA, asymptotic methods are required (Smith et al. 2001). An alternative analysis of the steady-state equations for the buffered diffusion of $Ca^{2+}$ involves linearizing the equations around the equilibrium concentrations of $Ca^{2+}$ and buffer, $\hat{c}_\infty$ and $\hat{b}_\infty$ (Naraghi and Neher 1997; Pape et al. 1998; Stern 1992; Neher 1998). Begining with the dimensionless equations, define $\delta\hat{c} = \hat{c} - \hat{c}_\infty$ and $\delta\hat{b} = \hat{b} - \hat{b}_\infty$. Substitute these expressions into (8.41) and (8.42) and drop the quadratic terms $\delta\hat{c}\delta\hat{b}$ to obtain

$$\hat{\varepsilon}_c \nabla^2_{\hat{\rho}} \delta\hat{c} - \left[ (1 + \hat{c}_\infty) \delta\hat{b} + \hat{b}_\infty \delta\hat{c} \right] = 0, \tag{8.58}$$

$$\hat{\varepsilon}_b \nabla^2_{\hat{\rho}} \delta\hat{b} - \left[ (1 + \hat{c}_\infty) \delta\hat{b} + \hat{b}_\infty \delta\hat{c} \right] = 0, \tag{8.59}$$

and associated boundary conditions

$$\lim_{\hat{\rho} \to \infty} \delta\hat{c} = 0, \quad \lim_{\hat{\rho} \to 0} \left( -\hat{\rho}^2 \frac{d\delta\hat{c}}{d\hat{\rho}} \right) = 1, \quad \lim_{\hat{\rho} \to \infty} \delta\hat{b} = 0, \quad \lim_{\hat{\rho} \to 0} \left( -\hat{\rho}^2 \frac{d\delta\hat{b}}{d\hat{\rho}} \right) = 0.$$

When this system of linear equations is solved and the result converted back to dimensional form, the following steady-state profiles for $Ca^{2+}$ and buffer result:

$$[Ca^{2+}]_i = [Ca^{2+}]_\infty + \frac{\sigma}{4\pi r (D_C + \gamma_\infty D_B)} \left[ 1 + \frac{\gamma_\infty D_B}{D_C} e^{-r/\lambda} \right], \tag{8.60}$$

$$[B] = [B]_\infty + \frac{\sigma\gamma_\infty}{4\pi r (D_C + \gamma_\infty D_B)} \left[ e^{-r/\lambda} - 1 \right], \tag{8.61}$$

where

$$\frac{1}{\lambda^2} = \frac{1}{\tau} \left( \frac{1}{D_B} + \frac{\gamma_\infty}{D_C} \right), \quad \frac{1}{\tau} = k^+ [Ca^{2+}]_\infty + k^-, \quad \gamma_\infty = \frac{K[B]_T}{(K + [Ca^{2+}]_\infty)^2}. \tag{8.62}$$

Convert (8.60)–(8.61) into dimensionless form to find expressions for $\delta\hat{c}$ and $\delta\hat{b}$ and show that these satisfy (8.58)–(8.59). Also show that when $\gamma_\infty$ is large (lots of buffer), (8.60) reduces to (8.47).

16. Substitute the spherical polar Laplacian (8.37) into (8.57) and show that the steady-state RBA can be found by integrating (Smith 1996).

# CHAPTER 9

# Biochemical Oscillations

## John J. Tyson

Biochemical and biophysical rhythms are ubiquitous characteristics of living organisms, from rapid oscillations of membrane potential in nerve cells to slow cycles of ovulation in mammals. One of the first biochemical oscillations to be discovered was the periodic conversion of sugar to alcohol ("glycolysis") in anaerobic yeast cultures (Chance et al. 1973). The oscillation can be observed as periodic changes in fluorescence from an essential intermediate, NADH; see Figure 9.1. In the laboratories of Britton Chance and Benno Hess, these oscillations were shown to arise from a curious property of the enzyme phosphofructokinase (PFK), which catalyzes the phosphorylation of fructose-6-phosphate to fructose-1,6-bisphosphate using ATP as the phosphate donor; Figure 9.1B. Although PFK consumes ATP, the glycolytic pathway produces more ATP than it consumes. To properly regulate ATP production, ATP inhibits PFK and ADP activates PFK. Hence, if the cell is energy "rich" (ATP high, ADP low), then PFK activity is inhibited, and the flux of sugars into the glycolytic pathway is shut down. As ATP level drops and ADP level increases, PFK is activated and glycolysis recommences. In principle, this negative feedback loop should stabilize the energy supply of the cell. However, because ATP is both a substrate of PFK and an inhibitor of the enzyme (likewise, ADP is both a product of PFK and an activator of the enzyme), the steady-state flux through the glycolytic pathway can be unstable, as we shall see, and the regulatory system generates sustained oscillations in all intermediates of the pathway.

JJT thanks Emery Conrad for help in writing this chapter.

**A**

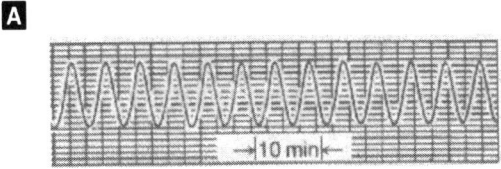

←10 min→

**B**

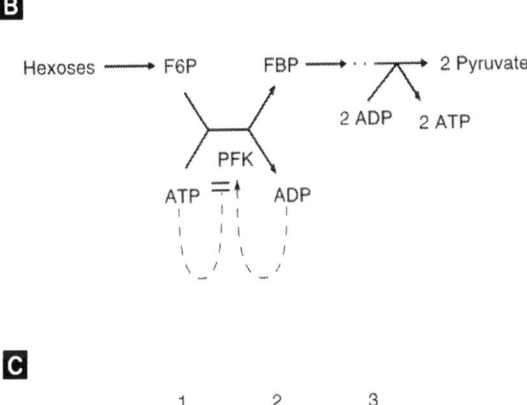

Hexoses ⟶ F6P          FBP ⟶ ··· ⟶ 2 Pyruvate

2 ADP    2 ATP

PFK

ATP ⇌ ADP

**Figure 9.1** (A) Sustained oscillations in NADH fluorescence in yeast cells *Saccharomyces cerevisiae*. Reprinted from Pye (1971). (B) The control properties of the enzyme phosphofructokinase (PFK) are thought to be responsible for the generation of oscillations in the glycolytic pathway. PFK catalyzes the conversion of fructose-6-phosphate (F6P) into fructose-1,6-bisphosphate (FBP), using ATP as phosphate-group donor. PFK activity is allosterically modulated by ATP (inhibitor) and ADP (activator). F6P is steadily supplied to PFK by a sugar source ("Hexoses") and FBP is steadily utilized in the production of metabolites ("Pyruvate"). ADP and ATP are also recycled by other metabolic processes. (C) Simplified mechanism. Reaction 1 is a steady supply of substrate Y for reaction 2, whose product X is removed by reaction 3. X activates the enzyme (PFK) catalyzing reaction 2. Roughly speaking, Y = F6P + ATP, X = FBP + ADP, "source" = hexoses, and "sink" = pyruvate.

**C**

source ⟶ Y ⟶ X ⟶ sink
          1      2      3

Another classic example of rhythmic behavior in biology is periodic growth and division of well-nourished cells. This phenomenon will be studied thoroughly in the next chapter, but for now, to illustrate some basic ideas, we focus on the periodic accumulation and degradation of cyclins during the division cycle of yeast cells; Figure 9.2. Kim Nasmyth and others have shown that these oscillations are intimately connected to dynamical interactions between CLN-type cyclins and CLB-type cyclins; Figure 9.2 (Nasmyth 1996). CLNs (in combination with a kinase subunit called CDC28) activate their own synthesis (product activation or "autocatalysis") and inhibit the degradation of CLBs. As CLBs accumulate, they inhibit the synthesis of CLNs, causing CLN-dependent kinase activity to drop and CLB degradation to increase. The mutual interplay of CLN and CLB generate the periodic appearance of their associated kinase activities, which drive the crucial events of the budding yeast cell cycle (bud emergence, DNA synthesis, mitosis, and cell division).

    The third example that we shall use in this chapter concerns periodic changes in physiological properties (physical activity, body temperature, reproduction, etc.) entrained to the 24 h cycle of light and darkness so prevalent to life on earth. These rhythms are not driven solely by the external timekeeper, because they persist under

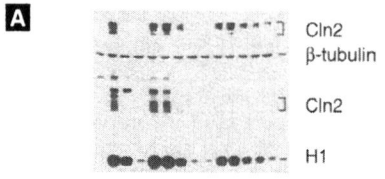

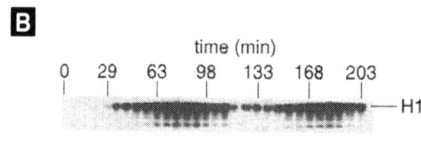

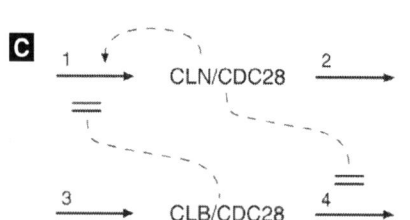

**Figure 9.2** Cyclin fluctuations during the cell cycle in budding yeast. (A) Samples are taken at increasing time points (left to right) from a synchronous culture of budding yeast cells and analyzed for CLN2/CDC28 protein content (top row) and CLN2/CDC28 kinase activity (last row, labeled "H1"). Reprinted from Tyers et al. (1993). β-tubulin serves as a control, nonoscillating, protein. (B) In a similar experiment, a synchronous culture of yeast cells is analyzed for CLB2/CDC28 activity (labeled "H1"; each column represents a 7 min increment). Reprinted from Surana et al. (1991). (C) Mechanism of cyclin oscillations. Budding yeast cells contain two classes of cyclins: CLN and CLB. These cyclins combine with a kinase partner (CDC28) to make active dimers. CDC28 subunits are always in excess, so dimer activity is limited by cyclin availability (i.e., cyclin synthesis and degradation, arrows 1-4 in the diagram). CLN/CDC28 activates the transcription factor that promotes CLN synthesis and inhibits the proteolytic enzymes that degrade CLB. In return, CLB/CDC28 inhibits the transcription factor that promotes CLN synthesis.

constant conditions of illumination and temperature; Figure 9.3A. Under constant conditions, the organism exhibits its own "endogenous" rhythm, which is close to, but not exactly, 24 h (hence "circadian," or "nearly daily"). The basic molecular mechanism of circadian rhythms has been uncovered only recently, by research in the laboratories of Michael Rosbash, Michael Young, Jay Dunlap, and others (Dunlap 1999). Central to the mechanism is a protein called PER, which, after being synthesized in the cytoplasm, moves into the nucleus to inhibit the transcription of its own mRNA; Figure 9.3B. Hence, the regulatory mechanism for PER synthesis contains a time-delayed negative-feedback loop (Figure 9.3C), which, as we shall see, is another common theme in biochemical oscillators.

# 9.1  Biochemical Kinetics and Feedback

In general, a biochemical reaction network is a schematic diagram of "boxes and arrows." Each box is a chemical species. Solid arrows represent chemical reactions producing or consuming a molecule. Dashed arrows represent control of a chemical reaction rate by a species that may not be a reactant or product of the reaction. Regulatory signals may be activatory (barbed end) or inhibitory (blunt end).

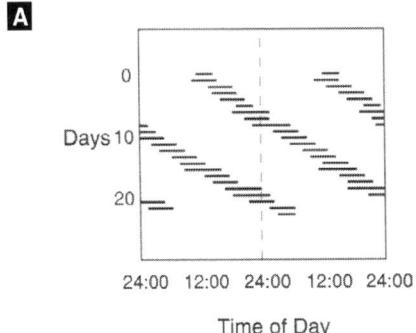

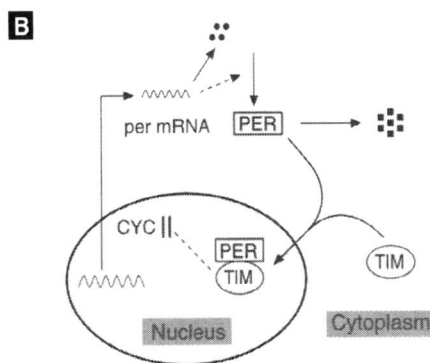

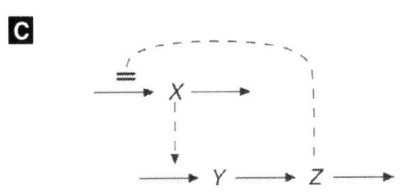

**Figure 9.3** (A) Endogenous circadian rhythm of activity. The black bars represent the sleep episodes of a human being isolated from all external temporal cues (variable light, temperature, etc.). The sleep episodes are plotted twice on each line to emphasize that the endogenous period of sleepiness is longer than 24 h. Data redrawn from Siffre (1975). (B) A protein called PER is known to play a crucial role in circadian rhythms in fruit flies and mice. The *per* gene is transcribed in the nucleus, with the help of two transcription factors, CLK and CYC. Then *per* mRNA is transported to the cytoplasm, where it codes for PER protein. PER protein is processed in the cytoplasm by phosphorylation and by binding to other proteins, such as TIM and CRY. Properly processed PER then moves into the nucleus, where it disrupts the binding of CLK and CYC, turning off the transcription of *per*. (C) A schematic representation of the negative-feedback loop in panel (B). This is Goodwin's classical mechanism for periodic protein expression, driven by feedback repression of transcription (Goodwin 1965).

To each reaction in a biochemical network is associated a rate law (with accompanying kinetic parameters). Hence, the network implies a set of rate equations of the form

$$\frac{dx}{dt} = v_{in1} + v_{in2} + \cdots - v_{out1} - v_{out2} - \cdots = f(x; p), \tag{9.1}$$

where $t$ = time, $x$ = concentration of species X, and $v_{in1}, v_{out1}, \ldots$ are the rates of the various reactions that produce and consume X. We lump together all these rate laws into a nonlinear function $f(x; p)$, where $p$ is a vector of kinetic parameters. In general, we can think of $x$ as a vector of concentrations of all the time-varying components in

**Table 9.1  Biochemical and Cellular Rhythms** Sources: Goldbeter (1996), Rapp (1979).

| Rhythm | Period |
|---|---|
| Membrane potential oscillations | 10 ms–10 s |
| Cardiac rhythms | 1 s |
| Smooth muscle contraction | seconds – hours |
| Calcium oscillations | seconds–minutes |
| Protoplasmic streaming | 1 min |
| Glycolytic oscillations | 1 min–1 h |
| cAMP oscillations | 10 min |
| Insulin secretion (pancreas) | minutes |
| Gonadotropic hormone secretion | hours |
| Cell cycle | 30 min–24 h |
| Circadian rhythms | 24 h |
| Ovarian cycle | weeks–months |

the reaction network, $x_i = [X_i]$, and $f$ as a vector-valued function. In component form,

$$\frac{dx_i}{dt} = f_i(x_1, x_2, \ldots, x_n; p_1, p_2, \ldots, p_r), \quad i = 1, \ldots, n. \tag{9.2}$$

A "steady-state" solution $(x_1^*, x_2^*, \ldots, x_n^*)$ of the kinetic equations satisfies the algebraic equations

$$f_i(x_1, x_2, \ldots, x_n; p_1, p_2, \ldots, p_r) = 0, \quad i = 1, \ldots, n. \tag{9.3}$$

Notice that steady-state concentrations depend on the parameter values $x_i^* = x_i^*(p_1, \ldots, p_r)$. Moreover, the algebraic equations are generally nonlinear, so there may be multiple steady-state solutions. Of course, since we are dealing with chemical reaction systems, we are interested only in solutions that satisfy $x_i \geq 0$ for all $i$.

In the theory of biochemical oscillations, based on rate equations of this sort, a crucial role is played by elements of the Jacobian matrix $\mathbf{J} = [a_{ij}]$, where $a_{ij} = \partial f_i / \partial x_j$ evaluated at a steady-state solution. The matrix $\mathbf{J}$ is a square matrix ($n \times n$) dependent on parameter values and the steady state under consideration. The diagonal elements of the Jacobian matrix of a chemical reaction system are usually negative, $a_{ii} < 0$, because $v_{\text{out}}$ terms in (9.1) are always proportional to the concentration of the substance being destroyed, so $\partial f_i / \partial x_i$ always has one or more negative terms. Positive contributions to $a_{ii}$ are rare because autocatalysis (reactions that produce $X_i$ at a rate that increases with $[X_i]$) is rare.

Common to biochemical networks are complex feedback loops, whereby the products of one reaction affect the rates of other reactions. A steady-state feedback loop can be defined as a set of nonzero elements of the Jacobian matrix that connect in a loop: $a_{ij} a_{jk} a_{kl} \cdots a_{mi} \neq 0$.

Representative feedback loops are illustrated in Figure 9.4.

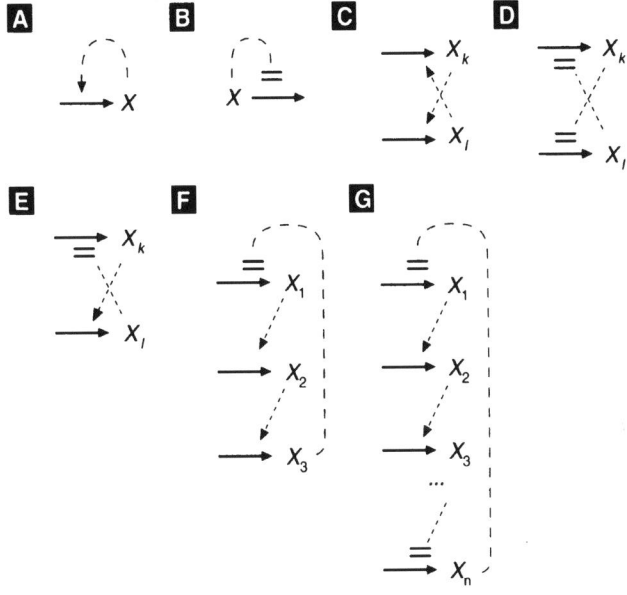

**Figure 9.4** Representative feedback loops.

(A) Autocatalysis ($a_{ii} > 0$), though rare, plays a major role in biochemical oscillations, as will become clear. We have already seen examples of autocatalysis in the mechanisms of glycolysis and yeast division; Figure 9.1 and Figure 9.2.

(B) Autocatalysis also occurs when a chemical decelerates the rate of its own destruction.

(C) Indirect autocatalysis occurs through a positive feedback loop ($a_{ij}$ and $a_{ji} > 0$), whereby $X_i$ activates the production of $X_j$, and $X_j$ returns the favor.

(D) A two-component positive-feedback loop is also created by a pair of antagonistic species ($a_{ij} < 0$ and $a_{ji} < 0$).

(E) A two-component negative-feedback loop ($a_{ij}a_{ji} < 0$) is created when $X_i$ activates the production of $X_j$, and $X_j$ inhibits the production of $X_i$.

(F) Longer negative feedback loops are common.

(G) Long feedback loops are either positive or negative, depending on the sign of the product $a_{1n}a_{n,n-1} \cdots a_{43}a_{32}a_{21}$.

Before we can speak definitively about the effects of feedback in biochemical reaction networks, we need to know how to characterize the rates of enzyme-catalyzed reactions that are subject to regulation by "distant" effectors, i.e., chemical species other than the reactants and products of the enzyme.

## 9.2   Regulatory Enzymes

Regulatory enzymes are usually multisubunit proteins with binding sites for reactants and products (on the catalytic subunits) and for activators and inhibitors (on the regulatory subunits). Although there are more sophisticated and accurate ways to characterize the binding of small molecules to multisubunit proteins, we shall limit ourselves to a straightforward generalization of the Michaelis–Menten equation from Section 4.7. For simplicity, consider a two-subunit enzyme (EE) that converts substrate (S) into product (P). The binding constant and turnover number of each subunit depend on whether the other (identical) subunit is bound to S or not. The mechanism can be written

EE + S ↔ EES        dissociation constant = $k_{-1}/k_1$,
EES → EE + P        rate constant = $k_3$,
EES + S ↔ SEES      dissociation constant = $k_{-2}/k_2$,
SEES → EES + P      rate constant = $k_4$.

As in Section 4.1, we can make rapid-equilibrium approximations on the enzyme–substrate complexes:

$$\frac{[EES]}{[EE]_T} = \frac{[S]/K_{m1}}{1 + ([S]/K_{m1}) + ([S]^2/(K_{m1}K_{m2}))},$$

$$\frac{[SEES]}{[EE]_T} = \frac{[S]^2/K_{m1}K_{m2}}{1 + ([S]/K_{m1}) + ([S]^2/(K_{m1}K_{m2}))},$$

where $[EE]_T = [EE]+[EES]+[SEES]$, $K_{m1} = (k_{-1}+k_3)/k_1$, and $K_{m2} = (k_{-2}+k_4)/k_2$. Here $K_{m1}$ is the Michaelis constant for the interaction of the enzyme with the first substrate to bind; $K_{m1}$ is inversely proportional to the affinity of the enzyme for the first substrate; $K_{m2}$ is inversely proportional to the affinity of the enzyme for the second substrate to bind.

Now it is easy to write an equation for the rate of the reaction:

$$v = \frac{d[P]}{dt} = -\frac{d[S]}{dt} = \frac{[EE]_T \left(\frac{[S]}{K_{m1}}\right)\left(k_3 + k_4\left(\frac{[S]}{K_{m2}}\right)\right)}{1 + \left(\frac{[S]}{K_{m1}}\right) + \left(\frac{[S]^2}{K_{m1}K_{m2}}\right)}. \tag{9.4}$$

We can distinguish two interesting limiting cases of (9.4).

*Hill equation*: $K_{m1} \to \infty, K_{m2} \to 0$, such that $K_{m1}K_{m2} = K_m^2$ = constant,

$$v = \frac{V_{max}([S]/K_m)^2}{1 + ([S]/K_m)^2}, V_{max} = k_4[EE]_T; \tag{9.5}$$

*Substrate inhibition*: $k_3 \to \infty, K_{m1} \to \infty, K_{m2} \to 0$, such that $K_{m1}K_{m2} = K_m^2$ = constant, $k_3K_{m2} \to \infty$, and $k_3[EE]_T K_m/K_{m1} = V_{max}$ = constant,

$$v = \frac{V_{max}[S]/K_m}{1 + ([S]/K_m)^2}. \tag{9.6}$$

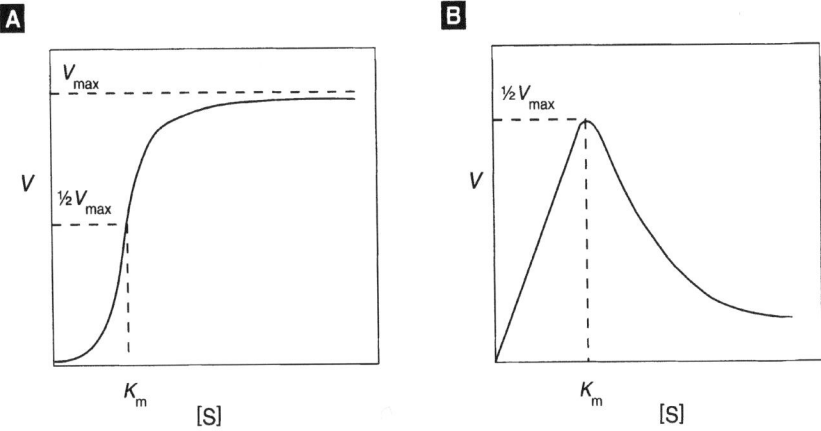

**Figure 9.5** Rate laws for activation and inhibition of multisubunit enzymes by cooperative binding.

Equations (9.5) and (9.6) are plotted in Figure 9.5.

Rate laws like (9.4) are said to express "cooperative" kinetics. The "empty" enzyme (EE) has low affinity for substrate ($K_{m1}$ large), but as the enzyme picks up its first substrate molecule, the two subunits change their conformation to a high affinity form ($K_{m2}$ small, such that $K_m = (K_{m1}K_{m2})^{1/2}$ is physiologically significant).

If the subunits do not exhibit cooperative binding, then $k_1 = 2k_2, 2k_{-1} = k_{-2}, k_4 = 2k_3$, and hence

$$v = \frac{V_{max}[S]}{K_m + [S]}.$$ (9.7)

That is, we regain the Michaelis–Menten rate law, with $V_{max} = 2k_3$ and $K_m = 2(k_{-1} + k_3)/k_1$.

Cooperative binding of ligands to a multisubunit enzyme is not limited to substrates. Other small molecules may bind to the enzyme and alter its catalytic properties (either its affinity for substrates or its rate of converting bound substrates into products). Such enzymes are called "allosteric" because in addition to substrate-binding sites, they have "other sites" for binding regulatory molecules that either activate or inhibit the enzyme. Allosteric proteins play crucial roles in the regulation of metabolic pathways, membrane transport, gene expression, etc. Relatively simple algebraic expressions for allosteric effects can be derived by the reasoning behind (9.4).

For instance, consider a tetrameric enzyme, with two catalytic subunits (EE) and two regulatory subunits (RR), which bind substrate (S) and ligand (L), respectively. In this case, the holoenzyme may exist in nine different forms. If each association–dissociation reaction is in rapid equilibrium at given concentrations of substrate and ligand, then

$$[E_T] = [EE_{00}] \left(1 + \frac{[S]}{K_{m1}} + \frac{[S]^2}{K_{m1}K_{m2}}\right) \left(1 + \frac{[L]}{K_{n1}} + \frac{[L]^2}{K_{n1}K_{n2}}\right)$$

where $[E]_T$ = total concentration of enzyme in all nine forms, $[EE_{00}]$ = concentration of holoenzyme unbound to substrate or ligand, $[S]$ = substrate concentration, $[L]$ = ligand concentration, $K_m$'s = Michaelis constants for substrate binding (e.g., $K_{m1} = k_{-1}/k_1$), and $K_n$'s = Michaelis constants for ligand binding (defined similarly). If, for example, the only form of the enzyme with significant catalytic activity is

$$\begin{bmatrix} SEES \\ LRRL \end{bmatrix},$$

then the rate of the reaction is

$$v = V_{max} \frac{[S]^2/(K_{m1}K_{m2})}{1 + [S]/K_{m1} + [S]^2/(K_{m1}K_{m2})} \cdot \frac{[L]^2/(K_{n1}K_{n2})}{1 + [L]/K_{n1} + [L]^2/(K_{n1}K_{n2})}. \qquad (9.8)$$

In this case, the ligand is an allosteric *activator* of the enzyme. For an allosteric *inhibitor*, the enzyme is most active when no ligand is bound to the regulatory subunits; hence

$$v = V_{max} \frac{[S]^2/(K_{m1}K_{m2})}{1 + [S]/K_{m1} + [S]^2/(K_{m1}K_{m2})} \cdot \frac{1}{1 + [L]/K_{n1} + [L]^2/(K_{n1}K_{n2})}. \qquad (9.9)$$

A comprehensive and accurate kinetic theory of allosteric enzymes (Rubinow 1980; Goldbeter 1996) is much more complicated than what has been presented, but (9.8) and (9.9) will serve our purposes in this chapter.

Finally, suppose EE and S above are not "multisubunit enzyme" and "substrate" but "dimeric transcription factor" and "ligand." The active form of the transcription factor promotes the expression of some gene, and the presence of ligand alters the distribution of transcription factor among its various forms: EE, EES, and SEES. Depending on which form of transcription factor is most active, the rate of gene expression can be activated and/or inhibited by ligand:

| ligand effect | active form | relative rate of gene expression |
|---|---|---|
| inhibition | EE | $\dfrac{1}{1+([S]/K_{m1})+([S]^2/K_{m1}K_{m2})}$, |
| activation | SEES | $\dfrac{[S]^2/K_{m1}K_{m2}}{1+([S]/K_{m1})+([S]^2/K_{m1}K_{m2})}$, |
| mixed | EES | $\dfrac{[S]/K_{m1}}{1+([S]/K_{m1})+([S]^2/K_{m1}K_{m2})}$. |

## 9.3  Two-Component Oscillators Based on Autocatalysis

First, we consider two minimal requirements for oscillations in chemical reaction systems.

1. If the "network" is absurdly simple, with only one time-varying component, then oscillations are, generally speaking, impossible. (Proof: For $x(t)$ to be periodic, $dx/dt$ must take on both positive and negative values at some values of $x$, which is impossible if $f(x; p)$ is a continuous single-valued function of $x$.) Thus, to understand biochemical oscillations, we must start with two-component networks, described by a pair of ODEs

$$\frac{dx}{dt} = f(x, y),$$
$$\frac{dy}{dt} = g(x, y), \qquad\qquad (9.10)$$

where $x$ and $y$ are (nonnegative) concentrations of the two components, and we have suppressed the dependence of $f$ and $g$ on parameters, for the time being.

2. Bendixson's negative criterion states that if $\partial f/\partial x + \partial g/\partial y$ is of constant sign in some region $R$ of the $x, y$ plane, then there can be no periodic solution of system (9.10) in $R$. Since $\partial f/\partial x$ and $\partial g/\partial y$ are usually negative (refer to our earlier discussion of the Jacobian matrix), Bendixson's criterion requires that a two-component chemical reaction system have an autocatalytic step ($\partial f/\partial x > 0$ or $\partial g/\partial y > 0$) in order to exhibit sustained oscillations.

In general, we can expect system (9.10) to have one or more steady-state solutions $(x^*, y^*)$, satisfying $f(x^*, y^*) = 0$ and $g(x^*, y^*) = 0$. The stability of such steady states is determined by the eigenvalues of the Jacobian matrix evaluated at the steady state

$$\mathbf{J} = \begin{bmatrix} f_x(x^*, y^*) & f_y(x^*, y^*) \\ g_x(x^*, y^*) & g_y(x^*, y^*) \end{bmatrix} = \begin{bmatrix} a_{11} & a_{12} \\ a_{21} & a_{22} \end{bmatrix}.$$

The eigenvalues of $\mathbf{J}$ are the roots of the characteristic equation $\lambda^2 - (a_{11} + a_{22})\lambda + (a_{11}a_{22} - a_{12}a_{21}) = 0$, namely,

$$\lambda_\pm = \frac{1}{2}\left[a_{11} + a_{22} \pm \sqrt{(a_{11} + a_{22})^2 - 4(a_{11}a_{22} - a_{12}a_{21})}\right].$$

For the steady state to be stable, Re $(\lambda)$ must be negative for both eigenvalues. If $a_{11}a_{22} - a_{12}a_{21} = \det(\mathbf{J}) < 0$, then $\mathbf{J}$ has one positive and one negative eigenvalue, and the steady state is a saddle point. If $\det(\mathbf{J}) > 0$ and $a_{11} + a_{22} = \text{tr}\,(\mathbf{J}) < 0$, then the steady state is stable, whereas, if $\det(\mathbf{J}) > 0$ and $\text{tr}\,(\mathbf{J}) > 0$, then the steady state is an unstable node or focus.

In general, the trace and determinant of $\mathbf{J}$ depend continuously on the kinetic parameters in $f$ and $g$. If by varying one of these parameters (call it $p_1$) we can carry $\text{tr}(\mathbf{J})$ from negative to positive values, with $\det(\mathbf{J}) > 0$, then the steady state loses stability at at $tr(\mathbf{J}) = 0$ (when $p_1 = p_{crit}$, say). At the bifurcation point, tr $(\mathbf{J}) = a_{11} + a_{22} = 0$, the eigenvalues are pure imaginary numbers, $\lambda_{\pm} = \pm i\omega$, $\omega = 2\sqrt{a_{11}a_{22} - a_{12}a_{21}}$. Close to the bifurcation point, i.e., $p_1 \approx p_{crit}$, small–amplitude limit–cycle solutions surround the steady state, and the period of oscillation is close to $2\pi/\omega$. We say that periodic solutions arise by a **Hopf bifurcation** at $p_1 = p_{crit}$ (Section A.5.2).

Biochemical oscillations usually arise by this mechanism, so we will consider first the requirements for Hopf bifurcation in a two-component network. In chemical reaction systems the diagonal elements of the Jacobian matrix are usually negative numbers, reflecting the various steps by which species X is transformed into something else. If both $a_{11}$ and $a_{22}$ are always negative, then $tr(\mathbf{J})$ never changes sign, and a Hopf bifurcation cannot occur. At least one of them must be positive for some values of the kinetic parameters. For a diagonal element to be positive, species X must be "autocatalytic;" i.e., with increasing [X], the rates of production of X increase faster than the rates of destruction. If $a_{11}$ and $a_{22}$ are of opposite sign, then $a_{12}$ and $a_{21}$ must also be of opposite sign in order for $det(\mathbf{J})$ to be positive. Thus we have two characteristic sign patterns for Jacobian matrices that typically produce Hopf bifurcations in chemical reaction systems with two time-dependent components:

$$\mathbf{J} = \begin{bmatrix} + & + \\ - & - \end{bmatrix} \text{ and } \mathbf{J} = \begin{bmatrix} + & - \\ + & - \end{bmatrix}. \tag{9.11}$$

Next, we describe the sorts of biochemical networks that produce these sign patterns and generate periodic solutions via Hopf bifurcations.

## 9.3.1 Substrate–Depletion Oscillator

The simplest oscillatory mechanism is probably the linear pathway in Figure 9.1C. Species Y is converted into X by an enzyme that is activated by its product. Hence, the production of X is autocatalytic; the reaction speeds up as [X] increases, until the substrate Y is depleted so much that the reaction ceases.

Using a rate law for the allosteric enzyme like those derived in Section 9.2, we can write a pair of ODEs describing this mechanism:

$$\frac{d[X]}{dt} = v_2[Y]\frac{\epsilon^2 + ([X]/K_n)^2}{1 + ([X]/K_n))^2} - k_3[X],$$

$$\frac{d[Y]}{dt} = k_1 - v_2[Y]\frac{\epsilon^2 + ([X]/K_n)^2}{1 + ([X]/K_n))^2}. \tag{9.12}$$

The rational function

$$\frac{\epsilon^2 + ([X]/K_n)^2}{1 + ([X]/K_n))^2}$$

represents activation of the allosteric enzyme by binding of product X to the enzyme's regulatory sites. The constant $\epsilon^2$ is the activity of the enzyme with no product bound relative to its activity when its regulatory sites are saturated by product molecules. We assume that product-binding to regulatory sites is highly cooperative, so that we can neglect the fraction of enzyme with only one product molecule bound. We shall also assume that $\epsilon \ll 1$.

It is convenient to define "dimensionless" variables $x = [X]/K_n$, $y = [Y]/K_n$, and $t' = k_3 t$, and a new variable $z = x + y$, and write system (9.12) as

$$\frac{dx}{dt'} = v(z - x)\frac{\epsilon^2 + x^2}{1 + x^2} - x,$$

$$\frac{dz}{dt'} = \kappa - x, \tag{9.13}$$

where $v = v_2/k_3$ and $\kappa = k_1/(k_3 K_n)$. The steady state for this model satisfies $x^* = \kappa$ and $v(z^* - \kappa) = \kappa(1 + \kappa^2)/(\epsilon^2 + \kappa^2)$. For the Jacobian matrix of system (9.13) at the steady state, it is easy to show that $\det(\mathbf{J}) = v(\epsilon^2 + \kappa^2)/(1 + \kappa^2) > 0$ and

$$\begin{aligned}
\mathrm{tr}\,(\mathbf{J}) &= -1 - v\frac{\epsilon^2 + \kappa^2}{1 + \kappa^2} + \frac{2\kappa^2(1 - \epsilon^2)}{(\epsilon^2 + \kappa^2)(1 + \kappa^2)} \\
&= -\frac{(1 + v)\kappa^4 - (1 - 3\epsilon^2 - 2v\epsilon^2)\kappa^2 + \epsilon^2(1 + v\epsilon^2)}{(\epsilon^2 + \kappa^2)(1 + \kappa^2)}.
\end{aligned} \tag{9.14}$$

It should be obvious from (9.14) that $\mathrm{tr}\,(\mathbf{J}) < 0$ and the steady state is stable if $\kappa$ is either close to 0 or very large. On the other hand, $\mathrm{tr}\,(\mathbf{J}) > 0$, and the steady state is unstable if $\kappa$ takes on intermediary values. For $\epsilon$ small, the steady state is unstable when $\kappa^2 - \epsilon^2 > 0$ and when $-(1 + v)\kappa^4 + \kappa^2 > 0$, i.e., when

$$\epsilon < \kappa < (1 - v)^{-1/2}, \tag{9.15}$$

which may, alternatively, be expressed as conditions on the input rate of Y:

$$\epsilon < \frac{k_1}{k_3 K_n} < \sqrt{\frac{k_3}{k_3 + v_2}}. \tag{9.16}$$

Between these limits the system executes stable limit cycle oscillations.

As we have often seen in this book, for two-component dynamical systems it is informative to plot nullclines in the phase plane. For system (9.13),

$$x\text{–nullcline: } z = x + \frac{x(1 + x^2)}{v(\epsilon^2 + x^2)},$$

$$z\text{–nullcline: } x = \kappa.$$

These nullclines are plotted in Figure 9.6. The $x$–nullcline has extrema at the roots of

$$(1 + v)x^4 - (1 - 3\epsilon^2 - 2v\epsilon^2)x^2 + \epsilon^2(1 + v\epsilon^2) = 0.$$

Notice that the steady state loses stability exactly when the steady state passes through the extrema of the $x$–nullcline.

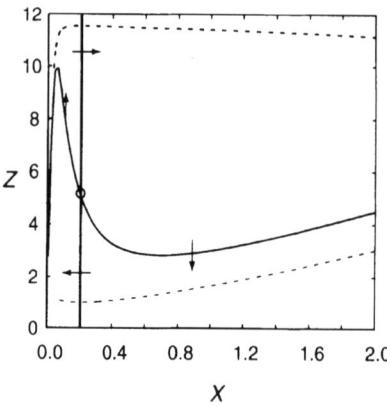

**Figure 9.6** Phase plane portrait of a typical substrate–depletion oscillator, (9.13). Parameter values: $\nu = 1, \kappa = 0.2, \epsilon = 0.05$. The solid lines are the x-nullcline (marked by short vertical arrows) and the y-nullcline (horizontal arrows). The nullclines intersect at an unstable steady state (o). The dashed line is the stable limit cycle solution (periodic orbit) of the dynamical system.

## 9.3.2   Activator–Inhibitor Oscillator

Looking back to (9.11), we see that there are two sign patterns consistent with Hopf bifurcation in a two-component biochemical reaction system. The first is illustrated by the substrate–depletion oscillator in Figure 9.1, and the second by the activator–inhibitor model in Figure 9.2. The ODEs describing Figure 9.2 are

$$\frac{d[X]}{dt} = \nu_1 \frac{\epsilon^2 + ([X]/K_m)^2}{1 + ([X]/K_m)^2} \cdot \frac{1}{1 + ([Y]/K_n)} - k_2[X],$$

$$\frac{d[Y]}{dt} = k_3 - \frac{k_4[Y]}{1 + ([X]/K_j)^2},$$

where X = CLN/CDC28 and Y = CLB/CDC28. In terms of dimensionless variables ($x = [X]/K_m, y = [Y]/K_n, t' = \nu_1 t/K_m$), these equations become

$$\frac{dx}{dt'} = \frac{\epsilon^2 + x^2}{1 + x^2} \cdot \frac{1}{1 + y} - ax,$$

$$\tau \frac{dy}{dt'} = b - \frac{y}{1 + cx^2}, \tag{9.17}$$

where $a = k_2 K_m/\nu_1$, $b = k_3/(k_4 K_n)$, $c = (K_m/K_j)^2$, and $\tau = \nu_1/(k_4 K_m)$.

Rather than analyze the stability of the steady state algebraically, which is difficult, we go directly to phase plane portraits of Figure 9.7. It is easy to find parameter values that produce either limit cycle oscillations or bistability.

Intuitively, the origin of the oscillations is clear. When Y is rare, X increases autocatalytically. Abundant X stimulates accumulation of Y (by inhibiting Y's degradation), which feeds back to inhibit the production of X. After X disappears, Y is also destroyed, and then X can make a comeback. Mechanisms like this one, whose Jacobian matrix

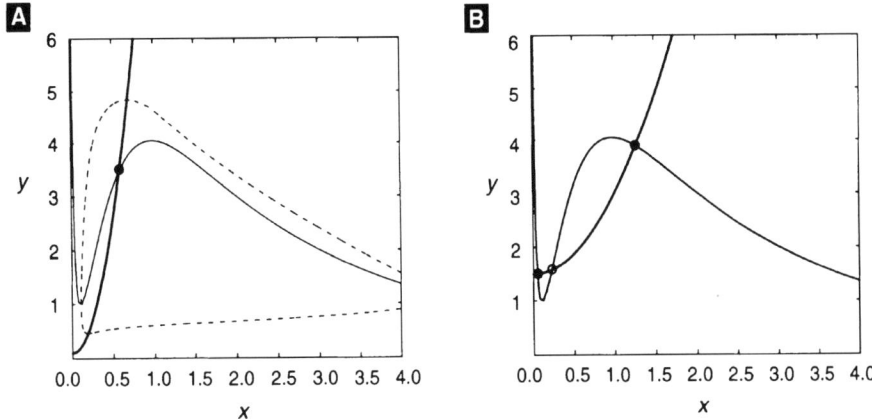

**Figure 9.7** Phase plane portrait of a typical activator–inhibitor system, (9.17). The solid lines are nullclines, intersecting at stable (•) or unstable (∘) steady states. The dashed line is a closed orbit of the dynamical system. (A) Oscillation, for $a = 0.1, b = 0.1, c = 100, \epsilon = 0.1, \tau = 5$. (B) Bistability, for $a = 0.1, b = 1.5, c = 1, \epsilon = 0.1, \tau = 5$.

has sign pattern

$$\begin{bmatrix} + & - \\ + & - \end{bmatrix},$$

are called activator–inhibitor models.

Two-component mechanisms with autocatalysis easily generate oscillations and bistabilty, as we have just seen. They also exhibit a rich structure of bifurcations to more complicated behavior (Boissonade and De Kepper 1980; Guckenheimer 1986).

## 9.4  Three-Component Networks Without Autocatalysis

In the previous section we have seen that two-component reaction systems can oscillate if they have autocatalysis ($a_{11}$ or $a_{22}$ positive) and negative feedback ($a_{12}$ and $a_{21}$ opposite sign). In this section we examine networks of three components ($x, y, z$) with Jacobian matrices of the form

$$\mathbf{J}_{\pm} = \begin{bmatrix} -\alpha & 0 & \pm\phi \\ c_1 & -\beta & 0 \\ 0 & c_2 & -\gamma \end{bmatrix},$$

where $\alpha, \beta, \gamma, c_1, c_2, \phi$ are all positive constants. Since the diagonal elements of the Jacobian are all negative, the system lacks autocatalysis. The Jacobian $\mathbf{J}_{+}$ describes a system with a positive feedback loop, and $\mathbf{J}_{-}$ one with a negative feedback loop.

### 9.4.1    Positive Feedback Loop and the Routh–Hurwitz Theorem

First, let us see whether Hopf bifurcations are possible in a system with a pure positive–feedback loop, i.e., a Jacobian of the form $\mathbf{J}_+$ at the steady state. Recall that the condition for a Hopf bifurcation is that the Jacobian matrix has a pair of purely imaginary eigenvalues $\lambda_\pm = \pm i\omega$. The eigenvalues of $\mathbf{J}_+$ are roots of the characteristic equation

$$
\begin{aligned}
G(\lambda) &= \lambda^3 + (\alpha + \beta + \gamma)\lambda^2 + (\alpha\beta + \beta\gamma + \gamma\alpha)\lambda + \alpha\beta\gamma - c_1 c_2 \phi \\
&= \lambda^3 + A\lambda^2 + B\lambda + C \\
&= 0.
\end{aligned}
$$

The roots of this equation can be characterized by the Routh–Hurwitz theorem. Let $G(\lambda_i) = 0$ for $i = 1, 2, 3$. Then Re $(\lambda_i) < 0$ for $i = 1, 2, 3$ if and only if (i) $A > 0$, (ii) $C > 0$, and (iii) $AB > C$. (For a proof, see Gantmacher (1959), p. 190.) Hence, in order for the steady state of the positive feedback loop to be unstable, we must insist that $C = \alpha\beta\gamma - c_1 c_2 \phi < 0$. In this case, $\mathbf{J}_+$ has at least one real positive root, call it $\lambda_1 > 0$. Then, $G(\lambda) = (\lambda - \lambda_1)H(\lambda)$, where $H(\lambda) = \lambda^2 + D\lambda + E$ and $D = A + \lambda_1 > 0, E = -C/\lambda_1 > 0$. From the quadratic formula, it follows that the two roots of $H(\lambda) = 0$ must have Re $(\lambda_i) < 0$. Hence, it is impossible for a steady state with Jacobian matrix $\mathbf{J}_+$ to undergo a Hopf bifurcation.

It is possible for a positive feedback loop to have multiple steady state solutions, with two stable nodes separated by a saddle point. At the saddle point, $\lambda_1 > 0$ and Re $(\lambda_i) < 0$ for $i = 2, 3$. For an example, see Exercise 9.6.

### 9.4.2    Negative Feedback Oscillations

The Jacobian $\mathbf{J}_-$ determines the stability of the steady state in a three-variable system with a pure negative feedback loop. In this case,

$$
G(\lambda) = \lambda^3 + (\alpha + \beta + \gamma)\lambda^2 + (\alpha\beta + \beta\gamma + \gamma\alpha)\lambda + \alpha\beta\gamma + c_1 c_2 \phi = 0,
$$

and the Routh–Hurwitz theorem implies that the steady state is unstable if and only if

$$
(\alpha + \beta + \gamma)(\alpha\beta + \beta\gamma + \gamma\alpha) < \alpha\beta\gamma + c_1 c_2 \phi.
$$

Furthermore, if equality holds, then $G(\lambda) = (\lambda + A)(\lambda^2 + B)$, so $G(\lambda)$ has conjugate roots on the imaginary axis at $\lambda = \pm i\sqrt{\alpha\beta + \beta\gamma + \gamma\alpha}$.

### 9.4.3    The Goodwin Oscillator

The quintessential example of a biochemical oscillator based on negative feedback alone (Figure 9.3C) was invented by Brian Goodwin ((Goodwin 1965; Goodwin 1966); see also Griffith (1968a)). The kinetic equations describing this mechanism are

$$
\frac{d[\mathrm{X}_1]}{dt} = \frac{v_0}{1 + ([\mathrm{X}_3]/K_\mathrm{m})^p} - k_1 [\mathrm{X}_1],
$$

$$\frac{d[X_2]}{dt} = v_1[X_1] - k_2[X_2],$$

$$\frac{d[X_3]}{dt} = v_2[X_2] - k_3[X_3].$$

Here $[X_1]$, $[X_2]$, and $[X_3]$ are concentrations of mRNA, protein, and end product, respectively; $v_0$, $v_1$, and $v_2$ determine the rates of transcription, translation, and catalysis; $k_1$, $k_2$, and $k_3$ are rate constants for degradation of each component; $1/K_m$ is the binding constant of end product to transcription factor; and $p$ is a measure of the cooperativity of end product repression.

Next we introduce dimensionless variables:

$$x_1 = \frac{v_1 v_2 [X_1]}{k_2 k_3 K_m},$$

$$x_2 = \frac{v_2 [X_2]}{k_3 K_m},$$

$$x_3 = [X_3]/K_m,$$

$$t' = \alpha t,$$

where $\alpha = (v_0 v_1 v_2)/(K_m k_2 k_3)$.

In terms of these new variables, the dynamical system becomes

$$\frac{dx_1}{dt'} = \frac{1}{1 + x_3^p} - b_1 x_1,$$

$$\frac{dx_2}{dt'} = b_2(x_1 - x_2),$$

$$\frac{dx_3}{dt'} = b_3(x_2 - x_3),$$

where $b_i = k_i/\alpha$.

Furthermore, to make the example easier, we shall assume that $b_1 = b_2 = b_3$. In this case, the dynamical system has a steady state at $x_1 = x_2 = x_3 = \xi$, where $\xi$ is the unique real positive root of $1/(1 + \xi^p) = b\xi$. The Jacobian matrix at this steady state is $J_-$, with $\alpha = \beta = \gamma = b, c_1 = c_2 = b$, and $\phi = (p\xi^{p-1})/(1 + \xi^p)^2 = bp(1 - b\xi) > 0$. Hence, the characteristic equation is $(b + \lambda)^3 + b^2\phi = 0$, whose roots are

$$\lambda_1 = -b - b\sqrt[3]{p(1 - b\xi)} < 0,$$

$$\lambda_{2,3} = -b + b\sqrt[3]{p(1 - b\xi)}\left[\cos(\pi/3) \pm i\sin(\pi/3)\right].$$

The steady state of Goodwin's model is unstable if Re $(\lambda_{2,3}) > 0$, i.e., if $-b + (b/2)\sqrt[3]{p(1 - b\xi)} > 0$. This condition is equivalent to $p(1 - b\xi) > 8$, or $b\xi < (p - 8)/p$. Hence, if $p$ (the cooperativity of end product repression) is greater than 8, then we can choose $k$ small enough to destabilize the steady-state solution of Goodwin's equations. At the critical value of $k$, when Re $(\lambda_{2,3}) = 0$, the steady state undergoes a

Hopf bifurcation, spinning off small-amplitude periodic solutions with period close to $2\pi/\text{Im}(\lambda_{2,3}) = 2\pi/(b\sqrt{3})$.

In Exercise 8 you are asked to generalize this derivation to negative feedback loops with an arbitrary number $n$ of components. You will find that the steady state is unstable when $b\xi < (p - p_{min})/p$, where $p_{min} = \sec^n(\pi/n)$. Notice that $p_{min} \to 1^+$ as $n \to \infty$, i.e., the minimum cooperativity of endproduct repression required for oscillations becomes small as the length of the feedback loop increases.

The analysis of Hopf bifurcations in Goodwin's model uncovers a number of problems with his negative–feedback mechanism for biochemical oscillations (Griffith 1968a). In a three-variable system (mRNA, protein, end product), the cooperativity of feedback must be very high, $p > 8$. Also, it is necessary, in this case, for the degradation rate constants of the three components to be nearly equal. If not, $p_{min}$ increases dramatically; e.g., if one of the $k_i$'s is tenfold larger than the other two, then $p_{min} = 24$. The value of $p_{min}$ can be reduced by lengthening the loop, but one must still ensure that the $k_i$'s are nearly equal.

Bliss, Painter and Marr (Bliss et al. 1982) fixed these problems by a slight modification of Goodwin's equations:

$$\frac{dx_1}{dt} = \frac{a}{1+x_3} - b_1 x_1,$$

$$\frac{dx_2}{dt} = b_1 x_1 - b_2 x_2,$$

$$\frac{dx_3}{dt} = b_2 x_2 - \frac{cx_3}{K+x_3}.$$

Notice that the feedback step is no longer cooperative ($p = 1$), and the uptake of end product is now a Michaelis–Menten function. The steady state of this system is $x_1^* = a/(b_1(1+\xi))$, $x_2^* = a/(b_2(1+\xi))$, $x_3^* = \xi$, where $\xi$ is the unique real positive root of $a/(1+\xi) = c\xi/(K+\xi)$. The stability of this steady state is determined by the roots of the characteristic equation $(b_1 + \lambda)(b_2 + \lambda)(\beta + \lambda) + b_1 b_2 \phi = 0$, where $\beta = cK/(K+\xi)^2$ and $\phi = a/(1+\xi)^2$.

The characteristic equation is hard to solve in this completely general case. In order to get a start on it, we make some simplifying assumptions. First, suppose that $K = 1$, so $\xi = a/c$. Next, suppose $b_1 = b_2 < c$, and choose $a = c\left(\sqrt{c/b_1} - 1\right)$, so that $\beta = b_1$ as well. In this case, $\phi = b_1 a/c$, and the characteristic equation becomes $(\lambda + b)^3 + b_1^3 a/c = 0$. The solutions of this characteristic equation are $\lambda_1 = -b_1\left(1 + \sqrt[3]{a/c}\right)$, $\lambda_{2,3} = -b_1 + b_1\sqrt[3]{a/c}[\cos(\pi/3) \pm i\sin(\pi/3)]$. The dynamical system has a Hopf bifurcation when Re$(\lambda_{2,3}) = 0$, i.e., when $a = 8c$. Hence, at the Hopf bifurcation, $c = 81b_1$ and $a = 8c = 648b_1$. If we set $b_1 = 0.2$, then the Hopf bifurcation occurs at $c = 16.2, a = 129.6$. At this Hopf bifurcation the period of oscillation is close to $2\pi/\text{Im}(\lambda_{2,3}) = 2\pi/(b_1\sqrt{3}) = 18$. Starting at this point, we can trace out the locus of Hopf bifurcations numerically as $a$ and $c$ vary at fixed $b_1$. See Figure 9.8.

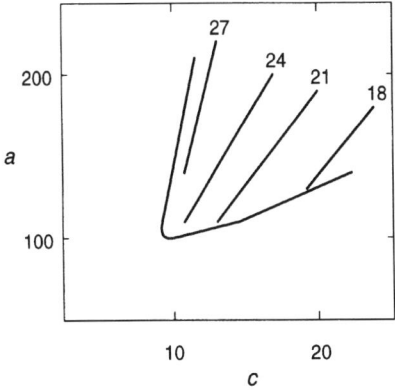

**Figure 9.8** Locus of Hopf bifurcations in the Bliss–Painter–Marr equations given in system (9.18), for $b_1 = b_2 = 0.2$. We also plot loci of constant period (18 – 27) within the region of limit cycle oscillations.

## 9.5 Time-Delayed Negative Feedback

In Goodwin's equations and Bliss–Painter–Marr's modified version, we assumed implicitly that there are no time delays in the processes of transcription, translation, or end product repression. However, there are surely some delays in transcription and translation associated with mRNA and protein processing in the nucleus and cytoplasm, respectively. And there are also bound to be delays in the feedback term, because the end product must move into the nucleus, bind with transcription factors, and interact with the "upstream" regulatory sites of the gene to affect its rate of transcription. If we lump all these delays together, we can write a delayed-differential equation for negative feedback:

$$\frac{d[X]}{dt} = \frac{a}{1 + (Z/K_m)^p} - b[X], \tag{9.18}$$

where $Z(t)$ is a functional of the past history of $[X](t)$. For a discrete time lag,

$$Z(t) = [X](t - \tau), \tag{9.19}$$

with $\tau$ =constant. For a distributed time lag,

$$Z(t) = \int_{-\infty}^{t} [X](s)G_c^n(t - s)ds, \text{ with } G_c^n(s) = \frac{c^{n+1}}{n!}s^n e^{-cs}. \tag{9.20}$$

The kernel $G_c^n(s)$ is plotted in Figure 9.9. In Exercise 11 you are asked to show that $G_c^n$ has a maximum at $s = n/c$. As $n$ and $c$ increase, with $n/c$ fixed, the kernel approaches a delta function, and the distributed time lag approaches the discrete time lag with $\tau = n/c$. In the same exercise, you are asked to prove that $d/ds(G_c^n(s)) = c[G_c^{n-1}(s) - G_c^n(s)]$, a fact that we shall presently put to good use.

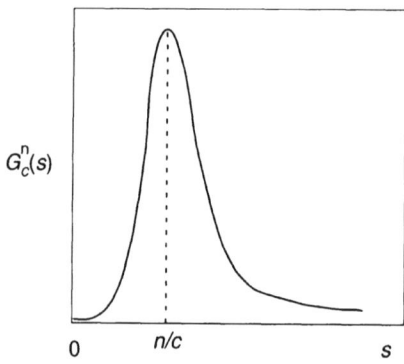

**Figure 9.9** The shape of the kernel $G_c^n(s)$ for distributed time lag in (9.20).

## 9.5.1 Distributed Time Lag and the Linear Chain Trick

Let us introduce a family of $Z_j$'s,

$$Z_j(t) = \int_{-\infty}^{t} [\text{X}](s) G_c^j(t-s) ds, \ j = 0, 1, \ldots, n.$$

Notice that $Z_n(t)$ is just the functional $Z(t)$ in (9.20). Differentiating the definite integral, we find that

$$\frac{dZ_j}{dt} = [\text{X}](t) G_c^j(0) + \int_{-\infty}^{t} [\text{X}](s) \frac{d}{dt} G_c^j(t-s) ds, \ j = 0, 1, \ldots, n.$$

Since $G_c^0(0) = c$ and $G_c^j(0) = 0$ for $j = 1, \ldots, n$, we see that

$$\frac{dZ_0}{dt} = c[\text{X}](t) + \int_{-\infty}^{t} [\text{X}](s)[-cG_c^0(t-s)] ds,$$

$$\frac{dZ_j}{dt} = \int_{-\infty}^{t} [\text{X}](s)[cG_c^{j-1}(t-s) - cG_c^j(t-s)] ds, \ j = 1, \ldots, n.$$

Hence, (9.18) with $Z$ defined in (9.20) can be written equivalently as a set of ODEs:

$$\frac{d[\text{X}]}{dt} = \frac{a}{1 + (Z_n/K_m)^p} - b[\text{X}],$$

$$\frac{dZ_0}{dt} = c([\text{X}] - Z_0),$$

$$\frac{dZ_j}{dt} = c(Z_{j-1} - Z_j), \ j = 1, 2, \ldots, n.$$

That is, the distributed time-delay model, with kernel $G_c^n(s)$, is identical to a Goodwin negative feedback loop of length $n + 2$. To see this, scale $[\text{X}], Z_0, \ldots, Z_n$ by $K_m$

and $t$ by $K_m/a$ to put these equations into the classical, dimensionless form of Goodwin's equations, (9.18). For simplicity, let $c = b$. Then, according to the results of Exercise 8, the loop has a Hopf bifurcation at $(bK_m)/a = (p - p_{min})/(\xi p)$, where $p_{min} = [\sec(\pi/(n+2))]^{n+2}$ and the dimensionless number $\xi$ is the unique real positive root of $\xi^{p+1} + \xi - a/(bK_m) = 0$.

As an example, suppose $n = 6$ and $p = 4$. In this case, $p_{min} = 1.884$ and the Hopf bifurcation occurs at $0.529 = (bK_m/a)\xi = 1/(1 + \xi^4)$, or $\xi = 0.9714$. Hence, the critical value of $b$ is $b_{crit} = 0.5446a/K_m$, and oscillations occur for $b < b_{crit}$. The period of oscillation close to the Hopf bifurcation is $2\pi/(b_{crit}\sqrt{3}) \approx 6.66K_m/a$.

### 9.5.2 Discrete Time Lag

For the case of a discrete time lag, we must solve the delay-differential equation

$$\frac{dx}{dt} = \frac{1}{1 + x(t - \tau)^p} - bx, \tag{9.21}$$

where $x$ and $t$ have been scaled to eliminate the parameters $a$ and $K_m$ from (9.18). (To do this, let $x = [X]/K_m$, dimensionless time $= at/K_m$, and dimensionless "b" $= bK_m/a$.) Then (9.21) has a steady–state solution $x^*$ satisfying $x^{p+1} + x - b^{-1} = 0$. To investigate the stability of the steady state, we use Taylor's theorem to expand (9.21) in terms of small deviations from the steady state, $y(t) = x(t) - x^*$,

$$\frac{dy}{dt} = -\phi y(t - \tau) - by(t) + \text{higher–order terms},$$

$$\phi = \frac{p(x^*)^{p-1}}{(1 + (x^*)^p)^2} = pb(1 - bx^*) = \frac{pb}{1 + (x^*)^{-p}}.$$

Looking for solutions of the form $y(t) = y_0 e^{\lambda t}$, we find that $\lambda$ must satisfy the characteristic equation $\lambda + b = -\phi e^{-\lambda \tau}$. At a Hopf bifurcation, the eigenvalue $\lambda$ must be purely imaginary; $\lambda = \pm i\omega$. Thus, for (9.21) to exhibit periodic solutions at a Hopf bifurcation, we must insist that

$$b = -\phi \cos(\omega \tau), \quad \omega = \phi \sin(\omega \tau). \tag{9.22}$$

From these equations we can determine the oscillatory frequency ($\omega$) and critical time delay ($\tau_{crit}$) at the onset of limit cycle oscillations:

$$\omega = \sqrt{\phi^2 - b^2} = b\sqrt{[p/(1 + (x^*)^{-p})]^2 - 1},$$

$$\tau_{crit} = \frac{\cos^{-1}\left(-[1 + (x^*)^{-p}]/p\right)}{b\sqrt{[p/(1 + (x^*)^{-p})]^2 - 1}}. \tag{9.23}$$

Because the domain of $\cos^{-1}$ is $[-1, 1]$, a necessary condition for Hopf bifurcation is $1 + (x^*)^{-p} < p$. For example, if $p = 4$, then $x^*$ must be $> (1/3)^{1/4} \approx 0.760$, which implies that $b = (x^{p+1} + x)^{-1} < 3^{5/4}/4 \approx 0.987$. If $b = 1/2$, then $x^* = 1$ and $\omega = \sqrt{3}/2$. Because $\cos(\omega \tau) = -b/\phi = -1/2, \tau = 4\pi\sqrt{3}/9$. Hence, for $b = 1/2$ and $p = 4$, small–amplitude

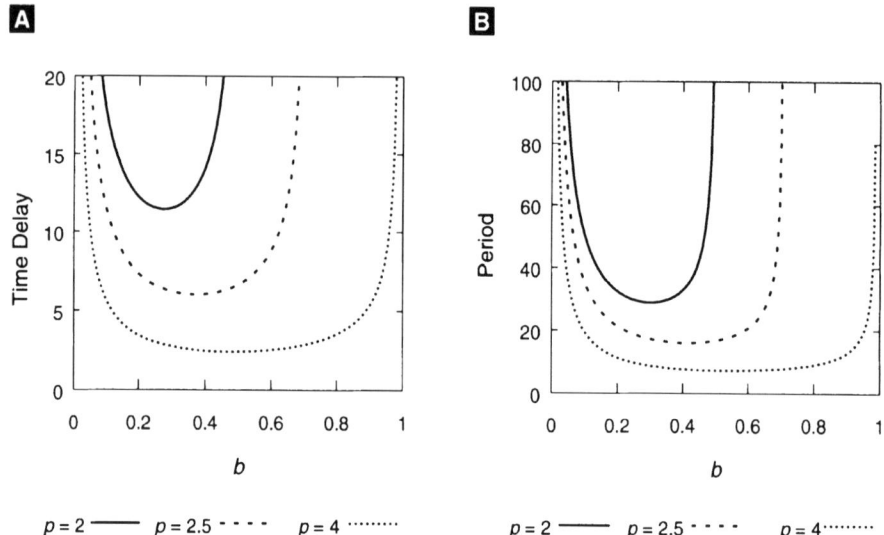

**Figure 9.10** Loci of Hopf bifurcations in (9.21). We plot $\tau_{crit}$ and period ($2\pi/\omega$) as functions of $b$, for three values of $p$: 2 (solid line), 2.5 (dashed line), and 4 (dotted line). Oscillatory solutions lie within the U-shaped domains. Notice that for $p = 4, b = 0.5$, as calculated in the text, $\tau_{crit} = 2.418$ and period = 7.255.

oscillations, with period $= 2\pi/\omega \approx 7.255$, bifurcate from the steady state as the time delay increases beyond 2.418. In Figure 9.10 we show how the characteristics of these Hopf bifurcations ($\tau_{crit}$ and period) depend on $b$, for fixed values of $p$.

## 9.6   Circadian Rhythms

Everyone is familiar with his or her own 24-hour sleep–wake cycle. Many other aspects of human physiology also exhibit daily rhythms, including body temperature, urine production, hormone secretion, and skin cell division. Such rhythms are observed in all kinds of plants, animals, and fungi, as well as unicellular organisms, and even cyanobacteria. Because these rhythms persist in the absence of external cues (light intensity, temperature, etc.), they reflect an endogenous oscillator within cells that runs at a period close to 24 h.

Biologists have long been puzzled by the molecular basis of circadian rhythms. Although a fundamental breakthrough was made by Konopka and Benzer in 1971 (Konopka and Benzer 1971), with their discovery of the *per* gene in *Drosophila* (mutations of which alter the endogenous circadian rhythm of affected flies), it was 25 years before the molecular details of the circadian oscillator began to become clear. We know now that PER protein inhibits transcription of the *per* gene, through a complicated pro-

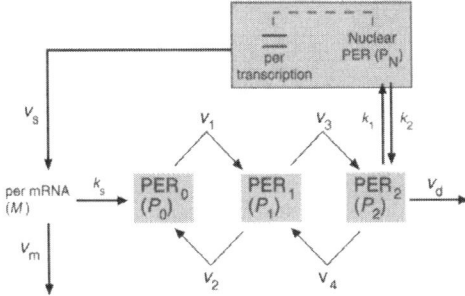

**Figure 9.11**  Goldbeter's model of circadian rhythms (Goldbeter 1995). PER protein is synthesized in the cytoplasm, where it is successively phosphorylated, as indicated by subscripts. The doubly phosphorylated form enters the nucleus and represses transcription of the *per* gene.

cess involving phosphorylation by DBT kinase, binding to TIM subunits, transport into the nucleus, and interaction with the transcription factors (CLK and CYC).

It is clear to all that the control system is dominated by a time-delayed negative–feedback loop, quite close in principle to Goodwin's original negative feedback oscillator. Numerous theoreticians have exploited the interesting nonlinear dynamics of delayed negative feedback in order to model certain characteristics of circadian rhythms. Ruoff and Rensing have explored the capabilities of Goodwin's Equation (9.18), with $p = 9$, to account for temperature compensation, entrainment, and phase resetting (Ruoff and Rensing 1996).

Goldbeter proposed a more complicated model, based loosely on Goodwin's idea, supplemented with reversible phosphorylation steps and nuclear transport; see Figure 9.11 (Goldbeter 1995). The kinetic equations describing this mechanism are

$$\frac{dM}{dt} = \frac{v_s}{1 + (P_N/K_I)^4} - \frac{v_m M}{K_{m1} + M},$$

$$\frac{dP_0}{dt} = k_s M - \frac{V_1 P_0}{K_1 + P_0} + \frac{V_2 P_1}{K_2 + P_1},$$

$$\frac{dP_1}{dt} = \frac{V_1 P_0}{K_1 + P_0} - \frac{V_2 P_1}{K_2 + P_1} - \frac{V_3 P_1}{K_3 + P_1} + \frac{V_4 P_2}{K_4 + P_2},$$

$$\frac{dP_2}{dt} = \frac{V_3 P_1}{K_3 + P_1} - \frac{V_4 P_2}{K_4 + P_2} - k_1 P_2 + k_2 P_N - \frac{v_d P_2}{K_d + P_2},$$

$$\frac{dP_N}{dt} = k_1 P_2 - k_2 P_N.$$

The basal parameter values are:

$v_s = 0.76\ \mu\text{M/h}, v_m = 0.65\ \mu\text{M/h}, v_d = 0.95\ \mu\text{M/h},$

$k_s = 0.38\ \text{h}^{-1}, k_1 = 1.9\ \text{h}^{-1}, k_2 = 1.3\ \text{h}^{-1},$

$V_1 = 3.2\ \mu\text{M/h}, V_2 = 1.58\ \mu\text{M/h}, V_3 = 5\ \mu\text{M/h}, V_4 = 2.5\ \mu\text{M/h},$

$K_1 = K_2 = K_3 = K_4 = 2\ \mu\text{M}, K_I = 1\mu\text{M}, K_{m1} = 0.5\ \mu\text{M}, K_d = 0.2\ \mu\text{M}.$

Figure 9.12A shows a numerical simulation of this system of ODEs, with a period close to 24 h. Figure 9.12B shows how the period of oscillation depends on $v_d$, the

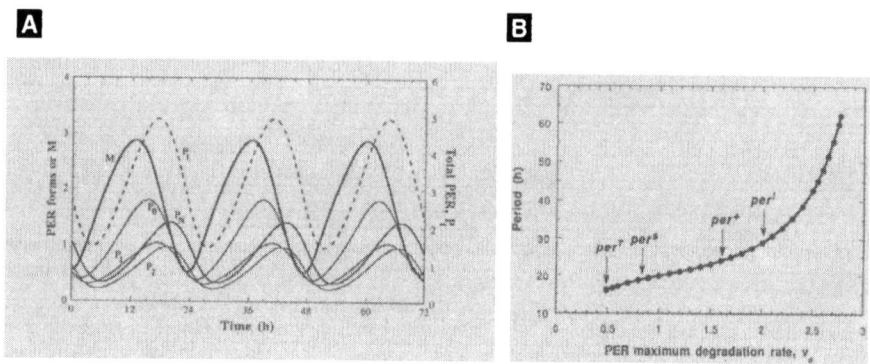

**Figure 9.12** Simulations of Goldbeter's model. (A) mRNA and protein concentrations as functions of time. (B) Period of oscillation as a function of $v_d$. Figures are reprinted from Goldbeter (1995).

degradation rate of PER. Goldbeter suggested that the short-period mutant of *per* (*per^S* has an autonomous period of 19 h) encodes a more stable form of PER, and the long-period mutant (*per^L* has an autonomous period of 28 h) encodes a less stable form of PER (Goldbeter 1995). In subsequent papers, Leloup and Goldbeter have studied temperature compensation and phase resetting in this model (Leloup and Goldbeter 1998; Leloup and Goldbeter 1999).

Perhaps the simplest model of circadian rhythms is negative feedback with discrete time delay described by (9.18) and (9.19), which has been explored in a recent paper by Lema, Golombek, and Echave (Lema et al. 2000). In this case, X is PER, $a$ = maximum rate of PER synthesis, $b$ = rate constant for PER degradation, $K_m$ = PER concentration at half-maximal synthesis rate, and $\tau$ = time delay (because the rate of synthesis of PER at the present moment depends on its cytoplasmic concentration some time in the past). (Lema et al. chose $a = 1$ h$^{-1}$, $b = 0.4$ h$^{-1}$, $K_m = 0.04$, $\tau = 8$ h, and $p = 2.5$.) In Section 9.5.2 we showed how to do stability analysis of the steady state of this delay-differential equation, in order to find the points of Hopf bifurcation to periodic solutions. Letting $x^*$ be the unique real positive root of $x^{p+1} + x - [a/(bK_m)] = 0$, we find that periodic solutions exist for $\tau > (K_m/a)\tau_{crit}$, with $\tau_{crit}$ (dimensionless) given by (9.23). The period of oscillation of the bifurcating solutions is close to $2\pi/\omega$, where $\omega\tau = \cos^{-1}\left(-[1 + (x^*)^{-p}]/p\right)$. For the parameter values chosen by Lema et al., $a/(bK_m) = 62.5$, $x^* = 3.21$, $\omega = 0.0344$, and $\tau_{crit} = 58.1$. Hence, for $\tau > 2.33$ h, their model oscillates with period close to $2\pi/\omega \approx \pi\tau \approx 7.32$ h. To get a period close to 24 h, they chose $\tau = 8$ h. In their paper Lema et al. explored entrainment by phase-resetting in response to light pulses, assuming that light interacts with PER dynamics at either the synthesis ($a$) or degradation ($b$) step.

Tyson and coworkers have taken a different approach, noting that phosphorylation of PER by DBT induces rapid degradation of PER, but multimers of PER and TIM are

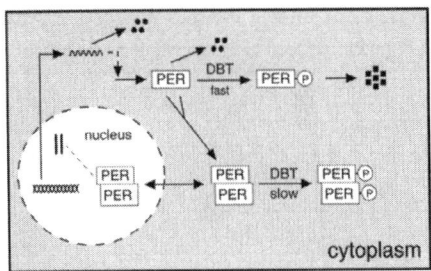

**Figure 9.13** An alternative model of circadian rhythms (Tyson et al. 1999). The circadian protein PER is synthesized in the cytoplasm, where it can form homodimers, PER:PER. (Multimeric protein complexes with TIM are neglected in this simple model.) PER monomers are rapidly phosphorylated by a protein kinase called DBT, and phosphorylated PER is rapidly degraded. PER dimers are only slowly phosphorylated by DBT.

not readily phosphorylated by DBT; see Figure 9.13 (Tyson et al. 1999). They describe this mechanism by three ODEs

$$\frac{dM}{dt} = \frac{v_{\mathrm{m}}}{1 + (P_2/A)^2} - k_{\mathrm{m}}M,$$

$$\frac{dP_1}{dt} = v_{\mathrm{p}}M - \frac{k_1 P_1}{J + P_1 + 2P_2} - k_3 P_1 - 2k_{\mathrm{a}}P_1^2 + 2k_{\mathrm{d}}P_2 + \frac{2k_2 P_2}{J + P_1 + 2P_2} + 2k_3 P_2,$$

$$\frac{dP_2}{dt} = k_{\mathrm{a}}P_1^2 - k_{\mathrm{d}}P_2 - \frac{2k_2 P_2}{J + P_1 + 2P_2} - 2k_3 P_2,$$

where

$M = [per\text{ mRNA}]$, $P_1 = [\text{PER monomer}]$, $P_2 = [\text{PER dimer}]$,

$v_{\mathrm{m}}, v_{\mathrm{p}}$ = rate constants for synthesis of mRNA and protein,

$k_{\mathrm{m}}, k_3$ = rate constants for non-specific degradation of mRNA and protein,

$k_1, k_2$ = rate constants for phosphorylation of monomer and dimer

$k_{\mathrm{a}}, k_{\mathrm{d}}$ = rate constants for association and dissociation of dimer,

$J, A$ = Michaelis constants for the binding of PER to phosphatase and transcriptional regulation factors.

If the dimerization reaction is in rapid equilibrium, then $P_1 = qP_{\mathrm{T}}$ and $P_2 = ((1 - q)/2)P_{\mathrm{T}}$, where $P_{\mathrm{T}} = P_1 + 2P_2 = [\text{total protein}]$, and

$$q = \frac{2}{1 + \sqrt{1 + 8KP_{\mathrm{T}}}},$$

where $K = k_a/k_d$ = the equilibrium binding constant for dimer formation. In this case, the system of three ODEs reduces to a pair of ODEs:

$$\frac{dM}{dt} = \frac{v_m}{1 + (P_2/A)^2} - k_{\mathrm{m}}M,$$

$$\frac{dP_{\mathrm{T}}}{dt} = v_{\mathrm{p}}M - \frac{k_1 P_1 + 2k_2 P_2}{J + P_{\mathrm{T}}} - k_3 P_{\mathrm{T}},$$

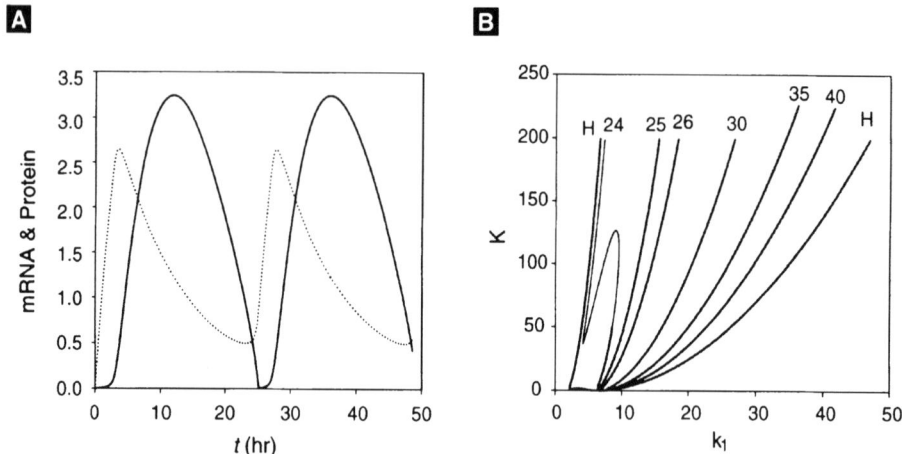

**Figure 9.14** Simulations of the model of Tyson et al. (1999). Basal parameter values: $v_m = 1$, $v_p = 0.5$, $k_m = k_3 = 0.1$, $k_1 = 10$, $k_2 = 0.03$, $J = 0.05$, $K = 200$, $A = 0.1$ (A) mRNA and protein concentrations as functions of time. (B) Locus of Hopf bifurcations (H) in dependence on $k_1$ and $K$. Also indicated are loci of limit cycles of constant period (24 h, 25 h, . . . , 40 h).

with $P_1$ and $P_2$ given as functions of $P_T$ by the equations above. Figure 9.14A shows a phase plane portrait for this system, with limit cycle oscillations of period close to 24 h. Figure 9.14B shows how these oscillations depend on $k_1$, the rate constant for phosphorylation of PER monomers by DBT, and $K$, the equilibrium binding constant for dimer formation.

# Suggestions for Further Reading

- Joseph Higgins published the first comprehensive theory of biochemical oscillators, and it is still valuable reading today (Higgins 1967).
- The earliest collection of articles on biological and biochemical oscillators was edited by Chance, Pye, Ghosh, and Hess, B. (Chance et al. 1973).
- For a comparative survey of cellular oscillators see (Berridge and Rapp 1979).
- Sol Rubinow's 1980 chapter is one of the best summaries of the kinetics of regulatory enzymes (Rubinow 1980).
- Although not easy reading, Albert Goldbeter's 1996 book *Biochemical Oscillators and Cellular Rhythms* is comprehensive, authoritative, and well motivated (Goldbeter 1996).

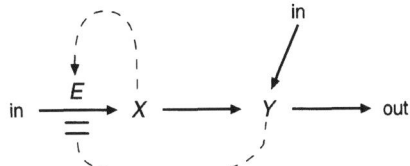

**Figure 9.15**   Exercise 1

## 9.7   EXERCISES

1.  Activator–inhibitor system. Consider the mechanism in Figure 9.15, which can be described by the (scaled) differential equations

$$\frac{dx}{dt} = \frac{a + bx^2}{1 + x^2 + ry} - x,$$

$$\frac{dy}{dt} = \epsilon(cx + y_0 - y).$$

Basal parameter values: $a = 1$, $b = 5$, $c = 4$, $r = 1$, $y_0 = 0$, $\epsilon = 0.1$.

(a)  Why is this called an activator–inhibitor system?

(b)  Draw a phase plane portrait (nullclines and typical trajectories) for the basal parameter values.

(c)  Vary $c$ and find the Hopf bifurcation points.

(d)  As you vary both $c$ and $y_0$, how many qualitatively different phase plane portraits can you find? Sketch them.

2.  Substrate-depletion system. Consider the mechanism in Figure 9.16, which can be described by the (scaled) differential equations

$$\frac{dx}{dt} = a - xy^2, \quad \frac{dy}{dt} = \epsilon + xy^2 - by$$

Basal parameter values: $a = 0.5$, $b = 1$, $\epsilon = 0.05$

(a)  Why is this called a substrate–depletion system?

(b)  Rewrite the ODEs in terms of $y$ and $z = x + y$.

(c)  Draw a phase plane portrait (nullclines and typical trajectories) for the basal parameter values.

(d)  Classify the stability of the steady state as $a$ varies.

3.  Substrate–inhibition oscillator (Murray 1981). Perhaps the simplest example of a biochemical oscillator is illustrated in Figure 9.17. The rate equations for this mechanism

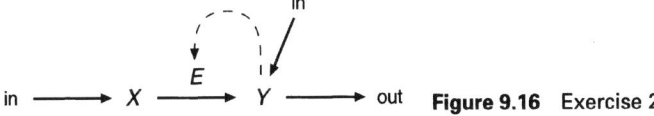

**Figure 9.16**   Exercise 2

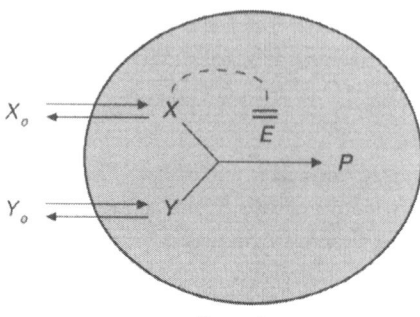

cell membrane

**Figure 9.17**   Exercise 3

are

$$\frac{dX}{dt} = a(X_0 - X) - \frac{vXY}{K_{m1} + X + (X^2/K_{m2})},$$

$$\frac{dY}{dt} = b(Y_0 - Y) - \frac{vXY}{K_{m1} + X + (X^2/K_{m2})},$$

where $a$, $b$ are membrane permeabilities, $v$ is a rate constant, and the $K_m$'s are dissociation constants of the enzyme–substrate complexes. Basal parameter values: $a = 1$, $b = 0.15$, $X_0 = 1$, $Y_0 = 6$, $v = 30$, $K_{m1} = 1$, $K_{m2} = 0.005$.

(a)   Plot nullclines and some characteristic trajectories.

(b)   Find points of Hopf bifurcation as $Y_0$ varies.

(c)   For $b = 1, X_0 = 1.5, Y_0 = 4$, plot nullclines and trajectories.

(d)   For $b = 1$, find the region of bistability in the $(X_0, Y_0)$ plane.

4.   Gene expression (Keller 1995). Consider a bacterial operon expressing two genes, *genX* and *genY*, as in Figure 9.18. The proteins X and Y form homodimers $X_2$ and $Y_2$, which then bind to the upstream regulatory sequence and affect the expression of *genX* and *genY*. $X_2$-binding stimulates gene expression, but $Y_2$-binding inhibits it. Assume that the homodimers are very stable and that they bind to the regulatory sequence with equal affinity.

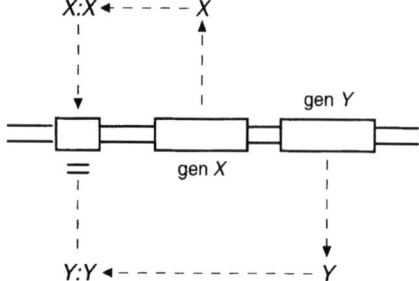

**Figure 9.18**   Exercise 4

(a)  Show that the mechanism can be described by a pair of (scaled) ODEs,

$$\frac{dx}{dt} = 1 + \frac{\alpha x^2}{1 + x^2 + y^2} - \beta x, \quad \frac{dy}{dt} = \epsilon\left(1 + \frac{\alpha x^2}{1 + x^2 + y^2}\right) - \gamma y,$$

where $\epsilon$ is the rate of expression of Y relative to that of X.

(b)  Construct a phase plane portrait for this system when $\alpha = 50, \beta = 10, \gamma = 1, \epsilon = 0.2$.

(c)  Would you describe this oscillator as activator–inhibitor or substrate–depletion?

(d)  Find the locus of Hopf bifurcations in the $(\beta, \gamma)$ parameter plane.

5.  Glycolysis. In Section 9.3.1 we studied a simple model of the glycolytic oscillator, system (9.12), and found that oscillations exist within a limited range of substrate injection rates, $k_{1\text{lo}} < k_1 < k_{1\text{hi}}$, where $k_{1\text{lo}} = \epsilon k_3 K_n$, and $k_{1\text{hi}} = (k_3 K_n)/(\sqrt{1 + v})$, with $v = v_2/k_3$ (See (9.16)).

(a)  Show that the period of oscillation close to the two Hopf bifurcation points is given by

$$T_{\text{lo}} = \frac{2\pi}{k_3 \epsilon \sqrt{2v}}, \quad T_{\text{hi}} = \frac{2\pi}{k_3}\sqrt{\frac{2 + v}{v}}.$$

(b)  Hess and Boiteux (Chance et al. 1973) reported that $k_{1\text{lo}} = 0.33$ mM/min, $T_{\text{lo}} = 8.6$ min, and $k_{1\text{hi}} = 2.5$ mM/min, $T_{\text{hi}} = 3.5$ min. Supposing that $v \ll 1$, show that $\epsilon \cong k_{1\text{lo}}/k_{1\text{hi}} = 0.13$ and $T_{\text{hi}}/T_{\text{lo}} \cong 2\epsilon \cong 0.25$, which is not too far from the observed ratio 0.4.

(c)  For the parameter values $\epsilon^2 = 0.017, K_n = 0.25$ mM, $v_2 = 1$ min$^{-1}$ and $k_3 = 10$ min$^{-1}$, use a numerical bifurcation package to compute $k_{\text{lo}}, T_{\text{lo}}, k_{1\text{hi}}$, and $T_{\text{hi}}$, and compare your results to the observations of Hess and Boiteux.

6.  Positive feedback on gene transcription (Griffith 1968b). Consider the simple case of a protein that activates transcription of its own gene, as in Figure 9.19. This mechanism is described by a pair of ODEs:

$$\frac{d[M]}{dt} = v_1 \frac{\epsilon^2 + ([P]/K_t)^2}{1 + ([P]/K_t)^2} - k_2[M], \quad \frac{d[P]}{dt} = k_3[M] - k_4[P].$$

(a)  How must the variables be scaled to write the ODEs in dimensionless form:

$$\frac{dx}{d\tau} = \frac{\epsilon^2 + y^2}{1 + y^2} - x, \quad \frac{dy}{d\tau} = \kappa(\sigma x - y)?$$

(b)  Assume that $\epsilon = 0.2, \kappa = 1$, and draw phase plane portraits for several values of $\sigma$.

(c)  Find the bifurcation values of $\sigma$.

(d)  Plot $y$ (the steady–state concentration of P) as a function of $\sigma$.

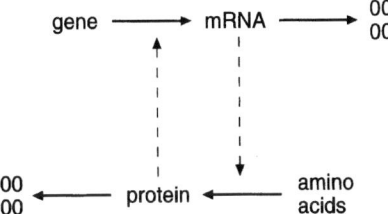

**Figure 9.19**  Exercise 6

7. Generalize the positive feedback system in the previous problem to a loop of arbitrary length.

   (a) Show that, when properly scaled, the steady state solution of the dynamical equations is given by the roots of

   $$\phi z = \frac{\epsilon^p + z^p}{1 + z^p}, \quad (*)$$

   where $\phi$ = constant, $p$ = an arbitrary exponent, and $z$ = the scaled concentration of end product.

   (b) Show that the system has saddle-node bifurcation points when $(*)$ is satisfied simultaneously with

   $$\phi = \frac{p z^{p-1}(1 - \epsilon^p)}{(1 + z^p)^2}. \quad (**)$$

   (c) For $\epsilon \ll 1$, show that $(*)$ and $(**)$ are satisfied simultaneously for

   $$z \approx (p - 1)^{1/p},$$
   $$\phi \approx p^{-1}(p - 1)^{(p-1)/p},$$

   and

   $$z \approx \epsilon(p - 1)^{-1/p},$$
   $$\phi \approx p \epsilon^{p-1}(p - 1)^{-(p-1)/p}.$$

   (d) Compute the saddle-node bifurcation points for $p = 2, 3, 4$, and compare to numerical results (pick some small value of $\epsilon$).

8. Goodwin's equations. Generalize the analysis in Section 9.4.3 to a negative–feedback loop with $n$ components,

   $$\frac{dx_1}{dt'} = \frac{1}{1 + x_n^p} - b_1 x_1, \quad \frac{dx_j}{dt'} = b_j(x_{j-1} - x_j), \ j = 2, 3, \ldots, n.$$

   Assume that $b_1 = b_2 = \cdots = b_n = b$, and show that the steady state $(x_1 = x_2 = \cdots = x_n = \xi)$ is unstable when $b\xi < (p - p_{min})/p$, where $p_{min} = \sec^n(\pi/n)$. Compute $p_{min}$ for $n = 4, 8, 16$.

9. Modify the Bliss-Painter-Marr equations given in Section 9.4.3 by replacing $a/(1 + x_3)$ with $a/(1 + x_3^p)$ in the first differential equation. Let $a = 1, b_1 = b_2 = 0.1, K = 1$. Plot the locus of Hopf bifurcation points in the $(p, c)$ parameter plane, for $0 < p < 10$.

10. Calcium-induced calcium release (Goldbeter et al. 1990). Consider a two-variable model for $Ca^{2+}$ oscillations in cells:

    $$\frac{dX}{dt} = v_0 + v_1\beta - V_2 + V_3 - k_4 X + k_5 Y,$$
    $$\frac{dY}{dt} = V_2 - V_3 - k_5 Y,$$

    where

    $$V_2 = v_2 \frac{X^2}{K_2^2 + X^2},$$
    $$V_3 = v_3 \frac{Y^2}{K_R^2 + Y^2} \cdot \frac{X^4}{K_A^4 + X^4}.$$

    In these equations $X = [Ca^{2+}]_{cytosolic}, Y = [Ca^{2+}]_{vesicular}$,

    $v_0$ = slow leak of $Ca^{2+}$ into the cytosol from extracellular fluid,

$v_1\beta$ = IP$_3$-induced release of Ca$^{2+}$ into the cytosol from intracellular stores,

$V_2$ = ATP-dependent Ca$^{2+}$ pump,

$V_3$ = Ca$^{2+}$-induced Ca$^{2+}$ release from storage vesicles,

$k_4$ = Ca$^{2+}$ elimination through the plasma membrane,

$k_5$ = Ca$^{2+}$ leak from storage vesicles.

Goldbeter et al. estimated the parameters to be

$$v_0 = 1 \ \mu M/s, v_1 = 7.3 \ \mu M/s, \beta = 0, v_2 = 65 \ \mu M/s, K_2 = 1 \ \mu M,$$
$$v_3 = 500 \ \mu M/s, K_R = 2 \ \mu M, K_A = 0.9 \ \mu M, k_4 = 10 \ s^{-1}, k_5 = 1 \ s^{-1}$$

(a) Draw a wiring diagram that corresponds to these equations.

(b) Draw a phase plane portrait for the basal parameter values.

(c) Increase $\beta$ (the IP$_3$ signal) from 0 to 1 (its maximum value), and find values that correspond to Hopf bifurcations.

(d) To simulate a pulse of IP$_3$, let $\beta(t) = \beta_0 e^{-\alpha t}$, with $\alpha = 0.25 \ s^{-1}$. Try $\beta_0 = 0.85$ and $0.95$. In each case, plot $X(t)$.

11. For the distributed time lag defined in (9.20), assume $n$ is an integer $\geq 1$ and show that

(a)
$$\int_0^\infty G_c^n(s)ds = 1$$

(b)
$$\frac{d}{ds}G_c^n(s) = c\left(G_c^{n-1}(s) - G_c^n(s)\right)$$

(c)
$$G_c^n(s) \text{ has a maximum at } s = n/c.$$

12. Periodic hemolytic anemia (Mackey 1996). Mackey has introduced a time-delayed negative-feedback model of an autoimmune disease that causes periodic crashes in circulating red blood cells (RBC)(see Figure 9.20). When RBC level in the blood is low, the cells produce a hormone, called erythropoietin, that stimulates the production of RBC precursor cells. After a few days, these precursors become mature RBCs and the production of erythropoietin is turned down. Hence, circulating RBC level is controlled by a negative–feedback system with time delay (maturation time). Mackey models this control system with equation (9.21):

$$\frac{dx}{dt} = \frac{1}{1 + x(t - \tau)^p} - bx,$$

where

[X] = concentration of circulating RBCs,

$a$ = maximal rate of production of mature RBCs,

$b$ = loss rate of RBCs from the blood stream,

$K_m$ = RBC concentration when production rate is cut in half,

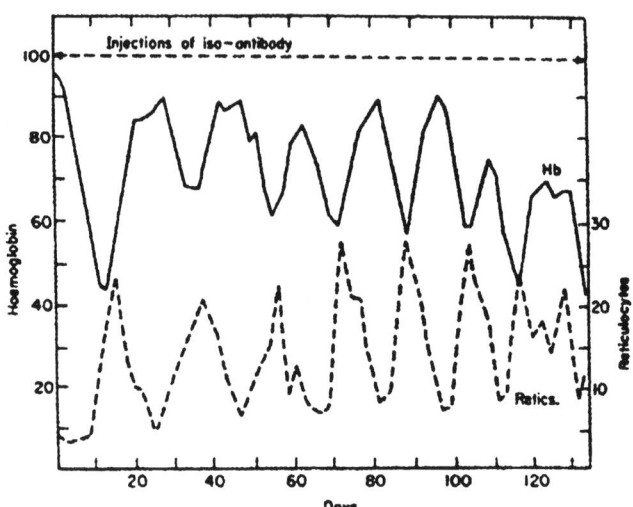

**Figure 9.20** Exercise 12. Laboratory induced autoimmune hemolytic anemia. Oscillations in circulating hemoglobin and reticulocyte counts in rabbit during constant application of red blood cell iso-antibody. From Kirk et al. (1968), as presented in Mackey (1996).

$\tau$ = time required for RBC precursors to develop into mature, circulating RBCs.

The nonlinear term in (9.21) captures the fact that current production of mature RBCs is a decreasing function of RBC concentration in the blood stream $\tau$ days ago; the parameter $p$ controls the steepness of this feedback function. Mackey estimated that in normal humans,

$$a = 7.62 \cdot 10^{10} \text{ cell kg}^{-1} \text{ day}^{-1}, \quad b = 2.31 \cdot 10^{-2} \text{ day}^{-1},$$
$$K_m = 2.47 \cdot 10^{11} \text{ cell kg}^{-1}, \quad \tau = 5.7 \text{ day}, \quad p = 7.6.$$

(a)   Calculate the steady–state level of circulating RBCs in a normal human, and show that this steady state is stable.

(b)   Periodic hemolytic anemia is an autoimmune disease that can be induced in rabbits by administration of RBC autoantibodies. The immune system destroys RBCs, thereby increasing parameter $b$ in the model. Treating $b$ as a bifurcation parameter, show that Mackey's model undergoes a Hopf bifurcation at $b = 5.12 \cdot 10^{-2}$/day and that the resulting oscillation has a period close to 20.6 day. Compare your results to the data in Figure 9.20.

(c)   Show that if $b$ is elevated beyond 0.27/day, then oscillations are lost by a reverse Hopf bifurcation.

(d)   Plot oscillation period and amplitude ($[X]_{max}$ and $[X]_{min}$) as functions of $b$ between the two Hopf bifurcations.

# CHAPTER 10

# Cell Cycle Controls

## John J. Tyson and Béla Novák

In recent years, molecular biologists have uncovered a wealth of information about the proteins controlling cell growth and division in eukaryotes. The regulatory system is so complex that it defies understanding by verbal arguments alone. To probe into the details of cell cycle control requires mathematical modeling of the type practiced in this book. We will start with an oversimplified model of the molecular controls of cell division in eukaryotes, in order to illustrate some of the basic principles involved. Then we will construct a more serious model of the budding yeast cell cycle, suitable for understanding wild-type and mutant cells. Next, a model of fission yeast introduces a new level of molecular controls. We finish up with a model of the "stripped down" cell cycles in early embryos, which leads into a discussion of whether the cell cycle is controlled by a limit cycle oscillator or not.

Before diving into the models, we need to answer two questions. What are the basic features of cell reproduction that we want to understand? And what are the molecular interactions that regulate these features?

## 10.1   Physiology of the Cell Cycle in Eukaryotes

The cell cycle is the sequence of events by which a growing cell duplicates all its components and divides into two daughter cells, each with sufficient machinery and information to repeat the process, Figure 1.6. The most important components are the cell's chromosomes, which contain linear DNA molecules in association with proteins. First, each DNA molecule must be accurately replicated, and then the two copies must

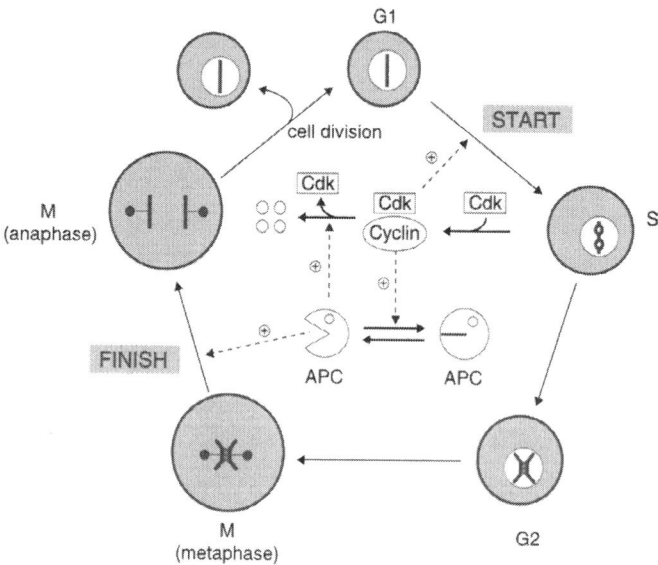

**Figure 10.1** The cell cycle. Outer ring illustrates the chromosome cycle. The nucleus of a new-born cell contains unreplicated chromosomes (represented by a single bar). At Start, the cell enters S phase and replicates its DNA (signified by replication bubbles on the chromosome). At the end of S phase, each chromosome consists of a pair of sister chromatids held together by tethering proteins. After a gap (G2 phase), the cell enters mitosis (M phase), when the repli-cated chromosomes are aligned on the metaphase spindle, with sister chromatids attached by microtubules to opposite poles of the spindle. At Finish, the tether proteins are removed so that the sister chromatids can be segregated to opposite sides of the cell (anaphase). Shortly thereafter the cell divides to produce two daughter cells in G1 phase. The inner icons repre-sent the fundamental molecular machinery governing these transitions. Start is triggered by a protein kinase, Cdk, whose activity depends on association with a cyclin subunit. Cdk activity drives the cell through S phase, G2 phase, and up to metaphase. Finish is accomplished by proteolytic machinery, APC, which destroys the tethers and cyclin molecules. In G1 phase, APC is active and Cdk inactive, because it lacks a cyclin partner. At Start, the APC must be turned off so that cyclins may accumulate. Cdk and APC are antagonistic proteins: APC destroys Cdk activity by degrading cyclin, and cyclin/Cdk dimers inactivate APC by phosphorylating some of its subunits.

be carefully segregated to the daughter cells at division. In eukaryotic cells, these two processes occur in the temporally distinct stages shown in Figure 10.1. During S phase, a new copy of each chromosome is synthesized. (The two identical DNA molecules are called sister chromatids.) Some time later, during M phase (mitosis), the sister chro-matids are separated so that each daughter cell receives a copy of each chromosome.

   DNA synthesis and sister chromatid separation (alternating S and M phases) make up the chromosome cycle of the cell. In parallel to it runs the growth cycle, whereby the cell's "hardware" (proteins, RNA, phospholipid bilayers, carbohydrates) is also duplicated and partitioned, more or less evenly, between daughters. During normal cell proliferation, these two cycles turn at the same rate, so that each round of DNA synthesis and mitosis is balanced by doubling of all other macromolecules in the cell. In this way, the DNA/protein ratio of the cell is maintained within advantageous limits. Of course, there are exceptions to this rule, such as oocytes, which grow very large without dividing, and fertilized eggs (embryos), which divide rapidly in the absence of growth. Nonetheless, the long-term viability of a cell line depends on balanced growth and division.

The chromosome cycle is usually subdivided into four phases (G1, S, G2, M), but it is better to think of it as two alternative "states" (G1 and S-G2-M) separated by two transitions (Start and Finish), as in Figure 10.1 (Nasmyth 1996). In G1, chromosomes are unreplicated and the cell is uncommitted to the replication–division process. At Start (the transition from G1 to S phase), a cell confirms that internal and external conditions are favorable for a new round of DNA synthesis and division, and commits itself to the process. The decision is irreversible; once DNA synthesis commences, it goes to completion.

During the process of DNA replication, sister chromatids are tethered together by specific proteins, called cohesins. As the mitotic spindle forms in M phase, microtubules from the spindle poles attach to chromosomes and pull them into alignment at the center of the spindle (metaphase). When DNA replication is complete and all chromosomes are aligned, the second irreversible transition of the cycle (Finish) is triggered. The cohesins are destroyed, allowing sister chromatids to be pulled to opposite poles of the spindle (anaphase). Shortly thereafter, daughter nuclei form around the segregated chromatids (telophase), and the incipient daughter cells separate.

These major events of the cell cycle must be tightly regulated. For instance, balanced growth and division is achieved in most cells by a size requirement for the Start transition. That is, cells must grow to a critical size before they can commit to chromosome replication and division. If this requirement is compromised by mutation, cells may become morbidly large or small. A second crucial regulatory constraint is to hold off the Finish transition if there have been any problems with DNA replication or chromosome alignment. Were anaphase to commence under such conditions, then daughter nuclei would not receive a full complement of chromosomes, which is usually a fatal mistake.

# 10.2 Molecular Mechanisms of Cell Cycle Control

Cell cycle events are controlled by a network of molecular signals, whose central components are cyclin-dependent protein kinases (Cdks). Cdks, when paired with suitable cyclin partners, phosphorylate many target proteins involved in cell cycle events (Figure 10.2A). For instance, by phosphorylating proteins bound to chromosomes at "origins of replication" (specific nucleotide sequences, where DNA replication can start), Cdks

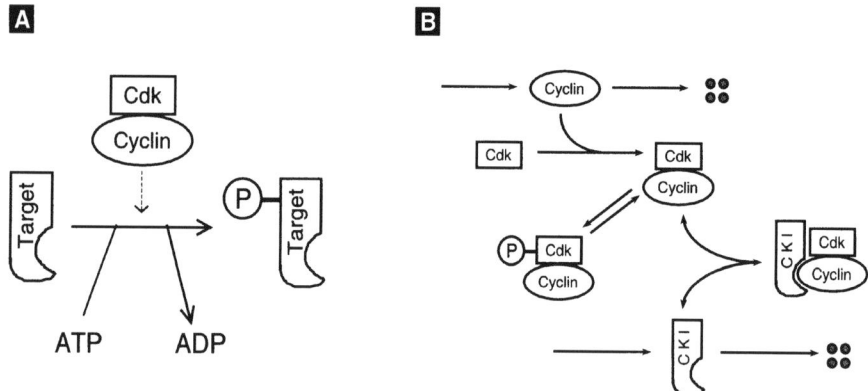

**Figure 10.2**  Cyclin–dependent kinase. (A) The role of a cyclin–dependent kinase (Cdk) is to phosphorylate certain target proteins using ATP as the phosphate donor. Cdk requires a cyclin partner in order to be active and to recognize proper targets. Cdk targets include proteins involved in DNA replication, chromosome condensation, spindle formation, and other crucial events of the cell cycle. (B) Cdk activity can be regulated in three ways: by availability of cyclin subunits, by phosphorylation of the Cdk subunit, and by stoichiometric binding to inhibitors (CKI = cyclin–dependent kinase inhibitor).

trigger the onset of DNA synthesis. By phosphorylating histones (proteins involved in DNA packaging), Cdks initiate chromosome condensation at the G2-M transition. Clearly, to understand the timing of these basic cell cycle events, one must understand the patterns of activation and inactivation of Cdks.

Cdk activities can be regulated throughout the cell cycle in many ways (Figure 10.2B). In principle, cells could regulate the availability of Cdk subunits, but this is uncommon; most Cdks are present in constant abundance throughout the cell cycle. Their activity is regulated, instead, by the availability of cyclin partners. Cyclin abundance is determined by the rates of cyclin synthesis and degradation, both of which can be regulated during the cell cycle, as we shall see. Secondly, Cdk/cyclin dimers can be put out of commission by binding a third partner, a stoichiometric inhibitor, generally referred to as a CKI (cyclin-dependent kinase inhibitor). CKIs come and go, because their synthesis and degradation rates are also cell-cycle regulated. Finally, Cdk activity can be inhibited by phosphorylation of a specific tyrosine residue, and the phosphorylation state of Cdk varies during the cell cycle as the activities of the tyrosine kinase (Wee1) and tyrosine phosphatase (Cdc25) fluctuate.

Because cells of higher eukaryotes contain many different Cdks and cyclins, "combinatorics" might play a major role in cell cycle progression, as the Cdk and cyclin subunits change partners. However, lower eukaryotes accomplish all the same basic tasks with many fewer components (one Cdk and 2-4 crucial cyclins), indicating that one Cdk is sufficient and that Cdk/cyclin holoenzymes can substitute for one another, to a large extent. Thus, progress through the cell cycle is not just a "square dance," with

Cdks and cyclins swapping partners to a steady rhythm, as some textbook diagrams might suggest, but rather a complex, nonlinear, dynamical system of interactions between Cdk/cyclin dimers and their regulatory agents: transcription factors, degradation machinery, CKIs, and tyrosine-modifying enzymes. Our task will be to understand the basic principles of this dynamical system, but first we need some more mechanistic details.

Nasmyth's two cell-cycle states shown in Figure 10.1, G1 and S-G2-M, are correlated with low and high Cdk activity, respectively. Cdk activity is low in G1 because its obligate cyclin partners are missing. Cyclin levels are low in G1 because cyclin mRNA synthesis is inhibited and cyclin protein is rapidly degraded. At Start, cyclin synthesis is induced and cyclin degradation inhibited, causing a dramatic rise in Cdk activity, which persists throughout S, G2, and M. The initial rise in Cdk activity is sufficient to initiate DNA replication, but further increase is required to drive cells into mitosis (Stern and Nurse 1996).

At Finish, a group of proteins, making up the anaphase-promoting complex (APC), is activated. The APC attaches a "destruction label" to specific target proteins, which are subsequently degraded by the cell's proteolytic machinery. The APC consists of a core complex of about a dozen polypeptides plus two auxiliary proteins, Cdc20 and Cdh1, whose apparent roles (when active) are to recognize specific target proteins and present them to the core complex for labeling. Activation of Cdc20 at Finish is necessary for degradation of cohesins at anaphase, and for activation of Cdh1. Together, Cdc20 and Cdh1 label cyclins for degradation at telophase, allowing the control system to return to G1. We must distinguish between these two different auxiliary proteins, because Cdc20 and Cdh1 are controlled differently by cyclin/Cdk, which activates Cdc20 and inhibits Cdh1.

## 10.3 A Toy Model of Start and Finish

A major challenge for theoretical molecular biologists is to explain the physiology of cell proliferation in a variety of unicellular and multicellular organisms in terms of their underlying molecular control systems. Of necessity, such connections will be made by ambitious computational models that reflect some of the inescapable complexity of real cell cycle controls. In order to design such models and understand how they work, we first need a solid grasp of the basic control principles of the cell cycle.

To this end, we draw attention to a simple theme that runs through the morass of molecular details. The irreversible transitions of the cell cycle (Start and Finish) are consequences of a hysteresis loop that derives from a fundamental antagonistic relationship between the central components of the machinery: The APC extinguishes Cdk activity by destroying its cyclin partners, whereas cyclin/Cdk dimers inhibit APC activity by phosphorylating Cdh1 (Figure 10.1). This antagonism creates two, alternative, stable steady states of the control system: a G1 state, with high Cdh1/APC activity and low cyclin/Cdk activity, and an S-G2-M state, with high cyclin/Cdk activity and low Cdh1/APC activity.

## 10.3.1   Hysteresis in the Interactions Between Cdk and APC

The biochemical reactions in the center of Figure 10.1 can be described by a pair of nonlinear ordinary differential equations (ODEs):

$$\frac{dX}{dt} = k_1 - (k_2' + k_2''Y)X, \tag{10.1}$$

$$\frac{dY}{dt} = \frac{(k_3' + k_3''A)(1 - Y)}{J_3 + 1 - Y} - \frac{k_4 mXY}{J_4 + Y}. \tag{10.2}$$

In these equations, $X$ and $Y$ are the concentrations of cyclin/Cdk dimers and active Cdh1/APC complexes, respectively; $m$ is cell "mass" (not to be confused with M for "mitosis"), and $A$ represents the concentration of a protein (to be identified later) that activates Cdh1 at Finish.

In writing these equations, we assume that cyclin molecules are synthesized in the cytoplasm, where they combine rapidly with an excess of Cdk subunits, and then the dimers move into the nucleus, where their effective concentration increases as the cell grows. Within the nucleus, the activity of cyclin/Cdk is proportional to $m \cdot X$. We also assume that APC cores are in excess, and that the total Cdh1 concentration is constant and scaled to 1. The $k$'s are rate constants, and the $J$'s are Michaelis constants.

The phase plane portrait for (10.1) and (10.2) is illustrated in Figure 10.3 The nullclines are described by simple algebraic equations:

$$\text{X nullcline: } X = \frac{\beta}{J_2 + Y},$$

$$\text{Y nullcline: } X = p \frac{(1 - Y)(J_4 + Y)}{Y(J_3 + 1 - Y)},$$

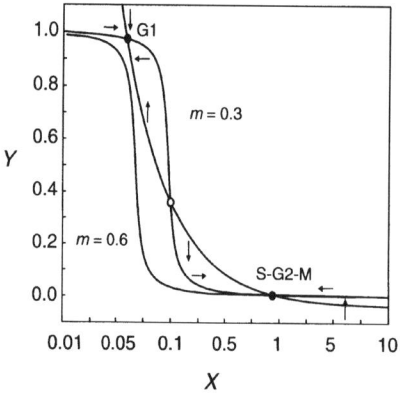

**Figure 10.3**  Phase plane portrait for the pair of nonlinear ODEs given in (10.1) and (10.2). Parameter values are given in Table 10.1. Curves are nullclines (see text) for $A = 0$, $m = 0.3$ and 0.6. Arrows indicate direction field for $m = 0.3$ only. For $m = 0.3$, the control system has three steady states: a stable node (**G1**) at $(X, Y) \approx (0.039, 0.97)$, a saddle point near (0.10, 0.36), and another stable node (**S-G2-M**) near (0.90, 0.0045). Suppose a newborn cell resides at **G1** (Cdh1 active and cyclin missing). As the cell grows ($m$ increases), the **G1** steady state is lost by a saddle-node bifurcation (at $m \approx 0.53$), and the control system is forced to the **S-G2-M** steady state.

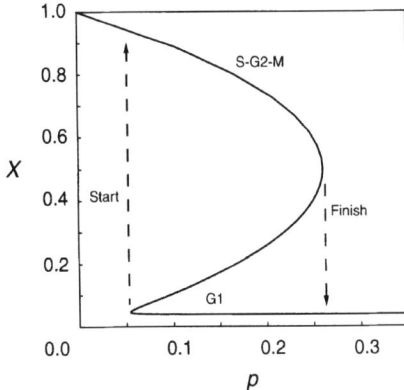

**Figure 10.4** Bifurcation diagram for (10.1) and (10.2). The steady–state concentration of $X = $ [cyclin/Cdk] is plotted as a function of the bifurcation parameter, $p = (k_3' + k_3''A)/(k_4 \cdot m)$. Other parameters: $\beta = \epsilon = 0.04$. Saddle–node bifurcations occur at $p_1 \approx 0.05418$ and $p_2 \approx 0.2604$.

where $\beta = k_1/k_2''$, $J_2 = k_2'/k_2''$, and $p = (k_3' + k_3''A)/(k_4 \cdot m)$ . The X-nullcline is a simple hyperbola. For $J_3 = J_4 \ll 1$, the Y-nullcline is a sigmoidal curve passing through $X = p$ at $Y = \frac{1}{2}$.

The control system has steady–state solutions wherever the nullclines intersect in Figure 10.4. The number of intersections depends on the value of $p$. For intermediate values of $p$ ($p_1 < p < p_2$), (10.1) and (10.2) have three steady states: two stable nodes separated by a saddle point. The stable nodes we refer to as **G1** (Cdh1 active, cyclin low) and **S-G2-M** (Cdh1 inactive, cyclin high), in boldface to distinguish the theoretician's stable steady state from the experimentalist's cell cycle phase. Progress through the cell cycle can be thought of as a tour around the hysteresis loop in Figure 10.4. For a small newborn cell in G1 phase (with $A \approx 0$ and $p \approx k_3'/k_4m > p_1$), the control system is attracted to the stable **G1** steady state. As the cell grows, $m$ increases and $p$ decreases. Eventually, $p$ drops below $p_1$, and the **G1** steady state disappears, forcing the control system to jump irreversibly to the **S-G2-M** steady state. High cyclin/Cdk activity initiates the processes of DNA synthesis and later mitosis, as the cell continues to grow. We assume that when DNA replication is complete and the chromosomes are properly aligned on the mitotic spindle, the parameter $A$ increases abruptly, forcing $p$ to increase above $p_2$. Consequently, the **S-G2-M** steady state is lost by a saddle–node bifurcation, and the control system jumps irreversibly back to the **G1** state. The cell divides ($m \rightarrow m/2$), $A$ decreases back to 0, and the control system returns to its starting condition.

In this toy model, the irreversible transitions of the cell cycle (Start and Finish) are the abrupt jumps of the hysteresis loop at the saddle-node bifurcation points. The **G1** $\rightarrow$ **S-G2-M** transition is driven by cell growth, and the reverse transition is driven by chromosome alignment on the mitotic spindle.

## 10.3.2 Activation of the APC at Anaphase

To fill out the picture in the previous section, we must identify the activator of Cdh1/APC and describe why $A$ increases abruptly at the metaphase $\rightarrow$ anaphase transition and de-

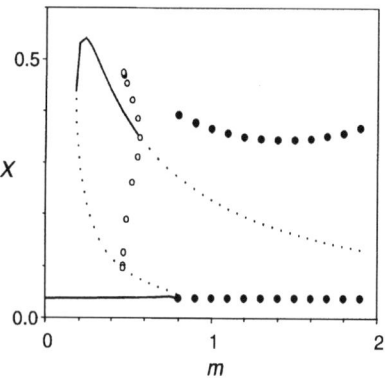

**Figure 10.5** Bifurcation diagram for (10.1)–(10.3). Asymptotic states of $X$ are plotted as functions of cell mass $m$. All other parameter values as specified in Table 10.1. Solid lines: stable steady states; dashed lines: unstable steady states; filled circles: stable limit cycles (maximum and minimum values); open circles: unstable limit cycles. (Notice that the bifurcation parameter $p$ in the previous figure is inversely proportional to $m$, so the hysteresis loop is flipped around.) There is a Hopf bifurcation on the upper branch of steady states (at $m \approx 0.57$), from which arises a branch of unstable limit cycles that disappear at an infinite–period saddle-loop bifurcation (at $m \approx 0.47$). A second branch of limit cycles at higher $m$ is stable. This branch originates at an infinite-period saddle-node-loop bifurcation (at $m \approx 0.79$) and terminates at another infinite–period bifurcation (at $m \approx 2.35$, not shown).

creases in G1 phase. The activator is a phosphatase (Cdc14) that removes from Cdh1 the inhibitory phosphate groups placed there by cyclin/Cdk. At the metaphase → anaphase transition, Cdc14 is activated indirectly by a complex pathway: cyclin/Cdk turns on Cdc20/APC, which destroys an inhibitor of Cdc14. To keep our toy model as simple as possible, we write an ODE for turnover of an "activator" whose production is promoted by cyclin/Cdk:

$$\frac{dA}{dt} = k_5' + k_5'' \frac{(mX)^n}{J_5^n + (mX)^n} - k_6 A. \tag{10.3}$$

Notice that cyclin ($X$), the activator ($A$), and Cdh1 ($Y$) are involved in a long, negative feedback loop: cyclin/Cdk turns on the activator, which indirectly activates Cdh1, which destroys cyclin subunits.

As shown in Figure 10.5 the three-component model (10.1)–(10.3) has a richer bifurcation diagram than the two-component model from Figure 10.4. The extended model still has two saddle-node bifurcations that separate a **G1** steady state ($X$ low) from an **S-G2-M** steady state ($X$ high), but now the upper steady state is unstable with respect to large–amplitude limit cycle oscillations generated by the negative feedback loop. These oscillations are "born" at a critical cell mass ($m \approx 0.79$) by an "infinite–period bifurcation," which is described in more detail in Exercise 3.

So far we have been assuming that cell mass is a parameter in our model. To complete this simple model of cell cycle controls, we reinterpret $m$ as a time-dependent variable and provide a differential equation for cell growth:

$$\frac{dm}{dt} = \mu m \left( 1 - \frac{m}{m_*} \right), \tag{10.4}$$

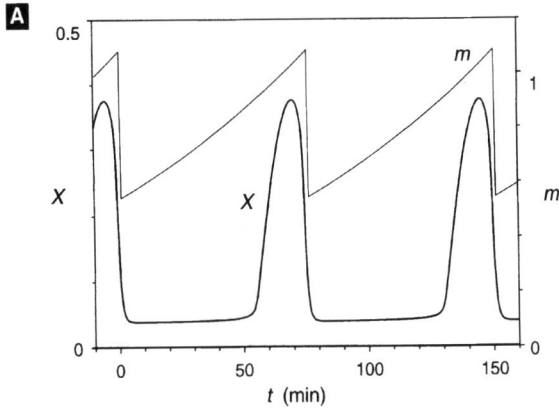

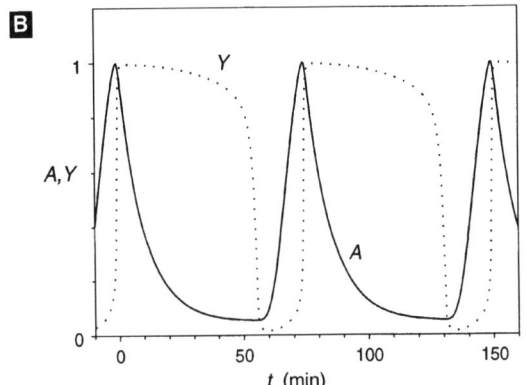

**Figure 10.6** Simulation of (10.1)–(10.4), with parameter values in Table 10.1. (A) Mass ($m$, scale on right) and [Cyclin/Cdk] ($X$, scale on left). Cell division occurs when $X$ crosses 0.1 from above. (B) [Cdc20] ($A$, solid line) and [Cdh1/APC] ($Y$, dashed line).

where $m_*$ is the maximum size to which a cell may grow if it does not divide, and $\mu$ is the specific growth rate when $m \ll m_*$. Our toy model consists of (10.1)–(10.4), with the proviso that $m \to m/2$ whenever the cell divides (i.e., when $X$ drops below some threshold level, taken to be 0.1). A typical simulation is presented in Figure 10.6. This simple model fulfills all the requirements of a functional eukaryotic cell cycle, with two irreversible transitions, Start (dependent on cell growth) and Finish (dependent on chromosome alignment). However, all organisms that have been studied in detail have additional layers of control on Cdk activity.

## 10.4    A Serious Model of the Budding Yeast Cell Cycle

A more realistic model of the basic cell cycle engine in eukaryotes is illustrated in Figure 10.7. The model is patterned after our understanding of the controls in budding yeast, but it applies to other organisms as well, as indicated in Table 10.2. The basic antagonism between cyclin/Cdk and Cdh1/APC is evident in the figure, as well as the role of Cdc20 in activating Cdh1 at Finish. (From now on, we denote the S-G2-M cyclin by CycB, to distinguish it from other types of cyclin molecules that play different roles in the cell cycle.) The figure introduces additional levels of control. Cdc20 is synthesized only during S-G2-M phase of the cell cycle, and it is activated abruptly at the metaphase-to-anaphase transition. A Cdc20-activatory signal is generated by the mitotic process itself (represented by the "intermediary enzyme" IE), and a Cdc20-inhibitory signal is generated by mitotic spindle abnormalities (through the "Mad" pathway). Further-more, CycB/Cdk is inhibited by a binding partner (CKI = "cyclin-dependent kinase inhibitor") that is prevalent in G1 phase of the cell cycle. The abundance of this sto-ichiometric inhibitor is controlled by phosphorylation reactions, which label CKI for rapid proteolysis. CKI can be phosphorylated by CycB/Cdk (note the antagonism be-tween these two components), but there is little or no CycB/Cdk activity in G1 cells. In order to leave G1 and enter S phase, the cell must produce a "starter kinase" (SK), whose job is to phosphorylate, and thereby remove, CKI. SK is a dimer of Cdk and a different type of cyclin (called Cln2 in budding yeast, Cig2 in fission yeast, and cyclin D in vertebrates). The starter kinase is not inhibited by CKI and not destroyed by Cdh1,

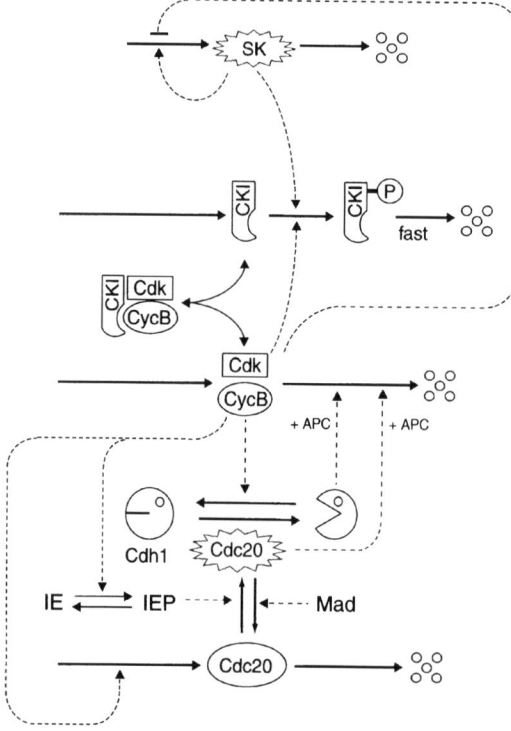

**Figure 10.7** The basic cell cycle en-gine in eukaryotic cells. The generic components in this mechanism cor-respond to specific gene products in well-studied organisms (see Table 10.2). Dynamical properties of this mechanism are determined by the set of kinetic equations given in system (10.5). A basal set of parameter val-ues, suitable for yeast cells, is given in Table 10.1. Notice that we have given Cdc20 some ability to degrade cyclin B. This well-known interaction enforces the negative feedback loop at exit from mitosis: CycB activates IE, which ac-tivates Cdc20, which degrades CycB directly, as well as activating Cdh1.

so when this alternative cyclin protein is produced in late G1, SK can help CycB/Cdk to overcome its enemies.

Figure 10.7 can be converted into a set of nonlinear ODEs modeled after equations (10.1)–(10.4):

$$\frac{dm}{dt} = \mu m \left(1 - \frac{m}{m_*}\right),$$

$$\frac{d[\text{CycB}]_\text{T}}{dt} = k_1 - (k_2' + k_2''[\text{Cdh1}] + k_2'''[\text{Cdc20}]_\text{A})[\text{CycB}]_\text{T},$$

$$\frac{d[\text{Cdh1}]}{dt} = \frac{(k_3' + k_3''[\text{Cdc20}]_\text{A})(1 - [\text{Cdh1}])}{J_3 + 1 - [\text{Cdh1}]} - \frac{(k_4 m[\text{CycB}] + k_4'[\text{SK}])[\text{Cdh1}]}{J_4 + [\text{Cdh1}]},$$

$$\frac{d[\text{Cdc20}]_\text{T}}{dt} = k_5' + k_5'' \frac{(m[\text{CycB}])^n}{J_5^n + (m[\text{CycB}])^n} - k_6[\text{Cdc20}]_\text{T},$$

$$\frac{d[\text{Cdc20}]_\text{A}}{dt} = \frac{k_7[\text{IEP}]([\text{Cdc20}]_\text{T} - [\text{Cdc20}]_\text{A})}{J_7 + [\text{Cdc20}]_\text{T} - [\text{Cdc20}]_\text{A}} - \frac{k_8[\text{Mad}][\text{Cdc20}]_\text{A}}{J_8 + [\text{Cdc20}]_\text{A}} - k_6[\text{Cdc20}]_\text{A},$$

$$\frac{d[\text{IEP}]}{dt} = k_9 m[\text{CycB}](1 - [\text{IEP}]) - k_{10}[\text{IEP}],$$

$$\frac{d[\text{CKI}]_\text{T}}{dt} = k_{11} - (k_{12}' + k_{12}''[\text{SK}] + k_{12}'''m[\text{CycB}])[\text{CKI}]_\text{T},$$

$$\frac{d[\text{SK}]}{dt} = k_{13}' + k_{13}''[\text{TF}] - k_{14}[\text{SK}],$$

$$\frac{d[\text{TF}]}{dt} = \frac{(k_{15}'m + k_{15}''[\text{SK}])(1 - [\text{TF}])}{J_{15} + 1 - [\text{TF}]} - \frac{(k_{16}' + k_{16}''m[\text{CycB}])[\text{TF}]}{J_{16} + [\text{TF}]}, \tag{10.5}$$

where $[\text{CycB}] = [\text{CycB}]_\text{T} - [\text{Trimer}]$. We have assumed that CKI/CycB/Cdk trimers are always in equilibrium with CKI monomers and CycB/Cdk dimers: $[\text{Trimer}] = K_{eq}[\text{CycB}][\text{CKI}] = K_{eq} \cdot ([\text{CycB}]_\text{T} - [\text{Trimer}]) \cdot ([\text{CKI}]_\text{T} - [\text{Trimer}])$, or

$$[\text{Trimer}] = \frac{2[\text{CycB}]_\text{T}[\text{CKI}]_\text{T}}{\Sigma + \sqrt{\Sigma^2 - 4[\text{CycB}]_\text{T}[\text{CKI}]_\text{T}}}, \tag{10.6}$$

where $\Sigma = [\text{CycB}]_\text{T} + [\text{CKI}]_\text{T} + K_{eq}^{-1}$.

A simulation of system (10.5), with the parameter values given in Table 10.1, is presented in Figure 10.8A. To understand how the control system works, let us "walk" our way through the cell cycle, starting in early G1 phase (around $t = 20$ min in the figure). In early G1, [CycB/Cdk] is low because CycB is rapidly degraded (Cdh1 is active), and what little CycB/Cdk there is in the cell is inactive because the cell is full of stoichiometric inhibitor (CKI). As the cell grows, starter kinase (SK) starts to increase because the transcription factor (TF) for its cyclin component is activated by increasing cell mass (the term $k_{15}'m$ in the ODE for [TF]). As [SK] increases, [CKI] decreases, because CKI is being phosphorylated by SK and subsequently degraded. When [CKI] drops below [CycB/Cdk] (at $t \approx 100$ min), then an active fraction of CycB/Cdk begins to assert itself by phosphorylating CKI and, more importantly, by phosphorylating and inactivating Cdh1. The latter effect stabilizes CycB and permits [CycB/Cdk] to

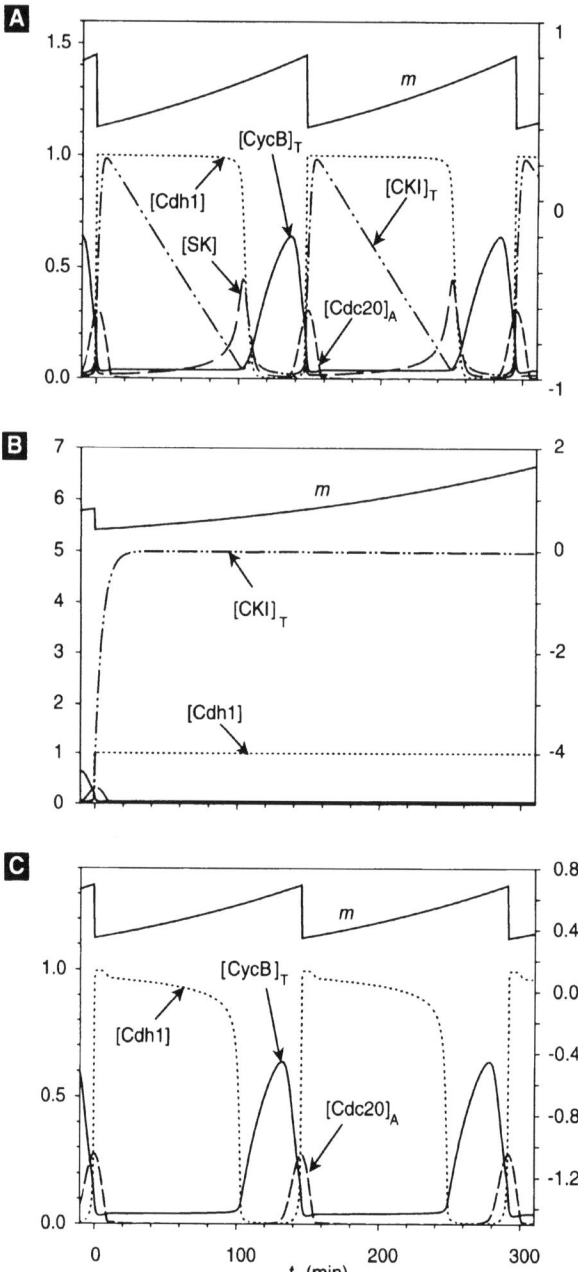

**Figure 10.8** The budding yeast cell cycle. (A) Wild–type cells: simulation of system (10.5), with parameter values in Table 10.1, except $\mu = 0.005\text{min}^{-1}$. (B) Mutant cells lacking SK ($k_{13}'' = 0$) block in G1 with copious CKI and active APC. (C) Mutant cells lacking SK and CKI ($k_{11} = k_{13}'' = 0$) are viable.

**Table 10.1  Parameter Values**

| Component | Rate constants (min$^{-1}$) | Dimensionless constants |
|---|---|---|
| CycB | $k_1 = 0.04, k_2' = 0.04, k_2'' = 1, k_2''' = 1$ | $[\text{CycB}]_{\text{threshold}} = 0.1$ |
| Cdh1 | $k_3' = 1, k_3'' = 10, k_4' = 2, k_4 = 35$ | $J_3 = 0.04, J_4 = 0.04$ |
| [Cdc20]$_T$ | $k_5' = 0.005, k_5'' = 0.2, k_6 = 0.1$ | $J_5 = 0.3, n = 4$ |
| [Cdc20]$_A$ | $k_7 = 1, k_8 = 0.5$ | $J_7 = 10^{-3}, J_8 = 10^{-3}, [\text{Mad}] = 1$ |
| IEP | $k_9 = 0.1, k_{10} = 0.02$ | |
| CKI | $k_{11} = 1, k_{12}' = 0.2, k_{12}'' = 50, k_{12}''' = 100$ | $K_{eq} = 1000$ |
| SK | $k_{13}' = 0, k_{13}'' = 1, k_{14} = 1,$ <br> $k_{15}' = 1.5, k_{15}'' = 0.05, k_{16}' = 1, k_{16}'' = 3$ | $J_{15} = 0.01, J_{16} = 0.01$ |
| Wee1 | $k_{\text{wee}}' = 0.01, k_{\text{wee}}'' = 1,$ <br> $V_{\text{awee}} = 0.25, V_{\text{iwee}}' = 0, V_{\text{iwee}}'' = 1$ | $J_{\text{awee}} = 0.01, J_{\text{iwee}} = 0.01$ |
| Cdc25 | $k_{25}' = 0.05, k_{25}'' = 5,$ <br> $V_{\text{a25}}' = 0, V_{\text{a25}}'' = 1, V_{\text{i25}} = 0.25$ | $J_{\text{a25}} = 0.01, J_{\text{i25}} = 0.01$ |
| $m$ | $\mu = 0.01$ | $m_* = 10$ |

rise rapidly. CycB-dependent kinase activity initiates DNA synthesis at about the same time ($t \approx 100 - 120$ min) that SK (Cln-dependent kinase activity) initiates a new bud. Meanwhile, CycB/Cdk turns off TF and [SK] drops. But now [CycB/Cdk] is sufficiently active to keep its enemies, CKI and Cdh1, in check. As CycB/Cdk drives the cell into mitosis, Cdc20 is synthesized (in its inactive form) and IE is phosphorylated. When $k_7[\text{IEP}]$ exceeds $k_8[\text{Mad}]$, Cdc20 is abruptly activated ($t \approx 140$ min). Cdc20 "tips the scales" in favor of Cdh1 and Sic1; CycB is destroyed, and the cell exits mitosis. When [CycB/Cdk] drops below 0.1, the cell divides ($m \rightarrow m/2$) and the process starts over again.

With this "serious" model of the yeast cell cycle we can simulate the behavior of mutant cells, in addition to wild-type cells. As shown in Figure 10.8B, mutants lacking SK ($k_{13}' = k_{13}'' = 0$) block in G1 with abundant CKI and active Cdh1, which is the phenotype of cells in which all Cln-cyclins are deleted ($cln1\Delta\ cln2\Delta\ cln3\Delta$). (Although Cln3 plays a different role than Cln1-2, it can serve as their backup; so it too must be deleted in order to observe the expected phenotype.) Because the only essential job of the Cln-cyclins is to remove Sic1 (CKI), the quadruple-deletion mutant $cln1\Delta\ cln2\Delta\ cln3\Delta\ sic1\Delta$ is viable; see Figure 10.8C. For a more thorough analysis of the budding yeast cell cycle, consult Chen et al. (2000).

# 10.5  Cell Cycle Controls in Fission Yeast

The fundamental regulatory proteins of the cell cycle engine are found in many different types of eukaryotic cells (Table 10.2), and these components are wired together in much the same way (Figure 10.7) in all organisms. Unlike budding yeast, most other organisms have an additional level of control of cyclin-dependent kinase activity by

**Table 10.2   Cell Cycle Regulatory Proteins in Yeasts and Vertebrates**

| Component | Budding Yeast | Fission Yeast | Frog Egg | Mammalian Cell |
|-----------|---------------|---------------|----------|----------------|
| Cdk | Cdc28 | Cdc2 | Cdc2 | Cdk1 |
| CycB | Clb1-6 | Cdc13 | Cyclin B | Cyclin B |
| Cdh1 | Cdh1 | Ste9 | Fizzy-related | Cdh1 |
| Cdc20 | Cdc20 | Slp1 | Fizzy | p55cdc |
| IE | Cdc5? | Plo1? | Plx1? | Plk1? |
| CKI | Sic1 | Rum1 | Xic1 | p27$^{Kip1}$ |
| SK | Cln1-2 | Cig2 | Cyclin E? | Cyclin D |

inhibitory phosphorylation of a tyrosine residue of the catalytic subunit (tyrosine-15 of Cdk). This phosphorylation (which is carried out by a tyrosine kinase called Wee1) is evident during S and G2 phases of the cell cycle: It permits DNA synthesis to occur, but it prevents entry into M phase. Before the cell can enter mitosis, the phosphate group must be removed from tyrosine-15, which is the job of a tyrosine phosphatase called Cdc25. Wee1 and Cdc25 are, in turn, targets of phosphorylation by CycB/Cdk: Wee1P is less active and Cdc25P more active than the unphosphorylated forms. Thus, Wee1 and CycB/Cdk are antagonistic proteins, whereas Cdc25 and CycB/Cdk are involved in a mutually enhancing feedback loop.

We can incorporate these regulatory signals into the model in the previous section by noticing that $[CycB]_T$ is now the sum of four forms, $[CycB/Cdk]$ + $[CycB/CdkP]$ + $[CKI/CycB/Cdk]$ + $[CKI/CycB/CdkP]$. Letting $[Trimer] = [CKI/CycB/Cdk]$ + $[CKI/CycB/CdkP]$ and $[PF] = [CycB/CdkP]$ + $[CKI/CycB/CdkP]$, we can write a differential equation for the phosphorylated forms (PF) that accounts for the actions of Wee1 and Cdc25:

$$\frac{d[PF]}{dt} = k_{wee}([CycB]_T - [PF]) - k_{25}[PF] - (k'_2 + k''_2[Cdh1] + k'''_2[Cdc20]_A)[PF], \quad (10.7)$$

where

$$k_{wee} = k'_{wee} + k''_{wee}G(V_{awee}, V'_{iwee} + V''_{iwee}[MPF], J_{awee}, J_{iwee})$$
$$k_{25} = k'_{25} + k''_{25}G(V'_{a25} + V''_{a25}[MPF], V_{i25}, J_{a25}, J_{i25}).$$

We have assumed that Wee1 and Cdc25 function as Goldbeter–Koshland ultrasensitive switches (see Exercise 1). The kinase that phosphorylates Wee1 and Cdc25 is CycB/Cdk (called here MPF, "M-phase promoting factor"). $V_{awee}$ and $V_{i25}$ refer to phosphatase activities that oppose MPF. To compute [MPF] from $[CycB]_T$, [Trimer], and [PF], we use the approximate expression (Exercise 6)

$$[MPF] \approx ([CycB]_T - [Trimer])(1 - ([PF]/[CycB]_T)). \quad (10.8)$$

Figure 10.9 shows a simulation of this model of the fission yeast cell cycle. Parameter values for the Wee1-Cdc25 interactions have been chosen so that size control in

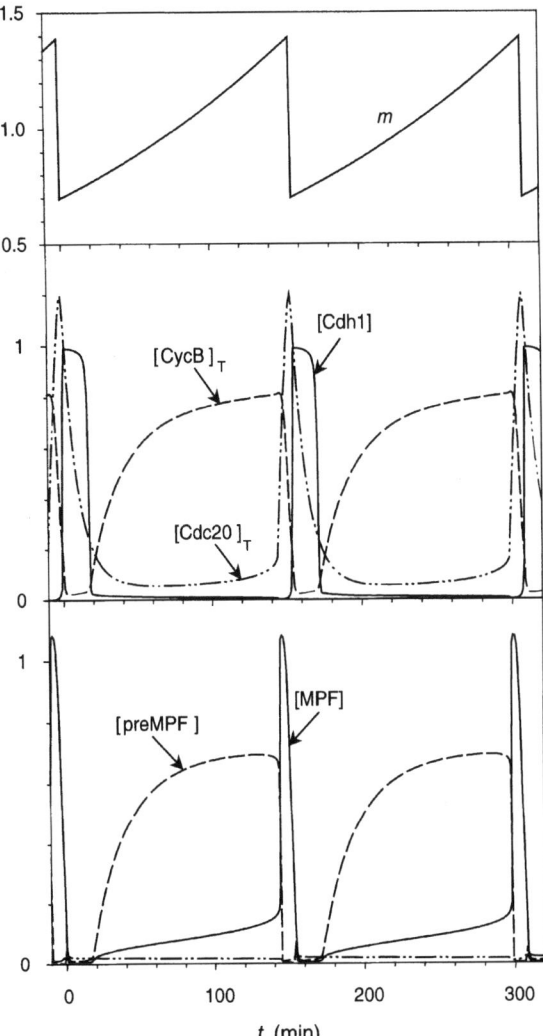

**Figure 10.9** The fission yeast cell cycle. Simulation of (10.5)–(10.8), with parameter values in Table 10.2, except $\mu = 0.005$ min$^{-1}$. This simulation represents wild–type cells, whereas Figure 10.8A represents $wee1^-$ mutant cells. Notice that, as observed, $wee1^-$ mutants are about half the size of wild-type cells and have a considerably longer G1 phase.

wild–type fission yeast occurs at the G2-M transition, rather than at G1-S. This is evident from the large size of cells at birth and the short duration of G1 phase. A $wee1\Delta$ mutant ($k''_{wee} = 0$) behaves much like Figure 10.8: Cells divide at about half the size of wild type and have a long G1 phase, exactly as observed. (The wee1 mutant, discovered by Paul Nurse in 1975, played a central role in unraveling the molecular machinery of the cell cycle.) It should be clear that $cdc25^-$ is a lethal mutation, but the double mutant $cdc25^-$ $wee1^-$ is viable and small (see Exercise 5).

## 10.6    Checkpoints and Surveillance Mechanisms

A basic job of the cell cycle engine shown in Figure 10.7 is to coordinate DNA synthesis and mitosis with overall cell growth. We have seen how cell size ($m$) might feed into the engine to ensure balanced growth and division. If cells are too small, the engine stops at a stable steady state (**G1** in Figure 10.8). Only when $m$ exceeds some critical value is this stable steady state lost by coalescence with an unstable steady state (saddle-node bifurcation). When the **G1** attractor is lost, the cell can proceed into S phase. In this way, Start can be controlled by cell size. In fission yeast, size control operates at the G2-M transition by the same principle: A stable G2 steady state (many inactive CycB/CdkP dimers) is lost when cells grow beyond a critical size, phosphorylated MPF is converted into active MPF, and the cell enters mitosis.

We believe that this is a general principle of cell cycle control. A checkpoint corresponds to a stable steady state of the cell cycle engine (no further progress). The checkpoint is lifted by changes in crucial parameters, carrying the control system across a bifurcation. The crucial parameters are controlled by surveillance mechanisms that monitor the internal and external milieus of the cell. For instance, if DNA synthesis stalls for any reason, an inhibitory signal suppresses mitosis until the genome is fully replicated. If DNA is damaged in G1 or G2 phases, other surveillance mechanisms suppress entry into S phase or M phase, respectively. If chromosome alignment on the metaphase plate is delayed for any reason, Mad protein inactivates Cdc20 and blocks progression from metaphase to anaphase.

## 10.7    Division Controls in Egg Cells

The physiology of animal eggs is quite different from that of yeast cells. In the ovary, oocytes grow very large without dividing. The cytoplasm is packed with supplies, and the nucleus is arrested in G2 phase (DNA replicated, lots of CycB/CdkP, low MPF activity). In response to hormone signals, a clutch of these "immature" oocytes leaves the G2-arrested state and proceeds through meiosis I and II. Frog eggs stop at metaphase of meiosis II (haploid complement of replicated chromosomes aligned on the spindle; high MPF activity). In this state, the "mature" oocyte awaits fertilization. Sperm entry triggers the egg to exit meiosis II; the sperm and egg nuclei replicate their DNA and fuse to form a diploid G2 nucleus. In the frog, the first mitotic cycle takes about 1 h and is followed by eleven rapid (30 min), synchronous, mitotic cycles without checkpoints. These cell cycles are not size-regulated (the cells get smaller at each division), they are not stopped by drugs that block either DNA synthesis or spindle formation, and there is little or no tyrosine phosphorylation of Cdk subunits during these cycles. In the fertilized egg, the checkpoints (stable steady states) are missing, and the cell cycle engine exhibits its capacity for free-running oscillation.

The cyclin/Cdk network controlling cell divisions in early embryos is a stripped-down version of Figure 10.7, lacking CKI and Cdh1. With CKI missing, SK has no role to

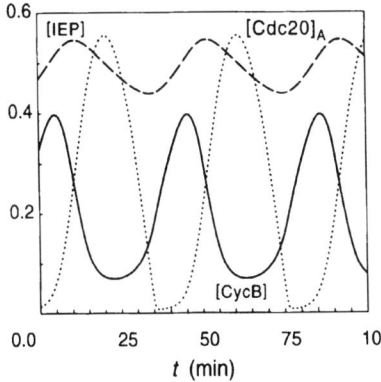

**Figure 10.10** Spontaneous oscillations of CycB/Cdk activity in early embryonic cells. Simulation of (10.9)–(10.11), with parameter values in Table 10.1. Period = 41 min.

play in the model. Also, Wee1 is inactivated, so we can neglect tyrosine phosphorylation reactions. All the antagonistic interactions are gone, leaving only the delayed negative feedback loop:

$$\frac{d[\text{CycB}]}{dt} = k_1 - (k_2' + k_2'''[\text{Cdc20}]_A)[\text{CycB}], \tag{10.9}$$

$$\frac{d[\text{IEP}]}{dt} = k_9[\text{CycB}](1 - [\text{IEP}]) - k_{10}[\text{IEP}], \tag{10.10}$$

$$\frac{d[\text{Cdc20}]_A}{dt} = \frac{k_7[\text{IEP}]([\text{Cdc20}]_T - [\text{Cdc20}]_A)}{J_7 + [\text{Cdc20}]_T - [\text{Cdc20}]_A} - \frac{k_8[\text{Mad}] \cdot [\text{Cdc20}]_A}{J_8 + [\text{Cdc20}]_A} \tag{10.11}$$

For simplicity, we assume that Cdc20 is a stable protein in the early embryo and set $[\text{Cdc20}]_T = 1$ in (10.11). Furthermore, $[\text{Mad}] = 1$, because the spindle assembly checkpoint is inoperative. Cell size ($m$) does not appear in these equations, because the embryo is not growing.

The system of equations given by (10.9)–(10.11) is a classical negative-feedback oscillator; see Chapter 9 and Exercise 7. As shown in Figure 10.10, it has limit cycle solutions, corresponding to spontaneous oscillations in activity of CycB/Cdk (usually called MPF in the frog-egg literature). After 12 rapid synchronous divisions, the frog egg undergoes an abrupt reorganization of the cell cycle (called the midblastula transition). Expression of zygotic genes provides the missing components of the cell cycle checkpoints. Consequently, the pace of cell division slows, as the controls described in Section 10.5 are put into place.

To describe the characteristic arrested states of frog oocytes we must add tyrosine phosphorylation of Cdk subunits to the negative-feedback oscillator above. So, to (10.9)–(10.11) we add

$$\frac{d[\text{MPF}]}{dt} = k_1 - k_{wee}[\text{MPF}] + k_{25}([\text{CycB}] - [\text{MPF}]) - (k_2' + k_2'''[\text{Cdc20}]_A)[\text{MPF}]$$

$$\tag{10.12}$$

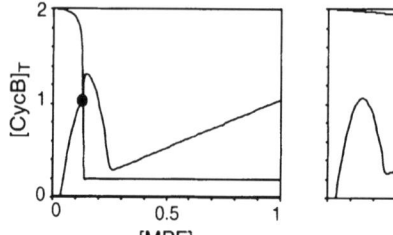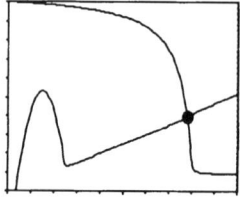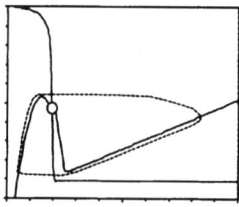

**Figure 10.11** Phase plane portraits for a model of frog oocytes: (10.9) and (10.12),with $[Cdc20]_A$ given by (10.13). Parameter values as in Table 10.1, except $k_1 = 0.06$, and $k_8$ varied. (A) Immature oocyte: $k_8 = 0.4$. Stable steady state with copious cyclin B, and MPF dimers in the inactive, tyrosine-phosphorylated form. (B) Mature oocyte: $k_8 = 0.8$. Stable steady state with copious cyclin B and high MPF activity. (C) Oscillating extract: $k_8 = 0.55$. Periodic oscillation of cyclin B level and MPF activity; period = 28 min.

where $k_{wee}$ and $k_{25}$ are defined along with (10.7), and [CycB]–[MPF] is just the concentration of tyrosine-phosphorylated dimers. In (10.10) we must replace [CycB] by [MPF], the active form of CycB/Cdk dimers.

The system given in (10.9)–(10.12) can be reduced to two variables and analyzed by phase plane methods, if we make pseudo-steady-state approximations to [IEP] and $[Cdc20]_A$. In this case, we are left with (10.9) and (10.12), and

$$[Cdc20]_A = G\left(\frac{k_7[MPF]}{(k_{10}/k_9) + [MPF]}, k_8, J_7, J_8\right). \tag{10.13}$$

Phase plane portraits for this model are illustrated in Figure 10.11. With proper choice of parameter values, one can observe stable G2 arrest (immature oocyte), stable metaphase arrest (mature oocyte), and stable limit cycle oscillations (reminiscent of MPF oscillations in frog egg extracts, which exhibit periodic tyrosine phosphorylation).

The full model given by (10.9)–(10.13), with both positive and negative feedback loops, has been studied by bifurcation theory (Borisuk and Tyson 1998).

# 10.8   Growth and Division Controls in Metazoans

In multicellular organisms, cell growth and division are under additional "social" constraints, because most somatic cells, though they find themselves bathed in a richly nutritious medium, are restrained from proliferating. Only if they receive specific "permission" from the body as a whole may these cells grow and divide. The permission slips include growth factors (small polypeptides secreted into the blood stream or interstitial fluids), and signals that reflect cell–cell contacts and adhesion to the extracellular matrix. Surveillance mechanisms monitor these signals and hold the cell in a resting state (alive but not proliferating) until conditions permit cell growth and division. If these surveillance mechanisms become mutated so that a cell loses crucial social constraints, it becomes transformed, in stages, to an invasive cancer, whose uncontrolled proliferation eventually interferes with some vital function and kills the organism.

Kohn (1999) has recently summarized our knowledge of the molecular signals controlling the cell cycle in mammals. The "wiring" diagram extends in fine print over four journal pages, and most people would agree that we are only beginning to unravel the details. How are we to make sense of a control system of such complexity? In this chapter we have seen that the molecular regulation of cell division can be understood in terms of some basic building blocks. Antagonistic interactions between CycB and Cdh1 and between CycB and CKI create the fundamental distinction between G1 and S-G2-M, starter kinases trigger the G1-S transition (Start), tyrosine phosphorylation of Cdk enforces a G2 checkpoint, and Cdc20 activation induces exit from mitosis (Finish). Within the complex wiring diagram of mammalian cell cycle controls, we can easily find all these basic building blocks. In other words, the generalized cell cycle control system in Section 10.3 applies equally well to mammalian cells as to yeast. Although the full mammalian control system has many extra bells and whistles, at its core lies a yeast-like cell cycle engine. What makes metazoans different from single-celled yeasts are the "social controls" on the brakes and accelerators of the engine.

## 10.9   Spontaneous Limit Cycle or Hysteresis Loop?

Almost by definition, the cell cycle is a periodic process, and for years this observation has tempted theoreticians to think of cell-cycle progression as a limit cycle solution of the underlying dynamical control system. By contrast, in this chapter we have been emphasizing a quite different picture: the cell cycle as an alternation between two self-maintaining states, **G1** (unreplicated DNA) and **S-G2-M** (DNA replication and mitosis). From a dynamical point of view, these two self-maintaining states are stable steady states of the kinetic equations describing the production of cyclin/Cdk activity and its destruction by cyclin proteolysis and CKI accumulation.

The control system is bistable because of the fundamental antagonism between Cdk and its "enemies." As expected for a dynamical system of this sort, bistability is observed only within a restricted region of parameter space; the boundaries of this region are parameter values where saddle-node bifurcations occur (e.g., Figure 10.4). The control system can be driven from one state to the other by parameter changes that carry the system across saddle-node bifurcation points. Because the stable state initially occupied by the cell (**G1**) is lost at the saddle-node bifurcation, the cell is forced to make an irreversible transition (Start) to the other stable state (**S-G2-M**). In general, the opposite transition (Finish) can be induced only by parameter changes that carry the system across a different boundary, where the **S-G2-M** state is lost and the system jumps irreversibly to **G1**. When traced out in a diagram like Figure 10.4, these parameter changes and state transitions create a "hysteresis loop."

The parameter changes that drive cells through Start and Finish are carried out by additional components of the control system, called "helper" molecules. The role of starter kinases is to inactivate Cdh1 and destroy CKI so that the cell can leave G1, and the role of Cdc20 is to activate Cdh1 and stabilize CKI so that the cell can reenter G1. The helpers do not participate in the antagonistic interactions: Starter kinases are not inhibited by CKI and not degraded by Cdh1/APC, and Cdc20 is not inhibited by cyclin/Cdk. Helper activity is only transient: It rises to induce a transition, but then falls back down in preparation for the reverse transition. Were the helper activity to stay high, it would impede the reverse transition. Mutations that interfere with the rise and fall of helper proteins are usually inviable or severely compromised in progress through the cell cycle.

If production and destruction of the helpers are included in the ODEs, converting former parameters into dynamical variables, don't we retrieve the notion of a limit cycle solution to the expanded equations? For instance, isn't Figure 10.8A a limit cycle solution to system (10.5)? Indeed, it is a stable periodic solution, but it lacks many of the properties that we usually associate with limit cycles. Our intuition about limit cycles has been honed on continuous, autonomous ODEs, but system (10.5) is discontinuous and nonautonomous, with $m = m(t)$ given by solution of (10.4) and the prescription that $m \rightarrow m/2$ whenever the cell exits mitosis. Hence, over a broad range of parameter values, the period of the cell cycle rhythm is identical to the mass-doubling time, $T_d \approx \ln 2/\mu$, and independent of all other kinetic parameters of the dynamical system. In our opinion, it is more profitable to think of $m$ not as a dynamical variable, but as an external parameter that drives the control network, system (10.5), back and forth between regions of stable steady states (Figure 10.5, when $m$ is small) and stable limit cycles (Figure 10.5, when $m$ is large).

The principle of balanced growth and division necessitates size control operating somewhere in the cell cycle. If all size controls are removed by mutation (e.g., in fission yeast, $wee1^{ts}$ removes size control at the G2 checkpoint and $rum1\Delta$ removes it at the G1 checkpoint), then the underlying limit cycle oscillation of the cyclin/Cdk control network is revealed. But it is fatal! The double mutation $wee1^{ts}$ $rum1\Delta$ is lethal: Because its division cycle runs faster than its growth cycle, this cell divides at ever-smaller size until it dies.

Fertilized eggs behave something like autonomous limit cycle oscillators, because all checkpoint requirements have been bypassed. But this is a temporary affair. At the midblastula transition, the rapid, synchronous, autonomous cycles of the early embryo disappear as checkpoints are reinserted in the control system. The egg replaces limit cycle oscillations by checkpoint-controlled progression (from one stable steady state to another), and it is high time theoreticians did the same!

## Suggested Reading

- *Molecular Biology of the Cell*, Chapter 17, is the best short introduction to the physiology, genetics, and biochemistry of the cell cycle (Alberts et al. 1994).
- *The Cell Cycle. An Introduction*, by Murray and Hunt, contains more details about the physiology, genetics, and biochemistry of the cell cycle (Murray and Hunt 1993).
- *At the heart of the budding yeast cycle* by Nasmyth argues for the notion that the cell cycle is an alternation between two self-maintaining states (G1 and S-G2-M) (Nasmyth 1996).
- *Molecular interaction map of the mammalian cell cycle control and DNA repair systems* by Kohn. Figure 6 has become an icon of the future challenges to the field of computational cell biology (Kohn 1999).
- *Chemical kinetic theory: understanding cell-cycle regulation* by Tyson et al. is a kinder, gentler introduction to modeling of the cell cycle (Tyson et al. 1996).

## 10.10  EXERCISES

1.  The trail-blazing paper by Goldbeter and Koshland (1981) showed that covalent modification of proteins (like the phosphorylation of Cdh1 by cyclin-dependent kinases) can generate abrupt, switch-like changes in activity of the modified protein. Let $Y$ be the concentration of the active form of the protein, $Y_P$ its inactive form, and assume that the total protein concentration is constant and scaled to 1 ($Y + Y_P = 1$). Then

$$\frac{dY}{dt} = \frac{V_a(1-Y)}{J_a + 1 - Y} - \frac{V_i Y}{J_i + Y}.$$

If inactivation occurs by phosphorylation, then $V_a$ and $J_a$ would be the activity and Michaelis constant of the phosphatase, and similarly $V_i$ and $J_i$ for the kinase. Clearly, the steady state activity of the protein is given by a quadratic equation, $AY^2 - BY + C = 0$, where $A = V_i - V_a$, $B = V_i - V_a + V_a J_i + V_i J_a$, $C = V_a J_i$.

(a)  Show that $Y = G(V_a, V_i, J_a, J_i) = (2C)/(B + \sqrt{B^2 - 4AC})$.

(b)  Why is it not a good idea (from a computational point of view) to write $G = (B - \sqrt{B^2 - 4AC})/2A$?

2.  The equations for the CycB and Cdh1 nullclines in Section 10.3.1, assuming $J_2 = J_3 = J_4 = \epsilon$, can be written

$$X = \frac{\beta}{\epsilon + Y}, \quad X = p\frac{(1-Y)(\epsilon + Y)}{Y(\epsilon + 1 - Y)}.$$

Saddle-node (SN) bifurcations occur when these two curves "touch" each other tangentially, i.e.,

$$\frac{\beta/p}{\epsilon + Y} = \frac{(1-Y)(\epsilon + Y)}{Y(\epsilon + 1 - Y)},$$

$$\frac{-\beta/p}{(\epsilon + Y)^2} = \frac{Y(\epsilon + 1 - Y)(1 - \epsilon - 2Y) - (1 - Y)(\epsilon + Y)(\epsilon + 1 - 2Y)}{Y^2(\epsilon + 1 - Y)^2}.$$

Show that if $\epsilon \ll 1$, these equations have two solutions: $Y \approx \epsilon$, $\beta/p \approx 4\epsilon$, and

$$Y = 1 - \sqrt{\epsilon} + \epsilon + O\left(\epsilon^{3/2}\right), \quad p/\beta = 1 + 2\sqrt{\epsilon} + O\left(\epsilon^{3/2}\right)$$

3. Equations (10.1)–(10.3) describe the interactions of cyclin/Cdk ($X$), Cdh1/APC ($Y$), and the activator ($A$) of Cdh1. We characterized the behavior of this system by a one-parameter bifurcation diagram in Figure 10.5. To gain some insight into the origin of the stable limit cycles in this diagram, it is useful to reduce (10.1)–(10.3) to two components, so that we can investigate the dynamics of the control system by phase plane techniques. To this end, we apply the "rapid equilibrium approximation" of Section 4.1 to Cdh1/APC to obtain $Y = G(k_3' + k_3''A, k_4mX, J_3, J_4)$ where $G$ is the Goldbeter-Koshland function defined in Exercise 1. In this case, (10.1)–(10.3) reduces to a pair of nonlinear ODEs:

$$\frac{dX}{dt} = k_1 - (k_2' + k_2''G(k_3' + k_3''A, k_4mX, J_3, J_4))X,$$

$$\frac{dA}{dt} = k_5' + k_5''\frac{(mX)^n}{J_5^n + (mX)^n} - k_6A.$$

(a) Draw phase portraits for this pair of ODEs for $m = 0.5$ and for $m = 1$. (All other parameters as in Table 10.1.)

(b) As $m$ decreases from 1, show that the period of the limit cycle solution becomes very long, until the limit cycle merges into a saddle-node loop at $m \approx 0.8$.

(c) Create a one-parameter bifurcation diagram similar to Figure 10.5 for the reduced system of ODEs.

(d) Using $k_2''$ as the second parameter, follow the saddle-node and Hopf bifurcation points in part (a) to create a two-parameter bifurcation diagram.

4. Figure 10.8 was constructed from system (10.5) and the parameter values in Table 10.1. You should verify the results in Figure 10.8A for wild-type budding yeast cells before continuing.

(a) Supposing DNA synthesis starts when CycB activity rises above 0.1 and cell division occurs when CycB activity falls below 0.1, determine the durations of G1 phase (time from birth to beginning of S phase) and S-G2-M phase (from onset of DNA synthesis to cell division), and determine cell size at birth, at onset of DNA synthesis, and at division. Check your results against Table 10.3.

(b) By simulation, confirm that $sk\Delta$ mutant cells are inviable (Figure 10.8B), but $sk\Delta\ cki\Delta$ double mutant cells are viable (Figure 10.8C). Fill in row 3 of Table 10.3.

(c) By simulation, show that $cki\Delta$ mutant cells are viable, and so are $cdh1\Delta$ mutants, but $cki\Delta cdh1\Delta$ double mutants are inviable. Fill in rows 4–6 of Table 10.3.

(d) Finish the rest of Table 10.3.

5. In the same way that you investigated budding yeast mutants in the previous problem, fill in the missing elements in Table 10.4 for fission yeast mutants.

6. Show that (10.8) is exact if

$$\frac{[CycB/CdkP]}{[CycB/Cdk]} = \frac{[CKI/CycB/CdkP]}{[CKI/CycB/Cdk]}.$$

Why should this condition be true?

7. Consider the negative–feedback oscillator (10.9)–(10.11) with $[Cdc20]_T = [Mad] = 1$.

(a) Show that to find the steady state one must solve a cubic equation in $z = [Cdc20]_A$:

$$\frac{k_1k_9}{k_2'''k_{10}} = \frac{k_8z(J_7 + 1 - z)(z + \epsilon)}{k_7(1 - z)(J_8 + z) - k_8z(J_7 + 1 - z)}, \quad \text{where } \epsilon = k_2'/k_2'''.$$

**Table 10.3  Budding Yeast Mutants**

| Genotype | Parameter Changes | Duration of G1 | Duration S-G2-M | Size at Birth | Size at DNA Syn. | Size at Division | Phenotype |
|---|---|---|---|---|---|---|---|
| Wild type | none | 108 | 39 | 0.40 | 0.69 | 0.87 | wild type |
| skΔ | $k_{13}'' = 0$ | - | - | - | - | - | block in G1 |
| skΔckiΔ | $k_{11} = k_{13}'' = 0$ | 105 | 42 | | | | viable |
| ckiΔ | $k_{11} = 0$ | | | | | | viable |
| cdhΔ | $k_2'' = 0$ | | | | | | viable |
| ckiΔcdhΔ | $k_{11} = k_2'' = 0$ | - | - | - | - | - | block in M |
| cdc20$^{ts}$ | $k_3'' = k_2''' = 0$ | - | - | - | - | - | block in M |
| Cdc20$^{OP}$ | $k_5' = 0.2$ | | | | | | viable |
| SK$^{OP}$ | $k_{13}' = 0.3$ | | | | | 0.75 | viable |
| SK$^{OP}$ | $k_{13}' = 0.4$ | - | - | - | - | - | inviable |
| CKI non-degradable | $k_{12}'' = 0$ $k_{12}''' = 0$ | | | | | | |
| Cdc20 non-degradable | $k_6 = 0.01$ | | | | | | |

Notes: The duration of S-G2-M in the model is too short, compared to experimental observations (~60 min).

(b)  Show that if $k_7 > k_8$ and both $J_7$ and $J_8$ are $\ll 1$, the steady state is given approximately by

$$[\text{Cdc20}]_A = \left(\frac{k_1 k_9}{k_2'' k_{10}}\right)\left(\frac{k_7 - k_8}{k_8}\right) - \frac{k_2'}{k_2''},$$

$$[\text{IEP}] = \frac{k_8}{k_7},$$

$$[\text{CycB}] = \frac{k_{10}}{k_9}\left(\frac{k_8}{k_7 - k_8}\right).$$

(c)  Show that the characteristic equation, which determines the stability of this steady state, is

$$\lambda\left[\lambda + \frac{k_1 k_9(k_7 - k_8)}{k_8 k_{10}}\right)\right]\left[\lambda + \frac{k_7 k_{10}}{k_7 - k_8}\right] + k_2'' k_8 k_{10} = 0.$$

(d)  At a Hopf bifurcation, $\lambda$ must be purely imaginary, $\lambda = \pm i\omega$. Show that the conditions for a Hopf bifurcation are

$$\omega^2 = \frac{k_1 k_7 k_9}{k_8} = \frac{k_2'' k_8^2 k_{10}^2 (k_7 - k_8)}{k_1 k_9(k_7 - k_8)^2 + k_7 k_8 k_{10}^2}.$$

(e)  With all the parameters fixed at their values in Table 10.1 except $k_9$, show that the conditions for a Hopf bifurcation are satisfied when $k_9 \approx 0.23$ and $\omega \approx 0.135$. What is the period of the limit cycle oscillation close to this Hopf bifurcation?

(f)  Verify the results of (e) numerically. Show that there is a second Hopf bifurcation at $k_9 \approx 0.042$.

### Table 10.4   Fission Yeast Mutants

| Genotype | Parameter Changes | Duration of G1 | Duration S-G2-M | Size at Birth | Size at DNA Syn. | Size at Division | Phenotype |
|---|---|---|---|---|---|---|---|
| Wild type | none | 16 | 136 | 0.70 | 0.76 | 1.40 | wild type |
| $wee1^-$ | $k''_{wee} = 0.01$ | | | 0.40 | | 0.80 | small |
| $wee1^-\,cki\Delta$ | $k_{11} = 0$, $k''_{wee} = 0.01$ | | | | | | |
| $cki\Delta$ | $k_{11} = 0$ | | | | | | wild type |
| $cdh\Delta$ | $k''_2 = 0$ | | | | | | viable |
| $cki\Delta\,cdh\Delta$ | $k_{13} = k''_2 = 0$ | | | | | | |
| $cdc25^-$ | $k''_{25} = 0.05$ | | | | | | |
| $wee1^-$ $cdc25^-$ | $k''_{wee} = 0.01$, $k''_{25} = 0.05$ | | | | | | viable |

Notes: There are two major discrepancies between the model and observations. (1) The simulated double mutant $wee1^-$ $cki\Delta$ is small but viable; whereas, in reality, these mutant cells become very small and die. (2) The simulated $cdc25^-$ mutant grows very large but eventually divides; whereas, in reality, these mutant cells block in G2.

(g)   If you are adept with numerical bifurcation software, follow these Hopf bifurcation points in two parameters, $k_9$ and $k'''_2$.

8.   Suppose you are informed of a primitive eukaryotic organism lacking Cdh1. To maintain itself in G1 phase, the organism relies only on the antagonism between CycB/Cdk and CKI. You model this control system by a pair of nonlinear ODEs:

$$\frac{dX}{dt} = k_1 - k'_2 Z - k''_2 W,$$

$$\frac{dY}{dt} = k_{11} - (k'_{12} + k''_{12} mZ)Y,$$

where $X = [\text{CycB}]_\text{T}$, $Y = [\text{CKI}]_\text{T}$, $Z = [\text{CycB}] = X - W$, and $W = [\text{Trimer}]$ is given by (10.6):

$$W = \frac{2XY}{X + Y + K + \sqrt{(X + Y + K)^2 - 4XY}}.$$

In this equation, $K^{-1}$ is the equilibrium constant for the binding of CKI to CycB/Cdk; hence, $K = 0.001$, according to Table 10.1. In writing the ODE for $X$, we have assumed that CKI binding renders CycB unstable ($k''_2 \gg k'_2$).

(a)   Draw phase portraits for this system. Show that cell growth can drive a Start transition.

(b)   To execute Finish, the organism, you are told, activates Cdc20, which degrades CycB at anaphase. Modify the ODEs to incorporate this signal, and show how Cdc20 activation returns the control system to G1.

# CHAPTER 11

# Modeling the Stochastic Gating of Ion Channels

**Gregory D. Smith**

In previous chapters we have seen several kinetic diagrams representing various molecular states and transitions between these states due to conformational changes and the binding or unbinding of ligands. Up to this point we have assumed a large number of molecules and written rate equations consistent with these transition-state diagrams. But how should we interpret a transition-state diagram when we are considering only a single molecule or a small number of molecules? The short answer to this question is that transition rates can be interpreted as transition probabilities per unit time.

## 11.1   Single–Channel Gating and a Two-State Model

The time course of voltage changes in a whole cell is the result of the average behavior of many individual channels. Our understanding of individual channel gating comes largely from experiments using the patch clamp technique (see Figure 11.1). For example, typical measurements from a so-called on-cell patch of T-type calcium currents in guinea pig cardiac ventricular myocytes are shown in the middle panel of Figure 11.2. The small current deviations in the negative direction indicate the opening of individual T-type calcium channels gating in response to a command membrane voltage stepped from $-70$ mV to $-20$ mV (top panel). Notice that two conductance states of the channel are visable: a closed state with no current flowing and an open state with unitary current of $\approx 10^{-12}$ amperes (1 picoampere or 1 pA). While transitions between these two conductance states are random in time, the mean of several hundred records (bottom panel of Figure 11.2) smooths out these current fluctuations and demonstrates that *on*

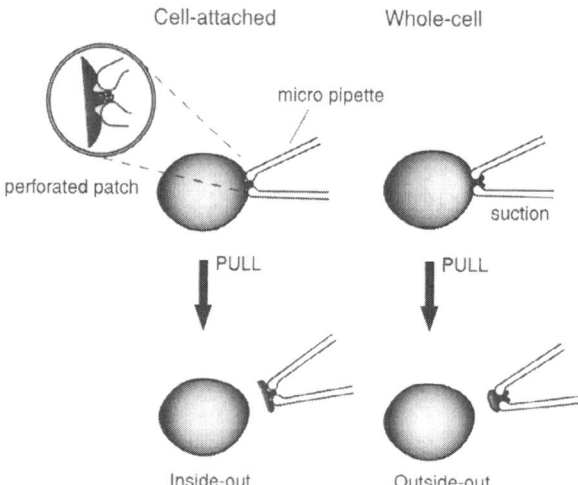

**Figure 11.1** Four methods of measuring electrical responses in cells with the patch clamp technique. In the patch technique, a pipette with an opening of $\approx 1~\mu$m is used to make a high–resistance ("gigaohm" = $10^9$ ohm) seal onto a cellular membrane. In the on-cell patch configuration all the current into the pipette flows directly through the patch, which can contain as few as one or two ion channels. In the whole–cell configuration, the patch is broken and a more accurate whole cell recording can be made as compared with a relatively leaky sharp electrode puncture. In a perforated patch configuation, an ionophore such as nystatin is introduced into the pipette in order to allow whole–cell-like access while minimizing exchange of the cell contents with the contents of the pipette. Alternatively, patches of membrane can be torn off, leading to inside-out and outside-out patches that can be studied in isolation. Adapted from Hille (2001).

*average* the stochastically gating channel activates and subsequently inactivates with time constants of $\approx 5$ ms and 50 ms, respectively. Interestingly, the average dynamics of individual T-type calcium channels is strikingly similar to "whole cell" measurements of the activation and inactivation of T-type calcium currents.

## 11.1.1 Modeling Channel Gating as a Markov Process

The stochastic gating of a single ion channel can be modeled as a continuous-time Markov process. Consider the simple transition-state diagram encountered first in Chapter 1, the kinetic scheme for ion channel with two states, one closed ($C$) and the other open ($O$),

$$C\,(\text{closed}) \overset{k^+}{\underset{k^-}{\rightleftharpoons}} O(\text{open}). \tag{11.1}$$

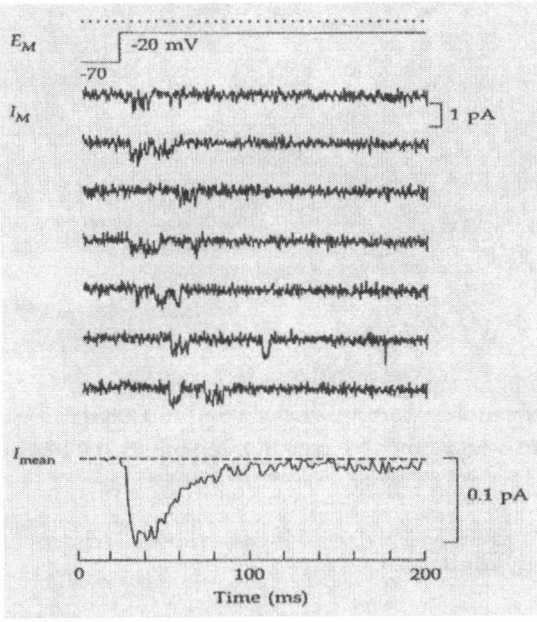

**Figure 11.2** On-cell patch clamp measurements of T-type calcium currents in guinea pig cardiac ventricular myocytes. The upper recordings show currents due to one (or a few) stochastically gating single channels when the command voltage is stepped from −70 mV to −20 mV. The lower plot is an average of several hundred such records that shows rapid activation followed by slow inactivation, proportional to macroscopic T-type calcium currents measured in whole–cell configuration. Reprinted from Hille (2001).

Define $s$ to be a random variable taking values $s \in \{C, O\}$ corresponding to these two states, and write Prob$\{s = i, t\}$ (or for short, $P_i(t)$) to represent the probability that $s(t) = i$; that is, the molecule is in state $i$ at time $t$. Because the molecule must always be in one of the two states, total probability must be conserved and we have

$$P_C(t) + P_O(t) = 1.$$

Now consider the possibility that the two-state ion channel is in state $C$ at time $t$. If this is the case, then the rate $k^+$ (e.g., with units of ms$^{-1}$) is related to the probability that in a short interval of time ($\Delta t$) the two-state ion channel will open. The relationship is given by

$$k^+ \Delta t = \text{Prob}\{s = O, t + \Delta t | s = C, t\}, \tag{11.2}$$

where $k^+ \Delta t$ is dimensionless (a pure number) and Prob$\{s = O, t + \Delta t | s = C, t\}$ is a short-hand notation for the probability, given that the channel is closed at time $t$, of a $C \to O$ transition occurring in the interval $[t, t + \Delta t]$. Multiplying by $P_C(t)$, the probability that the ion channel is indeed in state $C$, we find that $k^+ P_C(t) \Delta t$ is the probability that the transition $C \to O$ actually occurs.

The transition-state diagram (11.1) indicates two possible ways for the ion channel to enter or leave the closed state. Accounting for both of these, we have

$$P_C(t + \Delta t) = P_C(t) - k^+ P_C(t) \Delta t + k^- P_O(t) \Delta t.$$

Writing a similar equation relating $P_O(t + \Delta t)$ and $P_O(t)$ and taking the limit $\Delta t \to 0$ gives the system of ODEs

$$\frac{dP_C}{dt} = -k^+ P_C + k^- P_O, \tag{11.3}$$

$$\frac{dP_O}{dt} = +k^+ P_C - k^- P_O. \tag{11.4}$$

Because conservation of probability ensures that $P_C(t) = 1 - P_O(t)$, (11.3) can be eliminated to give

$$\frac{dP_O}{dt} = k^+ (1 - P_O) - k^- P_O.$$

Note that the similarity of this equation to (1.4), the kinetic equation derived in Chapter 1, is not accidental. The equation governing changes in probabilities for a single molecule always has the same form as the rate equation for a large number of molecules.

## 11.1.2  The Transition Probability Matrix

From our analysis of the two-state ion channel above, we know that for a channel closed at time $t$, $k^+ \Delta t$ is the probability that it undergoes a transition and opens in the time interval $[t, t + \Delta t]$, provided that $\Delta t$ is small. By conservation, we also know that the probability that the channel remains closed during the same interval is $1 - k^+ \Delta t$. Because a similar argument applies when the channel is open at time $t$, we can write the *transition probability matrix*

$$Q = \left[ \begin{array}{cc} \text{Prob}\{C, t + \Delta t | C, t\} & \text{Prob}\{C, t + \Delta t | O, t\} \\ \text{Prob}\{O, t + \Delta t | C, t\} & \text{Prob}\{O, t + \Delta t | O, t\} \end{array} \right] = \left[ \begin{array}{cc} 1 - k^+ \Delta t & k^- \Delta t \\ k^+ \Delta t & 1 - k^- \Delta t \end{array} \right], \tag{11.5}$$

where the elements of $Q_{ij}$ (row $i$, column $j$) correspond to the transition probability from state $j$ to state $i$, and conservation of probability ensures that all the columns sum to one, that is, for each column $j$,

$$\sum_i Q_{ij} = 1. \tag{11.6}$$

The transition probability matrix is especially useful when we write the current state of the channel as the vector

$$\vec{P}(t) = \left[ \begin{array}{c} \text{Prob}\{C, t\} \\ \text{Prob}\{O, t\} \end{array} \right]. \tag{11.7}$$

Using this notation, the state of the channel at $t + \Delta t$ is given by the matrix multiplication

$$\vec{P}(t + \Delta t) = Q\vec{P}(t). \tag{11.8}$$

For example, if the channel is known to be closed at time $t$, then

$$\vec{P}(t) = \begin{bmatrix} 1 \\ 0 \end{bmatrix},$$

and the distribution of probability after one time step is

$$\vec{P}(t + \Delta t) = \begin{bmatrix} 1 - k^+ \Delta t & k^- \Delta t \\ k^+ \Delta t & 1 - k^- \Delta t \end{bmatrix} \begin{bmatrix} 1 \\ 0 \end{bmatrix} = \begin{bmatrix} 1 - k^+ \Delta t \\ k^+ \Delta t \end{bmatrix}.$$

Applying (11.8) iteratively, we see that if the channel is closed at time $t$, the probability that it is closed or open at time $t + 2\Delta t$ is given by

$$\vec{P}(t + 2\Delta t) = Q\left[ Q\vec{P}(t) \right], = Q^2 \vec{P}(t)$$

or more generally,

$$\vec{P}(t + n\Delta t) = Q^n \vec{P}(t). \tag{11.9}$$

This iterative procedure can be used to calculate the evolution of the probability that the two-state channel is in an open or closed state. It amounts to using Euler's method to integrate (11.3) and (11.4).

## 11.1.3 Dwell Times

Using the transition probability matrix, it is possible to derive an expression for the average amount of time that the channel remains in the open or closed state, i.e., the open and closed *dwell times*. We have already seen that if a channel is closed at time $t$, the probability that it remains closed at time $t + \Delta t$ is $1 - k^+ \Delta t$. The probability that the channel remains closed for the following time step as well is thus $(1 - k^+ \Delta t)^2$. In general, we can write

$$\text{Prob}\{C, [t, t + n\Delta t] | C, t\} = (1 - k^+ \Delta t)^n. \tag{11.10}$$

This expression is actually much simpler than (11.9) because here we are insisting that the channel remain closed for the entire interval $[t, t + n\Delta t]$, while (11.9) accounts for the possibility that the channel changes states multiple times. If we define $\tau = n\Delta t$, we can rewrite (11.10) as

$$\text{Prob}\{C, [t, t + \tau] | C, t\} = \left( 1 - \frac{k^+ \tau}{n} \right)^n,$$

which is an approximate expression that becomes more accurate (for fixed $\tau$) as $\Delta t \to 0$ and $n \to \infty$ simultaneously. Taking this limit and using

$$\lim_{n \to \infty} \left( 1 - \frac{\alpha}{n} \right)^n = e^{-\alpha},$$

we obtain

$$\text{Prob}\{C, [t, t + \tau] | C, t\} = e^{-k^+ \tau}. \tag{11.11}$$

Thus, the probability that a channel closed at time $t$ remains closed until $t + \tau$ is a decreasing exponential function of $\tau$.

In order to complete our calculation of the closed dwell time for the two-state channel, we must consider the probability that a channel closed at time $t$ stays closed during the interval $[t, t+\tau]$ and then opens for the first time in the interval $[t+\tau, t+\tau+\Delta t]$. This probability is given by

$$\text{Prob}\{C, [t, t+\tau]|C, t\}\text{Prob}\{O, t+\tau+\Delta t|C, t+\tau\} = e^{-k^+\tau}k^+\Delta t.$$

Thus, the average closed time will be given by

$$\langle \tau_C \rangle = \int_0^\infty \tau e^{-k^+\tau}k^+ d\tau = \frac{1}{k^+},$$

where we have used

$$\int_0^\infty \tau e^{-t} dt = 1.$$

Similarly, the average open time of the two-state channel model is

$$\langle \tau_O \rangle = \int_0^\infty \tau e^{-k^-\tau}k^- d\tau = \frac{1}{k^-}.$$

## 11.1.4  Monte Carlo Simulation

The elements $Q_{ij}$ of the transition probability matrix represent the probability of making a transition from state $j$ to state $i$ in a time step of duration $\Delta t$. A simple method for simulating the transitions of a two-state channel is based on (11.6). Because conservation of probability ensures that each column of $Q$ will sum to unity, we can divide the interval $[0,1]$ into regions, each corresponding to a possible change of state (or lack of change of state). Next, we choose a random number $Y$ uniformly distributed on the interval $[0,1]$, and make a transition (or not) based upon the subinterval in which $Y$ falls. For example, let us return to the transition probability matrix for the two-state channel given by (11.5). If the current state is $O$ (open), then a transition to the closed state occurs if $0 \leq Y < k^-\Delta t$, while the channel remains open if $k^-\Delta t \leq Y \leq 1$, an interval of length $1 - k^-\Delta t$. Similarly, if the channel is closed, it remains closed if $0 \leq Y < 1 - k^+\Delta t$, and a transition to the open state occurs if $1 - k^+\Delta t \leq Y \leq 1$.

Several example simulations of stochastic gating of a two-state channel model using the Monte Carlo method are shown in Figure 11.3. By comparing open probabilities and dwell times in the three simulations shown, one can see how the transition probabilities $k^+$ and $k^-$ lead to distinct channel kinetics. In Exercise 10 the reader can use the Monte Carlo method to simulate a more complicated model, the four–state GLUT transporter model discussed in Section 3.1.

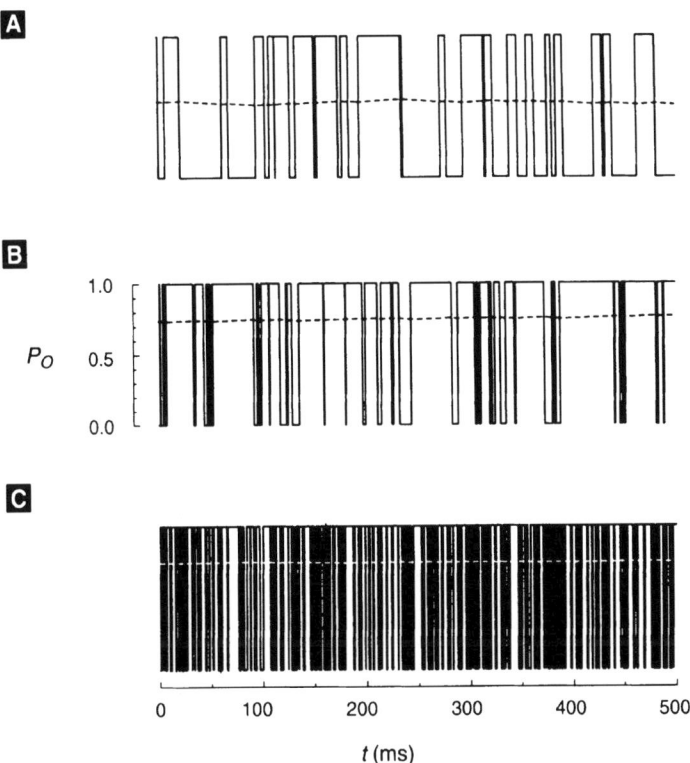

**Figure 11.3** A Monte Carlo simulation of the two-state ion channel. (A) $k^+ = 0.1$/ms, $k^- = 0.1$/ms, giving an equilibrium open probability (dotted lines) of 0.5. (B) $k^+$ changed to 0.3/ms, and now the equilibrium open probability is 0.75. (C) Transition probabilities increased by factor of 5 ($k^+ = 1.5$/ms, $k^- = 0.5$/ms). Note that average open time and average close time are shorter in this case, as evidenced by many more transitions between states.

## 11.1.5  Simulating Multiple Independent Channels

The gating of multiple independent channels can be simulated in one of several ways. An obvious possible method for simulating a small number of independent two-state ion channels is to implement $N$ Markov variables with identical transition probability matrices given by (11.5).

Under the assumption of identical and independent channels, an alternative method is to simulate a single Markov process that accurately tracks the number of open channels. Note that for an ensemble of $N$ two-state ion channels there are $N + 1$ possibilities for the number of open channels (i.e., $\{0, 1, 2, \ldots, N-2, N-1, N\}$) and thus $N + 1$ distinguishable states for the ensemble. If we label these states $S_0$ through $S_N$, we can write the transition-state diagram

$$
S_0 \underset{k^-}{\overset{Nk^+}{\rightleftharpoons}} S_1 \underset{2k^-}{\overset{(N-1)k^+}{\rightleftharpoons}} S_2 \underset{3k^-}{\overset{(N-2)k^+}{\rightleftharpoons}} \cdots \underset{(N-2)k^-}{\overset{3k^+}{\rightleftharpoons}} S_{N-2} \underset{(N-1)k^-}{\overset{2k^+}{\rightleftharpoons}} S_{N-1} \underset{Nk^-}{\overset{k^+}{\rightleftharpoons}} S_N, \tag{11.12}
$$

where the factors modifying the rate constants $k^+$ and $k^-$ account for combinatorics. For example, the transition probability $Nk^+$ leading out of state $S_0$ accounts for the fact that any one of $N$ closed channels can open (at rate $k^+$), resulting in one open channel and ensemble configuration $S_1$.

The transition probability matrix for the Markov process diagrammed in (11.12) is tridiagonal

$$
Q = \begin{bmatrix}
D_0 & k^- \Delta t & & & & & \\
Nk^+ \Delta t & D_1 & & & & & \\
& (N-1)k^+ \Delta t & & & & & \\
& & \ddots & & & & \\
& & & (N-1)k^- \Delta t & & & \\
& & & & D_{N-1} & Nk^- \Delta t & \\
& & & & k^+ \Delta t & D_N &
\end{bmatrix}, \tag{11.13}
$$

where the diagonal terms are such that probability is conserved and each column sums to 1:

$$
\begin{aligned}
D_0 &= 1 - Nk^+ \Delta t, \\
D_1 &= 1 - k^- \Delta t - (N-1)k^+ \Delta t, \\
D_{N-1} &= 1 - (N-1)k^- \Delta t - k^+ \Delta t, \\
D_N &= 1 - Nk^- \Delta t.
\end{aligned}
$$

The reader is encouraged to implement a simulation of a small number (e.g., $N = 4$) of identical and independent two-state channels.

## 11.1.6  Gillespie's Method

Both of the simulation methods described above involve iterating a transition probability matrix and can be quite cumbersome when the number of ion channels being considered is large. Fortunately, an alternative that works well for large $N$ has been devised (Gillespie 1977).

Consider a single two-state ion channel obeying the transition-state diagram (11.1) and recall that the probability that a channel closed at time $t$ remains closed until $t + \tau$ is an exponentially decreasing function of $\tau$ given by (11.11). Thus the closed dwell time ($\tau_C$) of the two-state channel is an exponentially distributed random variable; that is, the probability distribution function of $\tau_C$ is

$$\text{Prob}\{\tau < \tau_C \leq \tau + d\tau\} = k^+ e^{-k^+\tau} d\tau.$$

Similarly, the open dwell time ($\tau_O$) of the channel is an exponentially distributed random variable with probability distribution function

$$\text{Prob}\{\tau < \tau_O \leq \tau + d\tau\} = k^- e^{-k^-\tau} d\tau.$$

Thus, we can simulate a two-state ion channel by alternately choosing open and closed dwell times consistent with these distributions. If one has no subroutine for simulating an exponentially distributed random variable, simply choose a uniformly distributed random variable $U$ on the interval $[0,1]$ and use the relations

$$\tau_C = -\frac{1}{k_+} \ln U,$$

$$\tau_O = -\frac{1}{k_-} \ln U.$$

Gillespie's method is much faster computationally than the Monte Carlo methods described above. Furthermore, because there is no time step involved, the method is exact.

Gillespie's method becomes more involved when the transition-state diagram indicates that more than one possible transition contributes to the dwell time for a given state. This possibility is handled by first choosing an exponentially distributed random number for the dwell time that accounts for all of the possible transitions out of the current state (i.e., using the sum of the transition probabilities). After the length of the dwell time in the current state is thus determined, the destination state is selected by choosing a uniformly distributed random variable on an appropriately partitioned interval, a process similar to the selection of transitions during Monte Carlo simulation (see Section 11.1.4).

## 11.2 An Ensemble of Two-State Ion Channels

In the previous section we claimed that the equation governing changes in probabilities for a single molecule has the same form as the rate equation for a large number of molecules. This connection can be made more rigorous by specifying the number of molecules we are considering in advance. To simplify calculations we will again consider the two-state ion channel diagrammed in (11.1).

### 11.2.1 Probability of Finding N Channels in the Open State

Let us write $N$ as the number of molecules, and let $P_O(n, t)$ and $P_C(n, t)$ be the probabilities of having $n$ molecules in states $O$ and $C$, respectively. Because we will ultimately be interested in the statistics of current fluctuations, we will focus our attention on $P_O(n, t)$. In any case, the presence of $n$ open channels implies $N - n$ closed channels, i.e.,

$$P_C(n, t) = P_O(N - n, t) \qquad (0 \leq n \leq N).$$

Assume that all $N$ molecules are independent and consider a time interval $[t, t + \Delta t]$ short enough that only one molecule has appreciable probability of making a $C \to O$ or $O \to C$ transition. During this short time interval, there are four events that can influence $P_O(n, t)$, the probability that there are $n$ open channels. For example, it is possible that there are currently $n$ open channels, and during the time interval $[t, t + \Delta t]$ one of these channels closes. This probability is given by

$$\text{loss}_- = k^- n P_O(n, t) \Delta t,$$

where the parameter $k^-$ is the transition probability for $O \to C$, $P_O(n, t)$ is the probability that there were $n$ open channels to begin with, and the $n$ scales this probability to account for the fact that any one of the $n$ independent open channels can close with equivalent result. Similar reasoning leads to the expression

$$P_O(n, t + \Delta t) = P_O(n, t) + \text{gain}_+ - \text{loss}_+ + \text{gain}_- - \text{loss}_-, \qquad (11.14)$$

where

$$\text{gain}_- = k^-(n + 1)P_O(n + 1, t)\Delta t,$$
$$\text{loss}_+ = k^+(N - n)P_O(n, t)\Delta t,$$
$$\text{gain}_+ = k^+(N - n + 1)P_O(n - 1, t)\Delta t.$$

To give one more example, the $\text{gain}_+$ term in this equation represents a probability flux due to the possibility that there are $n - 1$ open channels and one of the closed channels opens. This transition probability is given by $k^+(N - n + 1)P_O(n - 1, t)\Delta t$, because any one of the $N - (n - 1) = N - n + 1$ closed channels can open with equivalent result.

Taking the limit $\Delta t \to 0$ of (11.14) gives the ordinary differential equation

$$\frac{d}{dt}P_O(n, t) = k^+(N - n + 1)P_O(n - 1, t) - k^+(N - n)P_O(n, t)$$
$$+ k^-(n + 1)P_O(n + 1, t) - k^- n P_O(n, t). \qquad (11.15)$$

This rather complicated expression is called a *master equation*. It actually represents $N + 1$ coupled ordinary differential equations, one for each $P_O(n, t)$ for $0 \leq n \leq N$ (all possible values for the number of open channels).

The equilibrium solution to the master equation is $N + 1$ time-independent probabilities $P_O^\infty(n)$, given by the binomial distribution

$$P_O^\infty(n) = \binom{N}{n} p^n (1 - p)^{N-n}, \qquad (11.16)$$

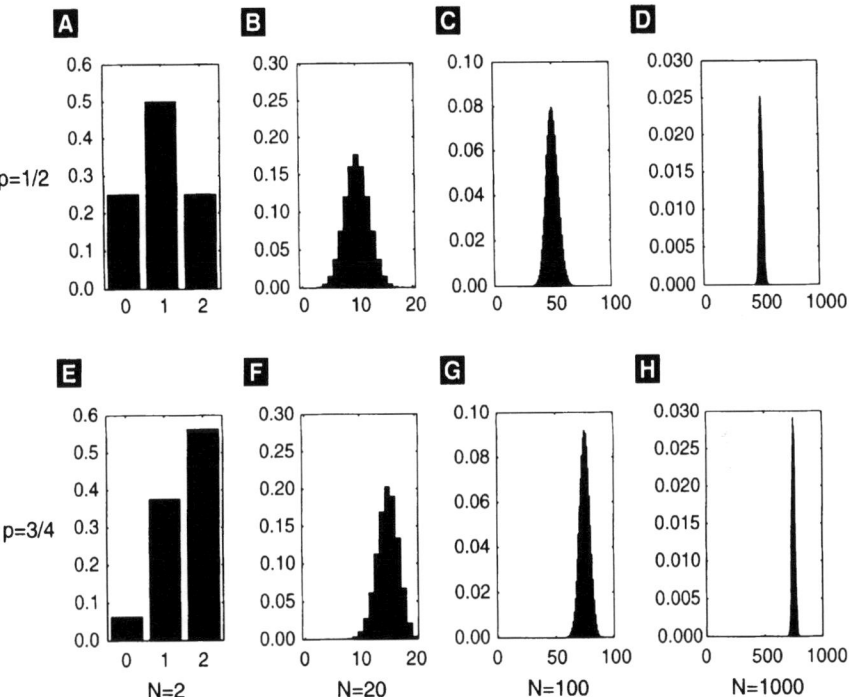

**Figure 11.4** For an equilibrium ensemble of $N$ two-state channels with open probability $p$, the likelihood of observing $n$ open channels is given by the binomial probability distribution (11.16), with parameters $N$ and $p$. The binomial probability distribution has mean $Np$, variance $Np(1-p)$, and coefficient of variation $[(1-p)/Np]^{1/2}$. Note that as the equilibrium open probability, $p$, increases the mean number of open channels shifts rightward. The $N^{-1/2}$ factor in the coefficient of variation is reflected in the narrowing of the distributions (from left to right).

where $p = k^+/(k^+ + k^-)$ and

$$\binom{N}{n} = \frac{N!}{n!(N-n)!}.$$

Although this may not be obvious at first, in Exercise 4 the mathematically inclined reader can use the method of substitution to confirm that the binomial distribution satisfies a time-independent version of (11.15).

Figure 11.4 shows several binomial probability distributions with parameters $N$ and $p$ varied. Given an ensemble of $N$ two-state channels, these distributions represent the equilibrium probability of finding $n$ channels in the open state. In the top row, the equilibrium open probability of $p = 0.5$ results in a centered distribution: The likelihood of observing $n$ open channels is equal to the likelihood of observing $N - n$

open channels. In the bottom row $p = 0.75$, and the enhanced likelihood that channels are open is evident in the rightward shift of the distributions.

## 11.2.2  The Average Number of Open Channels

The equilibrium solution to the master equation for the two-state channel given by (11.16) is the binomial distribution, and thus the average number of open channels at equilibrium is $\langle N_O \rangle_\infty = Np$. But what about the time-dependence of the average number of open channels? Because the average number of open channels is given by

$$\langle N_O \rangle = \sum_{n=0}^{N} n P_O(n, t), \tag{11.17}$$

we can find an equation for $d\langle N_O \rangle / dt$ by multiplying (11.15) by $n$ and summing. This gives

$$\frac{d\langle N_O \rangle}{dt} = k^+ \sum_{n=0}^{N} n(N - n + 1) P_O(n - 1, t) - k^+ \sum_{n=0}^{N} n(N - n) P_O(n, t)$$

$$+ k^- \sum_{n=0}^{N} n(n + 1) P_O(n + 1, t) - k^- \sum_{n=0}^{N} n^2 P_O(n, t). \tag{11.18}$$

In Exercise 6 the reader can show that this equation can be reduced to

$$\frac{d\langle N_O \rangle}{dt} = k^+ (N - \langle N_O \rangle) - k^- \langle N_O \rangle, \tag{11.19}$$

where

$$N - \langle N_O \rangle = \langle N_C \rangle. \tag{11.20}$$

Note that (11.19) is identical to the rate equation for a population of two-state channels derived by other means in Chapter 1. For the duration of this chapter we will refer to such an equation as an *average* rate equation. Also note that the equilibrium average number of open ($\langle N_O \rangle_\infty$) and closed ($\langle N_C \rangle_\infty$) channels can be found by setting the left–hand side of (11.19) to zero, that is,

$$\langle N_O \rangle_\infty = N \frac{k^+}{k^+ + k^-} = Np, \tag{11.21}$$

$$\langle N_C \rangle_\infty = N \frac{k^-}{k^+ + k^-} = N(1 - p), \tag{11.22}$$

in agreement with our knowledge of the mean of a binomial distribution.

If we divide (11.19) by the total number of channels $N$, we find the average rate equation for the fraction of open channels

$$\frac{d\langle f_O \rangle}{dt} = k^+ \left(1 - \langle f_O \rangle\right) - k^- \langle f_O \rangle, \tag{11.23}$$

where $\langle f_O \rangle = \langle N_O \rangle / N$, $\langle f_C \rangle = \langle N_C \rangle / N$, and (11.20) implies $\langle f_O \rangle + \langle f_C \rangle = 1$. The equilibrium fractions of open and closed channels are $\langle f_O \rangle_\infty = k^+ / (k^+ + k^-) = p$ and $\langle f_C \rangle_\infty = k^- / (k^+ + k^-) = 1 - p$. We thus see explicitly for a two-state channel that the master equation implies an average rate equation of the sort introduced in Chapter 2. This is true in general.

### 11.2.3 The Variance of the Number of Open Channels

One advantage of beginning with a master equation is that in addition to the average rate equation, an evolution equation for the variance in the number of open channels can be derived. The variance in the number of open channels is defined as

$$\sigma_{N_O}^2 = \langle (N_O - \langle N_O \rangle)^2 \rangle = \sum_{n=0}^{N} (n - \langle N_O \rangle)^2 \, P_O(n, t). \qquad (11.24)$$

Similarly, the variance in the number of closed channels is

$$\sigma_{N_C}^2 = \langle (N_C - \langle N_C \rangle)^2 \rangle = \sum_{n=0}^{N} (n - \langle N_C \rangle)^2 \, P_C(n, t). \qquad (11.25)$$

Again, we are ultimately interested in the statistics of current fluctuations, so we focus on $\sigma_{N_O}^2$. For the two-state channel under consideration, it is shown in Exercise 8 that these quantities are equal.

Beginning with (11.24) and the master equation (11.15), it can be shown that the variance $\sigma_{N_O}^2$ satisfies the ODE

$$\frac{d\sigma_{N_O}^2}{dt} = -2 \left( k^+ + k^- \right) \sigma_{N_O}^2 + k^+ \left( N - \langle N_O \rangle \right) + k^- \langle N_O \rangle. \qquad (11.26)$$

The equilibrium variance $\left( \sigma_{N_O}^2 \right)_\infty$ is thus given by steady states of this equation. Setting the left–hand side of this expression to zero, we obtain

$$\left( \sigma_{N_O}^2 \right)_\infty = N \frac{k^+ k^-}{\left( k^+ + k^- \right)^2} = N p (1 - p). \qquad (11.27)$$

From this equation it is clear that the equilibrium variance is proportional to $N$, the total number of channels. However, a relative measure of the variance known as the *coefficient of variation* is more meaningful. The coefficient of variation of the number of open channels, $CV_{N_O}$, is given by the ratio of the standard deviation $\sigma_{N_O}$ (the square root of the variance) to the mean $\langle N_O \rangle$. At equilibrium, we have

$$\left( CV_{N_O} \right)_\infty = \frac{\left( \sigma_{N_O} \right)_\infty}{\langle N_O \rangle_\infty} = \frac{1}{\sqrt{N}} \sqrt{\frac{k^-}{k^+}} = \sqrt{\frac{1 - p}{N p}},$$

where the last equality is in agreement with the mean and variance of a binomially distributed random variable being $Np$ and $Np(1 - p)$, respectively. From this expression it is clear that the equilibrium coefficient of variation for the number of open channels is inversely proportional to the square root of the number of channels $N$. Thus, in order to decrease this relative measure of channel noise by a factor of two, the number of channels must be increased by a factor of 4.

# 11.3    Fluctuations in Macroscopic Currents

When the voltage clamp technique is applied to isolated membrane patches, openings and closings of single ion channels can be observed. Recall the single-channel recordings of T-type $Ca^{2+}$ currents shown in the top panel of Figure 11.2. Importantly, the bottom panel of Figure 11.2 shows that when several hundred single-channel recordings are summed, the kinetics of rapid activation and slower inactivation of the T-type $Ca^{2+}$ current are evident. In this summed trace, the relative size of the fluctuations in the macroscopic current is much smaller than those observed in the single-channel recordings; however, the fluctuations in ionic current are still noticeable.

During voltage clamp recordings of large numbers of ion channels, stochastic gating leads to current fluctuations. For example, Figure 11.5B shows the time evo-

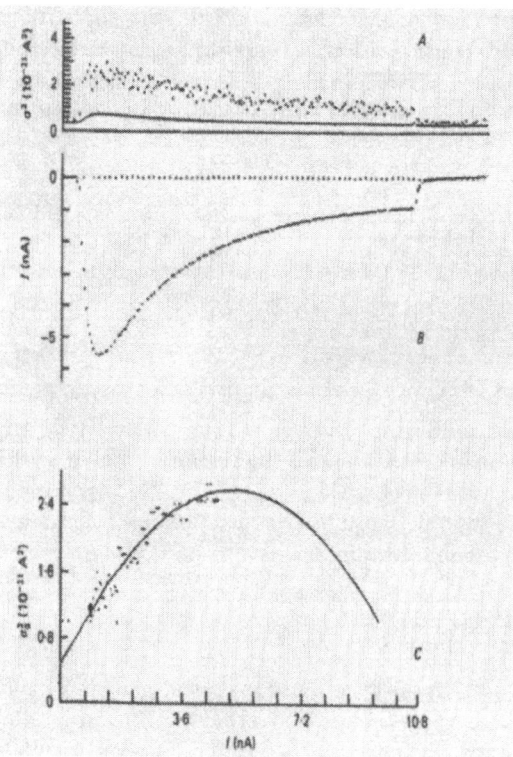

**Figure 11.5** Variance and mean sodium current measured from voltage-clamped single myelinated nerve fibers from *Rana pipiens* depolarized to −15 mV after 50 ms prepulses to −105 mV. (A) Variance arising from the stochastic gating of sodium channels (dots) and thermal noise (solid line). (B) Mean current. (C) After carefully accounting for contributions to the variance due to thermal noise, the parabolic relationship between variance and mean current suggests $N=20,400$ sodium channels at this node of Ranvier each with single channel conductance of $i_{unit} = 0.55$ pA. Reprinted from Sigworth (1980).

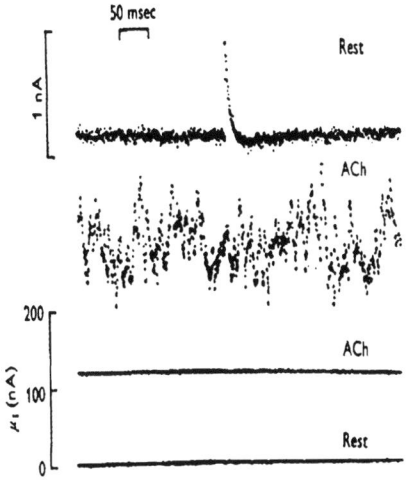

**Figure 11.6** Acetylcholine–produced current noise due to fluctuations in ionic conductance of voltage–clamped end-plates of *Rana pipiens* nerve–muscle preparation. Iontophoretic application of ACh resulted in an increase in mean current as well as variance. The second trace, labeled "Rest," also shows a spontaneous miniature end-plate current. Reprinted from Anderson and Stevens (1973).

lution of the mean sodium current measured from voltage-clamped single myelinated nerve fibers of *Rana pipiens* (frog) that were depolarized to $-15$ mV after 50 ms pre-pulses to $-105$ mV. After a careful accounting for contributions to the variance of the sodium current due to thermal noise (Figure 11.5A, solid line), the variance arising from the stochastic gating of sodium channels remains (Figure 11.5A, dots). Figure 11.6 shows macroscopic current fluctuations induced by the iontophoretic application of acetylcholine (ACh) to voltage–clamped end-plates of a *Rana pipiens* nerve-muscle preparation. Interestingly, iontophoretic application of ACh increased the variance of the end-plate current as well as the mean. While the first trace in Figure 11.6 shows a spontaneous miniature end-plate current (sharp peak), the phenomenon of interest is the 10-fold increase in variance observed throughout the duration of the second trace (as compared to the first).

In order to understand the relationship between fluctuations in macroscopic currents and the underlying single-channel kinetics, consider the statistics of ionic current implied by the two-state channel model presented in the previous section. In the simplest case, the unitary current of each two-state channel will be a random variable taking the value zero when the channel is closed or a fixed value $i_{unit}$ when the channel is open. That is, the unitary current will be a random variable $I_{unit}$ given by

$$I_{unit} = \begin{cases} i_{unit} = g_{unit} (V - V_{rev}) & \text{when open,} \\ 0 & \text{when closed,} \end{cases} \tag{11.28}$$

where $V$ is a fixed command voltage, $V_{rev}$ is the reversal potential for the single–channel conductance $g_{unit}$, and $i_{unit}$ is directly proportional to the conductance of the open channel. With these assumptions, it is straightforward to apply the results of Section

11.2 and derive the statistics of a fluctuating current that will result from $N$ two-state channels with unitary current given by (11.28). The fluctuating macroscopic current will be a random variable defined by

$$I_{\text{macro}} = N_O\, i_{\text{unit}} \qquad (0 \leq N_O \leq N),$$

where $N_O$ is the number of open channels (also a random variable). Because the macroscopic current is directly proportional to $N_O$, we can use (11.17) to find the equilibrium average macroscopic current

$$\langle I_{\text{macro}}\rangle_\infty = i_{\text{unit}} \langle N_O\rangle_\infty.$$

Similarly, the equilibrium variance in the number of open channels, $(\sigma_{N_O}^2)_\infty$, given by (11.27), determines the equilibrium variance of the macroscopic current

$$\left(\sigma_{I_{\text{macro}}}^2\right)_\infty = i_{\text{unit}}^2 \left(\sigma_{N_O}^2\right)_\infty.$$

Recall that if we write $p = k^+/(k^+ + k^-)$, the equilibrium mean and variance for the number of open channels are given by $\langle N_O\rangle_\infty = Np$ and $(\sigma_{N_O}^2)_\infty = Np(1-p)$. Thus, the equilibrium mean and variance for the macroscopic current are given by $\langle I_{\text{macro}}\rangle_\infty = i_{\text{unit}}Np$ and $(\sigma_{I_{\text{macro}}}^2)_\infty = i_{\text{unit}}^2 Np(1-p)$. Combining these expressions and eliminating $p$ gives

$$\left(\sigma_{I_{\text{macro}}}^2\right)_\infty = i_{\text{unit}}\langle I_{\text{macro}}\rangle_\infty - \langle I_{\text{macro}}\rangle_\infty^2/N, \qquad (11.29)$$

where both $\langle I_{\text{macro}}\rangle_\infty$ and $(\sigma_{I_{\text{macro}}}^2)_\infty$ are parameterized by $p$.

Equation (11.29) is the basis of a standard technique of membrane noise analysis whereby current fluctuations can be used to estimate the number of ion channels in a membrane patch. By repeatedly manipulating the fraction of open channels $p$ an estimate of $(\sigma_{I_{\text{macro}}}^2)_\infty$ as a function of $\langle I_{\text{macro}}\rangle_\infty$ is obtained. According to (11.29), the relationship will be parabolic with zero variance at $\langle I_{\text{macro}}\rangle_\infty = 0$ and $\langle I_{\text{macro}}\rangle_\infty = i_{\text{unit}}N$ and a maximum variance of $(\sigma_{I_{\text{macro}}}^2)_\infty = Ni_{\text{unit}}^2/4$ at $\langle I_{\text{macro}}\rangle_\infty = i_{\text{unit}}N/2$. In Figure 11.5C this technique was applied to voltage-clamped single myelinated nerve fibers from *Rana pipiens*. Equation (11.29) and a visual fit resulting in a maximum of $(\sigma_{I_{\text{macro}}}^2)_\infty = 2.5 \times 10^{-21}\, \text{A}^2$ at $\langle I_{\text{macro}}\rangle_\infty = 5\, \text{nA}$ suggests that $N = \langle I_{\text{macro}}\rangle_\infty^2/(\sigma_{I_{\text{macro}}}^2)_\infty = 10{,}000$ and $i_{\text{unit}} = 1\,\text{pA}$. However, after carefully accounting for contributions to the variance due to thermal noise, an adjusted fit (solid line in Figure 11.5C) gives $N = 20{,}400$ sodium channels at this node of Ranvier and a unitary current of $i_{\text{unit}} = 0.55\,\text{pA}$ (see Sigworth 1980).

In Figure 11.7 this technique was applied to end-plate conductance fluctuations of *Rana pipiens* nerve–muscle preparation. Here, the equilibrium variance of the macroscopic conductance $\left(\sigma_{g_{\text{macro}}}^2\right)_\infty$ is plotted against the mean conductance $\langle g_{\text{macro}}\rangle_\infty$, where the macroscopic conductance is related to the unitary conductance through $g_{\text{macro}} = Ng_{\text{unit}}$. Using (11.29) and the relations

$$\langle g_{\text{macro}}\rangle_\infty = \frac{\langle I_{\text{macro}}\rangle_\infty}{V - V_{\text{rev}}}, \qquad \left(\sigma_{g_{\text{macro}}}^2\right)_\infty = \frac{(\sigma_{I_{\text{macro}}}^2)_\infty}{(V - V_{\text{rev}})^2} \quad \text{and} \quad i_{\text{unit}} = g_{\text{unit}}\,(V - V_{\text{rev}}),$$

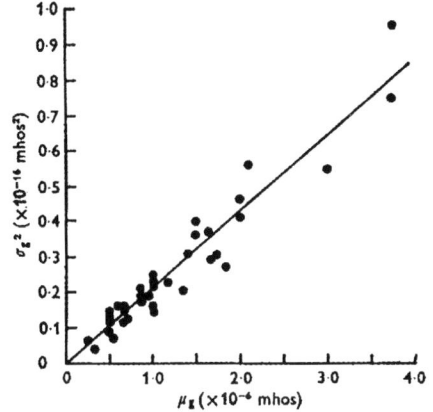

**Figure 11.7** Variance of conductance fluctuations as a function of mean end-plate conductance of *Rana pipiens* nerve–muscle preparation. Because the unitary conductance of end-plate channels is small, the relationship is linear and the slope of $0.19 \times 10^{-10}$ mho $= 19$ pS gives the single-channel conductance. Reprinted from Anderson and Stevens (1973).

the reader can confirm that this relationship is also expected to be parabolic, that is,

$$\left(\sigma^2_{g_{macro}}\right)_\infty = g_{unit}\langle g_{macro}\rangle_\infty - \langle g_{macro}\rangle^2_\infty/N.$$

However, because the unitary end-plate channel conductance of *Rana pipiens* nerve–muscle preparation is very small (quadratic term negligible), the relationship is nearly linear:

$$\left(\sigma^2_{g_{macro}}\right)_\infty = g_{unit}\langle g_{macro}\rangle_\infty.$$

Indeed, the slope of the line in Figure 11.7 gives a single channel conductance of 19 pS for the open end-plate channel.

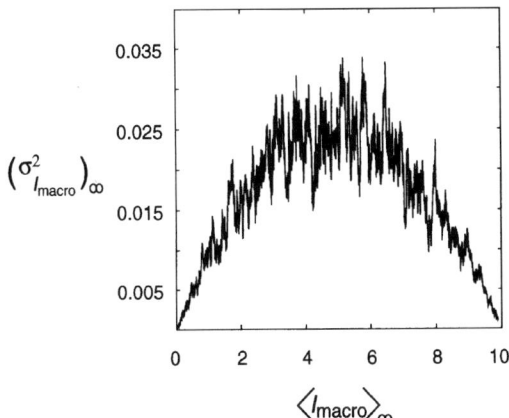

**Figure 11.8** The parabolic relationship between the variance and mean of ionic current through $N = 10,000$ two-state channels each with single-channel conductance of $i_{unit} = 0.01$ pA. The parabolic relationship between the variance $\left(\sigma^2_{I_{macro}}\right)_\infty$ and mean $\langle I_{macro}\rangle_\infty$ of current fluctuations is calculated from 100 simultaneously integrated Langevin equations.

## 11.4  Modeling Fluctuations in Macroscopic Currents with Stochastic ODEs

Figure 11.8 shows a simulation reproducing the parabolic relationship between the variance $\left(\sigma^2_{I_{\text{macro}}}\right)_\infty$ and mean, $\langle I_{\text{macro}}\rangle_\infty$, of current fluctuations due to the stochastic gating of ion channels. This simulation includes $N = 10,000$ identical two-state channels with unitary conductance of $i_{\text{unit}} = 0.01$ pA. A hundred trials were simultaneously performed and averaged to calculate the mean and variance as the open probability $p$ ranged from 0 to 1. Because the methods discussed in Section 11.1.4 would require the declaration of a Markov variable with 10,001 possible states, the reader may be wondering how this simulation was performed.

Indeed, in simulating the stochastic gating of large numbers of ion channels, Monte Carlo methods becomes impractical. However, when $N$ is large, fluctuations in macroscopic currents can instead be described using a stochastic ordinary differential equation, called a Langevin equation, that takes the form

$$\frac{df}{dt} = g(f) + \xi. \tag{11.30}$$

In this equation, the familiar deterministic dynamics given by $g(f)$ are supplemented with a rapidly varying random forcing term $\xi(t)$. Because $\xi$ is a random function of time, solving (11.30) often means finding a solution $f(t)$ that satisfies the equation for a particular instantiation of $\xi$. Alternatively, if the statistics of $\xi$ are given, we may be interested in deriving the statistics of the new random variable $f(t)$ that is formally defined by (11.30).

The most common fluctuating force to consider are the increments of a Wiener process. Similar to the unbiased random walk discussed in Chapter 12, a Wiener process $B(t)$ is a "Gaussian" random process that has zero mean

$$\langle B \rangle = 0 \tag{11.31}$$

and variance directly proportional to time,

$$\sigma_B^2 = \langle (B - \langle B \rangle)^2 \rangle = \langle B^2 \rangle = t. \tag{11.32}$$

Indeed, the instantiations of a Wiener process $B_1$ and $B_2$ shown in Figure 11.9 are similar to the random walks presented in Figure 12.4. Just as the increments of an unbiased random walk are $\pm\Delta x$ with equal probability, resulting in an increment with mean zero $\langle \Delta X \rangle = 0$, the increments of the numerical approximation to a Wiener process shown in Figure 11.9 are normally distributed with mean zero,

$$\langle \Delta B \rangle = 0.$$

In order to understand the variance of the increments of this simulated Wiener process, we must remember that unlike an unbiased random walk, a Wiener process is a continuous function of time, $B(t)$. A relevant statistic for the increments of a Wiener process is the two-time covariance or *autocorrelation function* $\langle \Delta B(t) \Delta B(t') \rangle$. Because

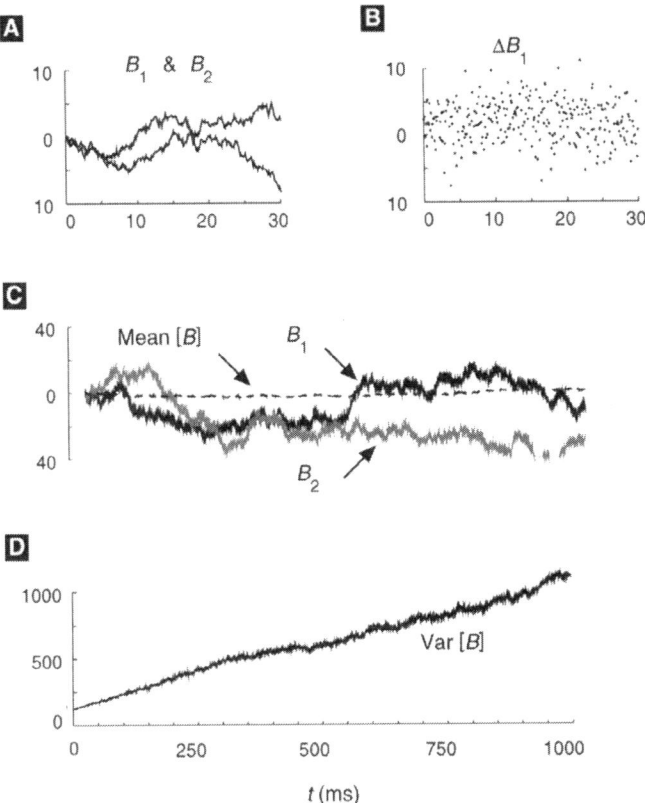

**Figure 11.9**  Two instantiations of a Wiener process, $B_1$ and $B_2$, have trajectories similar to the random walks shown in Figure 12.4. The mean of 100 trials is near zero, while the variance of 100 trials increases linearly with time. In these simulations the increments $(\Delta B)$ are normally distributed with mean zero and variance $1/\Delta t$, where $\Delta t$ is the integration time step.

nonoverlapping increments of a Wiener process are statistically independent and uncorrelated, $\langle \Delta B(t) \Delta B(t') \rangle = 0$ for $t' > t + \Delta t$ in Figure 11.9. In the limit as $\Delta t \to 0$ (i.e., for a "real" Wiener process), we might even write $dB(t)/dt = \xi(t)$, where $\langle \xi(t) \rangle = 0$ and

$$\langle \xi(t)\xi(t') \rangle = \delta\left(t - t'\right),\tag{11.33}$$

although technically this derivative does not exist. Equation (11.33) may appear unusual, especially if the reader is unfamiliar with the Dirac delta function, defined by $\delta(t) = 0$ for $t \neq 0$ and

$$\int_{-\infty}^{\infty} \delta(t)dt = 1.$$

A Wiener process $B(t)$ can be simulated by numerically integrating a piecewise constant approximation to the Wiener increment $\Delta B(t)$. The Wiener increment (Figure 11.9B) is a normally distributed random variable with zero mean that is held fixed during the time interval $[t, t + \Delta t]$ and updated after the integration time step is complete. Integrating this erratic function of time results in the Wiener trajectories of Figure 11.9A and Figure 11.9C. If we rewrite (11.33) to account for this piecewise constant approximation to the Wiener increments, we obtain

$$\langle \Delta B(t) \Delta B(t') \rangle = \begin{cases} 1/\Delta t, & t' \in [t, t + \Delta t], \\ 0 & \text{otherwise.} \end{cases}$$

Like the unbiased random walk, the variance of this simulated Wiener process is proportional to time as shown in Figure 11.9D. The reader can confirm through simulations that in order to achieve this macroscopic behavior, the variance of the Wiener increments $\Delta B(t)$ must be adjusted according to the integration time step (i.e., $\text{Var}[\Delta B] = 1/\Delta t$ gives $\text{Var}[B] = t$).

## 11.4.1   Langevin Equation for an Ensemble of Two-State Channels

In order to use a Langevin equation of the form of (11.30) to simulate a large number of ion channels, we must make an appropriate choice for both the deterministic function $g(f)$ as well as the statistics of the random variable $\xi$. Recalling the average rate equation for the dynamics of the open fraction of channels (11.23), we write

$$\frac{df_O}{dt} = k^+ (1 - f_O) - k^- f_O + \xi \tag{11.34}$$

$$= -\frac{f_O - \langle f_O \rangle_\infty}{\tau_f} + \xi, \tag{11.35}$$

where $f_O = N_O/N$ is a random variable, the fluctuating fraction of open channels, $\langle f_O \rangle_\infty = k^+/(k^+ + k^-)$, and $\tau_f = 1/(k^+ + k^-)$. For (11.35) to be meaningful, we must specify the statistics of $\xi$. An appropriate choice for $\xi$ is a fluctuating function of time that has zero mean,

$$\langle \xi(t) \rangle = 0,$$

and an autocorrelation function given by

$$\langle \xi(t) \xi(t') \rangle = \gamma \delta (t - t').$$

By methods of statistical physics beyond the scope of this book, $\gamma$ can be shown to be inversely proportional to $N$ and proportional to the sum of the rates of both the $O \rightarrow C$ and $C \rightarrow O$ transitions, that is,

$$\gamma(f_O) = \frac{k^+ (1 - f_O) + k^- f_O}{N}. \tag{11.36}$$

An appropriate choice for $\xi$ is thus $\xi = \sqrt{\gamma} \Delta B$, where the $\Delta B$ are the increments of a Wiener process.

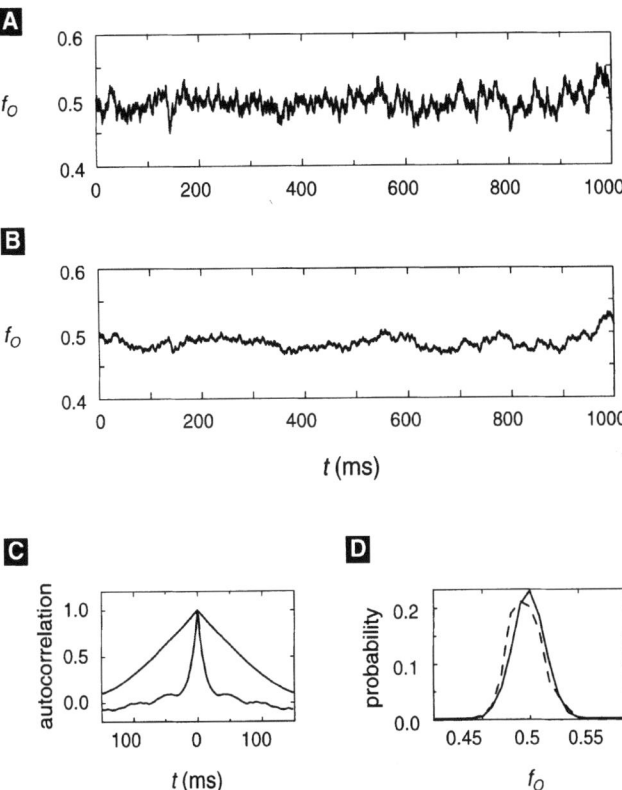

**Figure 11.10** (A,B) The open fraction $f_O$ of 1000 two-state ion channels simulated using a Langevin equation. Transition rates are ten times faster in (A) than (B) so that the time constant $\tau_f$ is 10 and 100 ms, respectively. (C) Numerically calculated autocorrelation function of $f_O$. (D) The equilibrium mean and variance of $f_O$ are approximately equal in the two cases.

Although we haven't fully justified this choice for $\xi$, we can check that this random variable and (11.35) define the random variable $f_O$ in a manner consistent with the work in previous sections. To do this we use the fluctuation–dissipation theorem (Keizer 1987; Gardiner 1997) from statistical physics that relates $\gamma$, which occurs in the correlation function of $\xi$, to the equilibrium variance of $f_O$. The relationship depends on the relaxation time constant $\tau_f$ and is given by

$$\gamma_\infty = \frac{2\left(\sigma_{f_O}^2\right)_\infty}{\tau_f}. \tag{11.37}$$

Using (11.27), and remembering that $\left(\sigma_{f_O}^2\right)_\infty = \left(\sigma_{N_O}^2\right)_\infty / N^2$, the reader can confirm that the last equality holds at equilibrium.

Stochastic simulations of the open fraction, $f_O$, of 1000 two-state ion channels calculated by integrating (11.35) are shown in Figure 11.10. The transition rates used were $k^+ = k^- = 0.05$/ms in panel (A) and $k^+ = k^- = 0.005$/ms in panel (B), giving time constants ($\tau_f$) of 10 and 100 ms and equilibrium open fraction $\langle f_O \rangle_\infty = p = 0.5$ in both cases. This difference in relaxation time constants is evident in the (normalized) autocorrelation functions compared in C. It can be shown that for an infinitely long simulation the autocorrelation functions for $f_O$ is

$$\langle f_O(t) f_O(t') \rangle = \left( \sigma_{f_O}^2 \right)_\infty e^{-|t-t'|/\tau_f}.$$

The narrower autocorrelation function in Figure 11.10C thus corresponds to the case with small time constant $\tau_f$. Note that although the time constant for relaxation to $\langle f_O \rangle_\infty = 0.5$ is faster in A than in B, the equilibrium variance $\left( \sigma_{f_O}^2 \right)_\infty$ shown in Figure 11.10D is approximately equal in the two cases, as expected according to (11.27).

## 11.4.2 Fokker–Planck Equation for an Ensemble of Two-State Channels

Rather than calculating trajectories for the fraction of open channels $f_O$ using a Langevin equation, an alternative is to calculate the evolution of the probability distribution function (PDF) for $f_O$. While the binomial distribution encountered in Section 11.2 is an example of a discrete probability distribution ($N_O$ takes on $N + 1$ discrete values), the Langevin equation for $f_O$ (11.35) implies that $f_O$ can take on any value on the interval $[0,1]$. Thus, the PDF for $f_O$ is continuous and defined as

$$P(f, t) \, df = \text{Prob}\{f(t) < f_O < f(t) + df\},$$

where conservation of probability gives

$$\int_0^1 P(f, t) \, df = 1. \tag{11.38}$$

We can write an evolution equation for $P(f, t)$, known as a Fokker–Planck equation, that corresponds to the Langevin description given by (11.35):

$$\frac{\partial P(f, t)}{\partial t} = -\frac{\partial}{\partial f} \left[ J_{\text{adv}}(f, t) + J_{\text{dif}}(f, t) \right]. \tag{11.39}$$

In this equation, $J_{\text{adv}}(f, t)$ is a probability flux due to advection (that is, transport) governed by the deterministic terms in (11.35):

$$J_{\text{adv}}(f, t) = -\frac{f - \langle f_O \rangle_\infty}{\tau_f} P(f, t). \tag{11.40}$$

In contrast, $J_{\text{dif}}(f, t)$ is a diffusive flux that accounts for the spread of probability induced by the random variable $\xi$. This diffusive probability flux is given by

$$J_{\text{dif}}(f, t) = -\frac{1}{2} \frac{\partial}{\partial f} \left[ \gamma(f) P(f, t) \right], \tag{11.41}$$

where $\gamma$ is given by (11.36). Rewriting (11.40) in terms of the total probability flux $J_{\text{tot}} = J_{\text{adv}} + J_{\text{dif}}$, we have

$$\frac{\partial P(f,t)}{\partial t} = -\frac{\partial J_{\text{tot}}(f,t)}{\partial f}, \tag{11.42}$$

with associated boundary conditions

$$J_{\text{tot}}(0,t) = J_{\text{tot}}(1,t) = 0$$

that imply no flux of probability out of the physiological range for $f_O$. An appropriate choice of initial conditions would be $P(f,0) = \delta(f - \langle f_O \rangle_\infty)$, implying that the system is known to be in equilibrium at $t = 0$.

Setting the left–hand side of (11.39) equal to zero, we see that the equilibrium probability distribution $P_\infty(f)$ solves $J_{\text{tot}}^\infty = 0$. That is,

$$-\frac{f - \langle f_O \rangle_\infty}{\tau_f} P_\infty(f) - \frac{1}{2}\frac{d}{df}\left[\gamma(f)P_\infty(f)\right] = 0. \tag{11.43}$$

This differential equation can be solved numerically. However, we obtain more insight by approximating $\gamma(f)$ by

$$\gamma_\infty = \frac{k^+\left(1 - \langle f_O \rangle_\infty\right) + k^-\langle f_O \rangle_\infty}{N}.$$

This procedure is valid when fluctuations of $f_O$ away from equilibrium, $\langle f_O \rangle_\infty$, are small (that is, when $N$ is large). If we make this approximation, it can be shown that the probability distribution

$$P_\infty(f) = A \exp\left[\frac{f\left(2\langle f_O \rangle_\infty - f\right)}{\gamma_\infty \tau_f}\right] \tag{11.44}$$

satisfies (11.43), where the normalization constant $A$ is chosen to satisfy conservation of probability (11.38). While this expression may not look familiar, when $\gamma_\infty$ is sufficiently small ($N$ sufficiently large), $P_\infty(f)$ is well approximated by the Gaussian

$$P_\infty(f) = \frac{1}{\sqrt{2\pi}\left(\sigma_{f_O}^2\right)_\infty} \exp\left[-\frac{\left(f - \langle f_O \rangle_\infty\right)^2}{2\left(\sigma_{f_O}^2\right)_\infty}\right].$$

At equilibrium $f_O$ will be a normally distributed random variable with mean $\langle f_O \rangle_\infty$ and variance $\left(\sigma_{f_O}^2\right)_\infty = \gamma_\infty \tau_f/2$, in agreement with (11.37). The distribution $P_\infty(f)$ is approximately Gaussian for $N = 1000$ as shown in Figure 11.10D.

# 11.5 Membrane Voltage Fluctuations

In Section 11.3 we discussed macroscopic current fluctuations experimentally observed in voltage clamp recordings and a membrane noise analysis. In this section we will

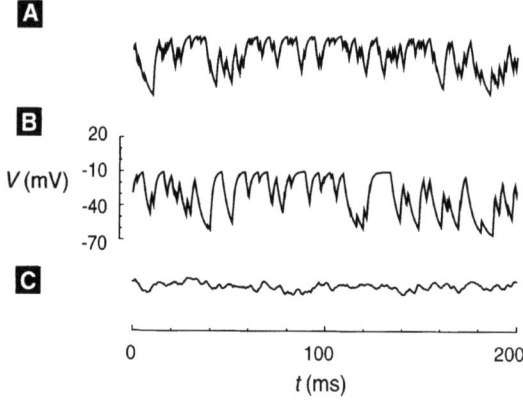

**A**

**B**

$V$ (mV)

20
-10
-40
-70

**C**

0         100         200

$t$ (ms)

**Figure 11.11** Membrane voltage fluctuations due to the stochastic gating of one or more sodium channels. (A,B) Single channel simulations. Transition probabilities are a factor of two slower in (B), leading to longer dwell times and fewer transitions as evidenced by 'kinks' in graph. (C) Twenty channels are simulated. As the number of sodium channels increases, the variance in membrane voltage decrease.

simulate electrical recordings in which the membrane potential is not clamped, but rather fluctuates under the influence of two-state ion channels. Although a misnomer, such measurements are referred to as *current clamp* recordings. For now we assume that ion channel gating is voltage-independent.

Simulations of membrane voltage fluctuations due to the stochastic gating of a single sodium channel obeying the transition-state diagram (11.1) are shown in Figure 11.11A and Figure 11.11B. The gating of the two-state channel is simulated using Monte Carlo methods, while membrane voltage is simultaneously calculated using the current balance equation

$$C\frac{dV}{dt} = -g_{\rm L}\left(V - V_{\rm L}\right) - g_{\rm Na}\left(V - V_{\rm Na}\right),\tag{11.45}$$

where $g_{\rm L}$ is leakage conductance with reversal potential $V_{\rm L} = -70$ mV, and $g_{\rm Na}$ (a random variable taking values of 0 or $g_{\rm Na}^{\max}$ depending on channel state) is a sodium conductance with reversal potential $V_{\rm Na} = 60$ mV. Because transition probabilities are a factor of two slower in Figure 11.11B, longer dwell times and fewer transitions are observed. For comparison, Figure 11.11C shows the result when twenty channels are simulated. Here $g_{\rm Na}$ takes on 21 possible values between 0 and $g_{\rm Na}^{\max}$ and as a consequence the variance in the fluctuating membrane voltage decreases.

Membrane voltage fluctuations such as those shown in Figure 11.11 can also be modeled by tracking the evolution of probability distribution functions (PDFs) for membrane voltage conditioned on the state of the ion channel. The governing equations are coupled partial differential equations each of which is similar to the advective component of the Fokker–Planck equations described above. We will refer to this method as the *ensemble density approach*.

If membrane voltage fluctuations are due to a single two-state sodium channel, there are two relevant conditional PDFs,

$$P_C(v,t)\,dv = \mathrm{Prob}\{v < V < v + dv | C, t\},$$

$$P_O(v,t)\,dv = \text{Prob}\{v < V < v + dv | O, t\},$$

where $P_C(v,t)$ and $P_O(v,t)$ are conditioned on the channel being closed or open, respectively. Conservation of probability implies

$$\int_{-\infty}^{\infty} P_C(v,t)\,dv = 1, \qquad \int_{-\infty}^{\infty} P_O(v,t)\,dv = 1.$$

The equations for the evolution of these conditional PDFs are

$$\frac{\partial}{\partial t} P_C(v,t) = -\frac{\partial}{\partial v} J_C(v,t) - k^+ P_C(v,t) + k^- P_O(v,t), \tag{11.46}$$

$$\frac{\partial}{\partial t} P_O(v,t) = -\frac{\partial}{\partial v} J_O(v,t) + k^+ P_C(v,t) - k^- P_O(v,t), \tag{11.47}$$

where $J_C(v,t)$ and $J_O(v,t)$ are advective probability fluxes due to membrane voltage obeying the current balance equation (11.45):

$$J_C(v,t) = -\frac{1}{C} [g_L (v - V_L)] P_C(v,t),$$

$$J_O(v,t) = -\frac{1}{C} [g_L (v - V_L) + g_{Na}^{max} (v - V_{Na})] P_O(v,t).$$

Notice that the sodium current term occurs only in $J_O(v,t)$, because if the channel is closed, $g_{Na} = 0$. The reaction terms that appear in (11.46) and (11.47) account for probability flux due to the stochastic gating of the sodium channel. Regardless of membrane voltage, the conditional probability $P_C(v,t)$ can decrease due to channel opening at a rate of $k^+ P_C(v,t)$ and increase due to closing of open channels at a rate of $k^- P_O(v,t)$. The reaction terms occur with opposite sign because any increase or decrease in the conditional probability $P_C(v,t)$ due to a channel gating implies commensurate change in $P_O(v,t)$.

The equilibrium conditional probability distribution functions for the membrane voltage, $P_C^{\infty}(v)$ and $P_O^{\infty}(v)$, calculated numerically from (11.46) and (11.47), are shown in Figure 11.12A. The simulation ran for 1 second, corresponding to approximately 1000 changes in channel state. As expected, $P_O^{\infty}(v)$ is shifted toward the right (depolarized $V$) relative to $P_C^{\infty}(v)$. The astute reader will note that the PDFs are not symmetric, indicating that when the channel is open probability advects toward $V_{Na}$ faster than it advects toward $V_L$ when the channel is closed; i.e., an open sodium channel leads to a smaller membrane time constant. In Figure 11.12B, the rate constants $k^+$ and $k^-$ are a factor of ten slower. In this case, more probability accumulates near both $V_L$ and $V_{Na}$.

## 11.5.1 Membrane Voltage Fluctuations with an Ensemble of Two-State Channels

The ensemble density formulation described above can be extended to the case where membrane voltage fluctuations are due to an ensemble of $N$ two-state channels. If we write $P_O(n,v,t)$ for the conditional probability density for membrane voltage given $n$

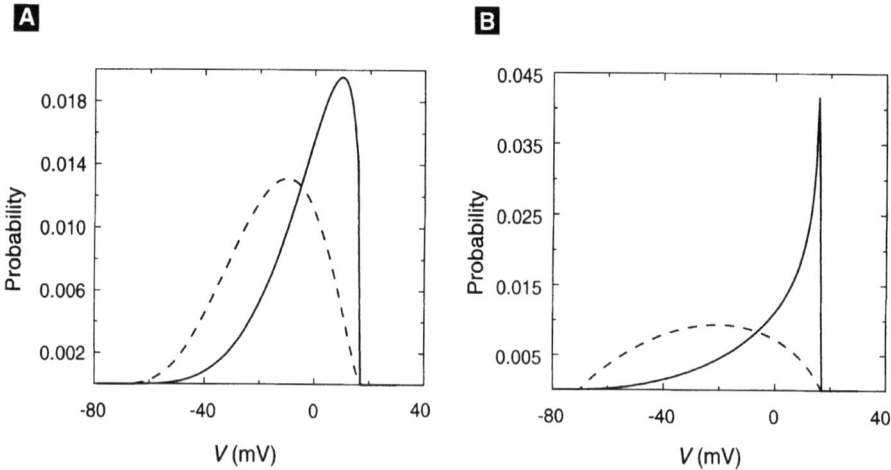

**Figure 11.12** Equilibrium conditional probability distribution functions (PDFs), $P_C^\infty(v)$ (dotted lines) and $P_O^\infty(v)$ (solid lines) for the membrane voltage conditioned on the state of a single two-state sodium channel. In (A), $k^+ = k^- = 1$/ms, while for (B) the transition probabilities are a factor of two slower.

open sodium channels, we have

$$\frac{\partial}{\partial t}P_O(n,v,t) = -\frac{\partial}{\partial v}J_O(n,v,t)$$
$$+ k^+(N-n+1)P_O(n-1,v,t) - k^+(N-n)P_O(n,v,t)$$
$$+ k^-(n+1)P_O(n+1,v,t) - k^- nP_O(n,v,t), \tag{11.48}$$

where the reaction terms are based on the master equation formulation presented in Section 11.2.1, $J_O(n,v,t)$ is given by

$$J_O(n,v,t) = -\frac{1}{C}\left[g_L(v-V_L) + g_{Na}^{max}\frac{n}{N}(v-V_{Na})\right]P_O(n,v,t),$$

and $g_{Na}^{max}$ is the sodium conductance when all $N$ channels are open. Note that (11.48) represents $N+1$ coupled partial differential equations, one for each $P_O(n,v,t)$, where $0 \leq n \leq N$.

Figure 11.13 shows equilibrium conditional PDFs $P_O^\infty(n,v)$ for membrane voltage fluctuations induced by 20 two-state sodium channels. These PDFs are calculated by numerically solving (11.48) until a steady state is achieved. Careful inspection of the figure shows that in the case of high $n$ (more open channels) the equilibrium distribution of membrane voltage is shifted toward $V_{Na}$. Note that these PDFs appear to be consistent with a binomial distribution for the total equilibrium probability for a given

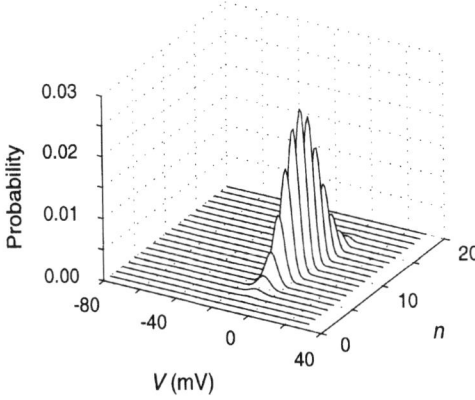

**Figure 11.13** Conditional probability distribution functions for membrane voltage fluctuations due to stochastic gating of 20 two-state sodium channels. Parameters as in Figure 11.12A.

value of $n$. That is,

$$P_O^\infty(n) = \int_{-\infty}^{\infty} P_O^\infty(n,v)\,dv$$

is in agreement with (11.16).

## 11.6 Stochasticity and Discreteness in an Excitable Membrane Model

Using the results of previous sections, we are prepared to explore the consequences of stochasticity and discreteness in an excitable membrane model. The deterministic Morris–Lecar model is

$$C\frac{dV}{dt} = I_{app} - g_L(V - V_L) - g_K w(V - V_K) - g_{Ca}m_\infty(V - V_{Ca}) \tag{11.49}$$

$$\frac{dw}{dt} = \frac{w_\infty - w}{\tau}, \tag{11.50}$$

where the activation function for the $Ca^{2+}$ current, $m_\infty(V)$; the activation function for the $K^+$ current, $w_\infty(V)$; and voltage-dependent time scale for activation of $K^+$ current, $\tau(V)$, are given in Chapter 2.

In (11.50), $w$ is usually thought to represent the fraction of open $K^+$ channels. However, we now understand that this differential equation is actually an average rate equation similar to (11.23). To be clear, let us write this deterministic average rate equation as

$$\frac{d\langle w \rangle}{dt} = \frac{w_\infty - \langle w \rangle}{\tau},$$

where $w$ (a random variable) represent the fraction of open $K^+$ channels. The reader can easily verify that this average rate equation corresponds to the two-state kinetic scheme

$$C\,(\text{closed}) \; \underset{\beta(V)}{\overset{\alpha(V)}{\rightleftharpoons}} \; O\,(\text{open}),$$

where $C$ and $O$ indicate closed and open states of the $K^+$ channel, and the voltage-dependent transition rates $\alpha(V)$ and $\beta(V)$ are given by

$$\alpha(V) = \frac{w_\infty}{\tau}$$

$$\beta(V) = \frac{1 - w_\infty}{\tau}.$$

This, in turn, implies that the equilibrium fraction of open $K^+$ channels is

$$\langle w \rangle_\infty = w_\infty(V) = \frac{\alpha}{\alpha + \beta},$$

and the time constant $\tau(V)$ is

$$\tau(V) = \frac{1}{\alpha + \beta}.$$

With these preliminaries, we can see that a Morris–Lecar simulation that includes channel noise due to a small number of $K^+$ channels could be performed with several Markov variables obeying the voltage-dependent transition probability matrix

$$Q = \begin{bmatrix} 1 - \alpha\Delta t & \beta\Delta t \\ \alpha\Delta t & 1 - \beta\Delta t \end{bmatrix}$$

for each channel. Alternatively, a larger collection of $N$ channels can be simulated by tracking only a single Markov variable, the number of open $K^+$ channels. In this case, the following tridiagonal transition probability matrix would be used:

$$Q = \begin{bmatrix} D_0 & \beta\Delta t & & & & \\ N\alpha\Delta t & D_1 & & & & \\ & (N-1)\alpha\Delta t & & & & \\ & & \ddots & & & \\ & & & (N-1)\beta\Delta t & & \\ & & & D_{N-1} & N\beta\Delta t \\ & & & \alpha\Delta t & D_N \end{bmatrix},$$

with diagonal terms $(D_0, D_1, \ldots, D_{N-1}, D_N)$ such that each column sums to unity.

## 11.6.1  Phenomena Induced by Stochasticity and Discreteness

Figure 11.14 and Figure 11.15 show Morris–Lecar simulations that include stochastic voltage-dependent gating of 100 $K^+$ channels. Spontaneous action potentials are

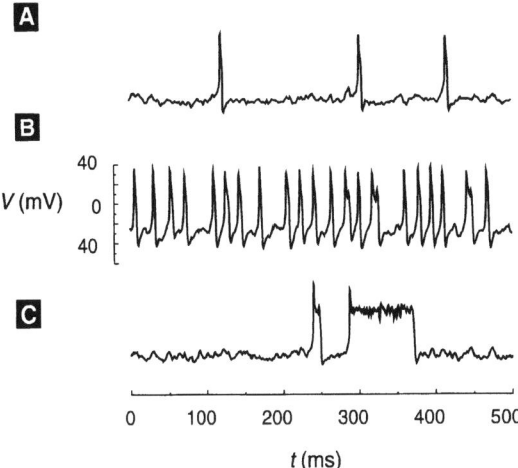

**Figure 11.14** Morris–Lecar simulations including stochastic gating of 100 $K^+$ channels. (A) Spontaneous excitability driven by channel noise is observed when $I_{app}$ = 10 and the deterministic model is excitable. (B) Stochastic oscillations are observed when $I_{app}$ = 12 and the deterministic model is oscillatory. (C) Stochastic bistability is observed when the deterministic model is bistable ($I_{app}$ = 12 and $v_3$ = 15 mV rather than standard value of 10 mV).

induced by this simulated channel noise (Figure 11.14A). We will refer to this phenomenon as "stochastic excitability," because it is understood as a sampling of the excitable phase space of the deterministic model made possible by membrane potential fluctuations due to the stochastic gating of $K^+$ channels. The $(V, w)$ phase plane trajectories for Figure 11.14 are shown in Figure 11.15. The discreteness and stochasticity of the $K^+$ gating variable $w$ allows trajectories to fluctuate around the fixed point of the deterministic model seen in the lower left of Figure 11.15. Occasionally, $K^+$ channels spontaneously inactivate ($w$ fluctuates toward 0) and a regenerative $Ca^{2+}$ current leads to an action potential. This type of spontaneous activity has been observed in stochastic versions of the Hodgkin–Huxley equations (Chow and White 1996; Fox 1997) and is thought to influence subthreshold membrane potential oscillations and excitability of stellate neurons of the medial entorhinal cortex of the hippocampal region (White et al. 1995; White et al. 2000).

In Figure 11.14B parameters are such that the deterministic model (as $N \to \infty$) is oscillatory. However, when $N = 100$ channel noise results in irregular oscillations. In Figure 11.14B stochastic bistability is observed. When parameters are chosen so that the deterministic model is bistable, channel noise allows the alternate sampling of two stable fixed points in the $(V, w)$ phase plane, a phenomenon known as basin hopping.

## 11.6.2 The Ensemble Density Approach Applied to the Stochastic Morris–Lecar Model

The ensemble density approach described in Section 11.4.2 can be applied to the stochastic Morris–Lecar model described above. The evolution equations for the

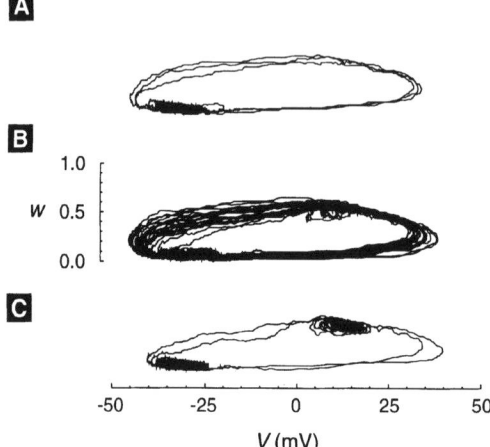

**A**

**B**

**C**

**Figure 11.15** Morris–Lecar simulations with stochastic gating of 100 $K^+$ channels shown in the $(V, w)$ phase plane. Panels correspond to (A) stochastic excitability, (B) oscillations, and (C) bistability shown in Figure 11.14.

conditional PDFs take the form

$$
\begin{aligned}
\frac{\partial}{\partial t} P_O(n, v, t) = &-\frac{\partial}{\partial v} J_O(n, v, t) \\
&+ \alpha(v)(N - n + 1) P_O(n - 1, v, t) \\
&- \alpha(v)(N - n) P_O(n, v, t) \\
&+ \beta(v)(n + 1) P_O(n + 1, v, t) - \beta(v) n P_O(n, v, t).
\end{aligned} \tag{11.51}
$$

This form is similar to (11.48) except that the transition probabilities are now voltage-dependent, and the probability fluxes $J_O(n, v, t)$ are given by the Morris–Lecar current balance equation (11.49). That is,

$$
J_O(n, v, t) = -\frac{1}{C} \left[ I_{\mathrm{app}} - g_{\mathrm{L}} (v - V_{\mathrm{L}}) - g_{\mathrm{K}}^{\max} \frac{n}{N} (v - V_{\mathrm{K}}) - g_{\mathrm{Ca}} m_\infty (v - V_{\mathrm{Ca}}) \right] P_O(n, v, t).
$$

Figure 11.16 shows equilibrium PDFs for the membrane voltage of the stochastic Morris–Lecar model conditioned on the number of open $K^+$ channels. These equilibrium PDFs are steady-state solutions to (11.51) and correspond to the three types of trajectories shown in Figure 11.15. The amount of time that trajectories spend in different regions of the $(V, w)$ phase plane is reflected in these distributions. It is clear from Figure 11.15A that the Morris–Lecar model exhibiting stochastic excitability spends a large proportion of time near the threshold for excitation.

## 11.6.3   Langevin Formulation for the Stochastic Morris–Lecar Model

To consider the behavior of the Morris–Lecar model under the influence of channel noise from a large number of $K^+$ channels, it is most convenient to use the Langevin formulation presented in Section 11.4.1. We do this by supplementing the rate equation for the average fraction of open $K^+$ channels, (11.50), with a rapidly varying forcing

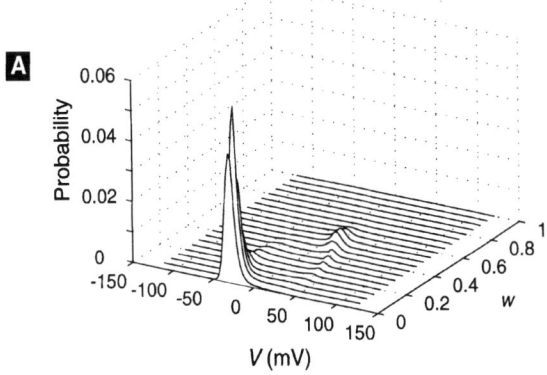

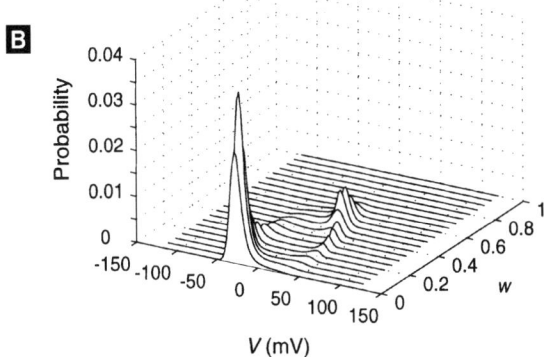

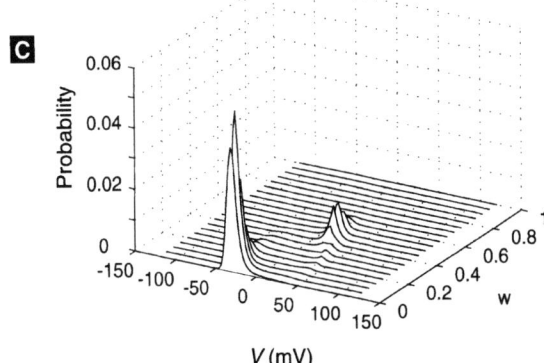

**Figure 11.16** Probability distribution functions for the membrane voltage of the stochastic Morris–Lecar model conditioned on the number of open $K^+$ channels. The equilibrium PDFs show evidence of stochastic excitability (A), oscillations (B), and bistability (C), corresponding to the trajectories shown in Figure 11.15.

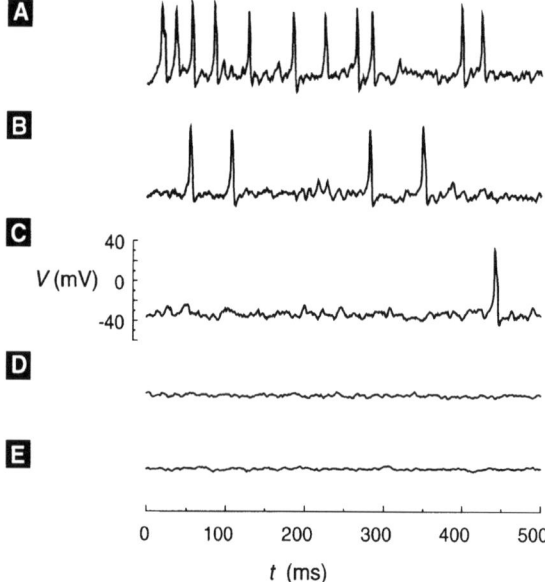

**A**

**B**

**C**

**D**

**E**

$V$ (mV)

40

0

-40

0     100     200     300     400     500

$t$ (ms)

**Figure 11.17** The stochastic Morris–Lecar model simulated using a Langevin equation for $w$, the fraction of open $K^+$ channels. As the number of $K^+$ channels is increased ($N = 25, 50, 100, 500,$ or 1000) spontaneous action potentials induced by stochastic gating are eliminated. For large $N$, the model is excitable, but essentially deterministic; i.e., fluctuations in $w$ are small and spontaneous action potentials are no longer observed without applied current.

term

$$\frac{dw}{dt} = \frac{w_\infty - w}{\tau} + \xi,$$

where $w$ is a random variable, $\langle \xi \rangle = 0$, and the autocorrelation function of $\xi$ is given by

$$\langle \xi(t)\xi(t') \rangle = \gamma(w, V)\delta\left(t - t'\right).$$

Following (11.36), $\gamma(w, V)$ is chosen to be

$$\gamma(w, V) = \frac{\alpha(1 - w) + \beta w}{N} = \frac{1}{N}\frac{(1 - 2w_\infty)w + w_\infty}{\tau}. \qquad (11.52)$$

Thus, $\xi$ is a random variable defined by $\xi = \sqrt{\gamma(w, V)}\Delta B$, where the $\Delta B$ are the increments of a Wiener process.

Figure 11.17 presents stochastic Morris–Lecar model simulations implemented using the Langevin formulation described above. Interestingly, the existence of stochastic excitability depends on the the number of $K^+$ channels included. When $N$ is relatively small ($N = 25, 50, 100$) membrane potential fluctuations are large, and spontaneous action potentials are frequent. However, when more $K^+$ channels are included ($N = 500, 1000$), the model becomes essentially deterministic. Although the model is still excitable, as $N \rightarrow \infty$ fluctuations in $w$ become smaller, and spontaneous action potentials are no longer observed.

# Suggestions for Further Reading

- *Handbook of stochastic methods for physics, chemistry, and the natural sciences*, G.W. Gardiner. This is an accessible introduction to stochastic methods including Markov systems, stochastic differential equations, Fokker–Planck equations, and master equations (Gardiner 1997).
- *Spontaneous action potentials due to channel fluctuations*, C.C. Chow and J.A. White. A theoretical and numerical analysis of the Hodgkin–Huxley equations when stochastic ion channel dynamics are included (Chow and White 1996).
- *Stochastic versions of the Hodgkin–Huxley equations*, R.F. Fox. A presentation of master equation and Langevin descriptions of the Hodgkin– Huxley equations with stochastic ion channel dynamics (Fox 1997).
- *Channel noise in neurons*, J.A. White, J.T. Rubinstein, and A.R. Kay. A good review of stochastic gating of voltage-dependent ion channels and channel noise in neurons (White et al. 2000).

## 11.7 EXERCISES

1. Consider a single GLUT molecule and the four states $(S_1, S_2, S_3, S_4)$ and transitions shown in Figure 3.1. Define $s$ to be a random variable taking values $s \in \{1, 2, 3, 4\}$ corresponding to these four states and write $P_i(t)$ to represent the probability that $s(t) = i$, that is, the molecule is in state $i$ at time $t$. Write an equation that indicates conservation of probability, that is, an equation resulting from the fact that the molecule must be in one of the four states.

2. Consider the possibility that the GLUT transporter (see previous exercise) is in state 1 at time $t$. What is the probability that in a short interval of time $[t, t + \Delta t]$ the GLUT transporter will associate with glucose, thereby transitioning out of state 1 and into state 2? What is the conditional probability, given that the channel is in state 1, of a $1 \to 2$ transition occurring during this interval?

3. Figure 3.1 and Table 3.1 indicate four possible ways for the GLUT transporter to enter or leave state 1. Accounting for all of these, we can write

   $$P_1(t + \Delta t) = P_1(t) - k_{12}[G]_{out}P_1(t)\Delta t - k_{14}P_1(t)\Delta t + k_{21}P_2(t)\Delta t + k_{41}P_4(t)\Delta t.$$

   Write three additional equations relating $P_i(t + \Delta t)$ and $P_i(t)$ for $i = \{2, 3, 4\}$ and take the limit $\Delta t \to 0$ to derive the a system of ODEs governing changes in probabilities for a single GLUT molecule being in states 1 through 4.

4. Use the method of substitution to confirm that the binomial distribution given by (11.16) is the steady state of the master equation for the two-state ion channel (11.15). Begin with a time-independent version of (11.15),

   $$0 = k^+(N - n + 1)P_O^\infty(n - 1) - k^+(N - n)P_O^\infty(n) + k^-(n + 1)P_O^\infty(n + 1) - k^- n P_O^\infty(n);$$

and subsitute $P_O^\infty(n-1)$, $P_O^\infty(n)$, and $P_O^\infty(n+1)$ according to (11.16). Remember that $p = k^+/(k^+ + k^-)$ and $1 - p = k^-/(k^+ + k^-)$. After some algebra you will find that in order to balance the $k^+$ and $k^-$ terms, one must have

$$(N - n + 1)\binom{N}{n-1} = n\binom{N}{n},$$

$$(n + 1)\binom{N}{n+1} = (N - n)\binom{N}{n},$$

as is indeed the case.

5. Using (11.17) as a guide, write an expression for the average number $\langle N_C \rangle$ of closed channels in terms of $P_C(n, t)$. Then show that $N - \langle N_O \rangle = \langle N_C \rangle$, as stated in (11.20). Hint: You will need to use $P_C(n, t) = P_O(N - n, t)$ and $\sum_{n=0}^{N} n P_O(N - n, t) = \sum_{n=0}^{N} (N - n) P_O(n, t)$.

6. Using the result of Exercise 5, confirm that (11.18) is equivalent to (11.19). In order to do so, you will need to show that

$$-\langle N_O \rangle = \sum_{n=0}^{N} n(n+1) P_O(n+1, t) - \sum_{n=0}^{N} n^2 P_O(n, t)$$

and

$$\langle N_C \rangle = \sum_{n=0}^{N} n(N - n + 1) P_O(n - 1, t) - \sum_{n=0}^{N} n(N - n) P_O(n, t).$$

7. Equation (11.18) is the master equation for the two-state channel with kinetic scheme given by (11.1). Derive the the master equation for the GLUT transporter shown in Figure 3.1.

8. Show that in the case of the two-state channel, the variances defined with respect to fluctuations in open channel number ($\sigma_{N_O}^2$) and closed channel number ($\sigma_{N_C}^2$), are equal. You will need some of the relations from Exercise 5 as well as (11.24) and (11.25).

9. Confirm (11.26), the equation for the time-dependence of the variance of the two-state channel. Hint: Differentiate (11.24) to obtain

$$\frac{d\sigma_{N_O}^2}{dt} = \sum_{n=0}^{N}\left[-2(n - \langle N_O \rangle)\frac{d\langle N_O \rangle}{dt} P_O(n, t) + (n - \langle N_O \rangle)^2 \frac{dP_O(n, t)}{dt}\right]. \qquad (11.53)$$

Now check to see whether the right–hand sides of (11.26) and (11.53) are equal. Use (11.19) and (11.15) as well as

$$\sum_{n=0}^{N}(n - \langle N_O \rangle)^2 (N - n + 1) P_O(n - 1, t) = \sum_{n=0}^{N-1}(n + 1 - \langle N_O \rangle)^2 (N - n) P_O(n, t),$$

$$\sum_{n=0}^{N}(n - \langle N_O \rangle)^2 (n + 1) P_O(n + 1, t) = \sum_{n=1}^{N}(n - 1 - \langle N_O \rangle)^2 n P_O(n, t).$$

10. Simulate the four-state GLUT transporter shown in Figure 3.1 by implementing a Markov process with transition probability matrix

$$Q = \begin{bmatrix} D_1 & k_{21}\Delta t & 0 & k_{41}\Delta t \\ k_{12}[G]_{\text{out}}\Delta t & D_2 & k_{32}\Delta t & 0 \\ 0 & k_{23}\Delta t & D_3 & k_{43}\Delta t \\ k_{14}\Delta t & 0 & k_{34}[G]_{\text{in}}\Delta t & D_4 \end{bmatrix},$$

where the diagonal entries $(D_1, D_2, D_3, D_4)$ are such that each column sums to 1.

11. Confirm the form of the time-invariant conditional PDFs shown in Figure 11.12.

12. Reproduce simulations from Figure 11.14, Figure 11.15, and Figure 11.17.

# CHAPTER 12

# Molecular Motors: Theory

Alex Mogilner, Timothy C. Elston, Hongyun Wang, and George Oster

Evolution has created a class of proteins that have the ability to convert chemical energy into mechanical force. Some of these use the free energy of nucleotide hydrolysis as fuel, while others employ ion gradients. Some are "walking motors," others rotating engines. Some are reversible; others are unidirectional. Could there be any common principles among such diversity?

The conversion of chemical energy into mechanical work is one of the main themes of modern biology. Biochemists characterize energy transduction schemes by free energy diagrams. But thermodynamics tells us only what *cannot* happen. Recent advances in laser trap and optical technology along with advances in molecular structure determination can augment traditional biochemical kinetic and thermodynamic analyses to make possible a more mechanistic view of how protein motors function. The result of these advances has been data that yield load–velocity curves and motion statistics for single molecular motors. This sort of data enables a more detailed, mechanistic level of modeling.

At first, the mechanics of proteins may seem counterintuitive because their motions are dominated by Brownian motion, the name given to the frequent changes in velocity of a macromolecule as it is buffeted about by random thermal motions of surrounding water molecules. In addition to "smearing out" deterministic trajectories, Brownian motion serves asan effective "lubricant," allowing molecules to pass over high energy barriers that would arrest a deterministic system. More subtly, it makes possible "uphill" motions against an opposing force by "capturing" occasional large thermal fluctuations.

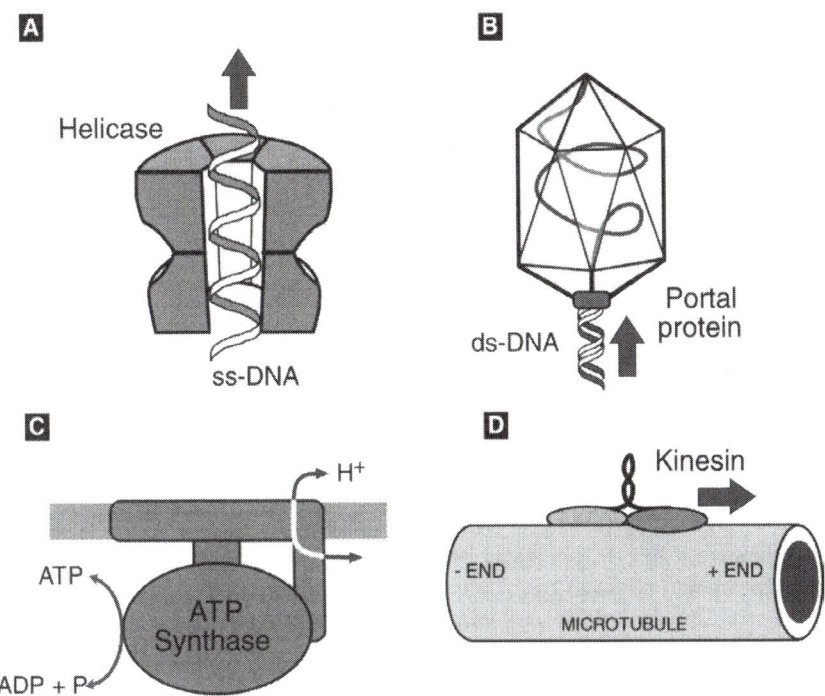

**Figure 12.1** Amazing variety of molecular motors: (A) Rotary motor DNA helicase translocates unidirectionally along the DNA strand using nucleotide hydrolysis as a "fuel." (B) Another rotary motor hydrolyzing ATP, bacteriophage portal protein, drives DNA in and out. (C) Reversible rotary motor ATP synthase either produces ATP using ion gradient or pumps protons hydrolyzing ATP. (D) Linear motor kinesin is a "walking enzyme." Utilizing chemical energy stored in ATP, it moves "head-over-head" toward the plus end of the microtubule "track." Some of these motors are discussed in this chapter.

In this chapter we will discuss protein motions on the molecular scale and derive a mathematical formalism to model such motions. To illustrate the formalism, in the Chapter 13 we will analyze (i) a "switch" controlling the direction of the bacterial flagellar motor, (ii) a polymerization ratchet, and (iii) a "toy" model related to the $F_0$ motor of ATP synthase. These models are simple enough to yield analytical as well as numerical results, and to illustrate many of the principles involved in mechanochemical energy transduction.

There is some ambiguity in what one calls a "motor." Here we take the narrow view that the principal–and proximate–function of a molecular motor is to convert chemical energy into mechanical force. This excludes, for example, ion pumps, which are surely protein machines that generate forces, but whose purpose is not force production. Chemical energy comes in various forms, for example, in transmembrane ion gradients and in the covalent bonds of nucleotides such as ATP and GTP, and the designs of motor proteins are tailored to each energy form.

Energy stored in one form frequently is converted into intermediate forms before being released as mechanical work. For example, a polymerizing actin filament or microtubule can generate a protrusive force capable of deforming a lipid vesicle or pushing out the leading edge of a cell (Honda et al. 1999; Fygenson et al. 1997; Dogterom and Yurke 1997). The energy source in this process is the free energy of binding monomers to the polymer tip. This energy is used to rectify the Brownian motion of the load against which the polymer is pushing. Strictly speaking, the force is generated by thermal fluctuations of the load, and the binding free energy is used to rectify its thermal displacements. Energy conversion here is relatively direct. However, in the acrosomal process of the *Limulus* sperm, thermal fluctuations are first trapped as elastic strain energy in the actin polymer by the binding of an auxiliary protein, scruin. Later, this strain energy is released to generate the force required to push the actin rod into the egg cortex (Mahadevan and Matsudaira 2000).

Many motors use nucleotide hydrolysis to generate mechanical forces, and it is frequently stated that the energy is stored in the $\gamma$-phosphate covalent bond. But releasing this energy to perform mechanical work can be quite indirect. The $F_1$ motor of ATP synthase uses nucleotide hydrolysis to generate a large rotary torque (Yasuda et al. 1998). However, the actual force–generating step takes place during the binding of ATP to the catalytic site; the role of the hydrolysis step is to release the hydrolysis products, allowing the cycle to repeat (Oster and Wang 2000a; Oster and Wang 2000b). In some motors, not all of the nucleotide binding energy is used immediately for force production; some energy is stored in elastic deformation of the protein to be released later as mechanical work. So energy transduction need not be a "pay as you go" process; deferred payments are permissible and common.

The bacterial flagellar motor and the $F_0$ motor of ATP synthase both use transmembrane ion gradients to generate a rotary torque (Berg 2000). Models of this process show how the chemical reaction of binding an ion onto a charged site creates an unbalanced electrostatic field that rectifies the Brownian motion of the motor and/or creates an electrostatic driving torque (Elston and Oster 1997; Elston et al. 1998). Although the proximal energy transduction process is a chemical binding event, the motion itself is produced by electrostatic forces and Brownian motion.

Thus a common theme in energy transduction is that *chemical reactions power mechanical motion using free energy* released during binding events, but the final producttion of mechanical force may involve a number of intermediate energy transductions.

The most important quality of molecular motors that distinguishes them from macroscopic motors is the overwhelming importance of thermal fluctuations. For this reason, all protein motors must be regarded as "Brownian machines." This means that carelessly applying macroscopic physics, where Brownian motion is negligible, to microscopic situations inevitably leads to incorrect conclusions. Therefore, we must begin our discussion by examining how to model molecular motions dominated by thermal fluctuations.

## 12.1  Molecular Motions as Stochastic Processes

### 12.1.1  Protein Motion as a Simple Random Walk

Generally, a stochastic process refers to a random variable that evolves in time. An example is a one-dimenisonal coordinate $x(t)$ locating a protein diffusing in an aqueous solution. We will begin by approximating the coordinate $x(t)$ by a discrete random variable. The rationale for this is twofold. Discrete random variables are conceptually simpler than their continuous counterparts. The results for the discrete case are applicable in studying continuous random processes because continuous random variables represent limiting behavior of their discrete counterparts. Our discussion is restricted to Markov processes. A Markov process is a mathematical idealization in which the future state of a protein is affected by its current state but is independent of its past. That is, the system has no memory of how it arrived at its current state. To a very good approximation, all systems considered in this text satisfy the Markov property. The mathematics involved in studying stochastic processes that are non-Markovian is considerably more complicated.

In the discrete model, a protein is initially started at $x = 0$. In each time interval $\Delta t$, it takes one step of length $\Delta x$ to the right with probability $\frac{1}{2}$ or to the left with probability $\frac{1}{2}$. Because the length of the step that the protein takes is always the same, this example is referred to as a simple random walk. Let $x_n$ denote the protein's position at time $t = n\Delta t$ and define the set of random variables $z_m$ with $m = 1, 2, \ldots, n$ to be independent and identically distributed with $\text{Prob}[z_m = 1] = \frac{1}{2}$ and $\text{Prob}[z_m = -1] = \frac{1}{2}$. Then we have

$$x_n = \Delta x(z_1 + z_2 + \cdots + z_n). \tag{12.1}$$

The collection of random variables $x = \{x_0, x_1, x_2, \ldots\}$ represents a spatially and temporally discrete stochastic process. In Exercise 1, (12.1) is used to verify that $\langle x_n \rangle = 0$ and $\text{Var}[x_n] = (\Delta x)^2 n = ((\Delta x)^2/\Delta t)t$. Here we use the notation $\langle \cdot \rangle$ to denote the average (expectation), and $\text{Var}[\cdot]$ to denote the variance: $\text{Var}[x] \equiv \langle x^2 \rangle - \langle x \rangle^2$. Note that the variance in $x$ grows linearly with time. This is a characteristic of diffusion; below we show in what sense the quantity $D = (\Delta x)^2/(2\Delta t)$ can be interpreted as a diffusion coefficient.

To fully characterize $x$ requires knowledge of the probability density for finding the particle at position $x_n$ after $k$ steps of size $\Delta x$: $p_k(n) = \text{Prob}[x_n = k\Delta x]$. Note that $p_k(n) = 0$ if $n < |k|$. This comes from the fact that the protein can take only one step per time interval. At any time $n\Delta t$, the total number of steps taken by the protein is $n = R_n + L_n$, where $R_n$ is the number of steps taken to the right and $L_n$ is the number of steps taken to the left. Clearly, $R_n$ is binomially distributed, like the number of "heads" in $n$ flips of a coin:

$$\text{Prob}[R_n = m] = \binom{n}{m}\left(\frac{1}{2}\right)^n = \frac{n!}{n!(n-m)!}\left(\frac{1}{2}\right)^n. \tag{12.2}$$

Using these definitions, $x_n$ is written as

$$x_n = \Delta x(R_n - L_n) = \Delta x(2R_n - n), \tag{12.3}$$

or equivalently

$$R_n = \frac{1}{2}\left(\frac{x_n}{\Delta x} + n\right). \tag{12.4}$$

Thus $x_n/\Delta x = k$ if and only if $R_n = \frac{1}{2}(n+k)$. Furthermore, $x_n/\Delta x$ must be even if $n$ is even and odd if $n$ is odd, since $R_n$ must be an integer. Therefore, we immediately find that the distribution for $x_n$ is

$$p_k(n) = \binom{n}{(k+n)/2}\left(\frac{1}{2}\right)^n \tag{12.5}$$

for $n \geq |k|$ and $k$ and $n$ either both even or both odd.

Note that in (12.1) $x_n$ is written as the sum of $n$ independent and identically distributed random variables. Therefore, the central limit theorem of probability theory guarantees that as $n$ gets large the distribution for $x_n$ becomes progressively closer to normal with $\langle x_n \rangle = 0$ and $\text{Var}[x_n] = ((\Delta x)^2/\Delta t)t = 2Dt$. That is,

$$\frac{p_k(n)}{\Delta x} \approx p(x,t) = \frac{1}{\sqrt{4\pi Dt}}\exp\left(-\frac{x^2}{4Dt}\right), \quad x = k\Delta x, \quad t = n\Delta t. \tag{12.6}$$

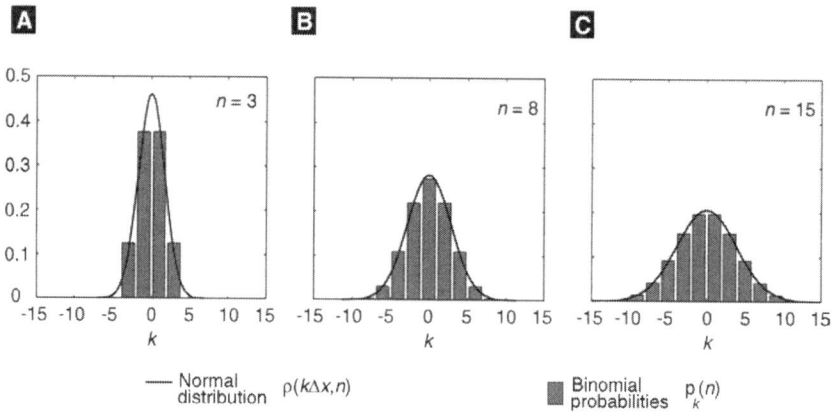

**Figure 12.2** In the limit of large $n$, the binomial distribution is well approximated by a normal distribution. In all three panels the bar graph represents the binomial probabilities and the solid line is the normal approximation. (A) n = 3, (B) n = 8, (C) n = 15.

Figure 12.2 shows the probability distribution for $x_n$ and the normal approximation for various values of $n$. By the time $n = t/\Delta t = 15$, the distribution of $x_n$ is close to normal, and the agreement gets better as $n$ is increased. Physically, the normal approximation amounts to a "coarse graining" of the process in which only length scales much larger than $\Delta x$ and time scales much larger than $\Delta t$ are resolved. In this limit the random variable $x_n$, which is discrete in both space and time, is approximated by $x(t)$, a random variable that is continuous in space and time. The value of $\Delta t$ can be approximated well by the "thermalization" time $\tau = 10^{-13}$ sec, described in Section 12.2.1. Thus, the continuous and discrete models of a protein's motion are equivalent at all time and distance scales of interest to us.

As an exercise, the reader is asked to verify that $p(x, t)$ satisfies the diffusion equation

$$\frac{\partial p(x, t)}{\partial t} = D \frac{\partial^2 p(x, t)}{\partial x^2}, \tag{12.7}$$

justifying our association of the quantity $(\Delta x)^2/(2\Delta t)$ with a diffusion coefficient.

## 12.1.2   Polymer Growth

Let us consider another example of a stochastic process: the number $N(t)$ of monomers in a polymerizing biopolymer. There is an important distinction between the stochastic variables $x(t)$ and $N(t)$. In the first example, $x(t)$ is a continuous random variable, since it can take on any real value. On the other hand, the number of subunits in a growing polymer is restricted to the positive integers, so that $N(t)$ is a discrete random variable.

Markov processes in which the random variable is discrete are often referred to as Markov *chains* because they can be represented as a sequence of jumps between discrete states. The simple random walk is an example of a spatially and temporally discrete Markov chain. As an example of a Markov chain in which time is continuous we consider a polymerizing biological polymer (filament), e.g., an actin filament or a microtubule. Figure 12.3A depicts the type of process we have in mind. In this example, two events change the length of the polymer by one monomer: polymerization and depolymerization. Mathematically, the state of the system is specified by a single number $N(t)$, the number of monomers in the filament at time $t$. The variable $N(t)$ is random, because we have no way to predict when the next polymerization or depolymerization event will occur. A diagram of the Markov chain for this process is shown in Figure 12.3B.

There are two equivalent, but conceptually different, levels at which stochastic processes can be studied. The first is at the level of individual sample paths or realizations of the process. To understand what is meant by a sample path, suppose that at $t = 0$ we start with three filaments that are each exactly 5 monomers long. As time goes on, we observe that the number of monomers in each filament instantaneously changes, or "jumps," by $\pm 1$ at random times. Figure 12.4 graphically illustrates this behavior. Even though each sample path starts with $N(0) = 5$, they all evolve differently, illustrating

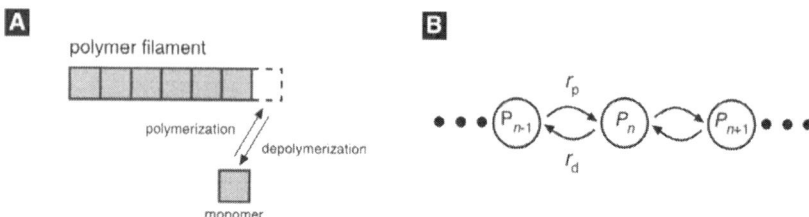

**Figure 12.3** A discrete Markov process (Markov chain). (A) A polymer filament grows by incorporating monomers from the solution onto its tip (polymerization). The process is stochastic. The monomer on the tip may dissociate from the filament into the solution (depolymerization). (B) A Markov chain model for the filament polymerization; $P_n$ represents the state of the filament when its length is $n$ monomers, $r_p$ is the polymerization rate, $r_d$ is the depolymerization rate. If $r_p > r_d$, the filament will grow over long times.

the randomness of the process. The sample paths of a large ensemble of such filaments can be used to determine the statistics of $N(t)$.

The second approach is to ask how the probability $p_n(t)$ of having exactly $n$ monomers in the filament at time $t$ changes in time. If $p_n(t)$ can be determined for all $t$ and $n$, then we have a complete characterization of the process. Both approaches are equally valid and are useful methods for studying stochastic processes. The advantages of staying at the level of sample paths are that in general it is easy to numerically generate single realizations of the process, and sample paths allow us to see the dynamics of the system. The advantage of working directly with the probability distribution is that it fully characterizes the system without the need to average over many sample paths to compute the statistics. Of course, there is no free lunch: Obtaining all this information comes at a computational price. Below we describe numerical techniques for treating both cases.

To begin our discussion we derive an equation that governs the evolution of $p_n(t)$. Let us assume that we know $p_n(t)$ for a specific value of $t$. At a slightly later time $t + \Delta t$,

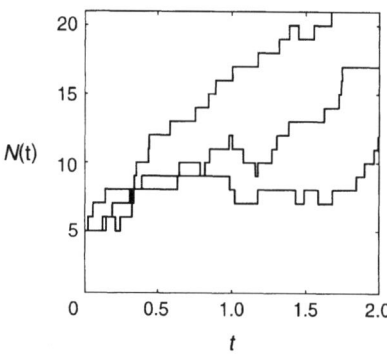

**Figure 12.4** Three sample trajectories of the tip of a growing filament. The polymerization process is stochastic with occasional depolymerization events.

we expect $p_n(t + \Delta t)$ to be equal to $p_n(t)$ plus a small correction. The key is to assume that $\Delta t$ is so small that the probability of two events in the interval $(t, t + \Delta t)$ is very unlikely. Here an event means polymerization or depolymerization. Then we can write

$$p_n(t + \Delta t) = \text{Prob}\left[N(t) = n \text{ and no event occurs in } (t, t + \Delta t)\right]$$
$$+ \text{Prob}\left[N(t) = n - 1 \text{ and polym. occurs in } (t, t + \Delta t)\right]$$
$$+ \text{Prob}\left[N(t) = n + 1 \text{ and depol. occurs in } (t, t + \Delta t)\right], \qquad (12.8)$$

where the right–hand side follows from the fact that the three events described in the square brackets are mutually exclusive. Next we make the reasonable assumption that the probability of polymerization or depolymerization is independent of the length of the filament. We also assume that these probabilities are proportional to $\Delta t$, and let $r_p \Delta t$ and $r_d \Delta t$ be the probabilities of polymerization and depolymerization, respectively, in $(t, t + \Delta t)$. Under these assumptions (12.8) can be written as

$$p_n(t + \Delta t) = p_n(t)(1 - (r_p + r_d)\Delta t) + p_{n-1}(t)r_p \Delta t + p_{n+1}(t)r_d \Delta t$$
$$= p_n(t) + \Delta t \left[r_p p_{n-1}(t) + r_d p_{n+1}(t) - (r_p + r_d)p_n(t)\right], \qquad (12.9)$$

where $(1 - (r_p + r_d)\Delta t)$ is probability of no event in $\Delta t$. There are two important points to be drawn from (12.9). First, it is clear that if we know $p_n(t)$ for all $n$ at a given time, then we have a mechanism for updating the probabilities at all later times. This illustrates the Markov property. Secondly, (12.9) represents a numerical algorithm for updating $p_n(t)$. That is, once a $\Delta t$ is chosen, we can write a computer program to generate $p_n(t + k\Delta t)$, where $k$ is positive integer. We now take the limit $\Delta t \to 0$ in (12.9):

$$\lim_{\Delta t \to 0} \frac{p_n(t + \Delta t) - p_n(t)}{\Delta t} = \frac{dp_n(t)}{dt} = -(r_p + r_d)p_n(t) + r_p p_{n-1}(t) + r_d p_{n+1}(t). \qquad (12.10)$$

Therefore, (12.9) is an algorithm for numerically solving the ordinary differential equation given by (12.10). This algorithm is called the forward Euler method, and is a very useful numerical tool that works adequately for many situations. However, problems may arise in using this scheme, as discussed below. Also, note that (12.10) can be interpreted as a chemical rate equation, so that $r_p$ and $r_d$ are the rates of polymerization and depolymerization, respectively.

## 12.1.3 Sample Paths of Polymer Growth

The next question we address is how to numerically generate sample paths that are consistent with (12.10). To analyze this problem consider the following experiment. At time $t = 0$, we start with a filament containing exactly $m$ monomers. That is, $p_m(0) = 1$. Next we watch the filament until the first event occurs (either polymerization or depolymerization). When this event occurs we record the time and start the experiment again. After doing this experiment many times, we find that the amount of time we must wait for the first event to occur is a random variable. Let us call it $T$. We are after the probability density $f_T(t)$ for $T$. Let $q(t)$ be the probability that no event has occurred in

$(0, t)$. Under the conditions of the experiment and from the derivation of (12.10), we have:

$$\frac{dq}{dt} = -(r_p + r_d)q. \tag{12.11}$$

Solving this equation, we obtain $q(t) = \exp(-(r_p + r_d)t)$. The probability $q(t)$ starts at 1 and decreases to 0 as time goes on. The probability that at least one event has occurred in $(0, t)$ is $1 - q(t)$. Hence, we can use $1 - q(t)$ to define a probability density funtion $f_T(t)$ for the *waiting time* distribution:

$$1 - q(t) = \text{Prob [Waiting time } T < t] = 1 - q(t) = \int_0^t f_T(t')dt'. \tag{12.12}$$

By differentiating (12.12), we obtain the relationship

$$f_T(t) = -\frac{dq(t)}{dt} = (r_p + r_d)\exp(-(r_p + r_d)t). \tag{12.13}$$

That is, the waiting time until the next event occurs has an exponential distribution with mean $1/(r_p + r_d)$. Thus, to produce realizations of $N(t)$, we need to be able to generate samples of an exponential random variable.

Most programming languages have built-in random number generators that produce numbers that are uniformly distributed between 0 and 1. If $R$ is such a random variable, its probability density function is $f_R(t) = 1$ in $[0, 1]$. The transformation that converts $R$ to an exponential random variable with mean $1/(r_p + r_d)$ is

$$T(R) = -\frac{1}{(r_p + r_d)} \ln R. \tag{12.14}$$

This can be verified mathematically as follows:

$$\int_0^t f_T(t)dt = \text{Prob } [T(R) < t] \tag{12.15}$$

$$= \text{Prob } [R > \exp(-(r_p + r_d)t)]$$

$$= 1 - \exp(-(r_p + r_d)t) = 1 - q(t). \tag{12.16}$$

Given this way to compute when the next transition occurs, the next thing we need to determine is whether polymerization or depolymerization takes place. Remember that in any time interval of length $\Delta t$ the probability of polymerization is $r_p \Delta t$. Likewise, the probability of depolymerization is $r_d \Delta t$. Therefore, given that an event has occurred, the probability that it was polymerization is

$$P \text{ [polymerization|an event occurred at } t] = \frac{r_p}{r_p + r_d}. \tag{12.17}$$

We may now generate sample paths of the stochastic process as follows: Start $N$ with a given value. Generate an exponentially distributed random number using (12.14). This determines when the next event takes place. To determine the type of event that occurred, generate a uniformly distributed random number $R_2$. If $R_2 < r_p/(r_p + r_d)$, then let $N \to N + 1$, otherwise $N \to N - 1$. Repeat the process and plot $N$ as a function of time. The trajectories shown in Figure 12.4 were generated in this way.

Before using this method to simulate protein motions, we briefly discuss the statistical behavior of polymer growth.

## 12.1.4 The Statistical Behavior of Polymer Growth

Since the intervals of time between events of monomers assembly and/or disassembly are random, one can measure only the statistical behavior of polymer growth, such as the average velocity of the polymer's tip, $\langle V \rangle = L \langle N \rangle / t$, where $L$ is the size of a monomer. Much useful information is buried in the statistical fluctuations about this mean velocity. One quantity that can be monitored as the polymer grows is the variance of the tip's displacement about the mean:

$$\text{Var}[x(t)] \equiv \langle x^2 \rangle - \langle x \rangle^2 = L^2(\langle N^2 \rangle - \langle N \rangle^2).$$

It is easy to show (see Exercise 4) that the average velocity of the polymer tip and the variance of its displacement are

$$\langle V \rangle = L(r_p - r_d), \quad \text{Var}[x(t)] = L^2(r_p + r_d)t.$$

Thus the variance grows linearly with time. In fact, a plot of $\text{Var}[x(t)]$ vs $t$ can be used to define an effective diffusion coefficient: $D_{\text{eff}} \equiv \text{Var}[x(t)]/2t$ (Wang et al. 1998). The coefficient $D_{\text{eff}}$ can be combined with the average velocity $\langle V \rangle$ to form a "randomness parameter" (Schnitzer and Block 1995)

$$r \equiv \frac{2D_{\text{eff}}}{L \cdot \langle V \rangle}. \tag{12.18}$$

As an example of the utility of this randomness parameter, let us consider the case whre there is no depolymerization: $r_d = 0$. Then, $\text{Var}[x(t)] = L^2 r_p t$, $D_{\text{eff}} = L^2 r_p / 2$, $\langle V \rangle = L r_p$, and $r = 1$. Now suppose that each polymerization event involves a sequence of reaction processes. Since chemical reactions are also stochastic processes, an additional variance is added to the spatial diffusion, so that the total variance will grow faster, and the randomness parameter is greater than 1. In this case, $1/r$ gives a lower bound on the number of reaction processes per step (i.e., $1/r <$ number of reaction processes per step) (Schnitzer and Block 1995).

Similar arguments are applicable to some "walking" motors, e.g., kinesin, that take a spatial step of constant size at random times. This time is determined by a sequence of hydrolysis reactions. If there is only one reaction, the walking motor is equivalent to the polymerizing filament, and is called a "Poisson stepper." Such a stepper is characterized by randomness parameter $r = 1$. In the next chapter we will show that the average velocity of a molecular motor is a function of the load force resisting the motor's advancement. The importance of considering the effective diffusion coefficient (or, equivalently, the randomness parameter) is that just as load–velocity data give information about the motor performance, load–variance data can provide independent estimates of model parameters (see, for example Peskin and Oster (1995)).

## 12.2 Modeling Molecular Motions

The botanist Robert Brown first observed Brownian movement in 1827. While studying a droplet of water under a microscope he noticed tiny specks of plant pollen dancing around. Brown first guessed, and later proved, that these were not motile, although at the time he had no clue as to the mechanism of their motion. It was not until Einstein contemplated the phenomenon 75 years later that a quantitative explanation emerged. In order to develop an intuition about molecular dynamics we begin with some simple remarks on Brownian motion of proteins in aqueous solutions.

### 12.2.1 The Langevin Equation

The radius of a water molecule is about 0.1 nm, while proteins are two orders of magnitude larger, in the range 2–10 nm. This size difference suggests that we can view the fluid as a continuum. A protein moving through the fluid is acted on by frequent and uncorrelated momentum impulses arising from the thermal motions of the fluid. We model these fluctuations as a time-dependent random "Brownian force" $f_B(t)$ whose statistical properties can be mimicked by a random number generator in a computer in a fashion described below. At the same time, the fluid continuum exerts on the moving protein a frictional drag force $f_d$ proportional to the protein's velocity: $f_d = -\zeta v$, where $\zeta$ is the frictional drag coefficient (see Section 12.6.1). Thus we can write Newton's law for the motion, $x(t)$, of a protein moving in a one-dimensional domain of length $L$:

$$\frac{dx}{dt} = v, \quad m\frac{dv}{dt} = -\zeta v + f_B(t), \quad 0 \le x(t) \le L. \tag{12.19}$$

The mass $m$ of a typical protein is about $10^{-21}$ kg, and the drag coefficient is about $10^{-7}$ pN·sec/nm.

If we multiply (12.19) by $x(t)$ and use the chain rule, we get

$$\frac{m}{2}\frac{d^2(x^2)}{dt^2} - mv^2 = -\frac{\zeta}{2}\frac{d(x^2)}{dt} + x \cdot f_B(t). \tag{12.20}$$

In order to see the consequences of (12.20) for molecular motions we first must average (12.20) over a large number of proteins so that the peculiarities of any particular trajectory are averaged out. We use the notation $\langle \cdot \rangle$ to denote this *ensemble* average:

$$\frac{m}{2}\frac{d^2\langle x^2 \rangle}{dt^2} - \langle mv^2 \rangle = -\frac{\zeta}{2}\frac{d\langle x^2 \rangle}{dt} + \langle x \cdot f_B(t) \rangle. \tag{12.21}$$

Next we take advantage of a central result from statistical mechanics called the *equipartition theorem* (Section 12.6.2), which states that each *degree of freedom* of a Brownian particle carries an average energy

$$\langle E \rangle = \frac{1}{2}k_B T \quad \text{[equipartition theorem]}, \tag{12.22}$$

where $k_B$ is Boltzmann's constant and $T$ the absolute temperature (Landau et al. 1980). Therefore, the second term in (12.21) is just twice the average kinetic energy of the protein: $\langle mv^2 \rangle = k_B T$. At room temperature, the quantity $k_B T \simeq 4.1$ pN·nm is the "unit" of thermal energy.

Because the random impulses from the water molecules are uncorrelated with position, $\langle x(t) \cdot f_B(t) \rangle = 0$. Introducing these two facts into (12.21) and integrating twice between $t = (0, t)$ with $x(0) = 0$, we obtain

$$\frac{d\langle x^2 \rangle}{dt} = \frac{2k_B T}{\zeta}(1 - e^{-t/\tau}), \quad \langle x^2 \rangle = \frac{2k_B T}{\zeta}[t - \tau(1 - e^{-t/\tau})], \tag{12.23}$$

where we have introduced the time constant $\tau = m/\zeta$.

For very short times $t \ll \tau$ we can expand the exponential in (12.23) to second order to obtain

$$\langle x^2 \rangle = \frac{k_B T}{m}t^2 \quad (t \ll \tau). \tag{12.24}$$

That is, at very short times the protein behaves like a ballistic particle moving with a velocity $v = \sqrt{k_B T/m}$. For a protein with $m \approx 10^{-21}$ kg [$= 10^{-18}$ pN·sec$^2$/nm], $v \approx 2$ m/s. However, in a fluid the protein moves at this velocity only for a time $\tau \approx m/\zeta = 10^{-13}$ sec, much shorter than any motion of interest in a molecular motor. During this short time the protein travels a distance $v \cdot \tau \sim 0.01$ nm before it collides with another molecule. This is only a fraction of a diameter of a water molecule, so the ballistic regime is very short–lived indeed! Very quickly, the kinetic energy of the protein comes into thermal equilibrium (is "thermalized") with the fluid environment. Thus when $t \gg \tau$, the exponential term disappears, and (12.23) becomes

$$\langle x^2 \rangle = 2\frac{k_B T}{\zeta}t \quad (t \gg \tau). \tag{12.25}$$

Einstein recognized that the frictional drag on a moving body is caused by random collisions with the fluid molecules, which is the same effect as the Brownian force $f_B(t)$ that gives rise to the diffusive motion of the body. Therefore, there must be a connection between the drag coefficient and diffusive motion. By comparing (12.25) to the relation we previously derived between the mean square displacement of a diffusing particle and its diffusion coefficient

$$\langle x^2 \rangle = 2Dt, \tag{12.26}$$

we arrive at the famous relation derived by Einstein in 1905:

$$D = k_B T/\zeta \quad \text{[Einstein relation]}, \tag{12.27}$$

where $D$ is the diffusion coefficient of the protein, typically $D \approx 10^7$ nm$^2$/sec. For diffusion in 2 and 3 dimensions the relation is

$$\langle \mathbf{x} \cdot \mathbf{x} \rangle = 2v \cdot Dt = 2v\frac{k_B T}{\zeta}t,$$

where $v = 2, 3$, respectively.

If an external force $F$ acts on the protein, this can be added to (12.19), so that the equation of motion for a protein becomes $\zeta \cdot dx/dt = F(x, t) + f_B(t)$ (the inertial term has been neglected; see Section 12.6.1). In general, forces acting on proteins can be characterized by a potential $F(x, t) = -\partial\phi(x, t)/\partial x$, so the equation of motion for a protein moving through a fluid becomes

$$\zeta\frac{dx}{dt} = -\frac{\partial\phi(x, t)}{\partial x} + f_B(t). \tag{12.28}$$

Equation (12.28) is frequently referred to as a *Langevin equation*, although this term more properly applies to the corresponding (12.19), which includes inertia.

## 12.2.2  Numerical Simulation of the Langevin Equation

Here we show how the stochastic algorithms developed above can be applied to a continuous Markov process describing a protein diffusing in water. We want a numerical algorithm for generating sample paths of equation (12.28). Let us integrate both sides of this equation over the interval $(t, t + \Delta t)$:

$$
\begin{aligned}
x(t + \Delta t) &= x(t) - \frac{1}{\zeta}\int_t^{t+\Delta t}\frac{\partial\phi(x, t')}{\partial x}dt' + \frac{1}{\zeta}\int_t^{t+\Delta t}f_B(t')dt' \\
&\approx x(t) - \frac{1}{\zeta}\frac{\partial\phi(x, t)}{\partial x}\Delta t + \frac{1}{\zeta}\int_t^{t+\Delta t}f_B(t')dt'.
\end{aligned}
\tag{12.29}
$$

In Section 12.6.3 we demonstrate that the way to include the effect from the Brownian force, $f_B(t)$, is to use the following numerical method for simulating (12.29):

$$x(t + \Delta t) \approx x(t) - \frac{1}{\zeta}\frac{\partial\phi}{\partial x}\Delta t + \sqrt{2D\Delta t}Z, \tag{12.30}$$

where $Z$ is a standard normal random variable, i.e., with mean 0 and variance 1. Many numerical software packages have built-in random number generators that will generate samples of a standard normal distribution. If one is not available, then a standard normal random variable $Z$ can be generated from two independent uniform random variables $R_1$ and $R_2$:

$$Z = -\sqrt{-2\ln R_1}\cos(2\pi R_2). \tag{12.31}$$

A derivation of this result is similar to the one presented above for generating an exponential random variable.

Although simulating (12.30) on a computer is easy (see Exercise 6), it is also easy to generate erroneous results, e.g., numerical instabilities that look much like random displacements due to Brownian motion, or currents that do not vanish at equilibrium. In order to derive a numerical method of simulating random motions that does not have these problems, we have to consider an alternative description of the molecular motion.

### 12.2.3 The Smoluchowski Model

Consider a protein moving under the influence of a constant external force, for example, an electric field. Because of Brownian motion, no two trajectories will look the same. Moreover, even a detailed examination of the path cannot distinguish whether a particular displacement "step" was caused by a Brownian fluctuation or the effect of the field. Only by tracking the particle for a long time and computing the average position vs. time can one detect that the diffusion of the particle exhibits a "drift velocity" in the direction of the force. Therefore, a better way to think about stochastic motion is to imagine a large collection of independent particles moving together. Then we can define the *concentration* of particles at position $x$ and time $t$ as $c(x, t)$ [#/nm], and track the evolution of this *ensemble*.

As the cloud of particles diffuses and drifts, we can write an expression for the *flux J* [#/area/time] of particles passing through a unit area; in one dimension $J_x$ has dimensions [#/sec]. The diffusive motion of the particles is modeled well by Fick's law: $J_x = -D\partial c/\partial x$. The external field exerts a force $F = -\partial\phi/\partial x$ on each particle that, in the absence of any diffusive motion would impart a drift velocity proportional to the field: $v = F/\zeta$. Thus the motion of the body is the sum of the Brownian diffusion and the field-driven drift: $J_x = -D\partial c/\partial x + v \cdot c$, which can be written in several ways (see also Section 12.6.4

$$
J_x = \quad \underbrace{-D\frac{\partial c}{\partial x}}_{\text{Diffusion flux}} \quad - \underbrace{\overbrace{\left(\frac{D}{k_B T} \cdot \frac{\partial \phi}{\partial x}\right)}^{\text{Drift velocity}} c}_{\text{Drift flux}} = -D\left(\frac{\partial c}{\partial x} + \frac{\partial(\phi/k_B T)}{\partial x} \cdot c\right)
$$

$$
= -\frac{1}{\zeta}\left(k_B T\frac{\partial c}{\partial x} + c\frac{\partial \phi}{\partial x}\right). \tag{12.32}
$$

At equilibrium the flux vanishes: $(k_B T/c_{eq})(\partial c_{eq}/\partial x) + \partial\phi/\partial x = 0$. Integrating with respect to $x$ shows that the concentration of particles at equilibrium in an external field $\phi(x)$ is given by the *Boltzmann distribution*:

$$
c_{eq} = c_0 e^{-\phi/k_B T} \quad \text{[Boltzmann distribution]}. \tag{12.33}
$$

Since the number of particles in the swarm remains constant, $c(x,t)$ must obey a *conservation law*. This is simply a balance on a small volume element $\Delta x$:

$$
\frac{\partial}{\partial t}(c\Delta x) = \text{net flux into } \Delta x = J_x(\text{in}) - J_x(\text{out}) = J_x(x) - J_x(x + \Delta x),
$$

or, taking the limit as $\Delta x \to 0$,

$$
\frac{\partial c}{\partial t} = -\frac{\partial J_x}{\partial x} \quad \text{[conservation of particles]}. \tag{12.34}
$$

Rather than focusing our attention on the swarm of particles, we can rephrase our discussion in terms of the *probability* of finding a *single* particle at $(x,t)$. To do this we normalize the concentration in (12.32) by dividing by the total population $p(x,t) \equiv c(x,t)/\left(\int_0^L c(x,t)dx\right)$. Inserting (12.32) expressed in terms of $p(x,t)$ into the conservation law (12.34) yields the *Smoluchowski equation*

$$\frac{\partial p}{\partial t} = D\left[\underbrace{\frac{\partial}{\partial x}\left(p\frac{\partial(\phi/k_BT)}{\partial x}\right)}_{\text{Drift}} + \underbrace{\frac{\partial^2 p}{\partial x^2}}_{\text{Diffusion}}\right] \qquad \text{[Smoluchowski equation].} \qquad (12.35)$$

Comparing this with the Langevin equation (12.28) shows that the Brownian force is replaced by the diffusion term, and the effect of the deterministic forcing is captured by the drift term.

We can nondimensionalize (12.35) by defining time and space scales. If the domain is $0 \le x \le L$, the spatial variable can be normalized as $x/L$. A time scale can be defined by $\tau = L^2/D$. Introducing the space and time scales, (12.35) can be written in dimensionless form as

$$\frac{\partial p}{\partial t} = \frac{\partial}{\partial x}\left(p\frac{\partial\phi}{\partial x}\right) + \frac{\partial^2 p}{\partial x^2}, \qquad (12.36)$$

where $t$ and $x$ are now dimensionless, and the potential $\phi$ is measured in units of $k_BT$.

Equation (12.36) must be augmented by appropriate boundary conditions specifying the value of $p(x=0,t), p(x=L,t)$, and $p(x,t=0)$, where $p(x,t)$ is defined on the interval $[0,L]$. These will depend on the system being modeled.

## 12.2.4  First Passage Time

A very useful quantity in modeling protein motions is the average time it takes for a diffusing protein to first reach an absorbing boundary located at $x = L$, starting from position $0 \le x \le L$ (Berg 1993; Weiss 1967). Denote the mean first passage time (MFPT) to position $L$ starting from position $x$ by $T(x,L)$. The equation governing $T(x,L)$ is derived as follows. A particle released at position $x$ can diffuse either to the right or to the left. After a time $\tau$, it covers an average distance $\Delta$, so that it is located at $x \pm \Delta$ with equal probability $\frac{1}{2}$. The MFPT to $L$ from the new positions are $T(x+\Delta,L)$ and $T(x-\Delta,L)$. The average value of $T(x,L)$ is just $T(x,L) = \tau+(1/2)[T(x+\Delta,L)+T(x-\Delta,L)]$. This equation can be rewritten in the form

$$\frac{1}{\Delta}\left(\frac{(T(x+\Delta,L)-T(x))}{\Delta} - \frac{(T(x)-T(x-\Delta,L))}{\Delta}\right) + \frac{2\tau}{\Delta^2} = 0.$$

Taking the limit as $\Delta \to 0$ and $\tau \to 0$ (so that $\Delta^2/2\tau$ = const) and recognizing that $\Delta^2/2\tau$ is just the diffusion coefficient $D$, the MFPT equation becomes

$$D\frac{\partial^2 T}{\partial x^2} = -1 \qquad \text{[MFPT equation].} \qquad (12.37)$$

$$\langle x^2 \rangle = 2Dt \quad \langle t \rangle = x^2/2D$$

The boundary conditions for this equation are simple. At an absorbing boundary, $T = 0$ (it takes no time to get there). At a reflecting boundary, $T$ is unchanged (i.e., a constant), so $\partial T / \partial x = 0$. For example, releasing a particle at a position $x$ with a reflecting boundary at $x = L$ and an absorbing boundary at $x = 0$, the MFPT is $T(x, L) = (2Lx - x^2)/2D$. The special case where $x = L$ (releasing the particle at the reflecting boundary) is just $T = L^2/2D$. Note the resemblance to the familiar equation $\langle x^2 \rangle = 2Dt$. This gives the mean squared distance diffused in time $t$, whereas the MFPT gives $\langle t \rangle = x^2/2D$, the mean time to diffuse a distance $x$. This suggests that the MFPT equation might be related to the Smoluchowski diffusion equation; in fact, they are adjoints of one another (see, for example, Lindenberg and Seshadri (1979)). We will use this result to compute the average velocity of the perfect Brownian ratchet in the next chapter.

## 12.3  Modeling Chemical Reactions

So far, we have paid attention exclusively to protein mechanics. To understand molecular motors, we have to consider chemical reactions, which supply the energy to drive molecular motors. Two of the most common energy sources are nucleotide hydrolysis and transmembrane protonmotive force. The former uses the energy stored in the covalent bond that attaches the terminal phosphate ($\gamma$-phosphate) to the rest of the nucleotide. The latter uses the electrical and entropic energy arising from a difference in ion concentrations across a lipid bilayer. Hydrolysis is a complicated process, still incompletely understood. Therefore, we will introduce the reaction model using the simple example of a positively charged ion (e.g., $H^+$) binding to a negatively charged amino acid: $H^+ + A^- \longleftrightarrow H \cdot A$. If we focus our attention on the amino acid, we see that it exists in two states: charged ($A^-$) and neutral ($H \cdot A \equiv A^0$), so that the neutralization reaction from the viewpoint of the amino acid is simply

$$A^- \underset{k^-}{\overset{k^+ \cdot [H^+]}{\rightleftharpoons}} A^0. \tag{12.38}$$

Here we use the chemists' convention of denoting concentrations in brackets: $k^+ \cdot [H^+]$ and $k^-$ are the forward and reverse rate constants; the forward rate constant depends on the ion concentration $[H^+]$, which we will treat as a constant parameter (i.e. we shorten our notation to $k^+ \cdot [H^+] \equiv k^*$, where $k^*$ is called a *pseudo–first–order* (or *pseudo–unimolecular*) rate constant).

The rate constants in reaction (12.38) conceal a great deal of physics, for the process of even as simple a reaction as this is quite complex at the atomic level. To model this reaction at a more microscopic level involves introducing additional coordinates to describe the process by which an ionic chemical bond is made and broken. These coordinates have a spatial scale much smaller than the motion of the motor itself (e.g., angstroms vs. nanometers), and a time scale much faster than any motion of the motor (picoseconds vs. microseconds). This is because all reactions involve a redistribution of electrons, and electrons, being very small, move very rapidly. Moreover, in all but the simplest cases, their movements are governed by quantum mechanics rather than classical mechanics. Nevertheless, it is instructive to use the Smoluchowski model to derive a more detailed expression for the rate constants. Deeper discussions can be found in Billing and Mikkelsen (1996), Warshel (1991), and Naray-Szabo and Warshel (1997).

The fundamental concept underlying the modeling of reactions is the notion of a "reaction coordinate," which we denote by $\xi$. In molecular dynamics simulations this is actually a 1-D path through a very high dimensional state space along which the system moves from reactants to products (Billing and Mikkelsen 1996; Warshel 1991; Naray-Szabo and Warshel 1997). For the reaction (12.38), $\xi(t)$ is the distance between the ion ($H^+$) and the amino acid charge ($A^-$). The spatial scale of this coordinate is much smaller (i.e. angstroms) than the spatial scale of the motor's motion, but we can imagine a "super–microscopic" view of the process as shown in Figure 12.5A, where we have plotted the free energy change $\Delta G$ during a reaction as a function of the reaction coordinate $\xi$. The reason for using free energy is that there are many "hidden" degrees of freedom that must be handled statistically, as will become clear presently. Here the chemical states of the amino acid, $A^-$ and $A^0$, are pictured as energy wells separated by barriers of heights $\Delta G_1$ and $\Delta G_2$, and whose difference in depth is $\Delta G$. The "transition state" (TrSt) is located at the top of the pass between the two wells.

For a fixed $H^+$ concentration, the forward chemical reaction $A^- \rightarrow A^0$ proceeds with a rate $k^* \cdot [A^-]$ [#/sec]. However, this rate is a statistical average over many "hidden" events. For a particular reaction to take place, the proton must diffuse to within a few angstroms of the amino acid charge so that the electrostatic attraction between them is felt. Moreover, if the amino acid is located within a protein, there will be steric diffusion barriers that must be circumvented before the two ions "see" each other electrostatically. (Actually, protons inside proteins move by "hopping" along strings of water molecules, or "water wires.") As the concentration of $H^+$ increases, there will be more "tries" at neutralization (i.e., hops from the left well to the right well).

Similarly, the reverse reaction $A^- \leftarrow A^0$ takes place when a thermal fluctuation confers enough kinetic energy on the proton to overcome the electrostatic attraction. Even then, the "free" proton will more often than not "jump" back and rebind to the amino acid, especially if the route between the solution and the amino acid is tortuous. Only when the proton manages a successful escape into solution (the left well) does it count in computing $k^-$.

The *net* flux over the barrier is

$$J_\xi = k^* \cdot [A^-] - k^- \cdot [A^0]. \tag{12.39}$$

After a long time the net flux between the two wells will vanish, $J_\xi = 0$, so that the population of neutral and charged sites will distribute themselves between the wells in a fixed ratio, which we denote by $K_{eq}$ (the equilibrium constant): $K_{eq} \equiv [A^0_{eq}]/[A^-_{eq}] = k^*/k^-$. If the transition state is high, then we can assume that population is apportioned between

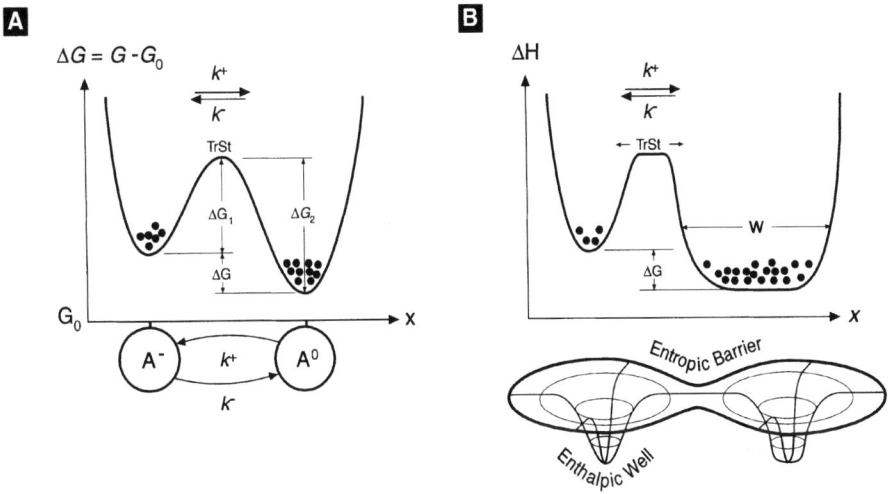

**Figure 12.5** (A) Free energy diagram illustrating the chemical reaction $A \longleftrightarrow B$ and the corresponding Markov model. The transition state TrSt is $\Delta G_1^{+}$ above the left well and $\Delta G_2^{+}$ above the right well. $\Delta G$ is the free energy difference between the well bottoms. The equilibrium distribution between the wells depends only on $\Delta G$. (B) The effect of entropic factors on the reaction $A \longleftrightarrow B$. Potential, rather than free, energy is shown as a function of $\xi$, effective one-dimensional reaction coordinate that involves concerted changes in both the chemical state and physical position along the path of the chemical reaction. The equilibrium populations in each well remain the same, but the transition rates between the wells are different due to the entropic effects of widening the transition state TrSt and the width W of the right well.

the two wells according to the Boltzmann distribution (12.33): $K_{eq} = \exp(\Delta G/k_B T)$, where $\Delta G$ is a free energy. The value of $\Delta G$ determines *how far the reaction goes*, but says nothing about the *rate* of the reaction. Now we know that $\Delta G = \Delta H - T\Delta S$ (cf. Section 12.6.4). The enthalpy term $\Delta H$ is due to the electrostatic attraction between the proton and the charged site. The entropic term $T\Delta S$ incorporates all the effects that influence the diffusion of the proton to the site and its escape from it, the "hidden coordinates." Thus we see that a thermodynamic equilibrium state $\Delta G = 0$ is a compromise between energy ($\Delta H$) and randomness ($T\Delta S$). The role of entropic factors is discussed further in Section 12.6.5.

There is one very significant effect in biochemical reactions that illustrates the importance of entropic effects: hydration. Before a charged ion can bind to the amino acid, it must divest itself of several "waters of hydration." This is because water, being a dipole, will tend to cluster about ions in solution, hindering them from binding to a charged site that is also insulated by its own hydration shell. Suppose for the sake of illustration that the energies binding the waters to the two reactants are just equal to the electrostatic energy of binding between the reactants. Binding seems unfavorable, since the ion will lose its translational and rotational degrees of freedom ($\approx 3k_BT$ according to the equipartition theorem, Section 12.6.2). The binding reaction can still proceed strongly because the liberation of the hydration waters is accompanied by a large entropy increase, since each water gains $\approx 3k_BT$ of rotational and translational energy, and so the term $-T\Delta S$ is strongly negative.

All of this means that the rate constants summarize the statistical behavior of a large number of "hidden" coordinates that are very difficult to compute explicitly, but may be easy to measure phenomenologically (see, for example Hanggi et al. (1990)). For our purposes, we shall adopt this phenomenological view of chemical reactions, and assume that the rate constants can be specified, so that the only entropic effect we need to deal with explicitly is the concentrations of the reactants, such as $H^+$ in (12.38). Therefore, we can treat reactions using Markov chain theory, as indicated by the 2-state model shown at the bottom of Figure 12.5A, whose equations of motion are

$$\frac{d[A^0]}{dt} = -\frac{d[A^-]}{dt} = \text{net flow over the energy barrier} = J_\xi = k^*[A^-] - k^-[A^0],$$

or in vector form

$$\frac{d}{dt}\mathbf{P} = \mathbf{J}_\xi = \mathbf{K} \cdot \mathbf{P}, \quad \mathbf{P} = \begin{pmatrix} p_- \\ p_0 \end{pmatrix}, \quad \mathbf{K} = \begin{pmatrix} k^* & -k^- \\ -k^* & k^- \end{pmatrix}. \tag{12.40}$$

Here $p_-$ and $p_0$ are the probabilities to have a negatively charged and neutral amino acid, respectively. In general, the reaction flux will have the form $\mathbf{J}_\xi = \mathbf{K}(\mathbf{P}) \cdot \mathbf{P}$, where the matrix $\mathbf{K}(\mathbf{P})$ is the matrix of transition rates, i.e., pseudo–first–order rate constants that may contain reactant concentrations that are held parametrically constant. Implicit in this formulation are the assumptions that (i) the actual reaction takes place instantaneously (electronic rearrangements are very fast), so that a substance remains in a chemical state for an exponentially distributed mean time before jumping (reacting) to another state; (ii) the transition out of a state depends only on the state itself, and not on any previous history.

## 12.4 A Mechanochemical Model

An important generalization is necessary to model molecular motors. We have spoken of the potential $\phi(x, t)$ that provides the deterministic forcing as an *external* force. However, a molecular motor $\phi(x, t)$ generally includes forces generated *internally* by the motor itself that drive the motor forward. Thus the potential term in (12.36) must be broken into two parts:

$$\phi(x, t) = \underbrace{\phi_I(x, t)}_{\text{internally generated forces}} + \underbrace{\phi_L(x, t)}_{\text{external load forces}},$$

where $\phi_I(x, t)$ is internally generated force and $\phi_L(x, t)$ is the external load force. A common situation is that of is a constant load force $F_L$ , in which case $\phi_L = F_L \cdot x$, so that $-\partial\phi/\partial x = -F_L$; i.e., the load force acts to oppose the motor's forward progress. The internally generated force potential will generally depend on the chemical state of the system. That is, the mechanical evolution of the system's geometrical coordinates governed by (12.36) is coupled to the chemical reactions described by a Markov chain (12.40). Each *chemical state* is characterized by its own probability distribution $p_k(x,t)$, where $k$ ranges over all the chemical states, and each chemical state is typically characterized by a separate driving potential $\phi_k(x,t)$. Thus there will be a Smoluchowski equation (12.36) for each chemical state, and these equations must be solved simultaneously to obtain the motor's motion.

For the neutralization reaction considered above, the total change in probability $p(x,\xi,t)$ is given by

$$\frac{\partial}{\partial t}\begin{pmatrix}p_1\\p_2\end{pmatrix} = \text{net flow in space} + \text{net flow along reaction coordinates}$$

$$= -\overbrace{\begin{pmatrix}(\partial/\partial x_1)J_{x_1}\\(\partial/\partial x_2)J_{x_2}\end{pmatrix}} + \overbrace{\begin{pmatrix}J_{\xi_1}\\J_{\xi_2}\end{pmatrix}}$$

$$= -D\begin{pmatrix}-(\partial/\partial x_1)[p_1\partial(\phi_1/k_BT)/\partial x_1 + (\partial p_1/\partial x_1)]\\-(\partial/\partial x_2)[p_2\partial(\phi_2/k_BT)/\partial x_2 + (\partial p_2/\partial x_2)]\end{pmatrix} + \begin{pmatrix}k^-p_2 - k^*p_1\\k^*p_1 - k^-p_2\end{pmatrix},$$

$$(12.41)$$

where the probability densities $p_i(x_i,t), i = 1, 2$, now keep track of the motion along the spatial and reaction coordinates, and $J_{\xi_1} = -J_{\xi_2}$ keeps track of flux along the reaction coordinate (since the reaction is of first order, i.e., has only two states). We can visualize the mechanochemical coupling by plotting the spatial and reaction coordinates as shown in Figure 12.6.

## 12.5  Numerical Simulation of Protein Motion

We return to the problem of simulating the protein's motion numerically. Equation (12.30) is a very useful and easy to implement numerical scheme. However, one of its

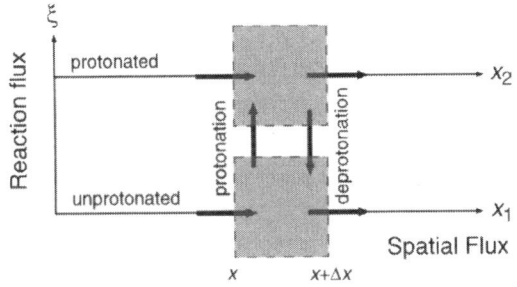

**Figure 12.6** The mechanochemical phase plane. A point is defined by its spatial and reaction coordinates $(x(t), \xi(t))$. The flow of probability in the spatial direction is given by the Smoluchowski model (12.35), and the flow in the reaction direction is given by the Markov model (12.40).

shortcomings is that it does not preserve the property of detailed balance. Detailed balance is the constraint placed on $c_{eq}(x)$ to ensure that systems in equilibrium do not experience a net drift. That is, when a system is in equilibrium, $J$ in (12.32) is required to be identically zero. Detailed balance ensures that the equilibrium density has a Boltzmann distribution.

It is important to understand the distinction between steady state and equilibrium. If we watch sample paths of $x(t)$, it is consistent for the trajectories to move with a mean velocity and for the system to have a steady state. In equilibrium the sample paths must not exhibit a mean velocity. It can be shown that (12.30) produces sample paths that show a net drift when the real system satisfies detailed balance (Elston and Doering 1995). Clearly, it is desirable to have an algorithm that preserves detailed balance in equilibrium and can be used to simulate both equilibrium and nonequilibrium processes.

## 12.5.1   Numerical Algorithm that Preserves Detailed Balance

To obtain an algorithm that has detailed balance built in, we convert the problem into a Markov chain and and use the procedures described above to simulate it numerically. The numerical algorithm given by (12.30) is based on the discretization of time. To convert the problem into a Markov chain requires that we discretize space. Let $x_n = (n - 1/2)\Delta x$ for $n = 0, \pm 1, \pm 2, \pm 3, \ldots$ be the discrete sites on which the protein can reside. Site $x_n$ is represented by the interval $[x_n - \Delta x/2, x_n + \Delta x/2]$. That is, when the protein is in the interval $[x_n - \Delta x/2, x_n + \Delta x/2]$, we treat it as being at site $x_n$. If the molecule is at site $x_n$, then it can jump to either $x_{n+1}$ or $x_{n-1}$. A diagram of this process is shown in Figure 12.7. The notation that we have adopted is that $F_{n+1/2}$ is the rate at which the protein jumps from $x_n$ to $x_{n+1}$ ($F$ refers to a "forward" jump). Similarly, $B_{n+1/2}$ is the rate at which the protein jumps from $x_{n+1}$ to $x_n$ ($B$ for "backward").

For small enough $\Delta x$, we have $p_n(t) \approx p(x, t)\Delta x$, where $p_n(t)$ is the probability that the protein is at $x_n$ at time $t$. The governing equation for $p_n(t)$ is

$$
\begin{aligned}
\frac{dp_n}{dt} &= -(B_{n-1/2} + F_{n+1/2})p_n + F_{n-1/2}p_{n-1} + B_{n+1/2}p_{n+1} \\
&= (F_{n-1/2}p_{n-1} - B_{n-1/2}p_n) - (F_{n+1/2}p_n - B_{n+1/2}p_{n+1}) = J_{n-1/2} - J_{n+1/2},
\end{aligned} \tag{12.42}
$$

where $J_{n+1/2}$ is the net flux between the points $x_n$ and $x_{n+1}$.

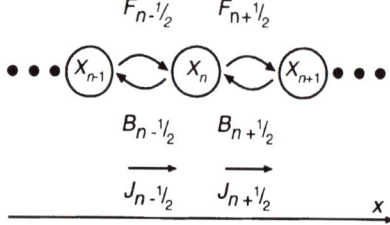

**Figure 12.7** The numerical discretization in spatial dimension. A continuous Markov process (the Langevin equation) is approximated by a discrete Markov process. The particle is restricted to a set of discrete sites ($x_n$) and is allowed to jump only to the neighboring sites ($x_{n-1}, x_{n+1}$). The site $x_n$ can be viewed as representing the interval $[x_n - \Delta x/2, x_n + \Delta x/2]$.

In addition to preserving detailed balance, our numerical scheme must approximate the actual dynamics of the protein. In Section 12.6.6 we demonstrate that the following jump rates preserve the mean drift motion as well as detailed balance:

$$F_{n+1/2} = \frac{D}{(\Delta x)^2} \cdot \frac{\Delta \phi_{n+1/2}/(k_B T)}{\exp\left(\Delta \phi_{n+1/2}/(k_B T)\right) - 1},$$

(12.43)

$$B_{n+1/2} = \frac{D}{(\Delta x)^2} \cdot \frac{-\Delta \phi_{n+1/2}/(k_B T)}{\exp\left(-\Delta \phi_{n+1/2}/(k_B T)\right) - 1},$$

(12.44)

where

$$\Delta \phi_{n+1/2} = \phi(x_{n+1}) - \phi(x_n).$$

(12.45)

## 12.5.2  Boundary Conditions

The algorithm described above must be complemented by boundary conditions. We discuss three types of boundary conditions: periodic, reflecting, and absorbing. In each case the total number of grid points within the interval is $M$, and $\Delta x = L/M$, where $L$ is the length of the spatial domain. The placement of the grid has been chosen such that $x_n = (n - 1/2)\Delta x$ for $n = 1, 2 \ldots, M$.

### Periodic

Periodic boundary conditions require that $p_{M+1}(t) = p_1(t)$ and $p_0(t) = p_M(t)$. Using these two equalities in (12.42) for $p_1(t)$ and $p_M(t)$ produces

$$\frac{dp_1}{dt} = -(B_{1/2} + F_{3/2})p_1 + F_{M+1/2}p_M + B_{3/2}p_2,$$

(12.46)

$$\frac{dp_M}{dt} = -(B_{M-1/2} + F_{M+1/2})p_M + F_{M-1/2}p_{M-1} + B_{1/2}p_1,$$

(12.47)

where we have also made use of the fact that $B_{1/2} = B_{M+1/2}$ and $F_{1/2} = F_{M+1/2}$.

### Reflecting

Figure 12.8A shows a reflecting boundary condition located midway between the grid points $M$ and $M + 1$. A reflecting boundary requires that

$$J_{M+1/2}(t) = F_{M+1/2}p_M(t) - B_{M+1/2}p_{M+1}(t) = 0.$$

(12.48)

We know that $p_{M+1}(t) = 0$, since it is located outside of the reflecting boundary and is inaccessible to the protein. Therefore, to enforce no flux through the boundary, we set $F_{M+1/2}^{\text{reflect}} = 0$. The equation for $p_M$ is then

$$\frac{dp_M}{dt} = -B_{M-1/2}p_M + F_{M-1/2}p_{M-1}.$$

(12.49)

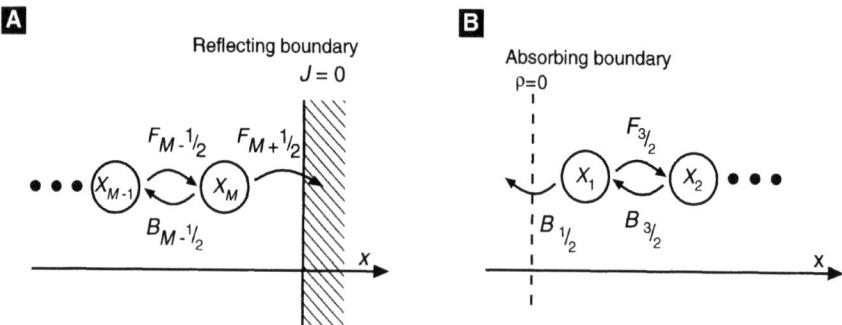

**Figure 12.8** Numerical treatments for two types of boundaries. (A) At a reflecting boundary, the particle is not allowed to jump through the boundary, and thus the flux through the boundary is zero. (B) At an absorbing boundary, the probability density is zero. Once the particle jumps out of the boundary, it should not be allowed to come back. However, in the numerical discretization, blocking the particle from coming back is not enough. The rate of the particle jumping out of the boundary has to be modified.

## Absorbing

If the protein reaches an absorbing boundary it is instantaneously removed from the solution. Therefore, the probability of finding the protein at an absorbing boundary is zero. Figure 12.8B illustrates an absorbing boundary at $x = 0$. Thus, we must enforce the condition $p(0, t) = 0$. In Section 12.6.7 we derive the appropriate jump rate at this boundary:

$$B_{1/2}^{\text{absorb}} = \frac{D}{(\Delta x)^2} \cdot \frac{\alpha^2}{\exp(\alpha) - 1 - \alpha}, \qquad \alpha = \frac{\phi_0 - \phi_1}{k_B T}. \qquad (12.50)$$

The equation for $p_1$ is then

$$\frac{dp_1}{dt} = -\left(B_{1/2}^{\text{absorb}} + F_{3/2}\right)p_1 + B_{3/2}p_2. \qquad (12.51)$$

It can be shown that this treatment of the absorbing boundary is accurate to second order in $\Delta x$ and that it preserves the velocity of a perfect Brownian ratchet subject to any load force (see Chapter 13).

Now we are ready to numerically integrate $p_n$. However, before we turn to examples of implementing the algorithm, we must address the issue of numerical stability and introduce an implicit method for the time integration as an alternative to Euler's method.

## 12.5.3 Numerical Stability

For notational convenience let $p_n(k\Delta t) = p_n^k$. Then Euler's method has the form

$$p_n^{k+1} \approx p_n^k - \Delta t \left[(B_{n-1/2} + F_{n+1/2})p_n^k + F_{n-1/2}p_{n-1}^k + B_{n+1/2}p_{n+1}^k\right]. \qquad (12.52)$$

Euler's method is called an explicit method because $p_n^{k+1}$ can be written explicitly in terms of $p_n^k$. Each time we use the above technique to update $p_n^k$, we introduce a small error due to the finite size of $\Delta t$ and round-off error. In the absence of round-off error, we can achieve any desired accuracy by decreasing $\Delta t$. However, there are some problems with this approach. Usually, the biggest problem is the amount of computer time required when we choose a very small $\Delta t$. However, the round-off error incurred in each step does not decrease with $\Delta t$; rather it accumulates. It is possible that if $\Delta t$ is too small, the total error is dominated by the round-off error. In that situation, the more steps we take, the larger the accumulated error. So a careful choice of time step is important.

Numerical stability is another issue with which we have to contend. That is, we do not want our numerical solutions to run off to $\pm\infty$, when the real solution is bounded for all time. It is possible to show that Euler's method is stable only if

$$\Delta t < \Delta t_c = \max_n \left( \frac{1}{F_{n+1/2} + B_{n+1/2}} \right), \tag{12.53}$$

where max in the above equation means to use the value of $n$ that produces the largest value of the quantity in the parentheses. Figure 12.9 illustrates this change in stability by using time steps slightly above and below $\Delta t_c$. To get an intuitive feel for this instability, let us consider one time step of the numerical scheme. At $t = 0$, we take $p_m^0 = 1$ and $p_n^0 = 0$ for $n \neq m$. From (12.52) we have

$$p_m^1 = \left[ 1 - \Delta t (B_{m-1/2} + F_{m+1/2}) \right] p_m^0. \tag{12.54}$$

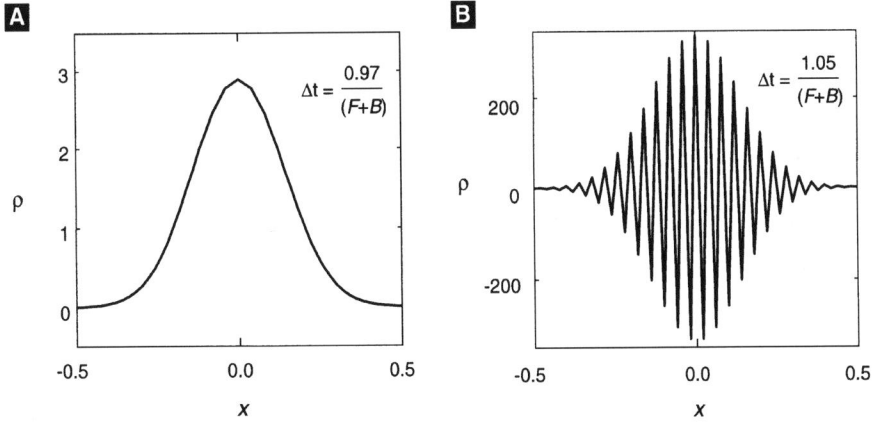

**Figure 12.9** Numerical stability/instability. (A) When the time step is slightly below the critical step size, the numerical solution is stable. (B) When the time step is slightly above the critical step size, the numerical solution is unstable.

It is clear that if $\Delta t > \Delta t_c$, then $p_m^1$ will be negative. Since $p_m^1$ is a probability, negative values clearly do not make sense. The condition on $\Delta t$ for stability is rather restrictive. Using the jump rates given in (12.84) and (12.85), it is possible to show that

$$\Delta t_c < \frac{(\Delta x)^2}{2D}. \tag{12.55}$$

This implies that in order to reduce the spatial step by a factor of 10 (which could be necessary to model accurately spatial fluctuations of the force, for example), the time step must be reduced by a factor of 100.

## 12.5.4  Implicit Discretization

We now improve upon Euler's method in two ways. First, we use a second–order algorithm that improves the accuracy of the solution for fixed $\Delta t$. Second, we choose an implicit method that is unconditionally stable. The implicit second–order algorithm we employ is called the Crank–Nicolson method. For a simple one–dimensional differential equation $dx/dt = h(x)$, the Crank–Nicolson method has the form

$$\frac{x^{k+1} - x^k}{\Delta t} = \frac{h\left(x^{k+1}\right) + h\left(x^k\right)}{2}. \tag{12.56}$$

For (12.42), this scheme becomes:

$$\frac{p_n^{k+1} - p_n^k}{\Delta t} = -\left(B_{n-1/2} + F_{n+1/2}\right) \cdot \frac{p_n^{k+1} + p_n^k}{2}$$

$$+ F_{n-1/2} \cdot \frac{p_{n-1}^{k+1} + p_{n-1}^k}{2} + B_{n+1/2} \cdot \frac{p_{n+1}^{k+1} + p_{n+1}^k}{2}. \tag{12.57}$$

If we now bring all the $p^{k+1}$ terms to the left-hand side and use the vector notation

$$\mathbf{p}^k = \begin{pmatrix} p_1^k \\ p_2^k \\ \vdots \\ p_M^k \end{pmatrix}, \tag{12.58}$$

(12.57) can be written in matrix form as

$$\mathbf{A}\,\mathbf{p}^{k+1} = \mathbf{C}\,\mathbf{p}^k, \tag{12.59}$$

where $\mathbf{A}$ and $\mathbf{C}$ are tridiagonal matrices with elements

$$A_{nn} = 1 + \frac{\Delta t}{2}(B_{n-1/2} + F_{n+1/2}), \quad A_{n,n-1} = -\frac{\Delta t}{2}F_{n-1/2}, \quad A_{n,n+1} = -\frac{\Delta t}{2}B_{n+1/2} \tag{12.60}$$

and

$$C_{nn} = 1 - \frac{\Delta t}{2}(B_{n-1/2} + F_{n+1/2}), \quad C_{n,n-1} = \frac{\Delta t}{2}F_{n-1/2}, \quad C_{n,n+1} = \frac{\Delta t}{2}B_{n+1/2}. \tag{12.61}$$

Equation (12.59) reveals why this method is called implicit. At each time step we must solve a linear set of coupled equations. Luckily, **A** is a sparse matrix, and it is therefore computationally fast to solve (12.59) for $\mathbf{p}^{k+1}$.

## 12.6  Derivations and Comments

### 12.6.1  The Drag Coefficient

The natural units of distance and force on the molecular scale are nanometers (1 nm = $10^{-9}$ m) and piconewtons (1 pN = $10^{-12}$ N), respectively. In these units, the viscosity of water at room temperature is $\eta \approx 10^{-9}$ pN·sec/nm². Then, a typical value for the hydrodynamic drag coefficient of a sphere of radius $R = 10$ nm is $\zeta = 6\pi\eta R \approx 10^{-7}$ pN·sec/nm. A dimensionless number that measures the ratio of inertial to viscous forces is the Reynolds number: $Re \equiv \rho v R/\eta$, where $\rho$ is the density of water ($10^3$ kg/m³ = $10^{-21}$ pN·sec²/nm⁴) (Happel and Brenner 1986; Berg 1993). Typical velocities of molecular motors are $v < 10^3$ nm/sec, so on the molecular scale, the Reynolds number is very small indeed: $Re \approx 10^{-8}$. This confirms our conjecture that we can safely ignore the inertial term in (12.19). If the fluid can truly be viewed as a continuum, then $\zeta$ can be computed from hydrodynamics (Happel and Brenner 1986). The frictional drag coefficient $\zeta$ depends on the particle shape and size as well as the fluid viscosity: $\zeta =$ (dimensionless geometric drag coefficient) × (size factor) × (shape factor). For a sphere, $\zeta = 6\pi\eta R$; drag coefficients for other shapes are given in Berg (1993), a good source of intuition on Brownian motion.

### 12.6.2  The Equipartition Theorem

Let us consider a collision of two particles of masses $m_1$ and $m_2$ with velocities $v_1$ and $v_2$ before the collision, and with velocities $v_1'$ and $v_2'$ after the collision, respectively. Conservation of energy and momentum guarantee conservation of the velocity of the center of mass after the collision, as well as of the absolute value of the relative velocity (Feynman et al. 1963). One of the central assumptions of statistical mechanics is that the velocities of the scattered particles are uncorrelated.

From this one can show that $\langle m_1 v_1'^2 \rangle = \langle m_2 v_2'^2 \rangle$. A more general result that can be derived from statistical mechanics is that each quadratic degree of freedom (e.g., linear or angular momentum) of a particle carries an average amount of energy $\langle E \rangle = k_B T/2$ (Reif 1965). Thus the mean kinetic energy of a point particle moving in the $x$ direction is $\langle m v_x^2/2 \rangle = k_B T/2$, or $\langle v_x^2 \rangle = k_B T/m$. If the particle is moving in a harmonic potential well (i.e., on a spring), its mean potential energy is $k\langle x^2 \rangle/2 = k_B T/2$. Thus the mean total energy is $\langle E \rangle = \langle E_{\text{kin}} \rangle + \langle E_{\text{pot}} \rangle = k_B T$.

### 12.6.3   A Numerical Method for the Langevin Equation

For (12.29) to be useful, we must specify the statistical properties of $f_B(t)$. Since $f_B(t)$ models the net effect of many protein–water interactions, from the central limit theorem it seems reasonable to assume that $f_B(t)$ is normally distributed. Physically, we also want $\langle f_B(t) \rangle = 0$, since a protein that is undergoing pure diffusion does not experience a net force. All that is left to fully characterize $f_B(t)$, is to specify its covariance $\mathrm{cov}[f_B(t)f_B(s)]$. Remember that a necessary condition for two random variables to be independent is that their covariance is zero. Since the motion of the water molecules is very fast as compared with the motion of the diffusing protein, we take $f_B(t)$ and $f_B(s)$ to be statistically independent whenever $t \neq s$. When $\phi = 0$, we should recover diffusive motion. That is, the variance of a Brownian particle started at $x(0) = 0$ is

$$\mathrm{Var}[x(t)] = \langle x(t)^2 \rangle = 2Dt. \tag{12.62}$$

We claim that an appropriate choice for the covariance is

$$\mathrm{cov}[f_B(t)f_B(s)] = \langle f_B(t)f_B(s) \rangle = 2k_B T \zeta \delta(t - s). \tag{12.63}$$

The Dirac delta function $\delta(t - s)$ in (12.63) is a mathematical concept that is best understood as the limit of a normal distribution centered at $s$ as the variance goes to zero:

$$\delta(t - s) = \lim_{\sigma \to 0} \frac{1}{\sqrt{2\pi\sigma^2}} \exp\left( -\frac{(t - s)^2}{2\sigma^2} \right). \tag{12.64}$$

This is illustrated in Figure 12.10. The only property of the Dirac delta function that we need is $\int_{-\infty}^{\infty} g(t)\delta(t - s)dt = g(s)$, which is easily understood when the Dirac delta function is interpreted as a sharply peaked probability density. We now have a complete description of $f_B(t)$. A random variable described as such is referred to as *Gaussian white noise*.

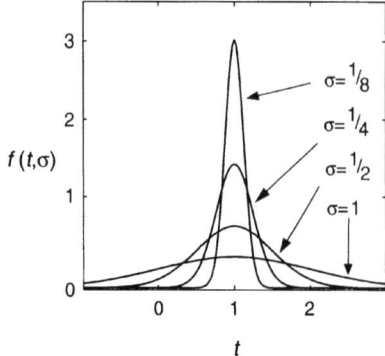

**Figure 12.10** The Dirac delta function can be viewed as the limit of a sequence of normal probability density functions as the standard deviation goes to zero.

Next we illustrate that our choice of $f_B(t)$ makes sense both mathematically and physically. Note that from (12.28) with $\phi = 0$ and $x(0) = 0$, we have

$$x(t) = \frac{1}{\zeta} \int_0^t f_B(t')dt', \tag{12.65}$$

from which it is clear that $\langle x(t) \rangle = 0$. Assume without loss of generality that $t > s$. One can show (from the theory of Dirac's delta function) that

$$\text{cov}[x(t)x(s)] = \frac{1}{\zeta^2} \left\langle \left( \int_0^t f_B(t')dt' \right) \left( \int_0^s f_B(t'')dt'' \right) \right\rangle$$

$$= 2D \int_0^t \int_0^s \delta(t' - t'')dt'dt'' = 2Ds, \tag{12.66}$$

which is consistent with (12.62) when $s = t$. Furthermore, since $f_B(t)$ is a normal stochastic variable, so is $x(t)$.

If we define the new random variable $W(t) = x(t)/\sqrt{2D}$, then $W(t)$ is a normal random variable characterized by

$$\langle W(t) \rangle = 0, \quad \text{cov}[W(t)W(s)] = \min(t, s), \tag{12.67}$$

where the min in (12.67) means to use the value of $t$ or $s$ that is smaller. The random variable $W(t)$ is referred to as a *Wiener process* and possesses some interesting mathematical properties, which we will not go into here. Note that from (12.29) what we need for our numerical algorithm is actually the incremental Wiener process defined as

$$\Delta W(t) = W(t + \Delta t) - W(t) = \frac{1}{\sqrt{2k_BT\zeta}} \int_t^{t+\Delta t} f_B(t')dt'. \tag{12.68}$$

Following a procedure similar to that used in (12.66), it is straightforward to show that $\Delta W(t)$ is a normal random variable with mean zero and standard deviation $\sqrt{\Delta t}$. It is also possible to show that all $\Delta W(t)$ and $\Delta W(s)$ are statistically independent for $t \neq s$. This gives us a way to generalize Euler's method to include Gaussian white noise. That is, a numerical method for simulating (12.28) is (12.30).

## 12.6.4 Some Connections with Thermodynamics

Note that the flux (12.32) can also be written as $J_x = -(c/\zeta)\partial/\partial x(k_BT \ln c + \phi)$. There can be many *steady states* characterized by a constant flux $J_x$ = const; one of these is the special case of *equilibrium*: $J_x = 0$. At equilibrium, one can define the quantity $\mu = (k_BT \ln c_{eq} + \phi)$ called the *chemical potential*. The equilibrium distribution of $c_{eq}(x)$ can be computed by setting the gradient in chemical potential to zero, so that $\mu$ =const; this is exactly equivalent to enforcing a Boltzmann distribution (12.33).

The chemical potential is also the free energy *per mole*, $G = \mu N$, where $N$ is the mole number. A mole is an Avogadro's number $N_a$ of objects (e.g., molecules), where $N_a = 6.02 \cdot 10^{23}$[#/mol]. At *equilibrium* we can define the *entropy* $S \equiv -k_BN \ln c_{eq}$ and the *enthalpy* $H = \phi N$. Then we arrive at the definition of the free energy $G = H - TS$. These definitions will prove useful when we discuss chemical reactions. Here we note simply that diffusion smoothes out the concentration, leading to an increase in entropy. Thus entropic increase accompanying the motion of the ensemble is handled by the Fickian diffusion term in the flux (12.32).

When the particles are charged (e.g., protons, $H^+$), then the chemical potential difference between two states, or across a membrane, is written as $\Delta \mu = \mu_2 - \mu_1 = (\phi_2 - \phi_1) + k_B T (\ln c_2 - \ln c_1) = \Delta \phi - 2.3 \, k_B T \, \Delta pH$. Here $pH = -\log_{10} c_{H^+}$, where $c_{H^+}$ is the proton concentration. The protonmotive force is defined as p.m.f. $= \Delta \mu / e = \Delta \psi - 2.3 \, (k_B T / e) \, \Delta pH$, where $e$ is the electronic charge, and $\Delta \psi = \Delta \phi / e$ (mV) is the transmembrane electric potential.

Consider the simple case where a protein motor is propelled by an internally generated motor force $f_M$ and opposed by a constant load force $f_L$. For example, $f_M = \Delta G / l$, where $l$ is the length of the power stroke and $\Delta G$ is the free energy drop accompanying one cycle of the chemical reaction that is supplying the energy to the motor. (This would be an ideal motor: It uses *all* of the chemical energy to produce a constant–force power stroke!) Then the Langevin equation (12.19) becomes

$$\frac{dx}{dt} = v, \quad m\frac{dv}{dt} = -\zeta v + f_M - f_L + f_B(t). \tag{12.69}$$

The diffusion equation associated with (12.69) for the probability density $p(x, v, t)$ is called the *Kramers equation*:

$$\frac{\partial p}{\partial t} = -\frac{\partial}{\partial x}(vp) - \frac{1}{m}\frac{\partial}{\partial v}\left[((f_M - f_L) - \zeta v)p - \left(\frac{k_B T}{m}\right)\frac{\partial p}{\partial v}\right]. \tag{12.70}$$

The Smoluchowski diffusion equation (12.35) is a special case of the Kramers equation; both are generically referred to as *Fokker–Planck equations* (Doi and Edwards 1986; Risken 1989; Gardiner 1997; Reif 1965). However, deriving (12.35) from (12.70) is not trivial: It requires a *singular perturbation treatment* that is beyond the scope of this chapter (Doering 1990; Risken 1989), but is discussed briefly in Appendix A.

We set $\partial p / \partial t = 0$ in (12.70) to look for the steady state. Multiplying by $v^2$ and taking the average by integrating over $x$ and $v$, and using the equipartition theorem, we obtain

$$0 = \underbrace{-\zeta \langle v^2 \rangle}_{1} + \underbrace{f_M \langle v \rangle}_{2} - \underbrace{f_L \langle v \rangle}_{3} + \underbrace{\frac{k_B T}{m}\zeta}_{4}. \tag{12.71}$$

At constant temperature, the terms in (12.71) (see also Figure 12.11) have the following interpretation:

- The rate at which the motor dissipates energy via frictional drag with the fluid.
- The rate at which energy is being absorbed by the motor from the chemical reaction.
- The rate of work done by the load force against the motor.

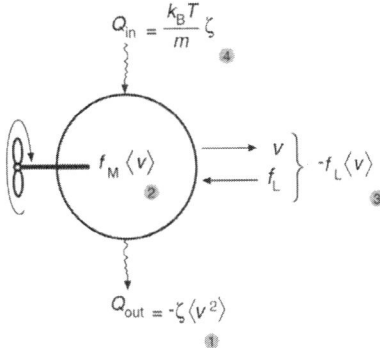

**Figure 12.11** Energy balance on a protein motor. $Q_{out}$ is the heat dissipated by the motion of the motor (term 1). $Q_{in}$ is the heat supplied to the motor by thermal fluctuations of the fluid (term 4). The rate of work done by the load force is $-f_L \cdot \langle v \rangle$ (term 3). The rate of work done by the chemical reaction driving the motor is $f_M \cdot \langle v \rangle$ (term 2).

- The rate at which the motor absorbs energy from the thermal fluctuations of the fluid.

Thus we see that when the chemical reaction is turned off, $\Delta G = 0$, the heat absorbed by the motor from the thermal environment (term 4) is just equal to the heat returned to the environment by frictional drag (term 1) in the absence of the load force. If the reaction driving the motor were *endothermic*, then it is possible for the motor to move by taking heat from the environment without violating the second law of thermodynamics.

## 12.6.5 Jumping Beans and Entropy

An analogy may make the role of entropic factors clearer. Imagine that the left enthalpic well in Figure 12.5B is filled with jumping beans whose hops are random in height and angle. We can vary the equilibrium populations of beans in each well without altering the height of the enthalpic barrier by simply increasing the width of the transition state or of one of the wells. This is shown in Figure 12.5B. Now a bean in the right well may execute many more futile jumps before hurdling the barrier: If it jumps from the right side of the well, it will fall back into the well even if its jump is high enough, or if it reaches the transition state it must diffuse (hop) along the plateau randomly with a high probability of hopping back into the right well. Both of these effects make it more difficult to escape from the right well, and so the equilibrium population there will increase, as will $K_{eq}$, the equilibrium population ratio.

The rate at which beans can pass the barrier from right to left will have the form

$$k^- = v \cdot \exp(\Delta G_2^{\ddagger}/k_B T)$$
$$= \underbrace{v}_{1} \cdot \underbrace{\exp(\Delta S/k_B)}_{2} \cdot \underbrace{\exp(-\Delta H/k_B T)}_{3} \cdot \underbrace{\exp(-\Delta F_L L/k_B T)}_{4},$$

where (1) $v$ is a frequency factor (number of jumps/unit time). For reactions that involve an atomic vibration, this is approximately $k_B T/\hbar$, where $\hbar$ is Planck's constant. For diffusion controlled reactions this is of order $D/L^2$, where $D$ is the diffusion coefficient and $L$ a characteristic dimension. (2) The entropic term $e^{\Delta S/k_B}$ accounts for geometric and "hidden variables" effects. (3) The enthalpic term $e^{-\Delta H/k_B T}$ is a free energy "payoff" for a successful jump; it accounts for the electrostatic and/or hydrophobic interactions. (4) If the reaction involves a mechanical step that is opposed by a load force, $F_L$, then the fourth term accounts for the penalty exacted by performing work against the load. All of these effects are contained in the kinetic rate constants and can estimated from more detailed models (Hanggi et al. 1990; Risken 1989).

Note that the height $\Delta G_2^{\ddagger}$ of the free energy barrier determines *how fast the reaction goes*. The exponent $\exp(\Delta G_2^{\ddagger}/k_B T)$ is called the *Arrhenius factor*. Because of this factor the reaction rate depends dramatically on the height of the free energy barrier.

## 12.6.6  Jump Rates

Here we determine the appropriate values of $F_{n+1/2}$ and $B_{n+1/2}$ in (12.42). Suppose the system, with proper boundary restrictions, attains an equilibrium as time goes to infinity: $p_n^{(eq)} = \lim_{t\to\infty} p_n(t)$. Then the rates should preserve the property of detailed balance. That is,

$$J_{n+1/2} = F_{n+1/2}p_n^{(eq)} - B_{n+1/2}p_{n+1}^{(eq)} = 0. \tag{12.72}$$

Making use of the equilibrium distribution given by (12.33), we have

$$\frac{F_{n+1/2}}{B_{n+1/2}} = \frac{p_{n+1}^{(eq)}}{p_n^{(eq)}} = \frac{p_{eq}(x_{n+1})}{p_{eq}(x_n)} = \exp\left(\frac{-\Delta\phi_{n+1/2}}{k_B T}\right), \tag{12.73}$$

where

$$\Delta\phi_{n+1/2} = \phi(x_{n+1}) - \phi(x_n). \tag{12.74}$$

Equation (12.73) is our first constraint on the jump rates.

Besides preserving detailed balance, our numerical scheme must of course approximate the actual dynamics of the protein. Let us consider the two simplest statistical properties of the random variable $x(t)$, namely the mean and the variance. To simplify the presentation, we make the assumption that $\phi(x) = -fx$. That is, our Brownian particle feels a constant force of strength $f$. For this problem (12.28) reduces to

$$\frac{dx}{dt} = \frac{f}{\zeta} + \frac{f_B(t)}{\zeta}. \tag{12.75}$$

Assuming $x(0) = 0$, the above equation can be integrated to produce

$$x(t) = \frac{f}{\zeta}t + \frac{1}{\zeta}\int_0^t f_B(s)\,ds. \tag{12.76}$$

Using (12.76) the mean and variance of $x(t)$ are found to be

$$\langle x(t)\rangle = \frac{f}{\zeta}t, \tag{12.77}$$

$$\text{Var}[x(t)] = 2Dt. \tag{12.78}$$

Remember that in the discrete version, $x(t) = \Delta x n(t)$. Since the force acting on the protein is constant, the forward and backward rates are independent of $n$. Therefore, we drop the subscripts and use $F$ and $B$. Using (12.42), it is straightforward to show that

$$\langle x(t) \rangle = \Delta x \langle n(t) \rangle = (F - B)t, \tag{12.79}$$

$$\text{Var}[x(t)] = (\Delta x)^2 \text{Var}[n(t)] = (\Delta x)^2 (F + B)t. \tag{12.80}$$

Equating the mean and the variance given in (12.77) and (12.78) with those of (12.79) and (12.80) gives us two more constraints on the jump rates. To summarize, we would like $F$ and $B$ to satisfy the following three equations:

$$\frac{F}{B} = \exp\left(\frac{f\Delta x}{k_B T}\right) \quad \text{[detailed balance]}, \tag{12.81}$$

$$(F - B)\Delta x = \frac{f}{\zeta} \quad \text{[mean]}, \tag{12.82}$$

$$(F + B)(\Delta x)^2 = 2D \quad \text{[variance]}. \tag{12.83}$$

Generally, it is impossible to satisfy three equations with two unknowns. Let us ignore the constraint on the variance for the time being. The rates that satisfy (12.81) and (12.82) are

$$F = \frac{D}{(\Delta x)^2} \cdot \frac{-f\Delta x/(k_B T)}{\exp\left(-f\Delta x/(k_B T)\right) - 1}, \tag{12.84}$$

$$B = \frac{D}{(\Delta x)^2} \cdot \frac{f\Delta x/(k_B T)}{\exp\left(f\Delta x/(k_B T)\right) - 1}. \tag{12.85}$$

Additionally, this set of jump rates satisfies (12.83) with an error of $O\left((\Delta x)^2\right)$. We point out that this choice of $F$ and $B$ is an improvement over the rates used by Elston and Doering (Elston and Doering 1995), since the mean is exactly preserved and $F$ and $B$ have finite values as $k_B T \to 0$.

In general, the force $f$ will not be constant, but will depend on $x$. In this case the jump rates will depend on $n$, and are given by (12.43) and (12.44).

## 12.6.7  Jump Rates at an Absorbing Boundary

To derive an appropriate jump rate at this boundary, we approximate (12.35) near $x = 0$ by

$$\frac{\partial p}{\partial t} = D \frac{\partial}{\partial x} \left(-\frac{f}{k_B T} p + \frac{\partial p}{\partial x}\right), \tag{12.86}$$

where $f = -(\phi_1 - \phi_0)/\Delta x$ is an approximation for $-\partial\phi/\partial x$ in $(0, \Delta x)$. To derive a second–order treatment of the boundary, we need only a first–order approximation for this derivative.

Next we assume that at any given time $p(x, t)$ in the interval $(0, \Delta x)$ is approximately at steady state. This assumption is valid because at small length scales diffusion is the dominant effect. The time scale for a particle with diffusion coefficient $D$ to diffuse a distance $\Delta x$ is $\Delta t_{\text{dif}} = (\Delta x)^2/2D$, which is proportional to $(\Delta x)^2$. The time scale for a flow with velocity $v$ to travel a distance of $\Delta x$ is $\Delta t_{\text{flow}} = \Delta x/v$, which is proportional to $\Delta x$. For small $\Delta x$, $\Delta t_{\text{diff}} \ll \Delta t_{\text{flow}}$. At small length scales, diffusion relaxes the system to steady state immediately after it is disturbed by the flow. Thus, at any given time, the local structure of the solution is given approximately by the steady-state solution. At an absorbing boundary, the steady-state assumption in $(0, \Delta x)$ can also be justified mathematically by examining (12.86) at $x = 0$. Because $p(0, t) = 0$, the left-hand side of (12.86) is exactly zero at $x = 0$. Therefore, we set the left-hand side of (12.86) to zero in the interval $(0, \Delta x)$ and solve the resulting ordinary differential equation subject to the following two conditions:

$$p(0) = 0, \quad \int_0^{\Delta x} p(x)dx = p_1. \tag{12.87}$$

The solution is

$$p(x) = p_1 \frac{\exp(fx/(k_B T)) - 1}{\frac{k_B T}{f}[\exp(f\Delta x/(k_B T)) - 1] - \Delta x}. \tag{12.88}$$

Using the above expression for $p(x)$, the flux is found to be

$$J = D\frac{f}{k_B T}p - D\frac{\partial p}{\partial x} = -p_1 \frac{D}{(\Delta x)^2} \cdot \frac{\alpha^2}{\exp(\alpha) - 1 - \alpha}, \quad \alpha = \frac{f\Delta x}{k_B T}. \tag{12.89}$$

In the numerical scheme, the flux at the boundary is

$$J_{1/2} = -p_1 B_{1/2}. \tag{12.90}$$

This equation reflects the fact that once the protein is absorbed, it does not return to the fluid. Comparing (12.89) with (12.90), we get (12.50).

## Suggestions for Further Reading

- *Elementary Applications of Probability Theory*, Henry Tuckwell. A good introduction to probability theory and stochastic processes with some applications to biology (Tuckwell 1995).
- *Random Walks in Biology*, Howard Berg. Introductory text on applying stochastic processes to cellular and molecular biological systems. This book is written for biologists (Berg 1993).
- *Handbook of Stochastic Methods*, Crispin Gardiner. A good reference that covers most important topics for studying stochastic processes. However, it is not a good book for learning the subject (Gardiner 1997).

## 12.7 EXERCISES

1. Use (12.1) to verify that $\langle x_n \rangle = 0$ and $\mathrm{Var}[x_n] = (\Delta x)^2 n = ((\Delta x)^2/\Delta t)t$.

2. Use (12.6) to compute $\partial p(x,t)/\partial t$ and $\partial^2 p(x,t)/\partial x^2$. Substitute your results for these two expressions into (12.7) to verify directly that $p(x,t)$ satisfies the diffusion equation.

3. Let $p_n(t)$ be the probability that a biological filament has $n$ subunits at time $t$. Assume that the depolymerization rate of the growing filament is zero. Write down the equations that govern $p_n(t)$. Assuming that $p_1(0) = 1$ solve the equations for $p_1(t)$, $p_2(t)$, and $p_3(t)$. Can you generalize your results for $p_n(t)$? Compute the first passage time density for the time it takes the filament to grow 4 subunits long.

4. Use (12.10) to verify that $\langle x(t) \rangle = L\langle N(t) \rangle = L(r_p - r_d)t$ and $\mathrm{Var}[x(t)] = L^2\mathrm{Var}[N(t)] = L^2(r_p + r_d)t$.

5. For a 10 nm sphere of mass $m = 10^{-21}$ kg moving in water, compute the thermalization time $\tau$.

6. Nondimensionalize (12.30) by choosing $L$ as the spatial scale and a characteristic time to diffuse across the domain $[-2L, 2L]$ as the time scale. Simulate the resulting equation on the computer for the double–well potential $\phi(x) = Ak_BT[(x/L)^4 - (x/L)^2]$, $A = 0.1, 1, 10$, and $x(0) = -L$ and no flux conditions at $x = \pm 2L$. Justify your choice of the time steps. Run simulations until (i) $t_{\mathrm{end}} = L^2/D$, (ii) $t_{\mathrm{end}} = 10L^2/D$, (iii) $t_{\mathrm{end}} = 100L^2/D$. Discuss the results.

7. Use the computer to plot the Boltzmann distributions in the case of the double well potential $\phi(x) = Ak_BT[(x/L)^4 - (x/L)^2]$, $-2L < x < 2L$, for $A = 0.1, 1, 10$. Discuss the results.

8. Solve equation (12.36) numerically on the interval $[0, 1]$ with potential $\phi(x) = x$, no flux boundary, and arbitrary initial conditions. Use any standard numerical method. Run the simulations until the transients die out. Compare the solutions with the corresponding Boltzmann distribution and discuss the results.

9. Derive (12.84) and (12.85) from (12.81) and (12.82).

10. Verify that the steady–state probability density given in (12.88) satisfies (12.86) and the conditions (12.87).

11. An Ornstein–Uhlenbeck process is characterized by the stochastic differential equation

$$m\frac{dV}{dt} = -\zeta V + f_B(t), \quad \langle f_B(t) \rangle = 0, \quad \mathrm{Cov}[f_B(t)f_B(s)] = 2k_BT\zeta\delta(t - s).$$

Write down the diffusion equation that corresponds to this process. Solve the diffusion equation for the equilibrium density. Using the numerical algorithm described in this chapter, generate sample paths for this process. Use the sample paths to generate a histogram of the particle velocities. Compare the histogram with the analytic equilibrium density.

12. Write a program to numerically solve the diffusion equation (nondimensionalize first!)

$$\frac{\partial p(x,t)}{\partial t} = D\left[ \frac{\partial}{\partial x}\left( A\sin\left(\frac{2\pi x}{L}\right) + F \right)\frac{p(x,t)}{k_BT} + \frac{\partial^2 p(x,t)}{\partial x^2} \right],$$

subject to periodic boundary conditions at $x = 0$ and $x = L$ and the initial condition $p(x,0) = \delta(x - L/2)$. Plot the distribution at various times to observe the relaxation to steady state. Use the steady-state distribution to compute the average velocity. Investigate how the average velocity changes as the parameters $F$, $A$, and $D$ are varied.

# CHAPTER 13

# Molecular Motors: Examples

Alex Mogilner, Timothy C. Elston, Hongyun Wang, and George Oster

## 13.1   Switching in the Bacterial Flagellar Motor

As an example of the use of the numerical algorithm developed in Chapter 12, we first consider a model for switching in the bacterial flagellar motor proposed by Scharf et al. (1998). Some bacteria, such as *Escherichia coli*, swim by spinning long helical flagella. Each cell has multiple flagella, all of which have the same handedness. When the flagella are spun in the counterclockwise (CCW) direction, they come together to from a bundle that propels the cell through the fluid. The motor that is responsible for flagella rotation is reversible. When spun in the clockwise (CW) direction, the flagella fly apart and the cell undergoes a tumbling motion. Addition of a chemical attractant causes the cell to suppress tumbling when moving toward this food source. One of the proteins in the signaling pathway is CheY. The binding of phosphorylated CheY to the portion of the motor located within the cytoplasm promotes CW rotation. To model motor reversals, the protein complex that forms the rotor is assumed to exist in two distinct conformational states that correspond to CW and CCW rotation. The binding of CheY decreases the free energy of the CW state, while at the same time increasing the free energy of the CCW by an equivalent amount. To capture this effect, the free energy of the rotor is assumed to have the following form (see Figure 13.1A):

$$G(x) = 4\Delta G_{nb} \left( \frac{x^4}{4} - \frac{x^2}{2} \right) - \frac{1}{2}\Delta Gx, \tag{13.1}$$

where $x$ is an appropriate reaction coordinate, $\Delta G = G(-1) - G(1)$ is the free energy difference between the CW and CCW states, and $\Delta G_{nb} = G(0) - G(-1)$ is the free

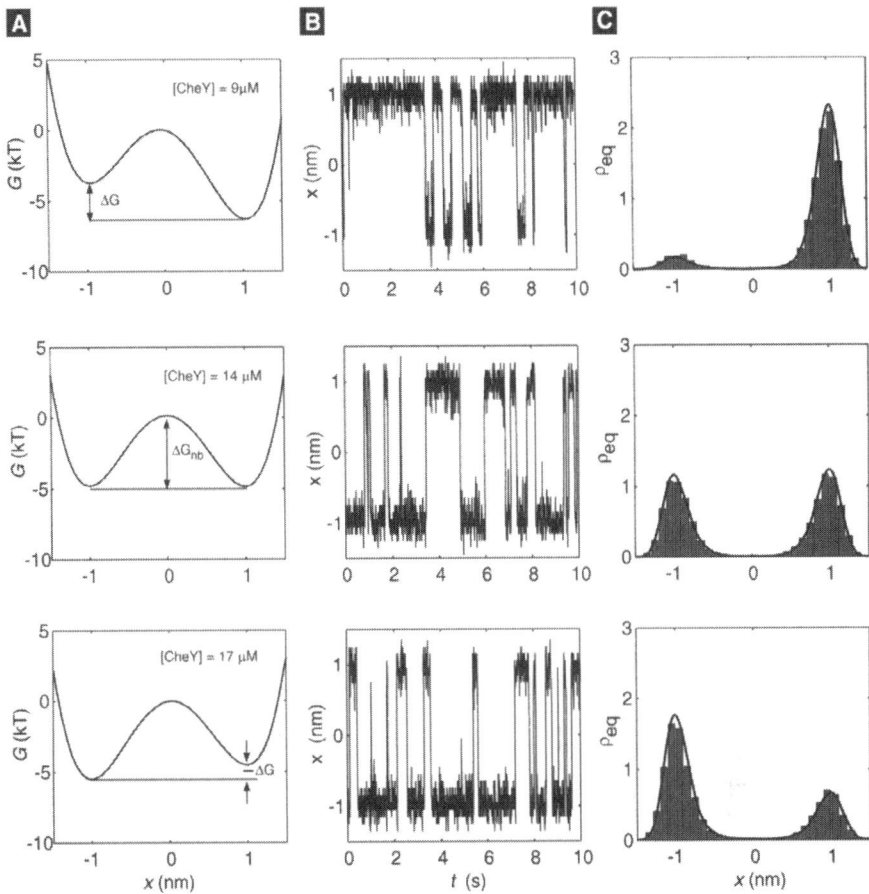

**Figure 13.1** (A) The free energy digram for the rotor at various levels of CheY concentration. (B) Time series generated from the three potentials shown in A. (C) The stationary distributions for the three potentials shown in A.

energy difference between the transition state and either the CW or CCW state when the CheY concentration is such that $\Delta G = 0$ (i.e., there is no bias toward CCW or CW rotation).

To model the chemical kinetics of CheY, we assume that the binding of CheY to the motor is a two–state processes with a single binding site being either empty or occupied. Let $p_E(t)$ be the probability that the site is empty at time $t$ and $p_O(t)$ be the probability that the site is occupied. The probabilities satisfy the following set of coupled equations:

$$\frac{dp_E(t)}{dt} = -k_{on}p_E(t) + k_{off}p_O(t), \tag{13.2}$$

$$\frac{dp_O(t)}{dt} = -k_{off}p_O(t) + k_{on}p_E(t), \tag{13.3}$$

where $k_{off}$ is the dissociation rate constant and $k_{on}$ is the rate at which CheY binds to the motor. From the law of mass action, $k_{on}$ should be proportional to the CheY concentration. That is, $k_{on} = k'_{on}[CheY]$, where $k'_{on}$ is a second–order rate constant and the brackets stand for concentration. If the concentration of CheY is held constant, then $p_E$ and $p_O$ will relax to their equilibrium values. These are found by solving (13.2) and (13.3) with the time derivatives set equal to zero. Doing this yields $p_O = [CheY]/(K_d + [CheY])$ where $K_d = k_{off}/k'_{on} = 9.1~\mu M$ is the dissociation constant.

There are approximately 26 binding sites on the motor, and the average number of occupied sites is $26p_O$. In the absence of CheY, $\Delta G = \Delta G_0 = 14k_B T$ and at saturating concentrations of CheY, $\Delta G = \Delta G_\infty = -9k_B T$. Therefore, the change in $\Delta G$ from low to high CheY concentrations is $\Delta\Delta G = \Delta G_0 - \Delta G_\infty = 23k_B T$, where the symbol $\Delta\Delta G$ indicates that we are talking about a change in the value of $\Delta G$. Thus each CheY contributes roughly $0.88k_B T$ toward changing the relative free energy of the CW and CCW states. These considerations lead to the following expression for $\Delta G$:

$$\Delta G = \Delta G_0 - \Delta\Delta G \frac{[CheY]}{K_d + [CheY]}. \tag{13.4}$$

Graphs of the free energy at various CheY concentrations are shown in Figure 13.1A, with two minima located roughly at $x = \pm 1$. Generally, we shall measure distance $x$ in nanometers (nm) and force in piconewtons (pN). In these units $k_B T = 4.1$ pN-nm at room temperature ($T = 298$ K). The force that arises from changes in free energy is $-\partial G/\partial x$. Therefore, the force vanishes at the minima. Additionally, if the conformation of the rotor is slightly displaced from either minimum, it experiences a force that moves it back toward that minimum. The force also vanishes at the local maximum located near $x = 0$. However, when the conformation of the rotor is slightly displaced from the origin, the force acts to move away from the rotor at $x = 0$ and toward one of the two minima. Thus, we expect the rotor to spend most of its time near the minima. To surmount the energy barrier between the minima requires a substantial thermal fluctuation.

The reaction coordinate $x(t)$, which determines the state of the rotor, can take on values anywhere between $\pm\infty$. Clearly, we cannot use an infinite interval in our numerical algorithm. However, since $G(x) \to \infty$ as $x \to \pm\infty$, there is a strong restoring force that drives the reaction coordinate back toward the origin when $|x|$ is large. This means that the probability of finding $x(t)$ at distances far from the origin is small, and ignoring large values of $|x|$ will not significantly affect our numerical solutions. For the parameters we shall consider, the interval $(-2, 2)$ is wide enough to ensure numerical accuracy. In practice, an appropriate interval can be determined by successively enlarging the length until the numerical results no longer change appreciably. At $x = \pm 2$ we enforce reflecting boundary conditions as described in Section 12.5.2. The diffusion equation for this process is

$$\frac{\partial p}{\partial t} = D\frac{\partial}{\partial x}\left(\frac{\partial G(x)}{\partial x}\frac{p}{k_B T} + \frac{\partial p}{\partial x}\right). \tag{13.5}$$

To find the equilibrium distribution for $p(x,t)$, the above equation is solved with $\partial p/\partial t = 0$. This yields

$$p_{eq}(x) = \frac{\exp\left(-\frac{G(x)}{k_B T}\right)}{\int_{-\infty}^{\infty}\exp\left(-\frac{G(y)}{k_B T}\right)dy}. \tag{13.6}$$

The diffusion coefficient $D$ in (13.5) represents an effective diffusion coefficient for the reaction coordinate that includes many microscopic effects. For all the results presented below, $D = 70$ nm$^2$/s and $\Delta G_{nb} = 5k_B T$. These values were chosen to be consistent with experimental observation that at 14 $\mu$M of CheY, motor reversals occur at an average rate of 2/s (we expand on this point below). Figure 13.1B shows time series generated by the three potentials shown in Figure 13.1A. The bistable nature of the system is clearly evident. The time series can be used to produce histograms of the reaction coordinate. An approximation for the distribution $p_n^{(s)}$ is constructed by dividing the number of points in each bin of the histogram by the total number of points in the time series. Then we estimate $p(x_n) \approx p_n^{(s)}/\Delta x$. Figure 13.1C shows distributions generated in this fashion. The solid lines are the exact results given by (13.6). As is clearly seen from the figure, the numerical algorithm accurately reproduces the equilibrium distribution.

In the discussion of Markov chains and diffusion in the previous chapter, we encountered the idea of a waiting or first passage time. This is a very important mathematical concept that comes up in many different biological contexts. At a CheY concentration of 14 $\mu$M the motor reverses roughly twice per second, and there is no bias toward CW or CCW rotation. To compute the switching rate, we must compute the average time for the system located at the reaction coordinate $x$ to surmount the energy barrier at $x = 0$. To this end, the reaction coordinate is started at $x = -1$ at $t = 0$ with an absorbing boundary at $x = 0$. Figure 13.2A shows the numerically generated probability density at various times. To generate this figure, 61 grid points were used. Note that the probability of finding the particle in the interval $(-\infty, 0)$ is continuously decreasing, due to the absorbing boundary. This probability can be used to determine the first passage time density $f(t)$ through the relation

$$f(t) = -\frac{d}{dt}\text{Prob}\left[-\infty < x(t) < 0\right] = -\frac{d}{dt}\int_{-\infty}^{0}p(x,t)dx$$

$$\approx -\frac{d}{dt}\sum_{n=1}^{M}p_n(t) = p_M F_{M+1/2}^{\text{abs}}, \tag{13.7}$$

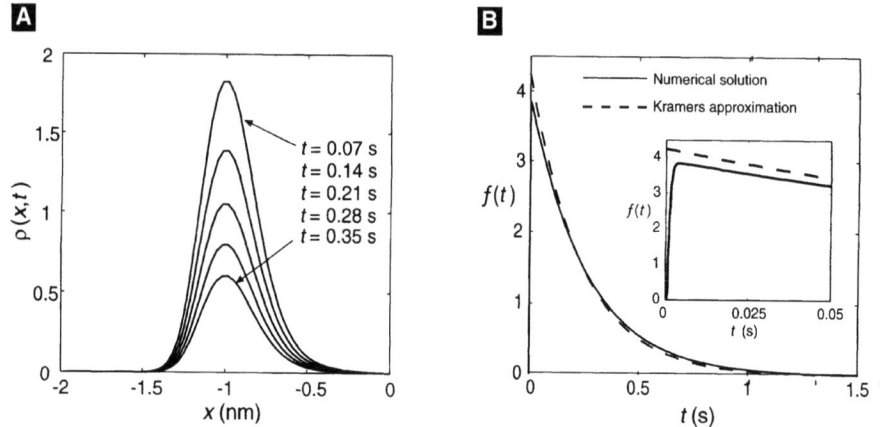

**Figure 13.2** Numerical results of the first passage time problem. The concentration of CheY has been chosen such that $\Delta G = 0$. At $t = 0$, the reaction coordinate is placed at $x = -1$. An absorbing boundary is placed at $x = 0$ and a reflecting barrier at $x = -2$. The first passage time is the time for the reaction coordinate to reach the origin. (A) Time evolution of the probability density. As time increases, the probability that the reaction coordinate remains inside the region decreases. (B) Probability density of the first passage time. Inset: an expanded view of the probability density near the origin showing the nonexponential nature of the distribution.

where the last equality follows from (12.42) and the absorbing boundary condition. Therefore, the numerical algorithm is well suited for computing first passage time densities. Figure 13.2B shows the first passage time density. The solid line is the numerical result. The dashed line is the Kramers approximation, which assumes that the process has an exponential distribution with mean first passage time

$$\text{MFPT} \approx \frac{k_B T \pi}{D \sqrt{G''(-1)|G''(0)|}} \exp\left(\frac{G(0) - G(-1)}{k_B T}\right)$$
$$= \frac{k_B T \pi}{D \sqrt{32 \Delta G_{nb}}} \exp\left(\frac{\Delta G_{nb}}{k_B T}\right). \tag{13.8}$$

A derivation of this result can be found in (Gardiner 1997). Note that the most significant factor in determining the mean first passage time is $\Delta G_{nb}$. The validity of the Kramers approximation depends on $\Delta G_{nb} \gg k_B T$. As shown in the inset of Figure 13.2B, the first passage time distribution is not exponentially distributed, since it must be equal to zero at $t = 0$. However, if we ignore this very short initial time interval, the distribution is approximated reasonably well with an exponential. Using the numerical distribution to compute the mean first passage time, we obtain MFPT = 0.253 s, and using Kramers approximation (13.8) we find MFPT = 0.236 s. An exact expression (Gardiner 1997) gives MFPT = 0.259 s. The agreement between this value and the numerical result given above provides evidence that the algorithm is faithfully reproducing the dynamics of the system. The switching rate is 1/(2 MFPT) = 1.98/s, where the factor of

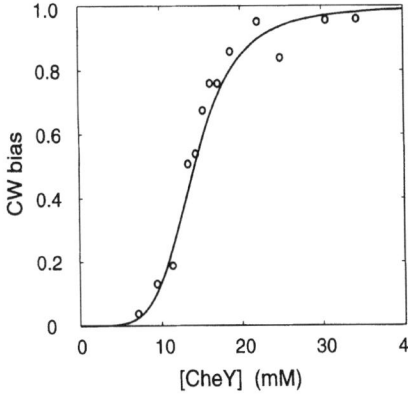

**Figure 13.3** The CW bias as function of CheY concentration. The solid line is the model prediction and the data points are from Scharf et al. (1998).

$\frac{1}{2}$ comes from the fact that half the time the reaction coordinate surmounts the barrier, it falls back into the well from which it started. This justifies our choice of $D$ and $\Delta G_{nb}$.

To test this simple model, we compute the CW bias as a function of CheY concentration. The CW bias is the fraction of time that the motor spends rotating in the CW direction. This can be computed by integrating (13.6) over the interval $(-\infty, 0)$. That is,

$$\text{CW bias} = \frac{\int_{-\infty}^{0} \exp\left(-\frac{G(x)}{k_B T}\right) dx}{\int_{-\infty}^{\infty} \exp\left(-\frac{G(y)}{k_B T}\right) dy}. \tag{13.9}$$

Figure 13.3 shows a comparison of the model's predictions to the experimental data of Scharf et al. (Scharf et al. 1998). The agreement between the data and the theoretical curve provides evidence to support the model's validity.

## 13.2 A Motor Driven by a "Flashing Potential"

The following process, called the *flashing ratchet* (Doering 1995; Doering 1998), is a paradigm for molecular motors. It is also a good application for the methods developed in Chapter 12. Imagine a protein driven by alternating its exposure to two potential energy profiles: $V_1$ (solid line) and $V_2$ (dashed line), as shown in Figure 13.4. The first potential is a piecewise linear asymmetric sawtooth potential, while the second potential is a constant. Thus, in the first potential, the protein is localized near a local minimum, while in the second potential the protein diffuses freely. While in either potential, the motion of the particle is given simply by $\zeta dx/dt = -dV_i/dx + f_B(t), i = 1, 2$. Switching between the potentials is governed by a chemical reaction (vertical arrows), which occurs with rate $k$.

Clearly, if the sawtooth potential is symmetric, the average displacement of the protein must be zero. However, in the case of the asymmetric potential shown in

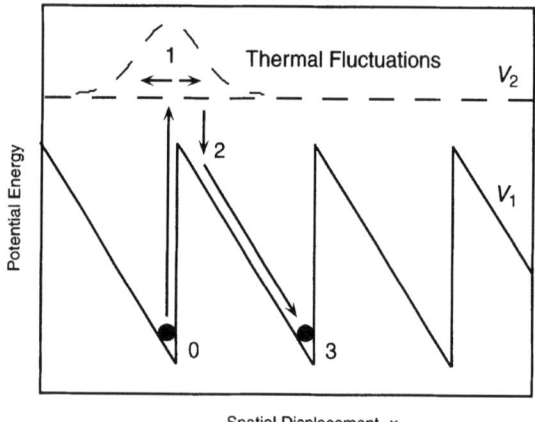

**Figure 13.4** The flashing ratchet. In the first potential, the protein is localized near a local minimum. Alternatively, the protein is free to diffuse. When the first potential is switched back on, the protein settles into the nearest local minimum.

Figure 13.4, the protein moves on the average to the right, although all steps are reversible. This phenomenon can be easily understood when the following inequalities are valid:

$$\frac{k_{\mathrm{B}}TL}{FD} \ll \frac{1}{k} \ll \frac{L^2}{D}. \tag{13.10}$$

Here $L$ is the wavelength of the sawtooth potential, and $F = -dV_1/dx$ is the slope of the first potential, i.e., the force driving the protein to the right in the first potential. The corresponding drift rate of the protein is $F/\zeta = FD/k_{\mathrm{B}}T$. The order of magnitude of the time for the protein to drift into a local minimum is $k_{\mathrm{B}}TL/FD$. The diffusion in the first potential and the protein's deviations from a local minimum can be neglected if the slope of the sawtooth potential is very steep. Quantitatively, this means that $k_{\mathrm{B}}T \ll FL$ (or $k_{\mathrm{B}}TL \ll FL^2$; compare with (13.10).

The first inequality (13.10)) means that the protein reaches a local minimum of the sawtooth potential well before this potential switches off. The second inequality (13.10) indicates that when the protein diffuses freely, it rarely can move farther than distance $L$ before the sawtooth potential is switched back on: the mean time between "flashes" of the potential, $1/k$, is much less than the characteristic diffusion time $L^2/D$. Thus by the time the sawtooth potential is switched off, the protein is in a local minimum. When the sawtooth potential is switched off, the protein diffuses with equal probability to the left and to the right, and will not diffuse very far compared to the period of the potential. If the protein diffuses to the left, then by the time the sawtooth potential is switched back on, the protein is in the *basin of attraction* of the same local minimum it started from. When the potential is on, the protein returns to the starting point of the cycle. However, if the diffusion to the right took place, the protein is in the basin of attraction of the next local minimum to the right. Thus, the protein either does not move, or moves the distance $L$ to the right, with equal probability. Said another way, the diffusion in the flat potential can be viewed as a spreading Gaussian distribution. The asymmetry of the sawtooth potential, when it is switched on, cuts a larger portion of the distribution into the next domain of attraction. The corresponding average rate of motion is $\langle V \rangle = Lk/4$ (the mean duration of a cycle is $2/k$, and on average, the system steps a distance $L$ per two cycles).

We considered the highly peculiar sawtooth potential with one of the slopes being infinitely steep. For more regular smooth potentials, the velocity of the flashing ratchet is computed analytically in the so–called fast and slow flashing limits (Doering 1995). In both of these limits, the protein advances very slowly. If the flashing is too fast, the protein does not have time to reach the local minimum, and the effect of asymmetry is lost. The protein is effectively exposed to the average potential, which does not support any steady movement. On the other hand, if the flashing is too slow, the freely diffusing protein moves too far from a local minimum, the information about its initial position is lost, and the average displacement becomes very small. The mean velocity of the flashing ratchet reaches a maximum $\approx Lk/4$ (for the smooth asymmetric potential, such that $\delta V \approx k_B T$) when $k \approx D/L^2$ (flashing frequency is of the same order of magnitude as the inverse time to diffuse over the potential's period).

In the general case, the average velocity of the flashing ratchet can be computed only numerically. Following the methods of the previous chapter, we can describe the ratchet by two coupled Smoluchowsky equations:

$$\frac{\partial p_i}{\partial t} = D \frac{\partial}{\partial x} \left[ \frac{\partial p_i}{\partial x} + \frac{\partial V_i / \partial x}{k_B T} p_i \right] + k(-p_i + p_j), \quad i = 1, 2, \; j \neq i. \tag{13.11}$$

These equations must be solved numerically (see Exercise 2) on the finite domain $[0, L]$ with periodic boundary conditions and normalization condition

$$\int_0^L (p_1(x, t = 0) + p_2(x, t = 0)) dx = 1.$$

When the probability distributions achieve steady state, the net current is

$$J = -D \left[ \frac{\partial^2 (p_1 + p_2)}{\partial x^2} + \frac{\partial V_1 / \partial x}{k_B T} p_1 \right],$$

from which the average velocity can be found: $\langle V \rangle = LJ$.

It is important to realize that in the process energy is consumed from the chemical reaction that drives the switching between the two potentials. The motion down the slope of the sawtooth potential $2 \rightarrow 3$ generates heat by frictional dissipation. Finally, if a small load force directed to the left is applied to the protein, the movement slows down. The load force is equivalent to tilting the potential to the left. Thus, the flashing ratchet is able to generate force, and has all characteristics of a molecular motor. However, there is no direct correspondence of the flashing ratchet mechanism to a real motor. In what follows, we consider two simple models of actual molecular motors.

## 13.3   The Polymerization Ratchet

Perhaps the simplest way to convert chemical energy into a mechanical force is by polymerizing a filament against a load force (see the previous chapter). Here the energy source is the free energy of binding of a monomer onto the tip of the polymer, $\Delta G_b$. If a polymer assembles against no resistance, the polymerization velocity (elongation rate) is simply

$$V_p = L(k_{on}M - k_{off}), \tag{13.12}$$

where $L$ (nm) is the size of the monomer, $M$ ($\mu$M) the monomer concentration, and $k_{on}$ ($1/\mu$M·s), $k_{off}$ (1/s) are the polymerization and depolymerization rate constants, respectively.

If an object is placed ahead of the growing polymer, there are two mechanisms by which the polymer can "push" the object: (i) by rectifying the Brownian motion of the object; (ii) by actively "pushing" against the object, i.e., a power stroke. First, we discuss the Brownian ratchet. We assume that the polymer is anchored at the left end and is perfectly rigid. The object has a diffusion coefficient $D = k_B T/\zeta$. For the moment, we neglect depolymerization ($k_{off} = 0$). In order for a monomer to bind to the end of the filament the object must open up a gap of size $L$ by diffusing away from the tip, and remaining there for a time $\approx (k_{on}M)^{-1}$ to allow a polymerization event to take place. In the limiting case when polymerization is much faster than diffusion, i.e., $k_{on}M \gg D/L^2$, we can consider the polymerization to happen instantaneously once a gap of size $L$ appears. Then the time for the load to diffuse a distance $L$ is simply the mean first passage time $\langle T \rangle = L^2/2D$. To cover $N$ such intervals takes $N \cdot \tau$ time units, so the average velocity is simply $\langle V_p \rangle = NL/(N \cdot \langle T \rangle) = 2D/L$. This is the speed of an ideal Brownian ratchet. Note that by reducing the size of the monomer $L$, the speed of the ratchet increases, since the likelihood of a thermal fluctuation of size $L$ increases exponentially as $L$ decreases. However, this is true only as long as our approximation holds: $k_{on}M \gg D/L^2$, or $L \gg \sqrt{D/(k_{on}M)}$. For smaller values of $L$ the polymerization reaction becomes the limiting factor, so that $V_p \approx L \cdot k_{on}M$ (cf. (13.12) with $k_{off} = 0$).

We can picture the situation as shown in Figure 13.5B: the object diffuses on a "staircase" sequence of identical free energy functions $\phi(x)$, each with a step height of the monomer binding free energy $\Delta G = -k_B T \ln(k_{on}M/k_{off})$. If a load force $F_L$ opposes the diffusive motion of the object, the potential becomes $\phi(x) - F_L x$. This corresponds to tilting the potential so that the object must diffuse "uphill," as shown in Figure 13.5B.

Including the depolymerization rate complicates the analysis considerably. However, a diffusion equation can be formulated that can be solved exactly when $k_{on}ML, k_{off}L \ll 2D/L$. In this regime, the approximate load–velocity relationship is given by the simple formula (Peskin et al. 1993)

$$V_p = L(k_{on}Me^{-f_L L/k_B T} - k_{off}). \tag{13.13}$$

That is, the polymerization rate in (13.13) is weighted by a Boltzmann factor where the exponent $F_L L/k_B T$ is the work done by the load in moving the object one step distance

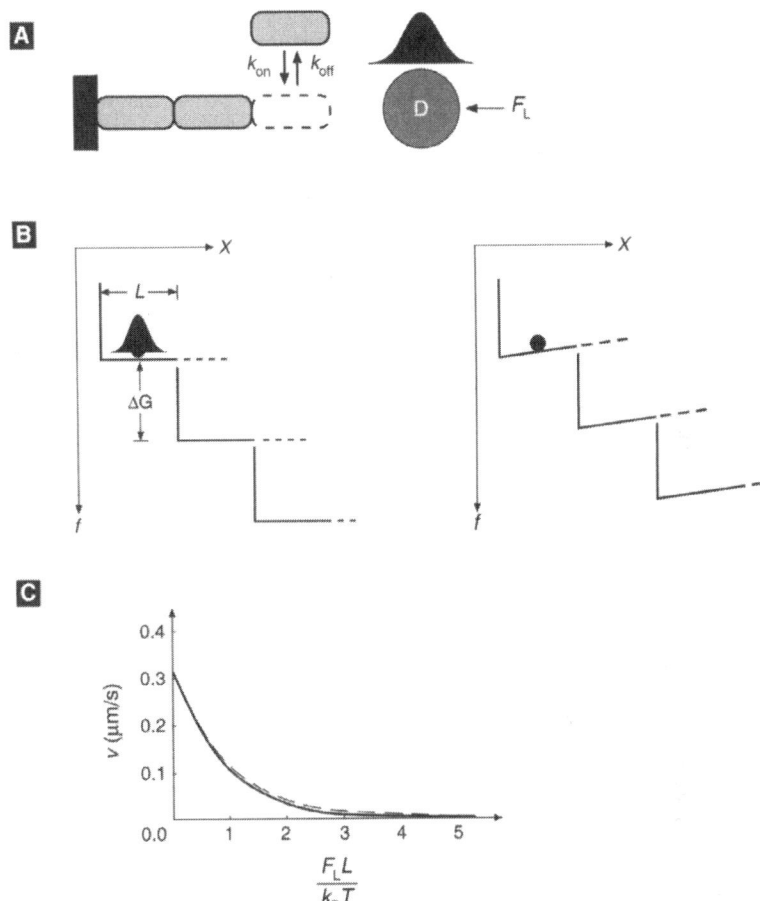

**Figure 13.5** The polymerization ratchet. (a) Monomers of length $L$ polymerize onto the end of a growing filament with rate constants $k_{on}$, $k_{off}$. An object with diffusion coefficient $D$ has its thermal motions rectified by the insertion of each new monomer. A load force $F_L$ opposes the the motion of the object to the right. (b) Free energy diagram of the polymerization process. The total free energy $\Delta G = \Delta G_b + F_L x$ where the binding free energy satisfies $\Delta G_b \gg k_B T$. Left panel: $F_L = 0$; right panel: $F_L > 0$ tilts the potential so that the object must diffuse uphill. (c) The load-velocity curve for the polymerization ratchet given by the approximation (13.13). The exact solution is shown by the dashed line (Peskin et al. 1993).

$L$. The stall load $F_s$ is reached when the work done in moving the object a distance $L$ is just equal to the free energy from the binding reaction, so that $V_p = 0$:

$$F_s = \frac{k_B T}{L} \ln \left[ \frac{k_{on} M}{k_{off}} \right]. \tag{13.14}$$

Note that without depolymerization, $k_{off} \to 0$, there is no finite stall load.

Variations and elaborations on the polymerization ratchet have been used to model a variety of cellular processes, including lamellipodial protrusion (Mogilner and Oster 1996), the polymerization of microtubules (Mogilner and Oster 1999), the propulsion of intracellular pathogens (Mogilner and Oster 1996), and the translocation of proteins (Simon et al. 1992; Peskin et al. 1993).

## 13.4   Simplified Model of the $F_0$ Motor

To futher illustrate the numerical formalism developed in the Chapter 12, we shall examine in detail a simplified model based on the principle of the ion-driven $F_0$ motor of ATP synthase (Elston et al. 1998; Dimroth et al. 1999). This enzyme uses electrochemical energy stored in a proton motive force across the inner membrane of mitochondria to produce ATP. This will illustrate many of the principles of mechanochemical energy conversion by proteins, but is sufficiently simple to analyze analytically and numerically. The motor is sketched schematically in Figure 13.6. It consists of two reservoirs separated by an ion–impermeable membrane. The reservoir on the left is acidic (high proton concentration) with concentration $c^{acid}$, and the reservoir on the right is basic (low concentration) with concentration $c^{base}$. The motor itself consists of two "parts": (i) a "rotor" carrying negatively charged sites spaced a distance $L$ apart that can be protonated and deprotonated; (ii) a "stator" consisting of a hydrophobic barrier that is penetrated by an apolar strip that can allow a protonated site to pass through the membrane, but will block the passage of an unprotonated site. (The height of the energy barrier blocking passage of a charge between two media with different dielectric constants $\epsilon_1$ and $\epsilon_2$ is $\Delta G \approx 200[(1/\epsilon_1) - (1/\epsilon_2)] \approx 45k_BT$ (Israelachvili 1992; Dimroth et al. 1999). This energy penalty arises from the necessity of stripping hydrogen–bonded water molecules from the rotor sites.)

Qualitatively, the motor works like this. Rotor sites on the acidic side of the membrane are frequently protonated. In this state (a nearly neutral dipole) the rotor can diffuse to the right, allowing the protonated site to pass through the membrane–stator interface to the basic reservoir. Once exposed to the low proton concentration in the basic reservoir, the proton quickly dissociates from the rotor site. In its charged state, the rotor site cannot diffuse backwards across the interface: Its diffusion is "ratcheted." We will show that thermal fluctuations will consistently drive the rotor to the right in Figure 13.6.

Thus a rotor site can exist in two states: unprotonated and protonated. To specify the mobility state of the rotor, we need to keep track only of the site immediately adjacent to the membrane on the acidic side. In its unprotonated state, the site adjacent to the membrane is immobilized, since it cannot pass into the stator, nor can it diffuse to the left, since the next rotor site on the basic side of the membrane is almost always deprotonated, and cannot diffuse to the left. (Of course, this depends on the thickness of the membrane being the same as the rotor spacing; this is unrealistic, but the full

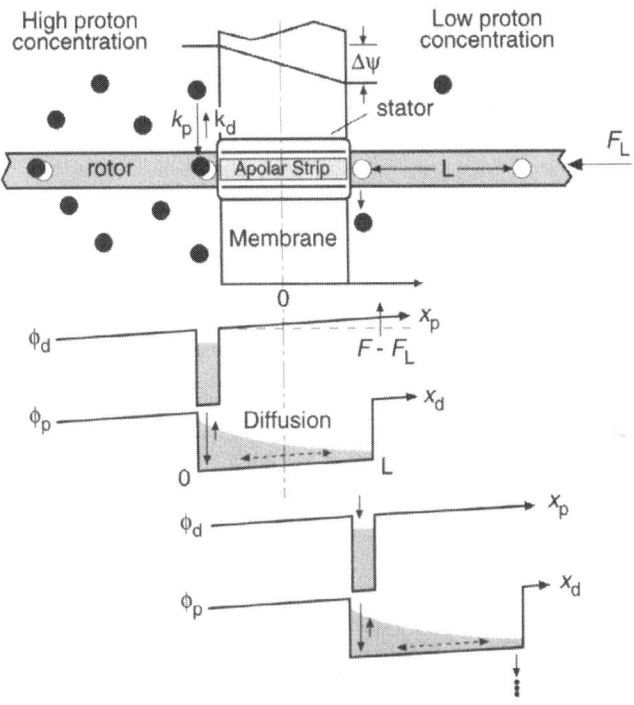

**Figure 13.6** Simplified model illustrating the principle of the $F_0$ motor. The state of the rotor can be pictured as a probability cloud (shown shaded) that drains from one potential to the next. $\phi_p$ is the potential seen by the site adjacent to the left side of the membrane in the protonated state, and $\phi_d$ is the potential in the deprotonated state. The potentials are tilted by an amout equal to the difference between the driving force ($F_i = e \cdot d\psi/dx$), and the load force, $F_L$. In the deprotonated state the site is immobilized, and in the protonated (almost neutral) state it can diffuse in the potential well $\phi_p$. In the fast diffusion limit, the probability cloud quickly settles into the exponential Boltzmann distribution inside $\phi_p$, which determines the probability of draining into the next deprotonated well, thus completing one step to the right.

model treated in the references does not have this constraint.) Thus the progress of the model can be pictured as a sequence of transitions between two potentials, as shown in Figure 13.6. When deprotonated, the rotor is immobilized in potential $\phi_d$, and when protonated, it can move in potential $\phi_p$. The effect of the load force $F_L$ is to tilt the potentials upward, so the motion in potential $\phi_p$ is "uphill" (i.e., the total potential when protonated can be written as $\phi_p(x) - F_L x$).

Below, we consider two limiting cases. In the first one, diffusion is much faster than the chemical reaction rates. In the second one, the diffusion time scale is comparable to the reaction rates in the basic reservoir. The first case can be treated *analytically*, while the second one will require *numerical* simulation.

## 13.4.1   The Average Velocity of the Motor in the Limit of Fast Diffusion

The model can be formulated mathematically in terms of the probability of the depro-tonated state $p_d(t)$ (non-dimensional), and the probability density of the protonated state $p_p(x,t)$ (1/nm). Here $x, 0 \leq x \leq L$, is the distance of the protonated site from the interface between the acidic reservoir and the membrane. The model equations have the form

$$\frac{dp_d}{dt} = \text{net deprotonated spatial flux} + \text{net reaction flux}$$

$$= J_{x_d} + J_\xi, \tag{13.15}$$

$$\frac{\partial p_p}{\partial t} = \text{net protonated spatial flux} - \text{net reaction flux}$$

$$= J_{x_p} - J_\xi, \tag{13.16}$$

where

$$J_{x_d} = 0,$$

$$J_{x_p} = D \frac{\partial}{\partial x} \left( \frac{\partial p_p}{\partial x} - \frac{F_I - F_L}{k_B T} p_p \right), \quad J_{x_p}(0) = J_{x_p}(L) = 0,$$

$$J_\xi = \text{deprotonation at acidic reservoir}$$

$$+ \text{deprotonation at basic reservoir}$$

$$- \text{net protonation at both reservoirs}$$

$$= k_d p_p(0) + k_d p_p(L) - \bar{k}_p p_d.$$

Protonation rates are proportional to hyrogen ion concentration on either side of the membrane, and the net protonation at both reservoirs is $\bar{k}_p = k_p c^{acid} + k_p c^{base}$. Note that the rates of protonation and deprotonation have the dimensions $\bar{k}_p$ (1/s) and $k_d$ (nm/s), respectively. We assume that the deprotonation rates are the same at both reservoirs.

First we nondimensionalize the model equations using the rescaled coordinate $(x/L) \to x$, the rescaled time $(k_d t/L) \to t$, and the following dimensionless parameters:

- Ratio of reaction to diffusion time scales: $\Lambda = D/k_d L$.
- Net work done in moving the rotor a distance $L$: $w = (F_I - F_L)L/k_B T$.
- Equilibrium constant: $\kappa = \bar{k}_p L/k_d$.

Here $F_I = e\Delta\psi/L$ is the electrical driving force, which is assumed to be constant. Substituting these variables and parameters into equations (13.15) and (13.16) gives

$$\frac{dp_d}{dt} = -\kappa p_d + p_p(0) + p_p(1), \tag{13.17}$$

$$\frac{\partial p_p}{\partial t} = \kappa p_d - p_p(0) - p_p(1) + \Lambda \frac{\partial}{\partial x} \left( \frac{\partial p_p}{\partial x} - w p_p \right), \tag{13.18}$$

where $x, t$ are now the nondimensional coordinate and time, respectively.

In many situations it turns out that diffusion is much faster than the chemical reaction rates: $\Lambda \gg 1$. (At $D \approx 10^7$ nm²/s, $L \approx 10$ nm, $k_d \approx 10^3$ nm/s, the order of magnitude of the parameter $\Lambda$ is $10^3$.) This means that the time between dissociation events is much longer than the time to diffuse a distance $L$, so the process is limited by the speed of the reactions, not by the diffusion of the rotor. In other words, the diffusive motion of the rotor is so fast that it achieves thermodynamic equilibrium, and so its displacement can be described by a Boltzmann distribution. In this case we can express the probability distribution as $p_p(x, t) = p_p(t) \cdot P(x)$, where $p_p(t)$ is the probability of the site being in the protonated state and $P(x)$ is the *equilibrium* spatial probability density describing the rotor's position relative to the stator.

We can obtain the steady–state Boltzmann distribution $P(x)$ from (13.18). First, we divide through by $\Lambda$ and take advantage of the fact that $\Lambda \gg 1$: All terms but the last are rendered negligible, so that the distribution of rotor positions in the protonated state potential well $\phi_p$ $(0 \le x \le 1)$ in Figure 13.6 is governed by

$$\frac{dP}{dx} - wP = 0.$$

The solution must be normalized to 1, since it represents a probability density. The result is:

$$P(x) = \left(\frac{w}{e^w - 1}\right) e^{wx}, \quad 0 \le x \le 1, \tag{13.19}$$

where the quantity in parentheses is the normalization factor. Thus, the rates of protonation at time $t$ are $p_p(0, t) = p_p(t) \cdot P(0)$ and $p_p(1, t) = p_p(t) \cdot P(1)$, respectively. Substituting this into (13.17) and imposing the conservation of probability, $p_d(t) + p_p(t) = 1$, we reduce the problem to the two–state Markov chain described by:

$$\frac{dp_d}{dt} = -\frac{dp_p}{dt} = -\kappa p_d + (P_+ + P_-)p_p, \tag{13.20}$$

where $P_+(w) = w/(1 - e^{-w})$, $P_-(w) = w/(e^w - 1)$. Therefore, the stationary probabilities are obtained directly by setting the time derivatives equal to zero and solving for $p_p$:

$$p_d(w, \kappa) = \frac{P_+ + P_-}{\kappa + P_+ + P_-}, \quad p_p(w, \kappa) = \frac{\kappa}{\kappa + P_+ + P_-}, \tag{13.21}$$

or, in the dimensional variables:

$$p_d(w, \kappa) = \frac{k_d P_+ + k_d P_-}{k_p(c^{acid} + c^{base})L + k_d P_+ + k_d P_-}, \tag{13.22}$$

$$p_p(w, \kappa) = \frac{k_p(c^{acid} + c^{base})L}{k_p(c^{acid} + c^{base})L + k_d P_+ + k_d P_-}. \tag{13.23}$$

The average velocity of the motor can be found using the following heuristic argument. The rotor effectively moves to the right either from the protonated state, when the proton is released to the basic reservoir with the effective rate $k_d P_+$, or from the deprotonated state, when the protonation takes place at the acidic reservoir with the effective rate $k_p c^{acid} L$. The corresponding effective rate of movement to the right is the sum of the corresponding rates weighted by the respective state probabilities: $\langle V_r \rangle = k_p c^{acid} L p_d + k_d P_+ p_p$. Similarly, the rotor effectively moves to the left either from the protonated state, when the proton is released to the acidic reservoir with the effective rate $k_d P_-$, or from the deprotonated state, when the protonation takes place at the basic reservoir with the effective rate $k_p c^{base} L$. The corresponding effective rate of movement to the left is the sum of the corresponding rates weighted by the respective state probabilities: $\langle V_l \rangle = k_p c^{base} L p_d + k_d P_- p_p$. The net average velocity $\langle V \rangle = \langle V_r \rangle - \langle V_l \rangle$, can be obtained by using the expressions for the state probabilities (13.22), (13.23) and some algebra:

$$\langle V \rangle (F_L) = \frac{k_p k_d L w (c^{acid} e^w - c^{base})}{(e^w - 1) k_p L (c^{acid} + c^{base}) + k_d w (e^w + 1)},$$

$$w = \frac{(F_I - F_L) L}{k_B T}. \tag{13.24}$$

As a check, note that when there is no load ($F_L = 0$), no membrane potential ($F_I = 0$), and no proton gradient ($c^{acid} = c^{base}$), the velocity vanishes, as it should. The load–velocity relationship given by (13.24) looks very similar to the one in the next limiting case, and is plotted in Figure 13.8C.

The *stall force* $F_s$ is reached when the load force just brings the motor to a halt ($c^{acid} e^w - c^{base} = 0$):

$$F_s = F_I + \frac{k_B T}{L} \ln \left[ \frac{c^{acid}}{c^{base}} \right]. \tag{13.25}$$

Since the electrical driving force satisfies $F_I = e \Delta \psi / L$, (13.25) can be written as an equilibrium thermodynamic relation in terms of the energy:

$$F_s \cdot L = e \Delta \psi - 2.3 k_B T \Delta pH. \tag{13.26}$$

This says that the reversible work done to move a rotor site across the membrane is equal to the work done by the electrical field plus the "entropic work" done by the Brownian ratchet. (The term "reversible" in this context means that the velocity is so slow (near stall) that we can neglect the viscous dissipation.) Dividing through by the unit charge $e$ gives the work per unit charge, which is just the protonmotive force discussed previously. One point about (13.26) is worth noting. Since the motor is working against a conservative load force, as the motor approaches stall its efficiency approaches 100%. For a motor working against a viscous load, a more sophisticated treatment is required (Oster and Wang 2000a).

It may seem from (13.24) that there is a definite average velocity of the motor in the limit $T \to \infty$. In other words, the motor continues to move in the absence of thermal fluctuation. The reason is that the solution of the Lengevin and Smoluchowski equations cannot be treated as a regular perturbation problem in the limit of low temperatures. This is a singlular perturbation problem, and the protein behavior at absolute zero temperature cannot be quantified as a simple limit of such behavior at low temperatures.

### 13.4.2  Brownian Ratchet vs. Power Stroke

In the last example the motion of the rotor was driven by a combination of Brownian motion and the membrane potential. The rotor diffusion is biased by the electrostatic forces that are switched off and on by the binding and dissociation of protons to the rotor sites. The membrane potential appears to drive the rotor unidirectionally without the aid of Brownian motion. However, this cannot happen without the binding and dissociation of protons, a stochastic process driven by thermally excited transitions. Thus even the "power stroke" depends on Brownian motion, so that setting $k_B T = 0$ in the model equations arrests the rotor motion. This is a fundamental distinction between molecular and macroscopic motors. The distinction between a motor driven by a "Brownian ratchet" and one driven by a "power stroke" may not be so clear in other systems.

In the polymerization ratchet model described above, the movement of the *load* is driven entirely by its Brownian motion. The chemical step simply rectifies, or biases, this motion (Peskin et al. 1993). By comparison, the $F_1$ motor in ATP synthase is driven by the hydrolysis of ATP. The conformational change in the protein that constitutes the power stroke is known: Binding of ATP to the catalytic site drives the change in protein shape that drives rotation. That is, the load is not driven significantly by Brownian motion; it sees the protein's conformational change as a "power stroke."

However, a closer look at how ATP binds to the catalytic site reveals that it is a multistep process involving the sequential annealing of hydrogen bonds between the protein and the nucleotide. Each step in this process is driven by Brownian motion, i.e., a thermally activated process as illustrated in Figure 12.5. Therefore, the power stroke itself can be viewed as a kind of Brownian ratchet, one that proceeds at a smaller length scale than the protein (ATP synthase is $\approx 10$ nm in diameter, while the catalytic site is $\approx 1$ nm). Thus the distinction between a process driven by a Brownian ratchet and by a power stroke can be largely a matter of size scale; a fuzzy boundary separates the two notions. In the extreme case where the motion of the *load* is due only to its diffusion, and the role of the chemical reaction is only to inhibit diffusion in one direction, we can say that the motor is a Brownian ratchet.

### 13.4.3  The Average Velocity of the Motor When Chemical Reactions Are as Fast as Diffusion

Next we consider a different limiting case: where the diffusion time scale is comparable to the reaction rates in the basic reservoir. In this case we have to change the mathematical formulation of the model. We make the simplifying assumption that the proton concentration in the acidic reservoir is so high that the binding sites on that side are always protonated. Now we define the right boundary of the membrane as the origin, and the distance between the membrane and the binding site nearest the membrane in the basic reservoir $x$ is always between 0 and $L$. The chemical state of the rotor is determined by the state of all the binding sites in the acidic reservoir. In general, if there are $N$ binding sites on this side, the total number of chemical states is $2^N$. However, for now we focus on the binding site nearest the membrane. In this case there are just two states: "off" if this site is unprotonated and "on" if it is protonated. In Chapter 12 the mechanochemistry of the motor was described by the following set of coupled diffusion equations:

$$\frac{\partial p_d}{\partial t} = D\frac{\partial}{\partial x}\left(\frac{F_L - F_I}{k_B T}p_d + \frac{\partial p_d}{\partial x}\right) - \bar{k}_p p_d + k_d p_p, \tag{13.27}$$

$$\frac{\partial p_p}{\partial t} = D\frac{\partial}{\partial x}\left(\frac{F_L - F_I}{k_B T}p_p + \frac{\partial p_p}{\partial x}\right) + \bar{k}_p p_d - k_d p_p, \tag{13.28}$$

where $p_p(x,t)$ and $p_d(x,t)$ (1/nm) are the probability densities for being at position $x$ and in the protonated and deprotonated states, respectively, at time $t$. The proton association and dissociation rates in basic reservoir are $\bar{k}_p$ (1/s) and $k_d$ (1/s), respectively. Note that dimensions of some of the model parameters and variables are different in this limit.

We can nondimensionalize these equations using the rescaled coordinate $(x/L) \to x$, the re-scaled time $k_d t \to t$, and the dimensionless parameters $\Lambda = (D/k_d L^2)$, $w = (F_I - F_L)L/k_B T$, and $\kappa = k_p/k_d$. The nondimensional equations have the form

$$\frac{\partial p_d}{\partial t} = \Lambda\frac{\partial}{\partial x}\left(w p_d + \frac{\partial p_d}{\partial x}\right) - \kappa p_d + p_p, \tag{13.29}$$

$$\frac{\partial p_p}{\partial t} = \Lambda\frac{\partial}{\partial x}\left(w p_p + \frac{\partial p_p}{\partial x}\right) + \kappa p_d - p_p. \tag{13.30}$$

Equations (13.29) and (13.30) are second–order partial differential equations. This means that four boundary conditions are required in order to have a mathematically complete description of the problem. One boundary condition is that $x = 0$ is reflecting:

$$\left[w p_d + \frac{\partial p_d}{\partial x}\right]_{x=0} = 0, \tag{13.31}$$

This takes into account the fact that an unprotonated site cannot pass back through the membrane. The remaining three boundary condtions require knowing the state of all the binding sites in the basic reservoir, which would necessitate solving a large number of coupled diffusion equations (one for each possible chemical state of the rotor). However, to keep things simple, we construct a reflecting boundary condition at $x = 1$ ($x$ is now measured in units of $L$) when the rotor is in the protonated state. That is,

$$\left[w p_p + \frac{\partial p_p}{\partial x}\right]_{x=1} = 0. \tag{13.32}$$

If the proton dissociation rate is fast and the proton concentration is low in the basic reservoir. we don't expect this artificial boundary condition to have much of an effect,

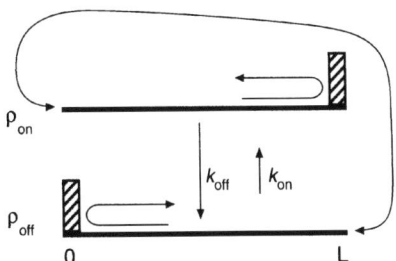

**Figure 13.7** In general, the chemical state of the motor is determined by all the binding sites in the right chamber. In the simplified version of the model only the binding site nearest the membrane is considered. The reflecting boundary condition in the off state at $x = 0$ is due to the fact that an unprotonated site cannot pass back through the membrane. The reflecting boundary condition in the on state at $x = L$ is artificial and is used to simplify the problem.

since in this limit the probability of the first site being occupied when $x = 1$ is very small, so that this boundary condition is rarely encountered.

When an unprotonated site moves to the right of $x = 1$, it brings a protonated site out of the membrane channel and into the region $0 < x < 1$. This protonated site becomes the new site that we follow. The state of the motor goes from unprotonated to protonated, and the coordinate of the motor goes from $x = 1$ to $x = 0$. Conversely, when a protonated site moves into the membrane, it brings an unprotonated site into the region $0 < x < 1$. This unprotonated site becomes the new site we follow. The state and the coordinate of the motor change accordingly. These considerations are illustrated in Figure 13.7 by the arrows connecting the right end of $p_d$ to the left end of $p_p$. The boundary conditions that model this situation are:

$$\left(wp_d + \frac{\partial p_d}{\partial x}\right)_{x=1} = \left(wp_p + \frac{\partial p_p}{\partial x}\right)_{x=0}, \quad p_d(1,t) = p_p(0,t), \tag{13.33}$$

which is the mathematical statement of the fact that the rotor in the off state at $x = 1$ is equivalent to the rotor in the on state at $x = 0$. Therefore, to implement these boundary conditions numerically, we make use of the periodic boundary condition discussed in the previous chapter. Note that there are two mechanisms for changing the chemical state of the rotor: movement of the rotor and chemical kinetics.

We are now in a position to use the numerical algorithm developed in Chapter 12 to approximate (13.29)–(13.33). The interval $(0, 1)$ is divided into $M$ segments. For each of the $M$ grid points there are two possible states of the rotor, off and on. Therefore, there are $2M$ possible states in the discrete approximation of the process. Let the first $M$ states correspond to the offstate, and the states $M + 1$ to $2M$ correspond to the onstate. The equations used in the numerical scheme for $1 < n < M$ are

$$\frac{dp_n}{dt} = (F_{n-1/2}p_{n-1} - B_{n-1/2}p_n) - (F_{n+1/2}p_n - B_{n+1/2}p_{n+1}) - \kappa p_n + p_{n+M},$$

$$\frac{dp_{n+M}}{dt} = (F_{n-1/2}p_{n-1+M} - B_{n-1/2}p_{n+M}) - (F_{n+1/2}p_{n+M} - B_{n+1/2}p_{n+1+M}) - p_{n+M} + \kappa p_n,$$

$$\tag{13.34}$$

and the equations used to implement the boundary conditions are

$$\frac{dp_1}{dt} = -(F_{3/2}p_1 - B_{3/2}p_2) - \kappa p_1 + p_{M+1}, \tag{13.35}$$

$$\frac{dp_M}{dt} = (F_{M-1/2}p_{M-1} - B_{M-1/2}p_M) - (F_{1/2}p_M - B_{1/2}p_{M+1}) - \kappa p_M + p_{2M}, \tag{13.36}$$

$$\frac{dp_{M+1}}{dt} = (F_{1/2}p_M - B_{1/2}p_{M+1}) - (F_{3/2}p_{M+1} - B_{3/2}p_{M+2}) - p_{M+1} + \kappa p_1, \tag{13.37}$$

$$\frac{dp_{2M}}{dt} = (F_{M-1/2}p_{2M-1} - B_{M-1/2}p_{2M}) - p_{2M} + \kappa p_M. \tag{13.38}$$

The potential used in $F_n$ and $B_n$ is $\phi(x) = (F_L - F_I)x$. Note that because of the chemical kinetics, the matrices $\mathbf{A}$ and $\mathbf{C}$ required for the numerical scheme are no longer tridiagonal. However, they are still sparse, so that solving (12.59) is not computationally expensive.

Let us discuss briefly how to calculate the protonation and deprotonation rates $\bar{k}_p$ and $k_d$. The chemical reaction is

$$\text{site}^- + \text{H}^+ \leftrightarrow \text{site} \cdot \text{H}. \tag{13.39}$$

At equilibrium, protonation and deprotonation balance. That is,

$$\bar{k}_p[\text{site}^-] = k_d[\text{site} \cdot \text{H}]. \tag{13.40}$$

Proton concentrations are generally reported as a pH value:

$$\text{pH} = -\log_{10}[\text{H}^+], \quad [\text{H}^+] = 10^{-\text{pH}}. \tag{13.41}$$

The higher the pH value, the lower the proton concentration. The $\text{pK}_a$ value of the binding site is calculated from the measured concentration values of $[\text{site} \cdot \text{H}]$ and $[\text{site}^-]$ as

$$\text{pK}_a = \text{pH} + \log_{10}\frac{[\text{site} \cdot \text{H}]}{[\text{site}^-]}. \tag{13.42}$$

Combining (13.40) and (13.42), we see that the rates $\bar{k}_p$ and $k_d$ are related to pH and $\text{pK}_a$ by

$$\frac{\bar{k}_p}{k_d} = 10^{\text{pK}_a - \text{pH}}. \tag{13.43}$$

Generally, $\bar{k}_p$ is limited by the rate at which protons diffuse to the binding site. In this limit, the association rate can be computed from the Smoluchowski formula (this rate is proportional to the proton concentration, $c^{\text{base}}$):

$$\bar{k}_p = \begin{pmatrix} \text{proton} \\ \text{concentration} \end{pmatrix} \cdot \begin{pmatrix} \text{absorption rate to a perfectly} \\ \text{absorbing disk of radius } r \end{pmatrix}$$

$$= \underbrace{0.6\,\text{nm}^{-3} \cdot 10^{-\text{pH}}}_{\text{protons/nm}^3} \cdot \underbrace{4\,r\,D_{\text{proton}}}_{\text{absorption rate}} .$$  (13.44)

Here $D_{\text{proton}}$ is the diffusion coefficient of protons. If we know $\bar{k}_\text{p}$, $k_\text{d}$ can be determined by (13.43). The table shown below lists typical parameter values for ATP synthase.

| Parameter Name | Parameter Value |
|---|---|
| Diffusion coefficient of the rod | $D = 10^4\ \text{nm}^2/\text{s}$ |
| pK$_\text{a}$ value of binding site | pK$_\text{a} = 6.0$ |
| pH of the right compartment | pH $= 6.0$ to $8.0$ (variable) |
| External loadforce on the rod | $F_\text{L} = 0$ to $3$ pN (variable) |
| Distance between binding sites | $L = 8$ nm |
| Diffusion coefficient of proton | $D_{\text{proton}} = 10^{10}\,\text{nm}^2/\text{s}$ |
| Absorbing radius of binding site | $r = 0.5\,\text{nm}$ |

Note that at these values of the model parameters, at pH $= 7$, the values of the nondimensional parameters are $\Lambda \approx 0.1, \kappa \approx 0.1, w \approx 1$.

We now have all the necessary information to use the numerical scheme. Figure 13.8A shows the relaxation of the marginal density $p(x,t) = p_\text{p}(x,t) + p_\text{d}(x,t)$ to steady state. The (dimensional) average velocity is a steady-state property of the system and is related to the total flux by the relations

$$\langle V \rangle = LJ = -LD \left[ \frac{F_\text{L} - F_\text{I}}{k_\text{B}T} \left( p_p^{(s)} + p_d^{(s)} \right) + \frac{\partial}{\partial x} \left( p_p^{(s)} + p_d^{(s)} \right) \right],$$  (13.45)

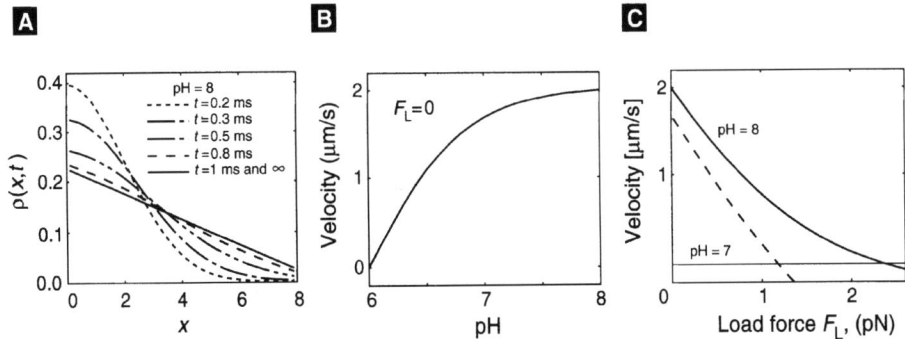

**Figure 13.8** Numerical results of the simplified $F_0$ motor. Here pH is the pH value of the right compartment. The external load force is $F_\text{L}$. In all the figures the number of grid points used in the simulations was $M = 32$. (A) Relaxation of the marginal density $\rho = \rho_{\text{rmon}} + \rho_{\text{off}}$ to steady state for pH $= 8$ and no external load. (B) Motor velocity as a function of pH with no external load. (C) Motor velocity as a function of the load force for pH $= 7$ and pH $= 8$.

where the superscript $s$ in the above equation stands for steady state. Once steady state is achieved, the average velocity $\langle V \rangle$ can then be computed from the relation

$$\langle V \rangle = L(F_{n-1/2}p_{n-1} - B_{n-1/2}p_n) + L(F_{n-1/2}p_{n-1+M} - B_{n-1/2}p_{n+M}) \qquad (13.46)$$

for $1 < n < M$. Typical results for the mean velocity are shown in Figure 13.8B and Figure 13.8C.

## 13.5 Other Motor Proteins

As we discussed in Chapter 12, there is such a variety of protein motors that no classification scheme can do justice to their diversity. However, for the purposes of discussion we can identify several physical properties that delineate classes of motors. The literature on molecular motors is vast, so we shall restrict our discussion here to those for which reasonably complete mathematical models exist as outlined in Table 13.5. Some of the most important characteristics of molecular motors are the following:

**Fuel**. The two most common energy sources for molecular motors are nucleotide hydrolysis (e.g., ATP, GTP) and transmembrane ion gradients. Certain specialized motors depend on stored elastic energy that has been captured during the assembly of the motor (e.g., spasmoneme and the acrosome of *Limulus* (Mahadevan and Matsudaira 2000)).

**Mechanical escapement**. Three common motor types are (i) rotary (e.g., the bacterial flagellar motor $F_0$ ATPase); (ii) linear motors that run along a "track," usually actin, microtubules, or nucleic acid polymers (e.g., myosin, kinesin, dynein, RNA polymerase); (iii) polymerization or depolymerization motors that directly push or pull against a load (e.g., the acrosome, cellular lamellipodia, the propulsive tail of *Listeria*). The latter category suggests a subclassification into those that operate in a continuous cycle (e.g., myosin, $F_1$), and motors that are "one-shot" processes; i.e., they function for but a single episode of polymerization, after which they are usually disassembled.

**Cooperative vs. "loners."** Because of its small "duty cycle" (i.e., attachment time to the load), myosin II must act in concert with many other partners to produce a continuous force on an actin filament. Myosin V and kinesin, however, have longer duty cycles, and so they can transport a vesicle without the cooperation of other motors.

These categories do not begin to delineate the variety of possibilities. However, one common event generally commences the transduction process between chemical energy and mechanical force. Because molecular motors can be viewed as enzymes, the binding of a substrate onto the motor initiates the transduction process. However, this does not really tell us much, since it is simply a restatement that chemical reactions (other than isomerizations) begin by combining substrates. The feature that distinguishes molecular motors from other enzyme reactions is that the binding event is directly or indirectly coupled to the creation of mechanical forces. For example, in $F_1$ ATPase, the binding of ATP to the catalytic site directly generates the power stroke. However, the coupling may not always be so direct. For example, binding of a proton to the $F_0$ rotor site switches off the local electrostatic field surrounding the rotor, permitting bidirectional diffusion. The binding of monomers to polymerizing actin captures thermal fluctuations in elastic strain, which is subsequently released to power protrusion.

Several points are important. First, the role of thermal fluctuations in all these processes is central, so that none could operate when $k_B T = 0$. Second, energy captured by binding or dissociation events can be stored and released later, and in a different location, to produce mechanical work. Third, the operation of every molecular motor depends on its specialized protein geometry, so that models of motors that ignore geometry are generally not useful to biologists.

Finally, we do not believe that the operation of molecular motors involves any novel physics or chemistry. However, the amazing variety of protein shapes requires that we treat each motor individually. General principles are not likely to provide more than philosophical comfort in understanding any particular motor. In the words of Katchalsky:

*"It is easier to make a theory of everything, than a theory of something"*

*–Aharon Katchalsky*

**Some Molecular Motors that Have Been Modeled Mathematically**

| Motor | References |
|---|---|
| Acrosome | (Mahadevan and Matsudaira 2000; Oster et al. 1982) |
| $F_1$ ATPase | (Oster and Wang 2000a; Wang and Oster 1998; Oosawa and Hayashi 1986) |
| $F_0$ ATPase | (Elston et al. 1998; Dimroth et al. 1999; Läuger 1991; Stein and Lauger 1990) |
| Bacterial flagellar motor | (Elston and Oster 1997; Läuger 1990; Berry 1993) |
| HSP70 | (Peskin et al. 1993; Simon et al. 1992; Elston 2000; Chauwin et al. 1998) |
| Kinesin | (Peskin and Oster 1995; Fox and Choi 2001; Derenyi and Vicsek 1996; Keller and Bustamante 2000) |
| Myosin | (Huxley and Simmons 1971; Huxley 1957; McMahon 1984; Smith and Geeves 1995) |
| Polymerization | (Peskin et al. 1993; Mogilner and Oster 1996; Mogilner and Oster 1999) |
| RNA polymerase | (Wang et al. 1998; Julicher and Bruinsma 1998) |

# Suggestions for Further Reading

- *Millennial musings on molecular motors*, R. Vale (Vale 2000).
- *The mechanochemistry of molecular motors*, D. Keller and C. Bustamante (Keller and Bustamante 2000).
- *Mechanics of Motor Proteins and the Cytoskeleton*, J. Howard (Howard 2001).

## 13.6  EXERCISES

1. In the example of switching in the bacterial flagellar motor, the first passage time density was calculated for the case $\Delta G = 0$. Write a program that numerically generates the first passage time density for the case when $\Delta G \neq 0$. Investigate how the distribution changes as a function of CheY concentration. Do you expect Kramers approximation for the mean first passage time to be valid for all values of CheY concentration?

2. Nondimensionalize equations (13.11) using the potentials

$$V_1(x) = k_B T[\cos(x/L) + 0.3\sin(2x/L)], \quad V_2 = 0.$$

   Use the parameters $L = 10$ nm, $D = 10^5$ nm$^2$/s, $10^2$/s $< k < 10^4$/s. Write a program that solves equations (13.11) numerically and compute the velocity of the flashing ratchet for various values of $k$ from the given range. Discuss the results.

3. Simulate two rigid filaments growing side by side against a "wall," which diffuses with the diffusion coefficient $D$. Assume that the rates of assembly and disassembly of the monomers onto the polymer tips are known. Use the computer to model this two-filament polymerization ratchet. Estimate the rate of growth in the absence of the load force and compare it with the average velocity of the one-filament polymerization ratchet. Compare the corresponding stall forces. Do the results depend on the mutual position of two filaments?

4. Consider the following model of a "walking" molecular motor that is roughly similar to kinesin. The motor walks a 1-D track with equidistant binding sites. The motor has two "legs" that either can be (i) attached to the adjacent binding sites, or (ii) attached by one foot, while the other diffuses freely between the binding sites adjacent to the attached foot. The rate of dissociation of the front foot from the track is different from that of the rear foot. Similarly, the rate of association of the diffusing foot to the site in front of the bound foot is different from the rate of association behind the bound foot. Describe this model with coupled Smoluchowski equations and demonstrate that it is mathematically equivalent to the model considered in Section 4.1.

5. Derive Langevin equations describing the model of the "walking" motor in the previous exercise. Simulate the motor's walk numerically. By changing the association/dissociation rates find conditions under which the motor would move on average to the right. Apply a load force to the free foot directed to the left and estimate the stall force numerically.

6. Estimate numerically the average velocity and effective diffusion coefficient of the walking motor at various values of the load force. Discuss the results.

7. Assume that the mechanochemical cycle of the "walking" motor is such that only the rear foot dissociates from the track and that the diffusing foot binds only to the site in front of the attached foot (i.e., the motor makes only forward steps). Describe how to find the ratio of the association and dissociation rates if you know from an experiment the average velocity and rate of growth of the displacement variance.

# Qualitative Analysis of Differential Equations

## G. Bard Ermentrout and Joel E. Keizer

Nonlinear ordinary differential equations are notoriously difficult or impossible to solve analytically. On the other hand, the solution to linear equations like those encountered in the kinetic model of the GLUT transporter can be expressed in terms of simple functions, and their behavior analyzed using standard results from linear algebra. In the first two sections, we cover some basic ideas in linear algebra and a review of power series. These are concepts that are needed in the sections that follow. In Section A.3 we summarize the main results for linear equations with two dependent variables. Although one often encounters models like the GLUT transporter that involve more than two variables, the basic ideas for two-variable ODEs carry over more or less unchanged for larger sets of linear equations. Thus the intuition gained from understanding simple two variable ODEs is enormously useful in understanding more complicated models. To help to develop this intuition, we introduce the notion of the *phase plane* and use phase plane analysis to help understand the solution of two-variable linear equations in Section A.4. Another reason for focusing on linear equations is that the stability of nonlinear ODEs can be understood by examining the behavior of small deviations around steady or oscillatory states. In Section A.4.2 we show how the properties of linearized equations can be used to understand stability of steady states for a system of nonlinear equations such as a membrane with a gated ion channel.

Although we have tried to include a good brief synopsis of the most important tools used in this book, more study on these topics may be necessary to fully appreciate some of the more complex mathematical concepts. For more information on these topics, please see the suggested readings listed at the end of this appendix.

# A.1 Matrix and Vector Manipulation

Matrices can be multiplied, added, multiplied by scalar numbers, and differentiated according to the rules of linear algebra. Here we summarize these results for the $2 \times 2$ matrices and two-component column vectors. The matrix $\hat{A}$ multiplying the vector $\mathbf{x}$ acts as a linear operator that produces a new vector, $\mathbf{z}$, according to the formula

$$\mathbf{z} = \hat{A}\mathbf{x} = \begin{pmatrix} a_{11} & a_{12} \\ a_{21} & a_{22} \end{pmatrix} \begin{pmatrix} x_1 \\ x_2 \end{pmatrix} = \begin{pmatrix} a_{11}x_1 + a_{12}x_2 \\ a_{21}x_1 + a_{22}x_2 \end{pmatrix}. \tag{A.1}$$

It can be verified using (A.1) that the identity matrix $\hat{I} = \begin{pmatrix} 1 & 0 \\ 0 & 1 \end{pmatrix}$ leaves vectors unchanged, i.e., $\mathbf{z} = \hat{I}\mathbf{x} = \mathbf{x}$. Matrices can be added together, as can vectors, using the rules

$$\hat{A} + \hat{B} = \begin{pmatrix} a_{11} + b_{11} & a_{12} + b_{12} \\ a_{21} + b_{21} & a_{22} + b_{22} \end{pmatrix} \quad \text{and} \quad \mathbf{x} + \mathbf{y} = \begin{pmatrix} x_1 + y_1 \\ x_2 + y_2 \end{pmatrix}. \tag{A.2}$$

To multiply either a matrix or a vector by a scalar $c$, each component is multiplied by $c$, e.g.,

$$c\hat{A} = c \begin{pmatrix} a_{11} & a_{12} \\ a_{21} & a_{22} \end{pmatrix} = \begin{pmatrix} ca_{11} & ca_{12} \\ ca_{21} & ca_{22} \end{pmatrix}. \tag{A.3}$$

Differentiation of matrices and vectors is also carried out on each component separately. Thus

$$d\mathbf{x}/dt = \begin{pmatrix} dx_1/dt \\ dx_2/dt \end{pmatrix}. \tag{A.4}$$

The *trace*, *determinant*, and *discriminant* are important scalars that characterize matrices and that appear in the solution to (A.15). We use the shorthand notation $tr\hat{A}$ for the trace of $\hat{A}$, $det\hat{A}$ for its determinant, and $disc\hat{A}$ for the descriminant. In terms of matrix elements they are defined as

$$tr\hat{A} = a_{11} + a_{22} \tag{A.5}$$
$$det\hat{A} = a_{11}a_{22} - a_{21}a_{12} \tag{A.6}$$
$$disc\hat{A} = (tr\hat{A})^2 - 4det\hat{A} \tag{A.7}$$

For example, for the matrix

$$\hat{A} = \begin{pmatrix} 1 & -1 \\ 3 & 6 \end{pmatrix}, \tag{A.8}$$

$tr\hat{A} = 7$, $det\,\hat{A} = 9$, and $disc\hat{A} = 13$.

The inverse of a matrix is the generalization of division by a number. The inverse of $\hat{A}$ is written as $\hat{A}^{-1}$ and is a matrix with the property that

$$\hat{A}\hat{A}^{-1} = \hat{A}^{-1}\hat{A} = \hat{I} \tag{A.9}$$

with $\hat{I}$ the identity matrix. The inverse of a matrix is useful in solving linear algebraic equations. For example, the solution of the linear equation

$$\hat{A}\mathbf{x} = \mathbf{y} \tag{A.10}$$

is

$$\mathbf{x} = \hat{A}^{-1}\mathbf{y}, \tag{A.11}$$

which can be verified by multiplying both sides of (A.10) on the left by $\hat{A}^{-1}$ and using (A.9). For a $2 \times 2$ matrix it is easy to verify by carrying out the matrix multiplication in (A.9) that if $\hat{A}$ is not *singular*, i.e., as long as $det\hat{A} \neq 0$, then

$$\hat{A}^{-1} = \frac{1}{det\hat{A}} \begin{pmatrix} a_{22} & -a_{12} \\ -a_{21} & a_{11} \end{pmatrix}. \tag{A.12}$$

## A.2   A Brief Review of Power Series

One of the most useful tehniques in applied mathematics is the method of power series expansion. The basic idea is that many functions can be expressed as a series in one or more variables. For example, the familiar exponential function can be written as

$$e^t = 1 + \frac{t}{1!} + \frac{t^2}{2!} + \cdots + \frac{t^n}{n!} + \cdots,$$

or more compactly as

$$e^t = \sum_{n=0}^{\infty} \frac{t^n}{n!},$$

where we define $0! = 1$. The series converges for all $t$ both real and complex. Given a function $f(t)$ and a point $t = t_0$, suppose that all the derivatives of $f$ at the point $t_0$ are defined. Then we can formally develop a power series approximation of the function $f$ around the point $t_0$. The formal power series is

$$f(t) = \sum_{n=0}^{\infty} f^{(n)}(t_0) \frac{(t - t_0)^n}{n!}. \tag{A.13}$$

Here $f^{(k)}(t_0)$ is the $k$th derivative of the function $f$ evaluated at the point $t_0$. That is, given the derivatives of a function at a point, we can approximate the function over some interval containing that point by using a series approximation. This series is called a

*Taylor series of $f$ about the point $t_0$.* When the point $t_0$ is 0 the series is often called a Maclaurin series.

If all of the derivatives of the given function exist at the point $t_0$, then the finite approximation to the Taylor series,

$$S_N(t) = \sum_{n=0}^{N} f^{(n)}(t_0) \frac{(t - t_0)^n}{n!},$$

is also defined for all $t$, since it is just a finite sum of polynomials. We say that the series *converges* for $t$ in some interval $I$ containing $t_0$ if the limit of $S_N(t)$ exists as $N \to \infty$ for all $t$ in $I$. The interval $I$ for which convergence is obtained is called the *interval of convergence* for the series. For the exponential series given above, the interval of convergence is the whole real line. Infinite series do not always converge on the whole line. For example, the geometric series

$$S(t) = 1 + t + t^2 + t^3 + \cdots + t^n + \cdots$$

converges for $|t| < 1$. A useful test for the convergence of a series of the form

$$S = \sum_{n=0}^{\infty} a_n$$

is the ratio test. Let $R_n = |a_{n+1}/a_n|$. If

$$\lim_{n \to \infty} R_n < 1,$$

then the series converges. Let us apply this to the exponential series above. Since $a_n = t^n/n!$, we have $|a_{n+1}/a_n| = t/(n+1)$, and the limit of this as $n$ goes to infinity is zero for any finite $t$, so that the series converges for all $t$.

Here are some examples. Let us find a series approximation for $f(0) = \sin(t)$ about $t = 0$. Note that $f(0) = 0, f'(0) = 1, f''(0) = 0, f'''(0) = -1$, and the higher derivatives just cycle among these numbers. That is, derivatives of order 1, 5, 9, etc., are equal to 1, those of order 3, 7, 11, etc., are equal to $-1$, and all others are zero. Thus

$$\sin(t) = t - \frac{t^3}{3!} + \frac{t^5}{5!} + \cdots + (-1)^m \frac{t^{2m+1}}{(2m+1)!} + \cdots.$$

The ratio $|a_{m+1}/a_m| = t^2/(2m+2)(2m+3)$, which tends to 0 as $m$ goes to infinity, so the sine series converges for all $t$.

You can similarly verify that

$$\cos(t) = \sum_{m=0}^{\infty} (-1)^m \frac{t^{2m}}{(2m)!}.$$

As a final example, consider the series for the square root function evaluated at $t = 1$. We have the following first few derivatives:

$$f(1) = 1, \quad f'(1) = \frac{1}{2}, \quad f''(1) = \frac{-1}{2}\frac{1}{2}, \quad f'''(1) = \frac{-3}{2}\frac{-1}{2}\frac{1}{2}.$$

Thus the $n$th derivative ($n > 1$) is

$$c_n = (-1)^{n+1} \frac{(2n - 3)(2n - 5) \cdots 1}{2^n}.$$

Thus,

$$\sqrt{(t)} = 1 + \frac{1}{2}(t - 1) + \sum_{n-2}^{\infty} (-1)^{n+1} \frac{c_n(t - 1)^n}{n!}.$$

We can apply the ratio test to this, noting that $|c_{n+1}/c_n| = n - \frac{1}{2}$, so that

$$R_n = |t - 1| \left( n - \frac{1}{2} \right) /(n + 1).$$

As $n \to \infty$ this ratio goes to $|t - 1|$. The interval of convergence satisfies $|t - 1| < 1$, or $0 < t < 2$.

## A.3   Linear ODEs

The simplest time-dependent differential equations to solve are linear in the dependent variables and of first order in the time. *First order* implies that only the first time derivative appears on the left–hand side of the equations, and linear implies that the right–hand side is a linear function of the dependent variables. The most general equations of this type in $n$ variables have the form

$$
\begin{aligned}
dx_1/dt &= a_{11}x_1 + a_{12}x_2 + \cdots + a_{1n}x_n + y_1, \\
dx_2/dt &= a_{21}x_1 + a_{22}x_2 + \cdots + a_{2n}x_n + y_2, \\
&\;\;\vdots \\
dx_n/dt &= a_{n1}x_1 + a_{n2}x_2 + \cdots + a_{nn}x_n + y_n.
\end{aligned}
\tag{A.14}
$$

Here we consider only the case where the $a_{ij}$ and $y_i$ are parameters that are independent of time. For simplicity, we focus in this chapter on the special case of two variables, which shares the main features of the more general case. Using the column vector and matrix notation introduced in Section A.1, we can write these equations concisely as

$$d\mathbf{x}/dt = \hat{A}\mathbf{x} + \mathbf{y} \tag{A.15}$$

with

$$\mathbf{x} = \begin{pmatrix} x_1 \\ x_2 \end{pmatrix}, \quad \mathbf{y} = \begin{pmatrix} y_1 \\ y_2 \end{pmatrix}, \quad \hat{A} = \begin{pmatrix} a_{11} & a_{12} \\ a_{21} & a_{22} \end{pmatrix}. \tag{A.16}$$

Using the rules for differentiation of vectors, matrix multiplication, and vector addition it is easy to verify that the vector equation (A.15), when written in terms of component vectors, is the special case of (A.14) for two variables, i.e.,

$$dx_1/dt = a_{11}x_1 + a_{12}x_2 + y_1, \tag{A.17}$$

$$dx_2/dt = a_{21}x_1 + a_{22}x_2 + y_2. \tag{A.18}$$

In this book we will be interested in equations for which there is a *steady–state* solution, $\mathbf{x}^{ss}$. In particular, we will be interested in what happens near these steady–state solutions. This is a solution that is independent of time, so that setting the left–hand side of (A.15) equal to zero and rearranging gives $\mathbf{y} = -\hat{A}\mathbf{x}^{ss}$. Using this expression we can eliminate $\mathbf{y}$ from (A.15) by defining $\mathbf{x}' = \mathbf{x} - \mathbf{x}^{ss}$, to get

$$d\mathbf{x}'/dt = \hat{A}\mathbf{x}'. \tag{A.19}$$

This has the same form as (A.15) with $\mathbf{y} = \mathbf{0}$, but now this equation concerns deviations, or perturbations, from the steady–state solution. In the next section we will show how to solve equations of this type using simple algebra.

## A.3.1   Solution of Systems of Linear ODEs

The simplest way to solve an equation like (A.15) is to use the component form of the equation to obtain a new equation that is of second order in time. For simplicity we focus on the special case that $\mathbf{y} = \mathbf{0}$ and introduce the notation $\dot{x}_1$ for the first time derivative and $\ddot{x}_1$ for the second, so that

$$\dot{x}_1 = a_{11}x_1 + a_{12}x_2, \tag{A.20}$$

$$\dot{x}_2 = a_{21}x_1 + a_{22}x_2. \tag{A.21}$$

Thus differentiating both sides of (A.20) with respect to time gives

$$
\begin{aligned}
\ddot{x}_1 &= a_{11}\dot{x}_1 + a_{12}\dot{x}_2 \\
&= a_{11}\dot{x}_1 + a_{12}(a_{21}x_1 + a_{22}x_2) \\
&= a_{11}\dot{x}_1 + a_{12}a_{21}x_1 + a_{22}(a_{12}x_2) \\
&= a_{11}\dot{x}_1 + a_{12}a_{21}x_1 + a_{22}(\dot{x}_1 - a_{11}x_1) \\
&= a_{11}\dot{x}_1 + a_{22}\dot{x}_1 - a_{11}a_{22}x_1 + a_{12}a_{21}x_1,
\end{aligned}
\tag{A.22}
$$

where in rewriting the right–hand side we have first used (A.21) to replace $\dot{x}_2$ and then used (A.20) to eliminate the term $a_{12}x_2$. Using the last equality in (A.22) and the definitions of the  trace and determinant of $\hat{A}$ in Section A.1 gives a second–order equation for $x_1$:

$$\ddot{x}_1 - \left(tr\hat{A}\right)\dot{x}_1 + \left(det\hat{A}\right)x_1 = 0. \tag{A.23}$$

Using similar manipulations, an identical second order equation can be derived for $x_2$:

$$\ddot{x}_2 - \left(tr\hat{A}\right)\dot{x}_2 + \left(det\hat{A}\right)x_2 = 0. \tag{A.24}$$

To solve (A.23), we try the exponential function $x_1(t) = c\exp(\lambda t)$ $(c \neq 0)$. Substituting this into the left–hand side of (A.23) gives

$$c\lambda^2 \exp(\lambda t) - c\lambda \exp(\lambda t)tr\hat{A} + c\exp(\lambda t)det\hat{A}. \tag{A.25}$$

Therefore, $c \exp(\lambda t)$ is a solution to (A.23) if

$$\lambda^2 - (tr\hat{A})\lambda + det\hat{A} = 0. \tag{A.26}$$

This is called the *characteristic equation* of the matrix $\hat{A}$. It is a quadratic equation in $\lambda$ with the well-known solution

$$\lambda_\pm = \frac{tr\hat{A} \pm \left(\left(tr\hat{A}\right)^2 - 4det\hat{A}\right)^{1/2}}{2}. \tag{A.27}$$

For example, if $\hat{A} = \begin{pmatrix} 1 & -1 \\ 3 & 6 \end{pmatrix}$, then $tr\hat{A} = 7$, $\det\hat{A} = 9$, and $\lambda_\pm = \left(7 \pm \sqrt{13}\right)/2$. As long as $\lambda_+ \neq \lambda_-$, then the solution to the characteristic equation gives two independent solutions to (A.23). In this case, because (A.23) is linear, it is easy to verify that the sum of these two solutions, $c_+ \exp(\lambda_+ t) + c_- \exp(\lambda_- t)$, is also a solution.

The argument of the square root in (A.27) is the *discriminant* of the matrix $\hat{A}$ defined in Section A.1. As long as $disc\hat{A} \neq 0$, then it is clear from (A.27) that there are two independent solutions for $x_1(t)$. Using the result in (A.23), it follows that the solution for $x_2(t)$ also has the same form. We write this concisely as

$$x_i(t) = b_{i1} \exp(\lambda_+ t) + b_{i2} \exp(\lambda_- t) \tag{A.28}$$

with $i = 1, 2$. The values of the constants $b_{1i}$ and $b_{2i}$ need to be chosen to satisfy the initial conditions. This is easily worked out, for example, for $x_1$. Recall that there are two initial conditions, $x_1(0)$ and $x_2(0)$, since there are two equations. Using (A.20) and (A.28) it follows that

$$x_1(0) = b_{11} + b_{12}, \tag{A.29}$$
$$\dot{x}_1(0) = a_{11}x_1(0) + a_{12}x_2(0) = b_{11}\lambda_+ + b_{12}\lambda_-. \tag{A.30}$$

Since $x_1(0), x_2(0), \lambda_+$, and $\lambda_-$ are known, (A.29) and the second equality in (A.30) provide two independent equations for the two unknowns $b_{11}$ and $b_{12}$. Solving these using elementary algebra gives

$$b_{11} = \frac{\dot{x}_1(0) - \lambda_- x_1(0)}{\lambda_+ - \lambda_-}, \tag{A.31}$$

$$b_{12} = \frac{-\dot{x}_1(0) + \lambda_+ x_1(0)}{\lambda_+ - \lambda_-}. \tag{A.32}$$

Since $\lambda_+ \neq \lambda_-$, the denominators of these equations are different from zero.

The time dependence of $x_1(t)$ is strongly dependent on the nature of the characteristic values. There are three possibilities that are determined by the sign of the discriminant and the trace. If $disc\hat{A} > 0$, then according to (A.27) the two characteristic values will be distinct real numbers, since for a matrix with real components, $tr\hat{A}$ is a real number. However, if $disc\hat{A} < 0$, then the roots will be conjugate complex numbers (if $tr\hat{A} \neq 0$), and will in fact be pure imaginary numbers (if $tr\hat{A} = 0$).

When the characteristic values are complex, (A.28) can be expressed in terms of sines, cosines, and exponentials. This follows from the representation of the exponential of a complex number $r + i\omega$ (with $i = \sqrt{-1}$) as

$$\exp(r + i\omega) = \exp(r)\exp(i\omega) = \exp(r)(\cos(\omega) + i\sin(\omega)). \qquad (A.33)$$

If we express the characteristic values in this fashion as $\lambda_\pm = r \pm i\omega$, then it is not difficult to show using (A.28)–(A.32) that

$$x_1(t) = \exp(rt)\left(x_1(0)\cos(\omega t) - \frac{(\dot{x}_1(0) - rx_1(0))}{\omega}\sin(\omega t)\right). \qquad (A.34)$$

Straightforward differentiation of this expression verifies that it satisfies the initial conditions and that it is identical to the expression in (A.28).

The solution to (A.23) is slightly different when $disc\hat{A} = 0$. In this case, according to (A.27) $\lambda_+ = \lambda_-$, and there is only a single characteristic value $\lambda = tr\hat{A}/2$. In this case, in addition to $c\exp(\lambda t)$ there is a second solution to (A.23), which is $c't\exp(\lambda t)$. This can be verified using the facts that $tr\hat{A} = 2\lambda$ and (since $disc\hat{A} = (tr\hat{A})^2 - 4det\hat{A} = 0$) that $det\hat{A} = \lambda^2$. Substituting these expressions for $tr\hat{A}$ and $det\hat{A}$ into (A.23) gives

$$\ddot{x}_1 - 2\lambda\dot{x}_1 + \lambda^2 x_1 = 0. \qquad (A.35)$$

It is easy to show then by substitution that $c't\exp(\lambda t)$ solves (A.35). Thus when $disc\hat{A} = 0$, the general solution to (A.23) is

$$x_1(t) = b_{11}\exp(\lambda t) + b_{12}t\exp(\lambda t). \qquad (A.36)$$

Using the initial conditions

$$x_1(0) = b_{11}, \qquad (A.37)$$
$$\dot{x}_1(0) = a_{11}x_1(0) + a_{12}x_2(0) = \lambda b_{11} + b_{12}, \qquad (A.38)$$

it is easy to show that in this case

$$b_{11} = x_1(0), \qquad (A.39)$$
$$b_{12} = \frac{a_{11} - a_{22}}{2}x_1(0) + a_{12}x_2(0). \qquad (A.40)$$

Although matrices with a vanishing discriminant are not typical, it is easy to construct specific examples, e.g., $\hat{A} = \begin{pmatrix} -3 & 5 \\ 0 & -3 \end{pmatrix}$. For this matrix $tr\hat{A} = -6$, $det\hat{A} = 9$, $disc\hat{A} = 0$, and $\lambda = -3$, and the solution for $x_1$ is easily found from (A.36)–(A.40) to be

$$x_1(t) = (x_1(0) + 5tx_2(0))\exp(-3t). \qquad (A.41)$$

## A.3.2 Numerical Solutions of ODEs

Although we have characterized the solutions to (A.20)–(A.21) analytically, it is just as easy to solve them numerically by creating a system of ODEs in a numerical solver.

Consider the matrix $\hat{A} = \begin{pmatrix} 1 & -1 \\ 3 & 6 \end{pmatrix}$, with characteristic values $\lambda_\pm = (7 \pm \sqrt{13})/2 = $
5.31 and 1.70. Since these values are distinct, we know that the solution is a sum of
two exponentials with positive exponents. Thus as long as $x_1(0) \neq 0$ *and* $x_2(0) \neq 0$, the
magnitude of $x_1$ and $x_2$ will increase exponentially with time. This is shown in Figure
A.1A, where the solution generated using a numerical package is plotted. Notice that
$x_1$ rapidly declines, whereas $x_2$ increases even more rapidly. The difference is due to
the coefficients of the two exponentials, which can be calculated explicitly from the
formulae in (A.31) and (A.32).

Figure A.1B shows another solution for the matrix $\hat{A} = \begin{pmatrix} -1 & -1 \\ 5 & 1 \end{pmatrix}$ for which

$disc\hat{A} = -16$ and therefore $\lambda_\pm = \pm 4i$. Since the characteristic values are imaginary,
(A.34) shows that the solution is a sum of sines and cosines, as shown in the figure. The
matrix $\hat{A} = \begin{pmatrix} -2 & -1 \\ 4 & 1 \end{pmatrix}$, on the other hand, has complex solutions with a real part

equal to $-\frac{1}{2}$. Thus the solution for this matrix will be a sum of sines and cosines multiplied by $\exp(-t)$. Numerical solution of the equations leads to the damped oscillations
shown in Figure A.1C.

## A.3.3   Eigenvalues and Eigenvectors

The characteristic values of $\hat{A}$ are also the *eigenvalues* corresponding to the *eigenvectors*
of the matrix. An eigenvector $e_\lambda$ of $\hat{A}$ has the property that

$$\hat{A}e_\lambda = \lambda e_\lambda, \tag{A.42}$$

where $\lambda$ is a number called the eigenvalue. In other words, the matrix $\hat{A}$ transforms an
eigenvector into a constant multiple of the eigenvector. This equation can be rewritten
in component form as

$$(a_{11} - \lambda)e_{1\lambda} + a_{12}e_{2\lambda} = 0,$$
$$a_{21}e_{1\lambda} + (a_{22} - \lambda)e_{2\lambda} = 0. \tag{A.43}$$

The only way to have a nonzero solution to this equation for $e_\lambda$ is that the determinant
of the coefficients on the left-hand side of (A.43) vanishes, i.e.,

$$det\hat{A} = \begin{pmatrix} a_{11} - \lambda & a_{12} \\ a_{21} & a_{22} - \lambda \end{pmatrix} = 0. \tag{A.44}$$

Expanding the determinant, one obtains the characteristic equation (A.26), which
shows that the eigenvalues are the same as the characteristic values of the matrix.

The eigenvectors of a matrix are defined only up to a multiplicative constant, since
if $e_\lambda$ satisfies (A.42), then so does $ce_\lambda$. As long as $\lambda_+ \neq \lambda_-$, then it is not difficult to verify

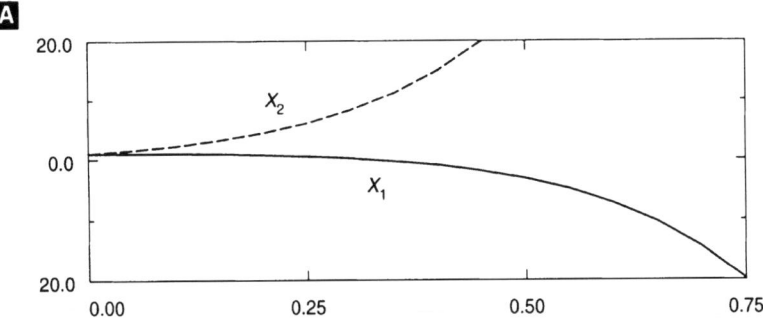

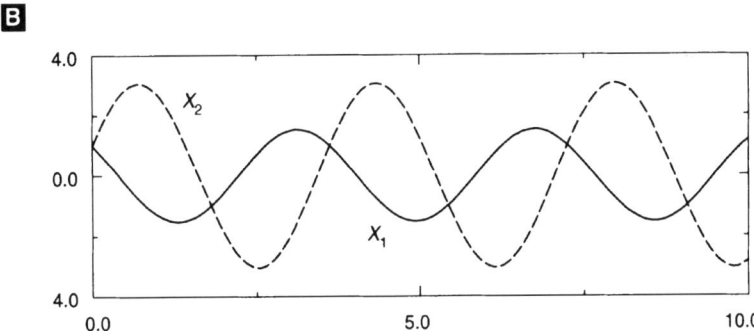

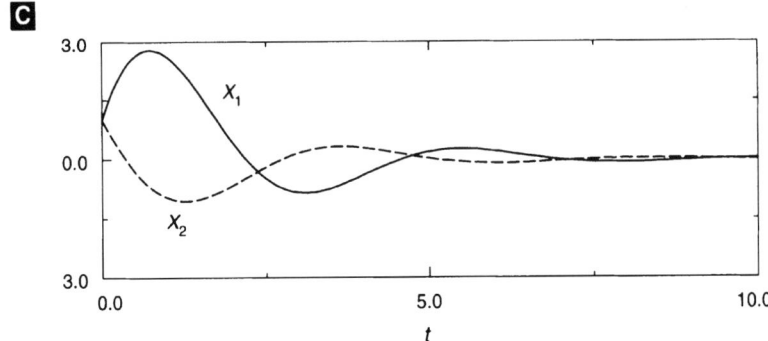

**Figure A.1** Solution of 2 × 2 linear equations using a numerical solver. Panels (A)–(C) give the time course of the solutions for the three matrices described in the text.

that the eigenvectors are given by the simple formula

$$\mathbf{e}_\lambda = \begin{pmatrix} 1 \\ (\lambda - a_{11})/a_{12} \end{pmatrix}.$$ 

(A.45)

For example, for the matrix $\hat{A} = \begin{pmatrix} 1 & 1 \\ 2 & 1 \end{pmatrix}$ $tr\hat{A} = 2$, $det\hat{A} = -1$, and $\lambda_\pm = 1 \pm \sqrt{2}$.

Applying (A.45), the eigenvectors are

$$\mathbf{e}_+ = \begin{pmatrix} 1 \\ \sqrt{2} \end{pmatrix} \text{ and } \mathbf{e}_- = \begin{pmatrix} 1 \\ -\sqrt{2} \end{pmatrix}. \tag{A.46}$$

A useful property of eigenvectors of $\hat{A}$ is that if $\mathbf{e}_\lambda$ is the initial condition for (A.20) and (A.21), then the solution is

$$\mathbf{x}(t) = \exp(\lambda t)\mathbf{e}_\lambda. \tag{A.47}$$

This result can be verified by differentiating the right–hand side of (A.47) to get $\dot{\mathbf{x}}(t) = \lambda \exp(\lambda t)\mathbf{e}_\lambda$ and then noticing that (A.42) implies that

$$\hat{A}\mathbf{x}(t) = \hat{A}\exp(\lambda t)\mathbf{e}_\lambda = \exp(\lambda t)\hat{A}\mathbf{e}_\lambda = \lambda \exp(\lambda t)\mathbf{e}_\lambda, \tag{A.48}$$

which shows that $\mathbf{x}(t)$ solves the equations. We apply this result in the following sections.

## A.4 Phase Plane Analysis

Obtaining a "solution" to first–order ODEs means that you have expressed all of the dependent variables as functions of time. In the case of the $2 \times 2$ linear equations in Section A.3, this means that we have the time series for $x_1$ and $x_2$. A great deal can be learned about these solutions by plotting the dependent variables as a function of time as done in Figure A.1. However, there are other ways of plotting solutions that give additional insight. For example, one can plot $\dot{x}_1$ versus time, or some function of $x_1$ and $x_2$ versus time. Perhaps the most useful plot is a *phase plane* plot, in which $x_2$ is plotted versus $x_1$ with time serving only as a parameter, as shown in Figure A.2 for the numerical solutions shown in Figure A.1. This type of plot represents the *trajectory* of the solution, just as the arc of a baseball thrown in the air is a trajectory in three dimensional space.

Technically, the phase plane (or phase space for more than two variables) is a Cartesian plane with coordinates $(x_1, x_2)$. Since the initial condition for the ODEs is arbitrary, any one of these points could be the initial point of a trajectory like those in Figure A.2. Continuing the analogy of phase space trajectories to the trajectory of a baseball, it makes sense to associate a *velocity* with the trajectory that goes through a point in phase space. This can be done directly using the differential equations, since the right–hand sides of the equations are explicit expressions for $\dot{x}_1$ and $\dot{x}_2$ as functions of $x_1$ and $x_2$. Thus for the matrix $\hat{A} = \begin{pmatrix} -2 & -1 \\ 4 & 1 \end{pmatrix}$ that gives rise to the trajectory in Figure A.2C, the $x_1$ component of the velocity at the point $(x_1, x_2)$ is $-2x_1 - x_2$, whereas the $x_2$ component of the velocity is $4x_1 + x_2$. For the initial point $(0.5, 0.5)$ of the

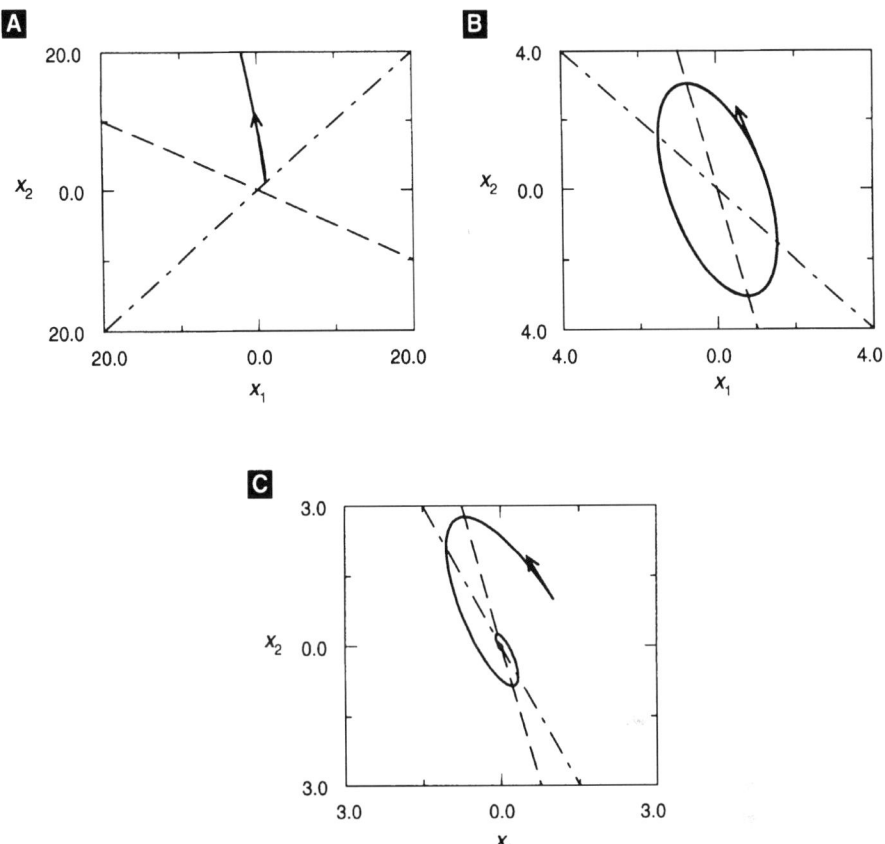

**Figure A.2** The three solutions in Figure A.1(A)–(C) represented here in phase plane plots in corresponding (A)–(C). The arrow represents the direction of the initial point on the trajectory, which is given by the full line. The dashed line is the $x_2$ nullcline, the broken dashed line is the $x_1$ nullcline, and their intersection is the steady state, which is unstable in (A), marginally stable in (B), and stable in (C).

trajectory in Figure A.2C the velocity vector at that point has components $(-1.5, 2.5)$. In the figure, the head of the arrow on the velocity vector indicates its direction, and the length is proportional to its magnitude. Just as the velocity of a baseball is parallel to its trajectory, so is the velocity vector in phase space parallel to its trajectory.

There are a number of important curves and points in the phase plane that are defined by the differential equations. *Isoclines* are lines in the phase plane where the time rate of change of a variable is constant. For example, for the matrix in the previous paragraph, the isoclines for $x_1$ are defined by $c = -2x_1 - x_2$, i.e., $x_2 = -2x_1 + c$, and the isoclines for $x_1$ are given by $x_2 = -4x_1 + c$, where $c$ is a constant. A particularly useful isocline is the *nullcline*, for which the time rate of change is zero, i.e., $c = 0$. So for this special case the nullclines are given by the straight lines through the origin, $x_2 = -2x_1$ and $x_2 = -4x_1$, shown in Figure A.2C. It is straightforward to show that the nullclines for the general $2 \times 2$ linear equations (A.17)–(A.18) are also straight lines. Since $\dot{x}_1 = 0$ on the $x_1$ nullcline, $x_1$ cannot decrease if the trajectory crosses the nullcline from the right and cannot increase if the trajectory crosses it from the left. This means, as can be verified by looking at Figure A.2, that the trajectory must cross the $x_1$ nullcline perpendicular to the $x_1$ axis. Similary, the trajectory crosses the $x_2$ nullcline perpendicular to the $x_2$ axis.

*Steady states* are defined as points in the phase space at which both $\dot{x}_1 = 0$ and $\dot{x}_2 = 0$. These points, which are also known as *singular points, equilibrium points*, or *stationary points*, have the property that neither variable changes as a function of time. They are determined graphically by the intersection of the nullclines. However, just because the variables do not change in time at a steady state does not mean that trajectories starting from nearby points will end up at the steady state. Three different situations are illustrated in Figure A.2. In panel A the steady state is at the origin, $(0, 0)$. However, the trajectory starting at $(0.5, 0.5)$ grows without bound. In panels B and C the steady states are also at the origin, but the trajectory in B circles the origin periodically, whereas in C it spirals into the steady state.

## A.4.1   Stability of Linear Steady States

As we saw in the preceding section, a steady state may or may not be an *attractor* for nearby trajectories; i.e., just because an initial condition is close to the steady state, it does not mean that after a time the trajectory will approach the steady state. However, when this is the case, the steady state is said to be *stable and attractive* or *asymptotically stable*. Three qualitatively different behaviors near steady state are illustrated by the solutions of the linear ODEs in Figure A.2. The matrix for the ODEs in panel A has positive eigenvalues, and the trajectory is repelled, not attracted, by the steady state. So the steady state in panel A is asymptotically unstable. In panel B the trajectory is circular and periodically returns to the initial condition $(0.5, 0.5)$. In this case the steady state is neither attractive nor repulsive and is said to be *neutrally stable*. Note that if the steady state of a nonlinear problem is determined to be neutrally stable by finding the eigenvalues of the linearized problem, we are not able to conclude *anything* about the stability of the steady state. Neutral stability is a borderline case, and the nonlinear parts of the equations can affect the stability in either direction. In these cases, more analysis must be done to determine the stability. Finally, in panel C the trajectory spirals into the steady state, which is stable and attractive.

The attentive reader may have noticed a correlation between the eigenvalues of the three matrices represented in Figure A.2 and the stability of the steady states. Indeed, unstable states of linear equations are characterized by at least one eigenvalue with a positive real part. If, in addition, both eigenvalues are positive, as in Figure A.2A, then the state is called an *unstable node*. An asymptotically stable state like that in Figure

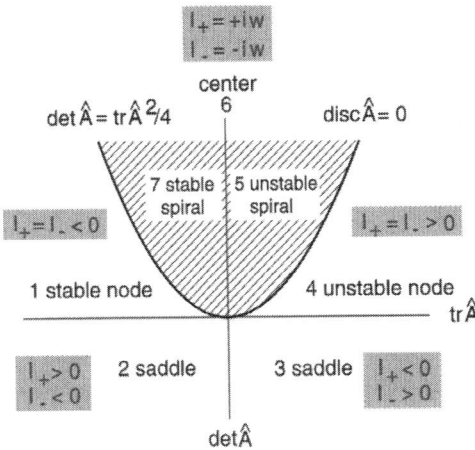

**Figure A.3** Graphical representation of the stability properties of 2 × 2 matrices. The trace is plotted on the $x$-axis and the determinant on the $y$-axis. The eigenvalues in the cross–hatched region are complex, and real elsewhere. These two regions are separated by the parabola $det\hat{A} = tr\hat{A}^2/4$ on which $disc\hat{A} = 0$. Seven regions with various degrees of stability are indicated by the sign of the eigenvalues.

A.2C, on the other hand, has negative real parts for all of its eigenvalues. Marginal (or neutral) stability occurs when the real part of a pair of eigenvalues vanishes, as is the case in Figure A.2B. A two–variable linear equation has only two eigenvalues, and a marginally stable steady state implies sinusoidal solutions. Another name for a marginal state for a two–variable system is a *center*.

Because we have at our disposal the analytical form of the solutions for $2 \times 2$ linear equations, it is possible to give a complete description of the stability of their steady states. Figure A.3 gives a graphical representation of the stability behavior of a matrix $\hat{A}$ as a function of the trace (plotted on the $x$-axis) and the determinant (plotted on the $y$-axis). The $tr\hat{A}$, $det\hat{A}$ plane in Figure A.3 is divided into seven distinct regions separated by the two axes and the parabola $det\hat{A} = tr\hat{A}^2/4$, which is the curve on which $disc\hat{A} = 0$. According to the expression for the characteristic values in (A.27), $\lambda_+ = \lambda_-$ on the parabola, and in the quadrant with $tr\hat{A} > 0$ the eigenvalues are both positive, whereas for $tr\hat{A} < 0$ both eigenvalues are negative. Marginal stability occurs when the real part of both eigenvalues is zero, i.e., when $tr\hat{A} = 0$ and $det\hat{A} > 0$, which occurs on the positive $y$-axis. Using (A.27) it is easy to verify that complex eigenvalues occur only in the cross–hatched region above the parabola (since $disc\hat{A} < 0$ there). In that region to the right of the $y$-axis, $tr\hat{A} > 0$, the eigenvalues have positive real parts, and the steady states are unstable spirals (region 5), whereas in region 7 the spirals are stable. When $tr\hat{A}$ and $det\hat{A}$ have values in regions 1 and 4, the steady state is a stable or unstable node, respectively. Below the $x$-axis (where $det\hat{A} < 0$) the steady states are unstable with the property that they have two real eigenvalues, one positive and one negative. Unstable states like this are called *saddle* points, because trajectories that start in the direction of the positive eigenvector recede from the steady state exponentially. Trajectories along the direction of the negative eigenvector move toward the steady state, also exponentially.

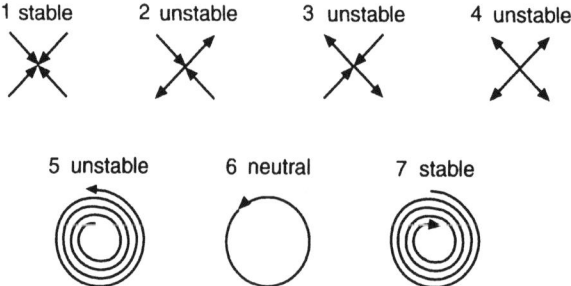

**Figure A.4** Schematic representation of phase space trajectories near the steady states in the seven regions shown in Figure A.3. Unstable states have trajectories that diverge from the steady state, whereas stable steady states have converging trajectories, and neutrally stable states are surrounded by closed trajectories. The states shown in 2 and 3 are saddle points, with both converging and diverging trajectories in the directions of the eigenvectors of the matrix.

Using Figure A.3 we can classify the qualitative behavior of phase plane trajectories for $2 \times 2$ linear ODEs based on the value of their trace and determinant. Representative trajectories are illustrated in Figure A.4 for each of the seven regions in Figure A.3. Region 1 is a stable node, and the two trajectories correspond to the directions of the two stable eigenvectors, which have velocity vectors directed at the steady state. Regions 2 and 3 are saddle points, with eigenvectors that move toward or away from the steady state, whereas the unstable node in region 4 has both eigenvectors moving away. Regions 5 and 7 have trajectories that spiral away from or toward the steady state. In region 6 the trajectories are circles, corresponding to sinusoidal oscillations.

## A.4.2 Stability of Nonlinear Steady States

What we have learned about stability of steady states for linear systems can be transferred partially to nonlinear ODEs. To be specific, let us consider a biological membrane with a gated ion channel. To do this we combine a model of ion gating with an expression that governs the membrane potential (see Chapter 1 and Chapter 2). For simplicity, we will consider only one conductance. If $n$ represents the gating variable and $V$ the voltage, then the two are coupled by the differential equations

$$C dV/dt = -gn(V - V_{\text{rev}}) + I_{\text{app}}, \tag{A.49}$$

$$dn/dt = -(n - n_{\infty})/\tau, \tag{A.50}$$

where $V_{\text{rev}}$ is the reversal potential. We assume that $n_{\infty}$ has the following voltage dependence:

$$n_{\infty} = \frac{1}{1 + \exp(-(V + V_{0.5})/S)} \tag{A.51}$$

with $V_{0.5}$ and $S$ positive constants. Equations (A.49) and (A.50) are both nonlinear due to the factor $n(V - V_{rev})$ in (A.49) and the voltage dependence of $n_\infty$ in (A.50).

To analyze the stability of the steady states of these equations we first must find the steady states by setting the right–hand sides of the equations equal to zero. This gives

$$gn^{ss}(V^{ss} - V_{rev}) = I_{app} \tag{A.52}$$

$$n^{ss} = n_\infty(V^{ss}), \tag{A.53}$$

which can be written as a single nonlinear equation to solve for $V^{ss}$:

$$\frac{I_{app}}{g} = \frac{V^{ss} - V_{rev}}{1 + \exp(-(V^{ss} + V_{0.5})/S)}. \tag{A.54}$$

This equation cannot be solved in closed form, and a much simpler way to locate the steady state is graphically in the $(V, n)$ phase plane using the nullclines. Setting the left–hand sides of (A.49) and (A.50) separately equal to zero and solving for $n$ as a function of $V$ gives

$$n = \frac{I_{app}}{g(V - V_{rev})} \quad (V\text{-nullcline}), \tag{A.55}$$

$$n = n_\infty = \frac{1}{1 + \exp(-(V + V_{0.5})/S)} \quad (n\text{-nullcline}). \tag{A.56}$$

The $V$- and $n$-nullclines are plotted in Figure A.5A, along with representative trajectories. Due to the nonlinearities in (A.49) and (A.50) the nullclines are curved rather than straight lines. This curvature influences the shape of the trajectories, which must

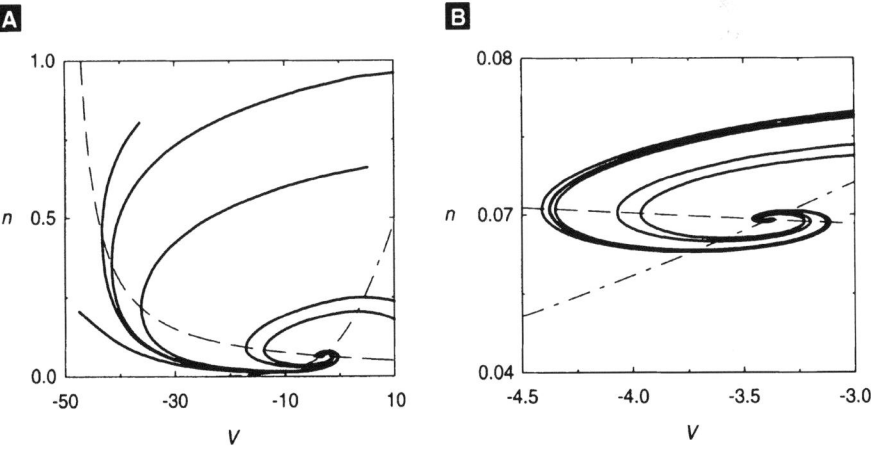

**Figure A.5**  Phase plane plots for (A.49)-(A.51) showing typical trajectories (full lines), the $V$-nullcline (dashed line), and the $n$-nullcline (broken dashed line). (B) is zoomed-in around the steady state, illustrating that the nullclines are approximately straight lines near the steady state.

cross the nullcline perpendicular to the axis of the variable. Close to the steady state, however, both nullclines become approximately straight lines, as is seen in Figure A.5B, which is the same phase plane as in Figure A.5A, but zoomed in around the steady state.

If we restrict the initial conditions for trajectories to be close to the steady state, then the nonlinear equations are well approximated by a $2 \times 2$ linear system. This can be seen in detail if we define as new variables $x_1 = V - V^{ss}$ and $x_2 = n - n^{ss}$, the deviations of the voltage and gating variable from their steady–state values. Since the steady–state values are constants, it follows that $dx_1/dt = dV/dt$ and $dx_2/dt = dn/dt$, so that we can use (A.49) and (A.50) to obtain differential equations for $x_1$ and $x_2$. In particular, if the initial conditions are close to the steady state, then we can substitute $V = V^{ss} + x_1$ and $n = n^{ss} + x_2$ into the right–hand sides of (A.49) and (A.50) and then use a Taylor series expansion in the small deviations $x_1$ and $x_2$. Explicitly:

$$
\begin{aligned}
dx_1/dt &= \left( g(n^{ss} + x_2)(V^{ss} + x_1 - V_{rev}) + I_{app} \right)/C \\
&= [gn^{ss}(V^{ss} - V_{rev}) + I_{app}]/C + (gn^{ss}x_1 + g(V^{ss} - V_{rev})x_2)/C \\
&\quad + gx_1x_2/C, \tag{A.57} \\
dx_2/dt &= -(n^{ss} + x_2 - n_\infty(V^{ss} + x_1))/\tau \\
&= -[n^{ss} - n_\infty(V^{ss})]/\tau + (dn_\infty/dV)^{ss}x_1/\tau - x_2/\tau \\
&\quad + \text{higher–order terms in } x_1. \tag{A.58}
\end{aligned}
$$

In the second equality in both (A.57) and (A.58) the terms in square brackets vanish because of the steady–state conditions in (A.52) and (A.52); the second terms are linear in $x_1$ and $x_2$; and the third terms are quadratic or of higher order in $x_1$ and $x_2$. Thus keeping the lowest–order terms gives the linear equations

$$
dx_1/dt = (gn^{ss}/C)x_1 + (g(V^{ss} - V_{rev})/C)x_2, \tag{A.59}
$$

$$
dx_2/dt = (dn_\infty/dV)^{ss}x_1/\tau - x_2/\tau. \tag{A.60}
$$

Once the elements of the matrix of this $2 \times 2$ linear equation have been evaluated, the behavior of the solution in a neighborhood of the steady state can be evaluated. This type of linear analysis, which gives information only about trajectories nearby the steady state, is called *linear stability analysis*.

The trajectories in Figure A.5B make it clear that the steady state is asymptotically stable, and according to the catalogue of possibilities in Figure A.4, the steady state is a stable spiral. It is also possible to find the steady states numerically and, in addition, determine the stability of the steady state by numerical evaluation of the eigenvalues. Combining the analytical tools developed in this chapter with the numerical tools available in various software packages, we are ready to explore the dynamics of a variety of cellular and neural dynamical systems in the remaining chapters.

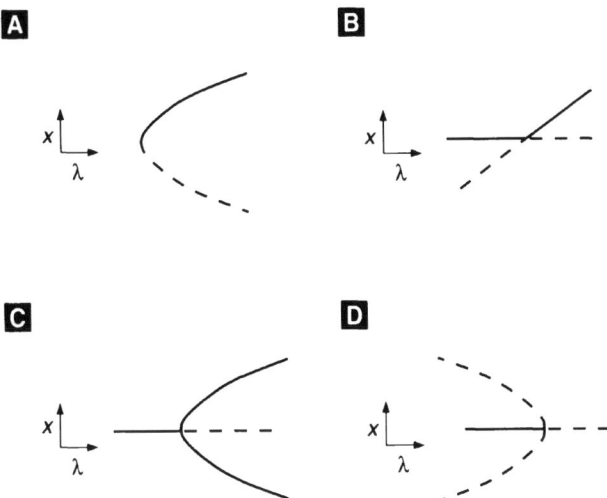

**Figure A.6** Bifurcations of new fixed points: (A) saddle–node bifurcation, (B) transcritcial bifurcation, (C) supercritical pitchfork bifurcation, (D) subcritical pitchfork bifurcation. Stable fixed points are solid, and unstable are dashed.

# A.5 Bifurcation Theory

In many systems of differential equations there are parameters that we would like to vary. As these parameters vary, we want to know whether the solutions to the equations remain similar in nature. For example, as current is injected into a cell, we want to know if the cell will remain at rest or whether some other phenomena that are qualitatively different will take place. The changes in the qualitative nature of solutions to differential equations as a parameter varies is called bifurcation theory. In this section we will review simple bifurcations from equilibrium of ordinary differential equations. Bifurcation from equilibrium solutions is intimately related to the stability of equilibria, a subject described earlier in this chapter. Suppose that we have found an equilibrium solution to a system of differential equations and study its stability as some relevant parameter varies. The stability is determined from the eigenvalues of the linearized system. There are two simple ways that stability can change as a parameter varies: (i) A real negative eigenvalue can cross through zero and become positive; (ii) a pair of complex conjugate eigenvalues with negative real parts crosses through the imaginary axis and becomes a pair of complex eigenvalues with positive real parts. In a fully nonlinear system these changes in stability will often lead to the appearance of new solutions to the differential equations. Because these are new branches of solutions that were not there previously, the system has undergone a qualitative change in behavior.

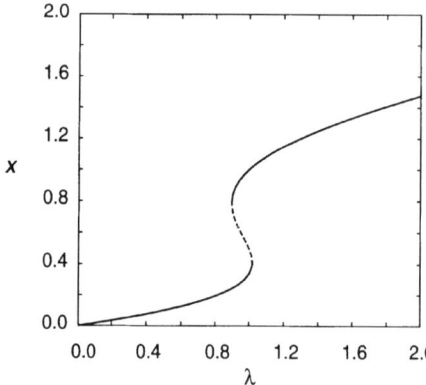

**Figure A.7** Numerically computed bifurcation diagram for the autocatalytic chemical model. Stable fixed points are solid, and unstable are dashed.

## A.5.1 Bifurcation at a Zero Eigenvalue

Consider the differential equation

$$\frac{dx}{dt} = \lambda - x^2, \tag{A.61}$$

where $\lambda$ is a parameter. For $\lambda < 0$ there are no real equilibria. However, if $\lambda > 0$, then there are two equilibrium solutions, $x = \pm\sqrt{\lambda}$. Consider the case $\lambda > 0$. The linearization about the positive fixed point is $-2\sqrt{\lambda}$. Thus, it is a stable fixed point. Note that as $\lambda$ tends to zero the eigenvalue of this $1 \times 1$ matrix goes to zero. Any time an eigenvalue of the linearization around an equilibrium point crosses zero, we can expect to see more than one fixed point in the neighborhood of the parameter. The graph of the equilibrium solution against the parameter along with the stability information is called a *bifurcation diagram*. Figure A.6A shows the bifurcation diagram for (A.61). This type of bifurcation is called a saddle node. The autocatalytic chemical model

$$\frac{dx}{dt} = \lambda - 6x + \frac{10x^2}{1 + x^2}$$

has two saddle-node bifurcations as the input $\lambda$ increases from 0. For $0 < \lambda < 0.9$ there is a single equilibrium point. At $\lambda \approx 0.9$ a new pair of equilibria appear at $x \approx 0.8$. As $\lambda$ continues to increase these new equilibria drift apart, and at $\lambda \approx 1.02$ the leftmost equilibrium merges with the middle one and disappears at $x \approx 0.4$. We can use a numerical package to draw a complete bifurcation diagram of this. Figure A.7 illustrates the complete bifurcation diagram. Note the two saddle-node bifurcations; for $\lambda$ between these two values there are three equilibria (two stable and one unstable), while for $\lambda$ outside the two values there is a unique stable equilibrium point. Techniques from nonlinear analysis can be used to show that every saddle-node bifurcation (no matter what the dimension of the system) is equivalent and can be transformed into (A.61).

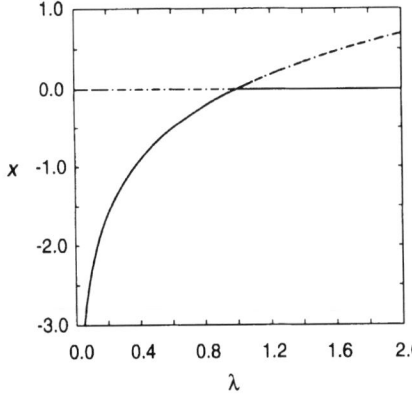

**Figure A.8** Numerically computed bifurcation diagram for the example transcritical bifurcation. Stable fixed points are solid, and unstable are dashed.

Consider next the differential equation

$$\frac{dx}{dt} = \lambda x - x^2. \tag{A.62}$$

In some model systems there is always a "trivial" equilibrium point, no matter what the parameter is. (In this case, 0 is always a solution.) For $\lambda < 0$, $x = 0$ is a stable equilibrium, and for $\lambda > 0$ it is unstable. The equilbrium point $x = \lambda$ is unstable (stable) for $\lambda < 0$ ($\lambda > 0$). Thus as $\lambda$ crosses zero the two fixed points "exchange stability." This is called a *transcritical* or *exchange of stability* bifurcation. Figure A.6B illustrates this bifurcation. For example, consider the system

$$\frac{dx}{dt} = x(1 - \lambda y), \quad \frac{dy}{dt} = e^{-x} - y.$$

Clearly, one fixed point is $(0, 1)$, and the Jacobian matrix for the linearization about this point is

$$J = \begin{pmatrix} 1 - \lambda & 0 \\ -1 & -1 \end{pmatrix}.$$

The eigenvalues are $-1$ and $-1 + \lambda$. Thus at $\lambda = 1$ we expect that there could be a bifurcation. It is not a saddle node since there always exists the trivial equilibrium $(0, 1)$. Since there are no additional symmetries in the problem (see below), it is likely a transcritical bifurcation. The diagram is shown in Figure A.8. As with the saddle-node bifurcation, all transcritical bifurcations can be transformed into (A.62) near the bifurcation.

Many biological and chemical systems are characterized by symmetries. In this case, the behavior as parameters vary is analogous to

$$\frac{dx}{dt} = x(\lambda \pm x^2). \tag{A.63}$$

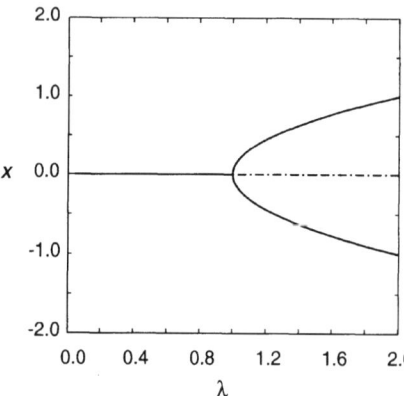

**Figure A.9** Numerically computed bifurcation diagram for the coupled system showing a pitchfork bifurcation. Stable fixed points are solid, and unstable are dashed.

As with the transcritical bifurcation, $x = 0$ is always a solution to this problem. For $\lambda < 0$, the fixed point 0 is stable, and for $\lambda > 0$ this trivial fixed point loses stability. At $\lambda = 0$ the linearized system has a zero eigenvalue. There can be *two* additional solutions depending on $\lambda$. Unlike the two bifurcations we previously described, the sign of the nonlinearity is important in this one. If we take the negative sign in (A.63), then the diagram in Figure A.6C is obtained. The new solutions are $x = \pm\sqrt{\lambda}$; they are both stable, and they occur for $\lambda > 0$. The branches open in the same direction as the trivial fixed point loses stability. This bifurcation is called a *supercritical pitchfork bifurcation*. If we take instead the positive sign for the nonlinearity in (A.63), then there are two solutions $x = \pm\sqrt{-\lambda}$, and they occur for $\lambda < 0$. As can easily be shown, they are both unstable. This is called a *subcritical pitchfork bifurcation*, since the branches open in the direction opposite from the change of stability of the trivial equilibrium point.

For example, consider the simple coupled system

$$\frac{dx}{dt} = -x + \lambda \frac{y}{1+y^2}, \quad \frac{dy}{dt} = -y + \lambda \frac{x}{1+x^2}.$$

It is easy to see that $x = y = 0$ is always a fixed point and that it is stable as long as $\lambda < 1$. At $\lambda = 1$ the Jacobian matrix has a zero eigenvalue, so we expect a bifurcation to occur. Figure A.9 shows that it is a supercritical pitchfork bifurcation. Every system that has a pitchfork bifurcation can be transformed into (A.63) near the bifurcation point. Unlike the saddle-node and the transcritical bifurcations, the details of the nonlinearity are crucial for determing the stability of the new branches of solutions.

## A.5.2  Bifurcation at a Pair of Imaginary Eigenvalues

Limit cycles and periodic solutions are extremely important in physiology. Thus, one is often interested in whether or not they occur in a given system. Unlike fixed points that can be found exactly or graphically, it is much more difficult to determine whether or not there are limit cycles in a system. There is one method that is arguably the best and perhaps only systematic method of finding parameters where there may be periodic solutions in any system of differential equations. The existence of periodic solutiuons emanating from a fixed point is established from the Hopf bifurcation theorem, which we now state.

**Hopf bifurcation theorem.** *Suppose that $X' = F(X, \lambda)$ has an isolated fixed point $X_0(\lambda)$. Let $A(\lambda)$ be the linearized matrix about this fixed point. Suppose that the matrix $A$ has a pair of complex conjugate eigenvalues $\alpha(\lambda) \pm i\omega(\lambda)$. Suppose the following conditions hold for some $\lambda_0$:*

1. $\alpha(\lambda_0) = 0$;
2. $\omega(\lambda_0) = \omega_0 > 0$;
3. $\nu \equiv d\alpha(\lambda)/d\lambda|_{\lambda=\lambda_0} \neq 0$;
4. $A(\lambda_0)$ *has no other eigenvalues with zero real part.*

*Then, the system contains an isolated limit cycle for $|\lambda - \lambda_0|$ small for either $\lambda > \lambda_0$ or for $\lambda < \lambda_0$. The magnitude of the limit cycle is proportional to $\sqrt{|\lambda - \lambda_0|}$, and the frequency is close to $\omega_0$. If $\nu > 0$ and the limit cycle exists for $\lambda > \lambda_0$ or if $\nu < 0$ and the limit cycle exists for $\lambda < \lambda_0$, then it is stable. Otherwise, it is unstable.*

Thus, the best way to try to find periodic solutions in a system of differential equations is to look for parameter values where the stability of an equilibrium is lost as a complex conjugate pair of eigenvalues crosses the imaginary axis. For a two-

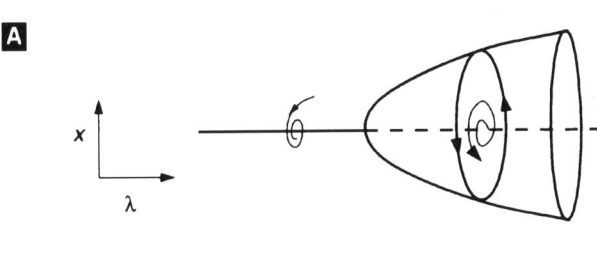

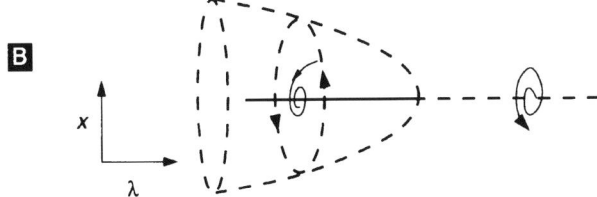

**Figure A.10** Illustration of the Hopf bifurcation of limit cycles. As the parameter changes, a branch of periodic solutions emerges from the fixed point. (A) Supercritical emergence of stable limit cycles. (B) Subcritical emergence of unstable periodic orbits.

dimensional system, this situation occurs when the determinant of $A$ is positive and the trace of $A$ changes from negative to positive. The following system illustrates the theorem:

$$\frac{dx}{dt} = \lambda x - y \pm x\left(x^2 + y^2\right),$$ (A.64)

$$\frac{dy}{dt} = \lambda y + x \pm y\left(x^2 + y^2\right).$$

Clearly, $(0,0)$ is always a fixed point. The eigenvalues of the linearization are $\lambda \pm i$, so that as $\lambda$ goes from negative to positive, there is a pair of eigenvalues with imaginary real part at $\lambda = 0$. If we convert (A.65) to polar coordinates, $x = r\cos\theta, y = r\sin\theta$ then we obtain

$$\frac{dr}{dt} = r\left(\lambda \pm r^2\right), \quad \frac{d\theta}{dt} = 1.$$

The equation for $r$ is just like (A.63), and thus the direction of bifurcation depends on the sign of the nonlinearity. We see that $r = \sqrt{\mp\lambda}$. Clearly, the solution to the $\theta$ equation is $\theta = t + C$, where $C$ is an arbitrary constant. We conclude that if the nonlinearity has a positive sign, then there is an unstable periodic solution for $\lambda < 0$ given by $(x(t), y(t)) = \sqrt{-\lambda}(\cos(t + C), \sin(t + C))$. If the nonlinearity has a negative sign, then the limit cycle exists for $\lambda > 0$, and it is stable. Figure A.10 illustrates the behavior for both cases. We remark that every system that undergoes a Hopf bifurcation can be transformed to (A.65).

As an example, we consider the Brusselator, a classic model for chemical oscillations:

$$\frac{dx}{dt} = a - (b + 1)x + x^2 y, \quad \frac{dy}{dt} = bx - x^2 y.$$

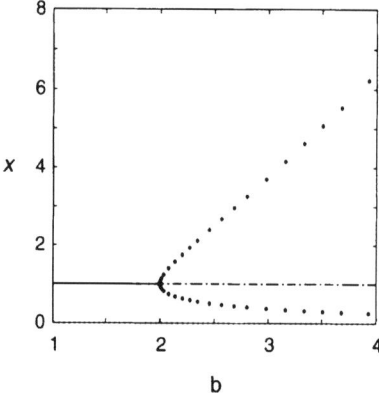

**Figure A.11** Numerically computed bifurcation diagram for the Brusselator as the parameter $b$ varies. Stable fixed points are solid, and unstable are dashed. Stable periodic orbits are filled circles.

The fixed points for this system are $(x,y) = (a, b/a)$, and the linearization about the fixed point is

$$A = \begin{pmatrix} b-1 & a^2 \\ -b & -a^2 \end{pmatrix}.$$

The determinant of $A$ is $a^2 > 0$. The trace is $b - 1 - a^2$. Thus, if $b$ is the parameter, then as $b$ increases past $1 + a^2$ there will be a Hopf bifurcation. The full bifurcation diagram is shown in Figure A.11.

## A.6  Perturbation Theory

As we have noted, nonlinear differential equations are not readily solved. In fact, even linear equations cannot always be solved in closed form if the coefficients are nonconstant in time. For this reason, one of the most powerful tools in applied mathematics is perturbation theory. In perturbation theory we look for very good approximate solutions. If some parameter in the equation is small, then a good initial approximation is to set it to zero. This can result in a simpler system of equations, which may be able to be solved. The idea is to assume that when the parameter is not zero, then we can use the simple case as a starting solution and expand the full solution in a power series in the small parameter. Typically, we need to expand the series to only one or two terms to see the dominant characteristics of the solution.

### A.6.1  Regular Perturbation

Let us first consider the general solution and then work some examples. Consider

$$\frac{dx}{dt} = f(x, \epsilon), \tag{A.65}$$

where $\epsilon$ is a small parameter. Suppose that we can solve the equation with $\epsilon = 0$; that is, we can find a solution $x_0(t)$ to

$$\frac{dx}{dt} = f(x, 0).$$

Formally, let us look for a solution of the form

$$x(t, \epsilon) = x_0(t) + \epsilon x_1(t) + \epsilon^2 x_2(t) + \cdots \tag{A.66}$$

and substitute this into (A.65). This leads to a sequence of equations,

$$\frac{dx_0}{dt} = f(x_0, 0),$$

$$\frac{dx_1}{dt} = D_x f(x_0, 0)x_1 + D_\epsilon f(x_0, 0),$$

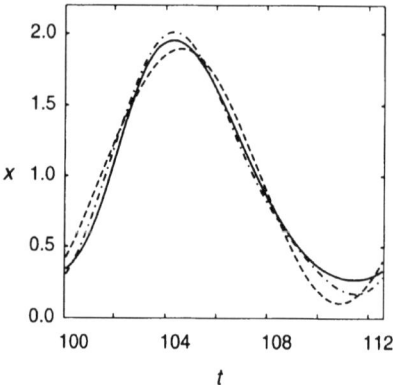

**Figure A.12** The true solution (solid lines) and the first two terms in the perturbation series for the linear time-dependent logistic equation.

$$\frac{dx_2}{dt} = D_x f(x_0, 0)x_2 + D_{x\epsilon}f(x_0, 0)x_1 + \frac{1}{2}(D_{xx}f(x_0, 0)x_1^2 + D_{\epsilon\epsilon}f(x_0, 0),$$

where $D_a$ is the derivative of $f$ with respect to $a$ evaluated at $x = x_0, \epsilon = 0$. Note that all but the first equation are linear. If the linear equation

$$\frac{dx}{dt} - D_x f(x_0, 0)x = g$$

is invertible, then we can continue this series method forever. As we will see later, when the equation is not invertible, then we run into problems, and other techniques are required. Another situation that can arise is that in which the small parameter multiplies $dx_k/dt$ for one of the variables $x_k$. We will also consider this type of perturbation below.

Let us look at a simple example. Consider the differential equation for population growth subject to periodic forcing:

$$\frac{dx}{dt} = x(1 + \epsilon \sin(\omega t) - x).$$

We are interested in the steady–state behavior; thus we want to find solutions that are periodic or constant. Obviously, $x = 0$ is a solution for any $\epsilon$, but this solution is of no interest, since it is unstable. When $\epsilon = 0$, another solution is $x = 1$. We will perturb from this solution:

$$x(t, \epsilon) = 1 + \epsilon x_1 + \epsilon^2 x_2 + \cdots.$$

Substituting this into the equation, we get

$$\frac{dx_1}{dt} = -x_1 + \sin(\omega t),$$

$$\frac{dx_2}{dt} = -x_2 + -x_1^2 + x_1 \sin(\omega t),$$

and so on. The $x_1$ equation has a periodic solution:

$$x_1(t) = \frac{\sin(\omega t) - \omega \cos(\omega t)}{1 + \omega^2}.$$

Thus, to order $\epsilon$,

$$x(t) = 1 + \epsilon \frac{\sin(\omega t) - \omega \cos(\omega t)}{1 + \omega^2}.$$

To do even better, we can go to the next order. A simple bit of calculus shows that

$$x_2(t) = \frac{2\omega^4 \cos(2\omega t) - 5\omega^3 \sin(2\omega t) - 4\omega^2 \cos(2\omega t) + \omega \sin(2\omega t)}{2 + 12\omega^2 + 18\omega^4 + 8\omega^6}.$$

Figure A.12 shows the numerical solution to the sample problem as well as the approximations $y_1(t) = 1 + \epsilon x_1(t)$ and $y_2(t) = y_1(t) + \epsilon^2 x_2(t)$ for $\epsilon = 1$ and $\omega = 0.5$. (For smaller values of $\epsilon$ and larger values of $\omega$ the approximation is much better.)

## A.6.2 Resonances

In many applied problems the general perturbation scheme described above breaks down. Typically, this arises when there is a family of solutions to the lowest–order perturbation and the linear equations that arise from higher–order perturbations are not invertible.

A typical example of this would be perturbation of eigenvalues of a matrix. For example, suppose that the matrix $A_0$ is simple and we can find the eigenvalues easily. We now ask what the eigenvalues of the matrix $B = A_0 + \epsilon A_1$ are. Suppose that $\lambda_0$ is an eigenvalue and $v_0$ is the corresponding eigenvector. That is,

$$A_0 v_0 = \lambda_0 v_0.$$

To find the eigenvalue of $B$ near $\lambda_0$ we suppose that both the eigenvalue and the eigenvector depend on $\epsilon$:

$$v(\epsilon) = v_0 + \epsilon v_1 + \cdots,$$
$$\lambda(\epsilon) = \lambda_0 + \epsilon \lambda_1 + \cdots.$$

Making the substitutions, we get

$$(A_0 - \lambda_0 I)v_1 = \lambda_1 v_0 - A_1 v_0 \equiv w. \tag{A.67}$$

There are two unknowns, $v_1$ and $\lambda_1$. However, the matrix $C = A_0 - \lambda_0 I$ is not invertible, so we cannot expect to solve this unless $\lambda_1$ is chosen so that $w$ is in the range of the matrix $C$. This condition uniquely determines the parameter $\lambda_1$. Then we can solve for $v_1$.

How do we know when a vector $w$ is in the range of a matrix $M$? The following theorem tells us precisely the conditions:

**Fredholm Alternative Theorem.** *The matrix equation*

$$My = w$$

*has a solution $y$ if and only if $w \cdot q = 0$ for every solution $q$ to the equation $M^* q = 0$. The matrix $M^*$ is the transpose complex conjugate of the matrix $M$.*

An analogous theorem holds for many other linear operators. Returning to (A.67), let $q_0$ be the solution to

$$C^T q_0 = 0, \qquad q_0 \cdot v_0 = 1.$$

Then the Fredholm alternative theorem implies that we must have

$$q_0 \cdot (\lambda_1 v_0 - A_1 v_0) = 0,$$

or

$$\lambda_1 = q_0 \cdot A_1 v_0.$$

Another classic example is to find a periodic solution to a weakly nonlinear differential equation. The  van der Pol oscillator is the standard example:

$$\ddot{x} + x = \epsilon \dot{x}(1 - x^2). \tag{A.68}$$

We seek periodic solutions to this problem. Expanding $x(t)$ in $\epsilon$,

$$x(t) = x_0(t) + \epsilon x_1(t) + \cdots,$$

and substituting into (A.68) we get

$$\ddot{x}_0 + x_0 = 0,$$
$$\ddot{x}_1 + x_1 = \dot{x}_0(1 - x_0^2).$$

The solution to the first equation is

$$x_0(t) = A \cos t + B \sin t.$$

Note that we can rewrite this as $x_0(t) = C \cos(t + \phi)$, where $\phi$ is a phase shift. Since the equation is autonomous, there is always an arbitrary phase shift, so we can set this to zero. In other words, we can assume $x_0(t) = A \cos t$, where $A$ is an arbitrary amplitude as yet unknown. The second equation is

$$\ddot{x}_1 + x_1 = -A \sin t (1 - A^2 \cos t).$$

This does not generally have a periodic solution. In fact, it is easy to solve explicitly (using a symbolic algebra program, like Maple). The key point is that the solution will be of the form

$$x_1(t) = P(t) + tQ(t),$$

where $P(t), Q(t)$ are periodic. Unless $Q(t) = 0$, the perturbed solutions $x_1(t)$ will not be periodic, so we must make $Q(t) = 0$. A simple calculation reveals that

$$Q(t) = A\frac{4 - A^2}{8} \cos t.$$

Thus, we choose $A = 2$, and to lowest order

$$x(t) = 2 \cos t.$$

## A.6.3  Singular Perturbation Theory

In many physiological systems there are vast differences in the time scales involved in the phenomena. For example, in a bursting neuron there is the period between bursts compared with the interspike interval of the action potential within a burst. Some variables may act much more slowly than other variables, while others act much more rapidly. Consider, for example, the simple linear differential equation

$$\epsilon\frac{dx}{dt} = y - x, \quad \frac{dy}{dt} = -x,$$

along with initial conditions $y = 1, x = 0$. We can easily solve this exactly using the methods of the previous section for any value of $\epsilon$. However, typically, in a real problem, the solutions are not so readily obtained. Let us suppose that we can set $\epsilon = 0$. Then we must have $0 = y - x$, or $x = y$. Thus our problem is now

$$\frac{dy}{dt} = -y, \quad y(0) = 1,$$

which has a solution $y(t) = \exp(-t)$. Furthermore, since $x = y$, we also have $x(t) = \exp(-t)$. Unfortunately, our "solution" does not satisfy the initial conditions $x(0) = 0$. Because we have reduced the order of the differential equation from 2 to 1, we cannot generally expect to find a solution for all initial conditions. This is why the problem is said to be *singular*.

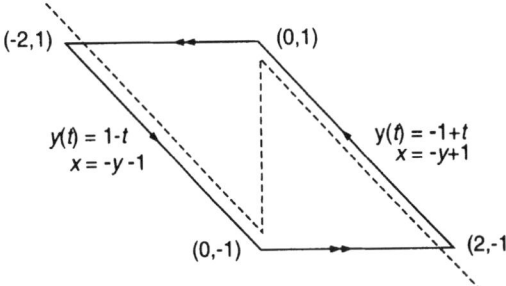

**Figure A.13**  $x$-nullclines for the relaxation oscillator example.

The way that we can fix this is to use a technique called matching. A complete description of matching goes well beyond this book, so we will just sketch this and another example. There are more examples throughout the text. The idea is to rescale time. Since the problems we are having occur at $t = t_0 = 0$, we introduce a new variable $\tau = (t - t_0)/\epsilon$. Under this change of variables our equation is

$$\frac{dX}{d\tau} = Y - X, \quad \frac{dY}{d\tau} = -\epsilon X.$$

(I have used capital letters to distinguish these solutions from the $t$-dependent solutions.) Now, we see that $Y$ is "slow" in the new time scale. Set $\epsilon = 0$. This means that $dY/d\tau = 0$, so that $Y$ is constant. The obvious constant to use is the initial value of $Y$, so we substitute $Y = 1$ into the $X$ equation:

$$\frac{dX}{d\tau} = 1 - X, \quad X(0) = 0.$$

The solution to this is $X(\tau) = 1 - \exp(-\tau)$. Thus we have two sets of solutions, $(x(t), y(t))$ and $(X(\tau), Y(\tau))$. The $(X, Y)$ solutions are valid for times near zero, and the $(x, y)$ are valid for larger times. Thus, to obtain the full solution, we add these two together and subtract the "common" part. To see what the common part is, we replace $\tau$ by $t/\epsilon$ in the $(X, Y)$ system and $t$ by $\epsilon\tau$ in the $(x, y)$ system. We take the limit as $\epsilon \to 0$ and obtain $(1, 1)$ for both sets of limits. This is the common part. Thus, our approximate solution is $(X(\tau) + x(t) - 1, Y(\tau) + y(t) - 1)$. Putting everything in terms of the original time, $t$, we obtain

$$x_c(t) = e^{-t} - e^{-t/\epsilon}, \quad y_c(t) = e^{-t}.$$

I close this section with another example that produces a singular nonlinear oscillator. The equations are

$$\epsilon\frac{dx}{dt} = -x + \text{sgn}(x) - y, \quad \frac{dy}{dt} = y + x,$$

where $\text{sgn}(x)$ is the signum function; it is $+1$ for $x > 0$ and $-1$ for $x < 0$. The nullclines are depicted in Figure A.13. For $\epsilon$ small, we expect that the solution will hug the $x$-nullcline, since we must have $-x + \text{sgn}(x) - y$ nearly zero. Setting $\epsilon = 0$ we must solve

$$-x + \text{sgn}(x) - y$$

for $x$ in terms of $y$. Unfortunately, for $y$ between $-1$ and $1$ there are two roots $x = -y \pm 1$. For the moment, let us pick $x = -y + 1$. We must have $-y + 1 > 0$, since our choice of $+1$ for $\text{sgn}(x)$ assumes that $x > 0$. Substituting this into the $y$ equation yields

$$\frac{dy}{dt} = 1,$$

so that $y(t) = y(0) + t$. Notice that as long as $y(t) < 1$, this is a valid solution, since $x > 0$. However, eventually $y(t)$ will exceed 1, and our root $x = -y + 1$ is no longer valid. So, what happens? Let $t_0$ be the time at which $y(t) = 1$. To see what happens, we must once again introduce a scaled time $\tau = (t - t_0)/\epsilon$. Then our equations are

$$\frac{dX}{d\tau} = -X + \text{sgn}(X) - Y, \quad \frac{dY}{d\tau} = \epsilon(Y + X).$$

Setting $\epsilon = 0$, this means that $Y$ must be constant. Since $y(t_0) = 1$, we will take $Y = 1$ as the constant. We must solve

$$\frac{dX}{d\tau} = -X + \text{sgn}(X) - Y, \quad X(0) = 0.$$

Note that for any $\tau > 0$, $X(\tau)$ is negative, so that $\text{sgn}(X) = -1$ and

$$X(\tau) = -2(1 - \exp(-\tau)).$$

This says that in the expanded time scale, $X(\tau)$ will drop from 0 down to $-2$. All the while, $Y$ is essentially constant at 1. Once $X$ has made the jump from 0 to $-2$, we can set $y = 1, x = -2$ and solve the $y(t)$ equation again. In this case, $x + y = -1$, since $\text{sgn}(x) = -1$, and we must solve

$$\frac{dy}{dt} = -1, \quad y(0) = 1.$$

The solution to this is $y(t) = 1 - t$. As above, this is valid only as long as $x = -y + \text{sgn}(x) = -y - 1$ is negative, that is, as long as $y(t) > -1$. Once $y(t)$ crosses $-1$, then $x$ will be positive, and we will have to jump back across to $x = +2$ keeping $y = -1$ constant again using the rescaled time. In retrospect, we see now that in the calculation on the right–hand branch (when $x > 0$) the correct initial condition for $y$ is $y(0) = -1$.

This completes the calculation of the singular trajectory. Figure A.13 illustrates this. We have the following:

$$y(t) = -1 + t, \quad x(t) = 2 - t \qquad \text{for} \quad 0 < t < 2,$$
$$y(t) = 1 - (t - 2), \quad x(t) = -2 + (t - 2) \quad \text{for} \quad 2 < t < 4,$$

in the normal time coordinates. In the scaled time coordinates, $x(t)$ jumps from 0 to $-2$ while $y = 1$, satisfying

$$x(t) = -2(1 - \exp((t - 2)/\epsilon)),$$

and from 0 to 2 while $y = -1$, satisfying

$$x(t) = 2(1 - \exp(-(t - 4)/\epsilon)).$$

The period of the oscillation is 4 to lowest order. The function $y(t)$ is continuous along the trajectory. The complete solution for $x(t)$ over one period is

$$x(t) = 2 - t - 2\exp(-t/\epsilon) \qquad \text{for} \quad 0 \le t < 2,$$
$$x(t) = t - 4 + 2\exp[(2 - t))/\epsilon)] \quad \text{for} \quad 2 \le t < 4.$$

# Suggested Readings

- *Mathematical Models in Biology*, Leah Edelstein-Keshet. This book details most of the mathematical techniques presented in this chapter, and contains a particularly good discussion of phase plane analysis, including linearization, stability, and qualitative analysis of systems of differential equations (Edelstein-Keshet 1988).
- *Applied Mathematics*, J. David Logan. This book covers a range of more advanced topics, in particular, perturbation and bifurcation theory (Logan 1997).

## A.7  EXERCISES

1. Using manipulations comparable to those used to obtain (A.23) show that $x_2$ in (A.18) also satisfies the second order equation (A.23).

2. Show by substitution that if $\mathbf{x}'$ and $\mathbf{x}$ are two different solutions to (A.19), then $c'\mathbf{x}' + c\mathbf{x}$ is also a solution.

3. Show by substition into (A.35) that $c't\exp(\lambda t)$ is a second solution to (A.23) when $disc\hat{A} = 0$. [Hint: Recall that $(t\exp(\lambda t)) = (1 + \lambda t)\exp(\lambda t)$; use this to show that $(t\ddot{\exp}(\lambda t)) = \lambda(2 + t)\exp(\lambda t)$.]

4. Show that any $2 \times 2$ matrix of the form $\hat{A} = \begin{pmatrix} a & b \\ 0 & a \end{pmatrix}$ with $a$ and $b$ arbitrary real numbers

   has $disc\hat{A} = 0$ and $\lambda = a$.

5. Use the solution (A.27) to the characteristic equation for $\hat{A}$, to show that $tr\hat{A} = \lambda_+ + \lambda_-$ and $det\hat{A} = \lambda_+ \lambda_-$.

6. Verify that the expression for the eigenvector of $\hat{A}$ given in (A.45) is correct by multiplying that expression by $\hat{A}$. Hint: You will need to use the fact that $det\hat{A} = \lambda^2 - (a_{11} + a_{22})\lambda$, which follows from (A.27).

7. Solve the general two-variable linear equations (A.17)–(A.18) numerically: Find $x_1(t)$ and $x_2(t)$ for the matrices $\hat{A} = \begin{pmatrix} 0 & -1 \\ 1 & 0 \end{pmatrix}$, $\hat{A} = \begin{pmatrix} 1 & -1 \\ 1 & 1 \end{pmatrix}$, and $\hat{A} = \begin{pmatrix} -2 & 1 \\ -3 & 1 \end{pmatrix}$ and $y_1 = y_2 = 0$. Determine the characteristic values of all three matrices and compare your numerical solutions to the solutions that you would expect based on the characteristic values. Explore how the solutions change when you change the values of $y_1$ and $y_2$.

8. Solve (A.17)–(A.18) numerically for the matrix $\hat{A} = \begin{pmatrix} 1 & -1 \\ 3 & 6 \end{pmatrix}$ and $y_1 = 1$ and $y_2 = 2$.

   Constrast your result with that in (A.41).

9. Make a phase plane plot of the solutions to the linear ODEs plotted as time series in Figure A.1A–C.

10. Show that the velocity vector for a point in phase space is parallel to the trajectory at the point. Hint: Calculate the slope of the trajectory $dx_2/dx_1$ using the ODEs.

11. Show that the nullclines for the general $2 \times 2$ linear equations (A.17) and (A.18) are linear.

12. Using the result in (A.47) verify the statement in Section A.4.1 that saddle point trajectories that start in the direction of the positive eigenvector grow away from the steady state exponentially, while those in the direction of the negative eigenvector approach the steady state exponentially.

13. Construct the solution to the following initial value problem, which arises in enzyme kinetics:

$$x' = 1 - xy, \quad \epsilon y' = -xy + 1 - y,$$

with the initial conditions $x(0) = 0, y(0) = 0$.

14. Find the periodic solution to

$$\epsilon x' = f(x) - y \quad y' = x,$$

where

$$f(x) = \begin{cases} -x - 2 & \text{for} \quad x < -1, \\ x & \text{for} \quad -1 \le x \le 1, \\ -x + 2 & \text{for} \quad x > 1. \end{cases}$$

15. Develop Taylor series for the following functions:

- $\cos(t)$ around $t = 0$

- $\ln(t)$ around $t = 1$.

- $\exp(t^2)$ around $t = 0$. Hint: Use the exponential series we have already determined.

# Solving and Analyzing Dynamical Systems Using XPPAUT

## G. Bard Ermentrout

Most of the examples and exercises in the book have been designed to be solvable with the ordinary differential equations solution and analysis package XPPAUT. One reason that we emphasize the use of XPPAUT rather than one of the other available packages is that XPPAUT is distributed at no cost and runs under both Unix and Windows environments. XPPAUT will also run under the new Macintosh operating system OSX with the appropriate Xwindows server. The second reason is that XPPAUT incorporates the bifurcation package AUTO, which is not included in other packages. The Windows version of XPPAUT, Winpp, uses a different bifurcation package, as explained below.

XPPAUT can be obtained from the web site of Bard Ermentrout, the developer. The site contains instructions for the installation of XPPAUT on various platforms, as well as a very useful tutorial. The web site is:

`http://www.math.pitt.edu/~bard/xpp/xpp.html`.

In addition, there is a full–length book describing the details of XPPAUT available (Ermentrout 2002). The tutorial in this appendix will introduce the reader to the main tools available in XPPAUT that are necessary to solve most of the exercises in this book.

# B.1 Basics of Solving Ordinary Differential Equations

## B.1.1 Creating the ODE File

Consider the simple linear differential equation system

$$\frac{dx}{dt} = ax + by,$$

$$\frac{dy}{dt} = cx + dy, \tag{B.1}$$

where $a, b, c, d$ are parameters. We will explore the behavior of this two-dimensional system using XPPAUT (even though it is easy to obtain a closed–form solution). To analyze a differential equation using XPPAUT, you must create an input file that tells the program the names of the variables and parameters, and defines the equations. By convention, these files have the file extension ode, and we will call them ODE files. Here is an ODE file for system (B.1):

```
# linear2d.ode
#
# right hand sides
dx/dt=a*x+b*y
dy/dt=c*x+d*y
#
# parameters
par a=0,b=1,c=-1,d=0
#
# some initial conditions
init x=1,y=0
#
# we are done
done
```

We have included some comments indicated by lines starting with #; these are not necessary but can make the file easier to understand. The rest of the file is fairly straight-forward. The values given to the parameters are optional; by default they are set to zero. The init statement is also optional. The minimal file for this system is

```
dx/dt=a*x+b*y
dy/dt=c*x+d*y
par a,b,c,d
done
```

In contrast to the more elaborate file, with the minimal file all parameters and intital conditions are set to zero. Use a text editor to type in the first file exactly as it is shown. Name the file linear2d.ode and save it. Note also that XPPAUT accepts other notation for equations, and you should not be surprised to see the more compact version used in Appendix C or in ODE files you might find on the XPPAUT web site or in the XPPAUT user's manual. For example, the minimal file could be written

```
x'=a*x+b*y
y'=c*x+d*y
par a,b,c,d
done
```

That's it! You have written an ODE file. The minimal steps are as follows:

- Use an editor to open a text file.
- Write the differential equations in the file; one per line.
- Use the par statement to declare all the parameters in your system. Optionally define initial conditions with the init statement.
- End the file with the statement done.
- Save and close the file.

**ODE FILE NOTES:** The equation reader is case-insensitive, so that AbC and abC are treated as identical. In statements declaring initial conditions and parameters, **do not** put spaces between the variable and the "=" sign and the number. XPPAUT uses spaces as a delimiter. Always write a=2.5 and **never** write a = 2.5.

## B.1.2    Running the Program

Run XPPAUT by typing

```
xpp linear2d.ode
```

The name of the exacutable, here xpp, might be different for your system. Use the name of your executable, along with all of the desired command line options (see on-line help for details). (*If you are using Winpp, click on the* **Winpp** *icon; then choose the file from the file selection dialog box.*)

Six windows will appear on the screen, or they may be iconified (depending on the command line options). If any of the windows appear "dead" or blank, iconify them manually and then uniconify them. Next time run XPPAUT without the -iconify command line option.

Menu commands will appear like this, ⎡Command⎤, and single–letter keyboard short-cuts will appear like this: ⎡**A**⎤. *Do not use the CapsLock key*; all shortcuts are lowercase. Every command can be accessed by a series of keystrokes. To make sure key clicks are interpreted correctly, click on the title bar of the window for which the shortcut is intended.

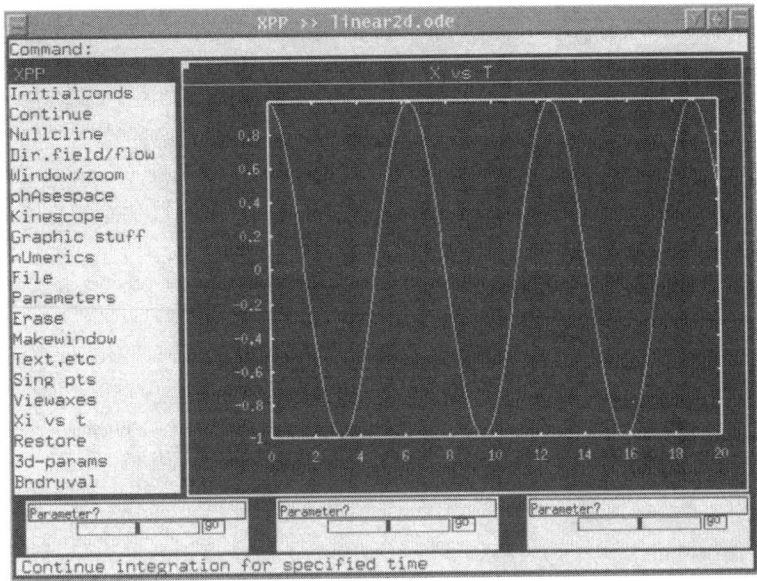

**Figure B.1**  The main XPPAUT window

## B.1.3  The Main Window

The **Main Window** contains a large region for graphics, menus, and various other regions and buttons. It is illustrated in Figure B.1. Commands are given either by clicking on the menu items in the left column with the mouse or tapping keyboard shortcuts. After a while, as you become more used to XPPAUT, you will use the keyboard shortcuts more and more. Both the full commands and the keyboard shortcuts are included here. In general, the keyboard shortcut is the first letter of the command unless there is ambiguity (such as `Nullcline` and `nUmerics`), and then, it is just the capitalized letter ($\boxed{\text{N}}$ and $\boxed{\text{U}}$, respectively). Unlike Windows keyboard shortcuts, the letter key alone is sufficient, and it is not necessary to press the Alt key at the same time. The top region of the **Main Window** is for typed input such as parameter values. The bottom of the **Main Window** displays information about various things as well as a short description of the highlighted menu item. The three little boxes with the words parameter are sliders to let you change parameters and initial data.

In addition to the **Main Window**, there are several other windows that appear. The **Equation Window**, shown in Figure B.2, allows you to see the differential equations that you are solving. We will describe the other windows as the tutorial progresses.

### Quitting the Program
To exit XPPAUT, click $\boxed{\text{File}}$ $\boxed{\text{Quit}}$ $\boxed{\text{Yes}}$ ($\boxed{\text{F}}$ $\boxed{\text{Q}}$ $\boxed{\text{Y}}$).

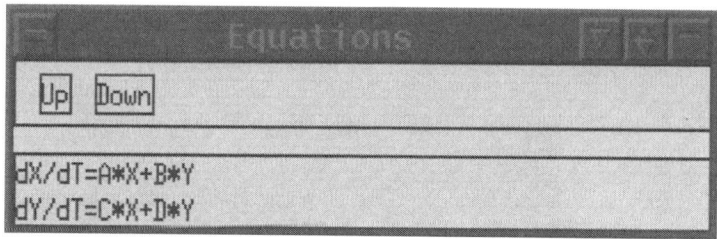

**Figure B.2**   The Equation Window.

## B.1.4   Solving the Equations, Graphing, and Plotting.

Here, we will solve the ODEs, use the mouse to select different initial conditions, save plots of various types, and create files for printing.

### Computing the Solution

In the **Main Window** you should see a box with axis numbers. The title in the window should say X vs T, which tells you that the variable X is along the vertical axis and T along the horizontal. The plotting range is from 0 to 20 along the horizontal and −1 to 1 along the vertical axis. When a solution is computed, this view will be shown. Click on $\boxed{\texttt{Init Conds}}$ $\boxed{\texttt{Go}}$ ( $\boxed{\textbf{I}}$ $\boxed{\textbf{G}}$ ) in the **Main Window**. A solution will be drawn followed by a beep. As one would expect given the differential equations, the solution looks like a few cycles of a cosine wave.

### Changing the View

To plot Y versus T instead of X, just click on the command $\boxed{\texttt{Xi vs t}}$ $\boxed{\textbf{X}}$ and choose Y by backspacing over X, typing in Y, and typing $\boxed{\textbf{Enter}}$.

   Many times you may want to plot a phase plane instead, that is, X vs. Y. To do this, click on $\boxed{\texttt{Viewaxes}}$ $\boxed{\texttt{2D}}$ ( $\boxed{\textbf{V}}$ $\boxed{\textbf{2}}$ ), and a dialog box will appear. Fill it in as follows:

| X-axis: X | Xmax: 1 |
|-----------|---------|
| Y-axis: Y | Ymax: 1 |
| Xmin: -1  | Xlabel: |
| Ymin: -1  | Ylabel: |

Click on OK when you are done. (Note that you could have filled in the labels if you had wanted, but for now, there is no reason to.) You should see a nice elliptical orbit in the window. This is the solution in the phase plane (cf. Figure B.3).

### Phase Plane Shortcuts

There is a very simple way to view the phase plane or view variables versus time. Look at the **Initial Data Window** (Figure B.4). You will see that there are little boxes next to the

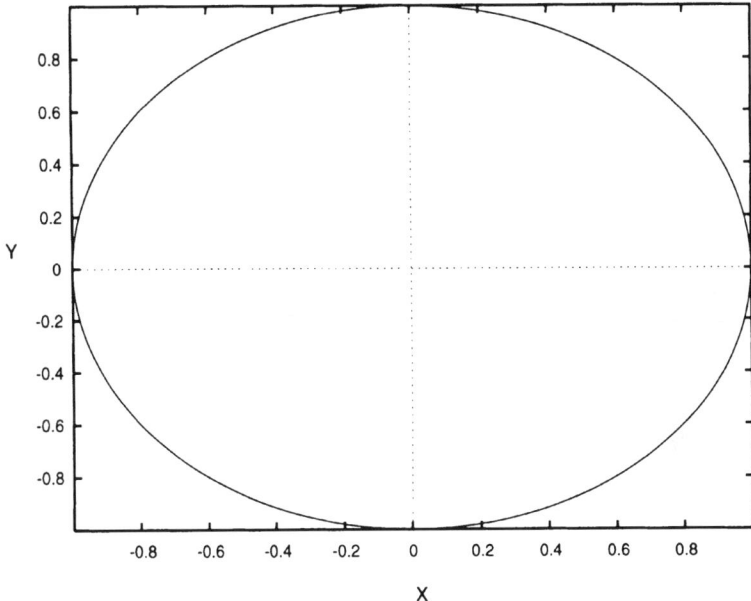

**Figure B.3**  Phase plane (X vs Y) for the linear 2D problem (B.1).

variable names. Check the two boxes next to X and Y. Then at the bottom of the **Initial Data Window**, click the XvsY button. This will plot a phase plane and automatically fit the window to contain the entire trajectory. This is a shortcut and does not give you the control that the menu command does. (For example, the window is always fit to the trajectory, and no labels are added or changed. Nor can you plot auxiliary quantities with this shortcut.) To view one or more variables against time, just check the variables you want to plot (up to 10) and click on the XvsT button in the **Initial Data Window**.

You should have a phase plane picture in the window. (If not, get one by following the above instructions or using the shortcut.) Click on $\boxed{\text{Init Conds}}\,\boxed{\text{Mouse}}$ ($\boxed{\text{I}}\,\boxed{\text{M}}$.) Use the mouse to click somewhere in the window. You should see a new trajectory drawn. This, too, is an ellipse. Repeat this again to draw another trajectory. If you get tired of repeating this, try $\boxed{\text{Init Conds}}\,\boxed{\text{mIce}}$ ($\boxed{\text{I}}\,\boxed{\text{I}}$), which, being "mice," is many mouses. Keep clicking in the window. When you are bored with this, click either outside the window or tap the escape key, $\boxed{\textbf{Esc}}$.

Click on $\boxed{\text{Erase}}$ and then $\boxed{\text{Restore}}$ ($\boxed{\textbf{E}}\,\boxed{\textbf{R}}$). Note that all the trajectories are gone except the latest one. XPPAUT stores only the latest one. There is a way to store many of them, but we will not explore that for now.

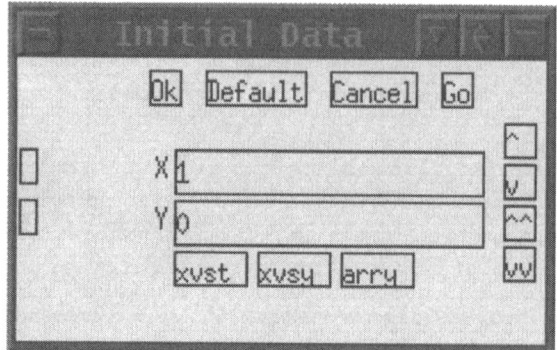

**Figure B.4** The initial conditions window.

## B.1.5  Saving and Printing Plots

XPPAUT does not directly send a picture to your printer. Rather, it creates a PostScript file that you can send to your printer. If you do not have PostScript capabilities, then you probably will have to use the alternative method of getting hard copy outlined below. (Note that Microsoft Word supports the import of PostScript and Encapsulated PostScript, but can only print such pictures to a PostScript printer. You can download a rather large program for Windows called GhostView which enables you to view and print PostScript on nonPostScript printers. Linux and other UNIX distributions usually have a PostScript viewer included.)

Here is how to make a PostScript file. Click on `Graphics` `Postscript` (`G` `P`), and you will be asked for three things: (i) Black and White or Color (ii) Landscape or Portrait; (iii) and the Fontsize for the axes. Accept all the defaults for now by just clicking `Enter`. Finally, you will be asked for a file name. The File Selector box is shown in Figure B.5. You can move up or down directory trees by clicking on the <>; choose files by clicking on them; scroll up or down by clicking on the up/down arrows on the left or using the arrow keys and the PageUp/PageDown keys on the keyboard; change the wild card; or type in a file name. For now, you can just click on `Ok` and a PostScript plot will be created and saved. The file will be called `linear2d.ode.ps` by default, but you can call it anything you want.

Once you have the PostScript file, you can type

```
lpr filename
```

on UNIX. In Windows, if your computer is hooked up to a PostScript printer, then you can print from a viewing application or type

```
copy filename lpt1
```

from the command line (if available).

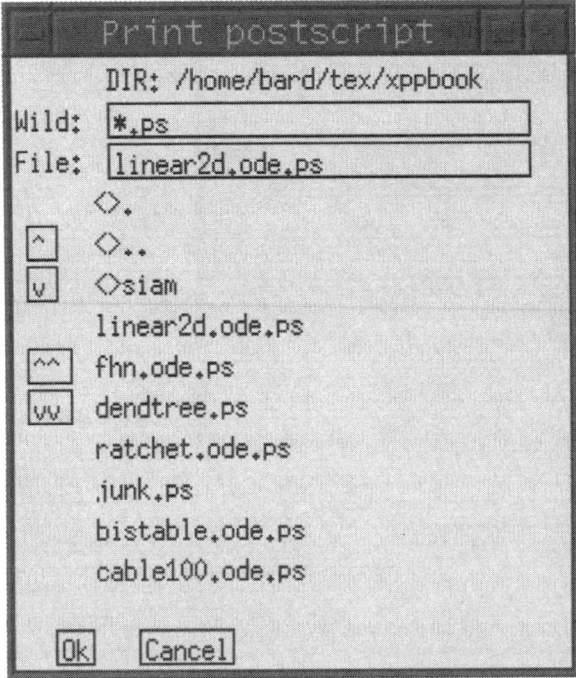

**Figure B.5** File selector.

## Other Ways to get Hard Copy

Another way to get hard copy that you can import into documents is to grab the image from the screen. In Windows, click on $\boxed{\texttt{Alt+PrtSc}}$ after making the desired window active. Paste this into the MSPaint accessory and then use the tools in Paint to cut out what you want. Pasting into Microsoft Word is useful for generating reports with added text. Alternatively, you can download a number of programs that let you capture areas of the screen. In the UNIX environment, you can capture a window using xv, an excellent utility that is free and available for most UNIX versions. All of the screen shots in this tutorial were captured with xv. Finally, you can capture the screen (or a series of screen images) with the $\boxed{\texttt{Kinescope}}\boxed{\texttt{Capture}}$ command and then write these to disk with the $\boxed{\texttt{Kinescope}}\boxed{\texttt{Save}}$ command. This produces a series of GIF files that are usable by many software packages.

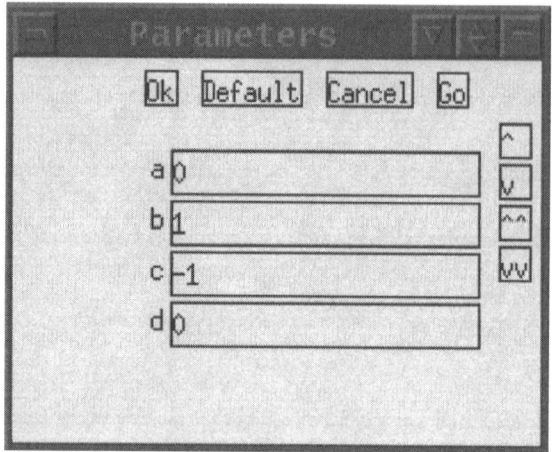

**Figure B.6** The parameter window.

## Getting a Good Window

If you have computed a solution and do not have a clue about the bounds of the graph, let XPPAUT do all the work. Click on `Window/zoom` `(F)it`, and the window will be resized to a perfect fit. The shortcut is `W` `F` and you will likely use it a lot!

## B.1.6  Changing Parameters and Initial Data

There are many ways to vary the parameters and initial conditions in XPPAUT. We have already seen how to change the initial data using the mouse. This method works for any $n$-dimensional system as long as the current view is a phase plane of two variables. Here are two other ways to change the initial data:

- From the main menu click on `Init Conds` `New` and manually input the data at the prompts. You will be prompted for each variable in order. (For systems with hundreds of variables, this is not a very good way to change the data!)
- In the **Initial Data Window** you can edit the particular variable you want to change. Just click in the window next to the variable and edit the value. Then click on the `Go` button in the **Initial Data Window**. If there are many variables, you can use the little scroll buttons on the right to go up and down a line or page at a time. If you click the mouse in the text entry region for a variable, you can use the `PageUp`, etc., keys to move around. Clicking `Enter` rolls around in the displayed list of initial conditions. The `Default` button returns the initial data to those with which the program started. If you do not want to run the simulation, but have set the initial data, *you must* click on the `Ok` button in the **Initial Data Window** for the new initial data to be recognized.

There are many ways to change parameters as well. Here are three of them:

- From the **Main Window**, click on $\boxed{\texttt{Parameters}}$. In the command line of the **Main Window**, you will be prompted for a parameter name. Type in the name of a parameter that you want to change. Click on $\boxed{\texttt{Enter}}$ to change the value and $\boxed{\texttt{Enter}}$ again to change another parameter. Click on $\boxed{\texttt{Enter}}$ a few times to get rid of the prompt.
- In the **Parameter Window** (shown in Figure B.6) type in values next to the parameter you want to change. Use the scroll buttons or the keyboard to scroll around. As in the **Initial Data Window**, there are four buttons across the top. Click on $\boxed{\texttt{Go}}$ to keep the values and run the simulation; click on $\boxed{\texttt{Ok}}$ to keep the parameters without running the simulation. Click on $\boxed{\texttt{Cancel}}$ to return to the values since you last pressed $\boxed{\texttt{Go}}$ or $\boxed{\texttt{Ok}}$. The $\boxed{\texttt{Default}}$ button returns the parameters to the values when you started the program.
- Use the little sliders (Figure B.7). We will attach the parameter d to one of the sliders. Click on one of the unused parameter sliders. Fill in the dialog box as follows:

| Parameter: d |
| Value: 0 |
| Low: -1 |
| High: 1 |

and click $\boxed{\texttt{Ok}}$. You have assigned the parameter d to one of the sliders and allowed it to range between $-1$ and 1. Grab the little slider with the mouse and move it around. Watch how d changes. Now click on the tiny $\boxed{\texttt{go}}$ button in the slider. The equations will be integrated. Move the slider some more and click on the $\boxed{\texttt{go}}$ button to get another solution.

## B.1.7   Looking at the Numbers: The Data Viewer

In addition to the graphs that XPPAUT produces, it also gives you access to the actual numerical values from the simulation. The **Data Viewer** shown in Figure B.8 has many buttons, some of which we will use later in the book. The main use of this is to look at the actual numbers from a simulation. The independent variable occupies the leftmost column, and the dependent variables fill in the remaining windows. Click on the top of the **Data Viewer** to make it the active window. The arrow keys and the $\boxed{\textbf{PageUp}}$,

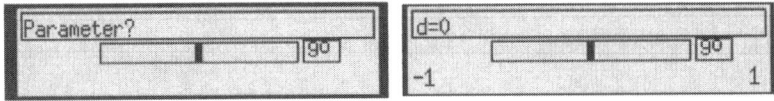

**Figure B.7**   Left: Unused parameter slider. Right: parameter slider used for d.

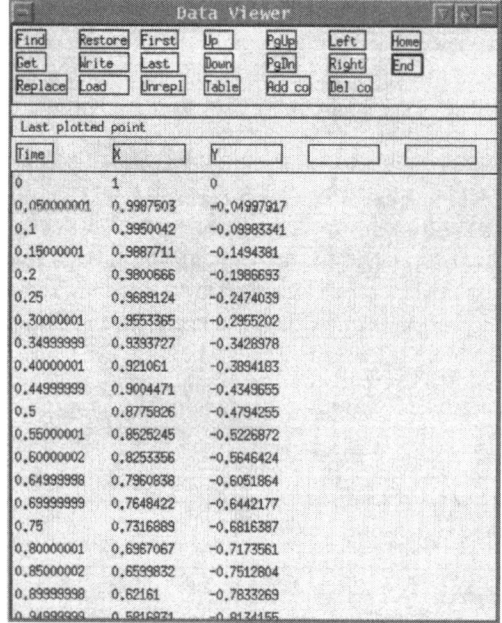

**Figure B.8**   The Data Viewer.

**PageDown** , **Home** , and **End** keys (as well as their corresponding buttons) do all the obvious things. Left and right keys scroll horizontally, a useful feature if you have many variables. Here we mention three buttons of use:

Find brings up a dialog box prompting you for the name of a column and a value. If you click on Ok , XPPAUT will find the entry that is closest and bring that row to the top. You can find the maximum and minimum, for example, of a variable.

Get loads the top line of the **Data Viewer** as initial data.

Write writes the entire contents of the browser to a text file that you specify.

## B.1.8   Saving and Restoring the State of Simulations

Often you will have a view, a set of parameters, and initial data that you want to keep. You can save the current state of XPPAUT by clicking on File Write set ( **F** **W** ) in the **Main Window**. This brings up a file selection box. Type in a file name; the default extension is .set. The resulting file is an ASCII file that is human and computer readable. The first and last few lines look like this:

```
## Set file for linear2d.ode on Fri Aug  4 13:53:31 2000
2    Number of equations and auxiliaries
4    Number of parameters
```

```
# Numerical stuff
1     nout
40    nullcline mesh

. . . . . . .


RHS etc ...
dX/dT=A*X+B*Y
dY/dT=C*X+D*Y
```

Once you quit XPPAUT, you can start it up again and then use the File Read set to load up the parameters, etc., that you saved.

Now you should quit the program. We will look at a nonlinear equation next, find fixed points, and draw some nullclines and direction fields. To quit, click on File Quit Yes (F Q Y).

## B.1.9  Important Numerical Parameters

XPPAUT has many numerical routines built into it, and thus there are many numerical parameters that you can set. These will be dealt with in subsequent sections of the book where necessary. However, the most common things you will want to change are the total amount of time to integrate and the step size for integration. You may also want to change the method of integration from the default fixed–step Runge–Kutta algorithm. To alter the numerical parameters, click on nUmerics (U), which produces a new menu. This is a top–level menu, so you can change many things before going back to the main menu. To go back to the main menu, just click on the [Esc]-exit or tap Esc. There are many entries in the numerics menu. The following four are the most commonly used:

Total sets the total amount of time to integrate the equations. (Shortcut: T.)

Dt sets the size of the time step for the fixed step size integration methods and sets the output times for the adaptive integrators. (Shortcut: D.)

Nout sets the number of steps to take before *plotting* an output point. Thus, to plot every fourth point, change Nout to 4. For the variable step size integrators, this should be set to 1.

Method sets the integration method. There are currently 13 available. (Shortcut: M.) They are described in the user manual.

When you are done setting the numerical parameters, just click on Esc-exit or tap the Esc key.

## B.1.10  Command Summary: The Basics

Initialconds Go computes a trajectory with the initial conditions specified in the **Initial Data Window** (**I** **G**).

Initialconds Mouse computes a trajectory with the initial conditions specified by the mouse. Initialconds m(I)ce lets you specify many initial conditions (**I** **M** or **I** **I**).

Erase erases the screen (**E**).

Restore redraws the screen (**R**).

Viewaxes 2D lets you define a new 2D view (**V** **2**).

Graphic stuff Postscript allows you to create a PostScript file of the current graphics (**G** **P**).

Kinescope Capture allows you to capture the current view into memory, and Kinescope Save writes this to disk.

Window/zoom (F)it fits the window to include the entire solution (**W** **F**).

File Quit exits the program (**F** **Q**).

File Write set saves the state of XPPAUT(**F** **R**).

File Read set restores the state of XPPAUT from a saved .set file (**F** **R**).

# B.2  Phase Planes and Nonlinear Equations

Here we want to solve a nonlinear equation. We will choose a planar system, since there are many nice tools available for analyzing two-dimensional systems. A classic model is the FitzHugh–Nagumo equations, which are used as a model for nerve conduction. The equations are

$$\frac{dv}{dt} = Bv(v - \beta)(\delta - v) - Cw + I_{\text{app}}, \tag{B.2}$$

$$\frac{dw}{dt} = \epsilon(v - \gamma w),$$

with parameters $I_{\text{app}}, B, C, \beta, \delta, \epsilon, \gamma$. Here we will use $I_{\text{app}} = 0, B = 1, C = 1, \beta = .1, \delta = 1, \gamma = 0.25$, and $\epsilon = .1$. Let us write an ODE file for this:

```
# Fitzhugh-Nagumo equations
dv/dt=B*v*(v-beta)*(delta-v)-Cw+Iapp
dw/dt=epsilon*(v-gamma*w)
par Iapp=0,B=1,C=1,beta=.1,delta=1,gamma=.25,epsilon=.1
@ xp=V,yp=w,xlo=-.25,xhi=1.25,ylo=-.5,yhi=1,total=100
@ maxstor=10000
done
```

We have already seen the first four lines: (i) lines beginning with a # are comments, (ii) the next two lines define the differential equations, and (iii) the line beginning with par defines the parameters and their default values. The penultimate line beginning with the @ sign is a directive to set some of the options in XPPAUT. These could all be done within the program, but this way everything is all set up for you. Details of these options are found in the user manual. For the curious, these options set the $x$-axis (xp) to be the v variable, the $y$-axis (yp) to be the w variable, the plot range to be $[-.15, 1.25] \times [-.5, 1]$, and the total amount of integration time to be 100. The last option, @ maxstor=10000, is a very useful one. XPPAUT allocates enough storage to keep 4000 time points. You can make it allocate as much as you want with this option. Here we have told XPPAUT to allocate storage for 10000 points. Type this in and save it as fhn.ode.

## B.2.1  Direction Fields

Run the file by typing xpp fhn.ode. The usual windows will pop up. One of the standard ways to analyze differential equations in the plane is to sketch the *direction fields*. Suppose that the differential equation is

$$\frac{dx}{dt} = f(x,y), \qquad \frac{dy}{dt} = g(x,y).$$

The phase plane is divided into a grid, and at each point $(x,y)$ in the grid a vector is drawn with $(x,y)$ as the base and $(x+sf(x,y), y+sg(x,y))$ as the terminal point, where $s$ is a scaling factor. This so-called direction field gives you a hint about how trajectories move around in the plane. XPPAUT lets you quickly draw the direction field of a system. Click on Dir.field/flow (D)irect Field ( **D** **D** ) and then accept the default of 10 for the grid size by clicking Enter . A bunch of vectors will be drawn on the screen, mainly horizontal. They are horizontal because $\epsilon$ is small so that there is little change in the w variable. The length of the vectors is proportional to the magnitude of the flow at each point. At the head of each vector is a little bead. If you want to have pure direction fields that do not take into account the magnitude of the vector field, just click on Dir.field (S)caled Dir. Fld ( **D** **S** ) and use the default grid size. (We prefer pure direction fields, but this is a matter of taste.)

Click on Initialconds m(I)ce to experiment with a bunch of different trajectories. Note how the vectors from the direction field are tangent to the trajectories. See Figure B.9.

## B.2.2  Nullclines and Fixed Points

As discussed in earlier chapters, a powerful technique for the analysis of planar differential equations and related to the direction fields is the use of *nullclines*. Nullclines are curves in the plane along which the rate of change of one or the other variable is zero. The $x$-nullcline is the curve where $dx/dt = 0$, that is, $f(x,y) = 0$. Similarly, the $y$-nullcline is the curve where $g(x,y) = 0$. The usefulness of these curves is that they break the plane

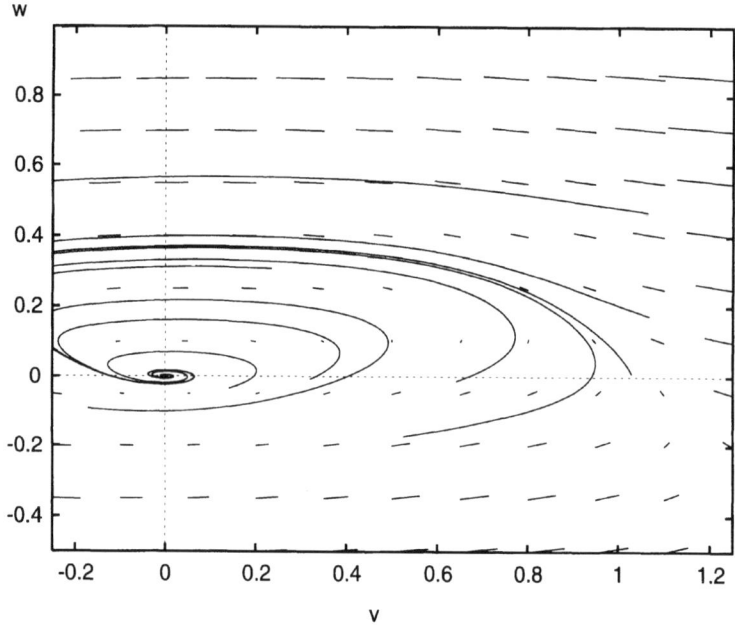

**Figure B.9**   Direction fields and some trajectories for the FitzHugh–Nagumo equations.

up into regions along which the derivatives of each variable have a constant sign. Thus, the general direction of the flow is easy to determine. Furthermore, any point where they intersect represents a fixed point of the differential equation.

XPPAUT can compute the nullclines for planar systems. To do this, just click on `Nullcline` `New` ( **N** **N** ). You should see two curves appear: a red one representing the $V$-nullcline and a green one representing the $W$-nullcline. The green one is a straight line, and the red is a cubic. They intersect just once: There is a single fixed point. Move the mouse into the phase plane area and hold it down as you move it. At the bottom of the **Main Window** you will see the $x$ and $y$ coordinates of the mouse. The intersection of the nullclines appears to be at $(0,0)$.

The stability of fixed points is determined by linearizing the system of equations about them and then finding the eigenvalues of the resulting linear matrix. XPPAUT will do this for you quite easily. XPPAUT uses Newton's method to find the fixed points and then numerically linearizes the system about them to determine stability. To use Newton's method, a decent guess needs to be provided. For planar systems this is easy to do; it is just the intersection of the nullclines. In XPPAUT fixed points and their stability are found using the `Sing pts` command, since "singular points" is a term sometimes used for fixed points or equilibrium points. Click on `Sing pts` `Mouse` ( **S** **M** ) and move the mouse to near the intersction of the nullclines. Click the button, and a message box will appear on the screen. Click on No, since we do not need the eigenvalues. A new window will appear that contains information about the fixed points. The stability is shown at the top of the window.

The nature of the eigenvalues follows: **c+** denotes the number of complex eigenvalues with positive real part; **c-** is the number of complex eigenvalues with negative real part; **im** is the number of purely imaginary eigenvalues; **r+** is the number of positive real eigenvalues; and **r-** is the number of negative real eigenvalues. Recall that a fixed point is linearly stable if all of the eigenvalues have negative real parts. Finally, the value of the fixed points is shown under the line. As can be seen from this example, there are two complex eigenvalues with negative real parts: the fixed point is $(0,0)$. ( XPPAUT reports a very small nonzero fixed point due to numerical error.) Integrate the system using the mouse, starting with initial conditions near the fixed point. (In the **Main Window**, tap $\boxed{\text{I}}\,\boxed{\text{I}}$.) Note how solutions spiral into the origin, as is expected when there are complex eigenvalues with negative real parts.

For nonplanar systems of differential equations you must provide a direct guess. Type your guess into the **Initial Data Window** and click on $\boxed{\text{Ok}}$ in the **Initial Data Window**. Then from the **Main Window**, click on $\boxed{\text{Sing Pts}}\,\boxed{\text{Go}}$ ($\boxed{\text{S}}$,$\boxed{\text{G}}$).

Change the parameter I from 0 to 0.4 in the **Parameter Window** and click on $\boxed{\text{Ok}}$ in the **Parameter Window**. In the **Main Window** erase the screen and redraw the nullclines: $\boxed{\text{Erase}}\,\boxed{\text{Nullclines}}\,\boxed{\text{New}}$ ($\boxed{\text{E}}$ $\boxed{\text{N}}$ $\boxed{\text{N}}$). The fixed point has moved up. Check its stability using the mouse ($\boxed{\text{Sing pts}}$ $\boxed{\text{Mouse}}$). The fixed point should be $(0.1, 0.4)$. Use the mouse to choose a bunch of initial conditions in the plane. All solutions go to a nice limit cycle. That is, they converge to a closed curve in the plane representing a stable periodic solution.

We can make a nice picture that has the nullclines, the direction fields, and a few representative trajectories. Since XPPAUT keeps only the last trajectory computed, we will "freeze" the solutions we compute. We can freeze trajectories automatically or one at a time, and we will do the former. Click on $\boxed{\text{Graphic stuff}}\,\boxed{\text{(F)reeze}}\,\boxed{\text{(O)n freeze}}$ ($\boxed{\text{G}}$ $\boxed{\text{F}}$ $\boxed{\text{O}}$) to permanently save computed curves. Up to 26 can be saved in any window. Frist we use the mouse to compute a bunch of trajectories. Draw the direction fields by clicking $\boxed{\text{Dir.field/flow}}\,\boxed{\text{(D)irect Field}}$ ($\boxed{\text{D}}$ $\boxed{\text{D}}$).

We can label the axes as follows: Click on $\boxed{\text{Viewaxes}}\,\boxed{\text{2D}}$ ($\boxed{\text{V}}$ $\boxed{\text{2}}$), and the 2D view dialog will come up. Change nothing but the labels (the last two entries), and put V as the **Xlabel** and w as the **Ylabel**. Click on $\boxed{\text{Ok}}$ to close the dialog. Finally, since the axes are confusing in the already busy picture, click on $\boxed{\text{Graphic stuff}}\,\boxed{\text{aXes opts}}$ ($\boxed{\text{G}}$ $\boxed{\text{X}}$) and in the dialog box change the 1's in the entries **X-org(1=on)** and **Y-org(1=on)** to 0's to turn off the plotting of the X and Y axes. Click $\boxed{\text{Ok}}$ when you are done.

To create a PostScript file, follow $\boxed{\text{Graphic stuff}}\,\boxed{\text{(P)ostscript}}$ ($\boxed{\text{G}}$ $\boxed{\text{P}}$) and accept all the defaults. Name the file whatever you want and click on $\boxed{\text{Ok}}$ in the file selection box. Figure B.10 shows the version that we made. Yours will be slightly different. If you want to play around some more, turn off the automatic freeze option,

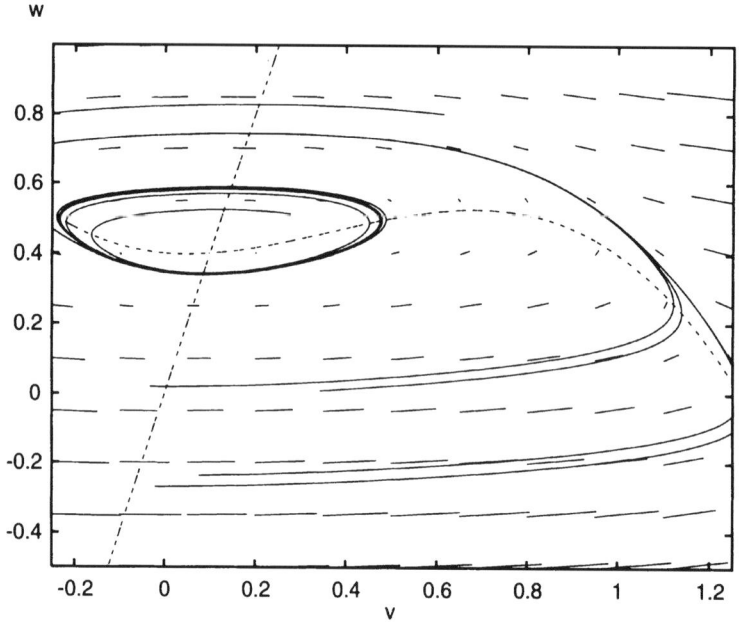

**Figure B.10** Nullclines, direction fields, trajectories for Iapp=0.4 in the Fitzhugh-Nagumo equations.

Graphic stuff Freeze Off freeze (**G** **F** **0**), and delete all the frozen curves, Graphic stuff Freeze Remove all (**G** **F** **R**).

## B.2.3  Command Summary: Phase Planes and Fixed Points

Nullcline New draws nullclines for a planar system (**N** **N**).

Dir.field/flow (D)irect Field draws direction fields for a planar system (**D** **D**).

Sing pts Mouse computes fixed points for a system with initial guess specified by the mouse (**S** **M**).

Sing pts Go computes fixed points for a system with initial guess specified by the current initial conditions (**S** **G**).

Graphic stuff Freeze On Freeze will permanently keep computed trajectories in the current window (**G** **F** **0**).

Graphic stuff Freeze Off Freeze will toggle off the above option (**G** **F** **0**).

Graphic stuff Freeze Remove all deletes all the permanently stored curves
( G | F | R ).

Graphic stuff aXes opts lets you change the axes ( G || X ).

Viewaxes 2D allows you to change the 2D view of the current graphics window and to label the axes ( V || 2D ).

# B.3   Bifurcation and Continuation

Once we have found the critical points of a system of interest, we can then embark on a continuation and bifurcation analysis of the solutions, as we mentioned in Appendix A. Continuation analysis describes how solutions to differential equations evolve over parameters, while bifurcation analysis refers, in particular, to how solutions appear and disappear as parameters are varied. One of the main strengths of XPPAUT is that it provides a convenient interface to many of the features found in the AUTO package for continuation/bifurcation analysis. AUTO remains one of the best such packages. However, the stand-alone versions of AUTO require coding compilation of the equations with the FORTRAN computer language. Note that AUTO currently only is available for the Unix version of XPPAUT. The Windows version uses a continuation package called LOCBIF, and slight differences from the procedures outlined below are explained in the user's manual on the web site. While AUTO is powerful even as implemented in XPPAUT, it is not foolproof. The results you obtain should be viewed with a critical eye. Bifurcation analysis is discussed in more depth in an excellent book by Kuznetsov (Kuznetsov 1998). It is useful to understand that the AUTO features available in XPPAUT are independent of the other tools, but that parameters and fixed points are exchanged back and forth. Bifurcation diagrams can be imported into XPPAUT for plotting.

## B.3.1   General Steps for Bifurcation Analysis

- AUTO bifurcation analysis must start from a fixed (or singular) point or from a periodic orbit. Get rid of transients and find a stable fixed point by by integrating several times: Click Initialconds Go and then click Initialconds Last several times. For limit cycles, get a good estimate of the period and integrate one full period only.
- Bring up the **AUTO window** by selecting File Auto ( F | A ).
- Use the Parameter function to choose the parameters that will be varied.
- Use the Axes Hilo function to select the parameters to be plotted and the range over which they will be varied. A one–parameter bifurcation diagram must be completed before a second parameter can be varied.
- Use the Numerics function to define direction, step size, etc.

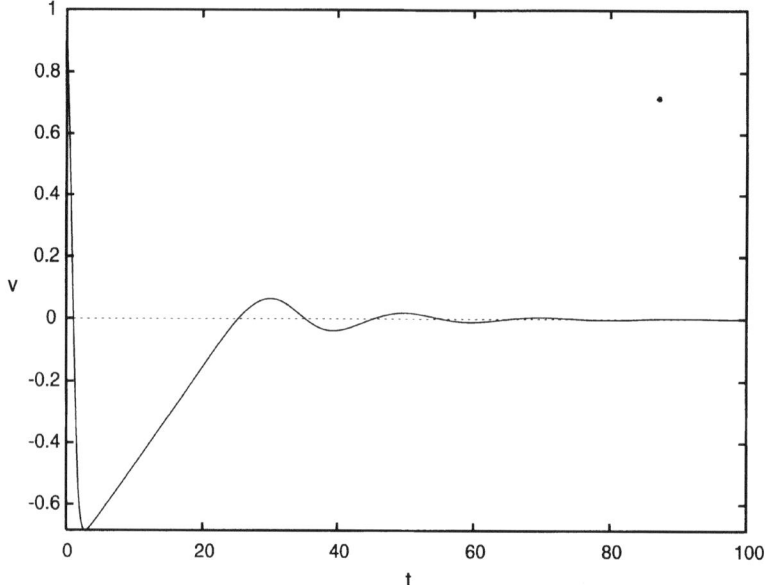

**Figure B.11** FH-N system with `Iapp=0`.

- ⎡Run⎤ the analysis for ⎡Steady state⎤ or ⎡Periodic⎤.

It is more difficult for AUTO to start from a periodic solution, and so some further assistance is required. First, a good approximation of the period must be determined from the data browser (use ⎡Find⎤ in conjunction with a large number to determine the maximum) or by measuring the peak-to-peak period with the mouse. Integrate over just that period by adjusting ⎡nUmerics⎤⎡Total⎤ to the period length. After starting AUTO and selecting parameters and bounds, choose ⎡Run⎤⎡Periodic⎤. AUTO may still fail, and further adjustments to numerical parameters or a better approximation of the period may be necessary.

## B.3.2   Hopf Bifurcation in the FitzHugh–Nagumo Equations

We will continue with the FitzHugh-Nagumo example and explore the bifurcation structure of this system. Using the FH-N equations from earlier, add the lines

```
v(0)=1,
w(0)=1,
```

to the ODE file to set initial conditions. Now we are ready to begin the analysis:

1. Start up XPPAUT with `Iapp=0`. Next, run the ODE file, ploting x vs t as discussed above. You should see $v$ oscillate a bit and go to zero, as seen in Figure B.11. Now

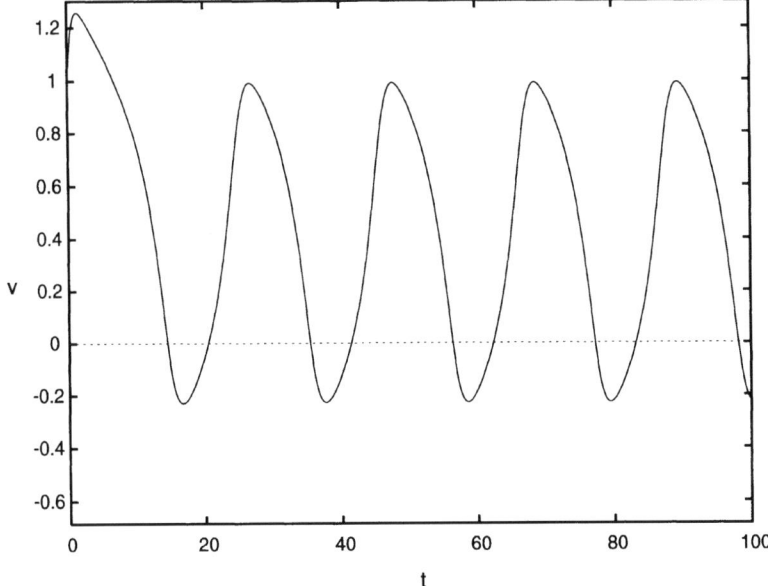

**Figure B.12**   FH-N system with Iapp=1.5.

run the simulation with Iapp=1.5. You should see the periodic solution shown in Figure B.12. We want to understand how this change ocurrs as the parameter Iapp is changed.

2. To set up the bifurcation analysis run the simulation again with Iapp=0: Click Initialconds Go and then click Initialconds Last several times. This will run the simulation until it is really at the steady state.

3. The next step is to bring up the **AUTO window** by selecting File Auto ( **F** **A** ). Once the **AUTO window** is present, make sure that Iapp is listed as Par1 under Parameter . It should already be there if you typed the file in as written. If not, select Iapp.

4. Set up the graphics axes with Axes Hilo . Fill in the dialog box as follows:

| Y-axis: V |
|---|
| Main Parm: Iapp |
| Xmin: -0.5 |
| Ymin: -3 |
| Xmax: 0.5 |
| Ymax: 0.5 |

5. Set up the Numerics , and change only Par Min=0 and Par Max=3.5.

6. To begin, click ⌑Run⌑ ⌑Steady state⌑. The beginning of the diagram should appear with four points labeled as in Figure B.13. The bold line represents a stable steady state, and the faint line represents an unstable steady state.

7. Now choose ⌑Grab⌑. This will allow you to see what the four marked points in the diagram represent. A cross appears on the plot at marker 1, and below the plot, a description will be present. Lab is the label for the point (1), Ty is the type of point. EP stands for End point, where we started computing. Iapp is the value of the parameter for that location of the graph.

8. Move the cursor over the (2). Under Ty should be the label HP, for Hopf bifurcation. Point (3) should also be a Hopf bifurcation, and (4) will be the other end point.

9. With the cross back on (2), press ⌑**Enter**⌑. This "grabs" that point as the beginning of a new calculation. Click ⌑Run⌑ again. This time, the pop-up screen is labeled as Hopf Pt, and we will choose ⌑Periodic⌑ to follow the periodic orbits as Iapp changes. You should get something like Figure B.14. The darkened circles show that the periodic orbits are stable. Open circles represent an unstable periodic solution. Note that there are upper and lower points in the plot for the periodic solution, showing the maximum and minimum values (of v) that the solution attains.

10. By using ⌑grab⌑ again, we can go to the periodic orbits, and their period will be shown below the plot.

11. We can save the plot using the ⌑File⌑ menu as discussed above.

## B.3.3   Hints for Computing Complete Bifurcation Diagrams

- Be sure to start at a fixed point or clearly defined limit cycle. As discussed above, get rid of transients and find a stable fixed point by integrating several times: Click

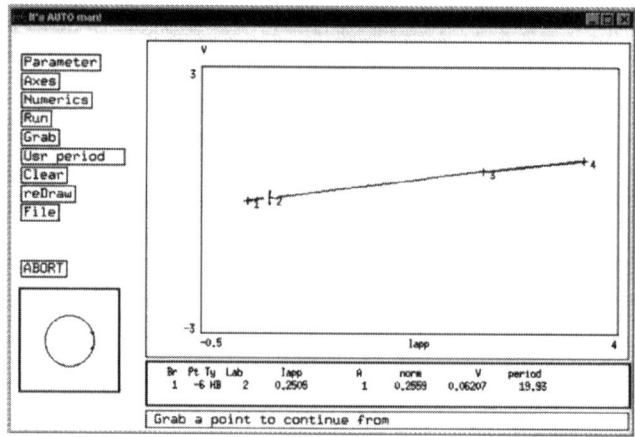

**Figure B.13**   Initial bifurcation diagram and AUTO window for the FH-N system.

Initialconds Go and then Initialconds Last several times. For limit cycles, get a good estimate of the period and integrate one full period only.

- Learn to navigate the diagram efficiently. Tab jumps to special points, Axes Fit reDraw will fit the entire diagram to the page, and Axes Zoom magnifies a given area.
- AUTO will try to follow all branches of fixed points. However, AUTO may need some assistance. Grab special points and Run in different directions by changing the sign of **Ds** in the numerics dialog box.
- Try to find the periodic solution from all Hopf points.
- Be sure also to change **Ds** for two parameter bifurcations.
- An initial **MX** label indicates that auto has failed and you may not have provided a good fixed point or periodic orbit.
- To erase the diagram and start again with different parameters, Grab the initial starting point and destroy the diagram with File Reset diagram.
- If AUTO fails to continue, try making **dsmin** smaller; for periodic orbits and boudary value problems make **ntst** larger.
- If AUTO clearly misses a bifurcation point, make **dsmin** smaller and recompute.

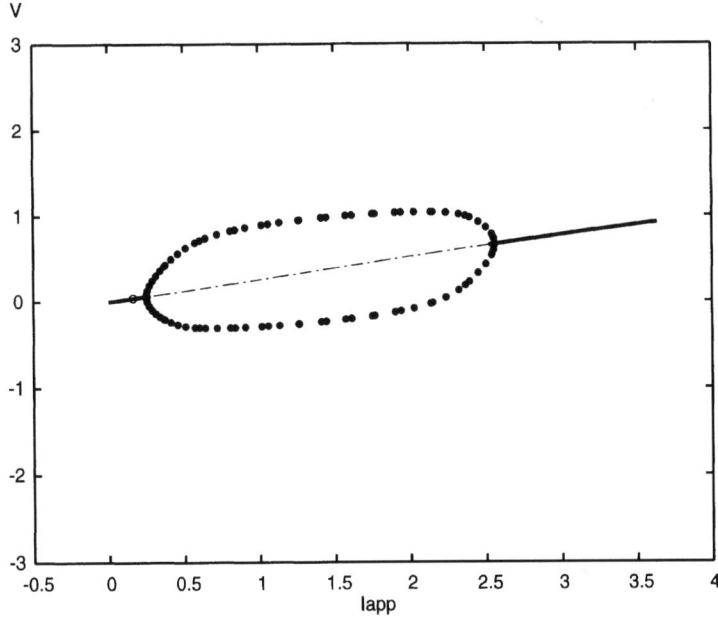

**Figure B.14** Final bifurcation diagram the FH-N system.

## B.4    Partial Differential Equations: The Method of Lines

XPPAUT doesn't have any way to solve PDEs other than by discretizing space and producing a series of ODEs using the method of lines. However, one does not have to write all the differential equations down, one at a time. There are ODE file shortcuts that make this easy to do. There is also a nice way of plotting the space–time behavior of a one dimensional PDE. We will go through one quick example here. Consider the PDE

$$\frac{\partial v}{\partial t} = f(v, w) + D\frac{\partial^2 v}{\partial x^2},$$

$$\frac{\partial w}{\partial t} = g(v, w),$$

where $f, g$ are the kinetics for the FitzHugh–Nagumo model or some other model. For simplicity, we assume Neumann boundary conditions (see Chapter 7). This system can be discretized with the method of lines, yielding the following system of ODEs:

$$\frac{dv_0}{dt} = f(v_0, w_0) + D(v_1 - v_0),$$

$$\frac{dv_j}{dt} = f(v_j, w_j) + D(v_{j+1} - 2v_j + vj - 1), \quad j = 1, \ldots, N-1,$$

$$\frac{dv_N}{dt} = f(v_N, w_N) + D(v_{N-1} - v_N),$$

$$\frac{dw_j}{dt} = g(v_j, w_j), \quad j = 0, \ldots, N.$$

We will use the FitzHugh–Nagumo kinetics and make an ODE file of the discretized system. Note that we have changed the parameter B=4, so that the system is more excitable (see Chapter 7):

```
# fitzhugh-nagumo action potential
f(v,w)=B*v*(v-beta)*(delta-v)-Cw+Iapp
g(v,w)=epsilon*(v-gamma*w)
par Iapp=0,B=1,C=1,beta=.1,delta=1,gamma=.25,epsilon=.1
par D=.5
dv0/dt=f(v0,w0)+D*(v1-v0)
dv[1..49]/dt=f(v[j],w[j])+D*(v[j+1]-2*v[j]+v[j-1])
dv50/dt=f(v50,w50)+D*(v49-v50)
dw[0..50]/dt=g(v[j],w[j])
@ total=200,dt=.25,meth=qualrk,tol=1e-6
@ xhi=200,yp=v20
done
```

XPPAUT actually expands this to 100 differential equations, and the variables are named v0,w0,v1,w1 and so on up to v50,w50. You must always use the letter "j" for the index. We have told  XPPAUT to use a quality step size (adaptive) Runge–Kutta

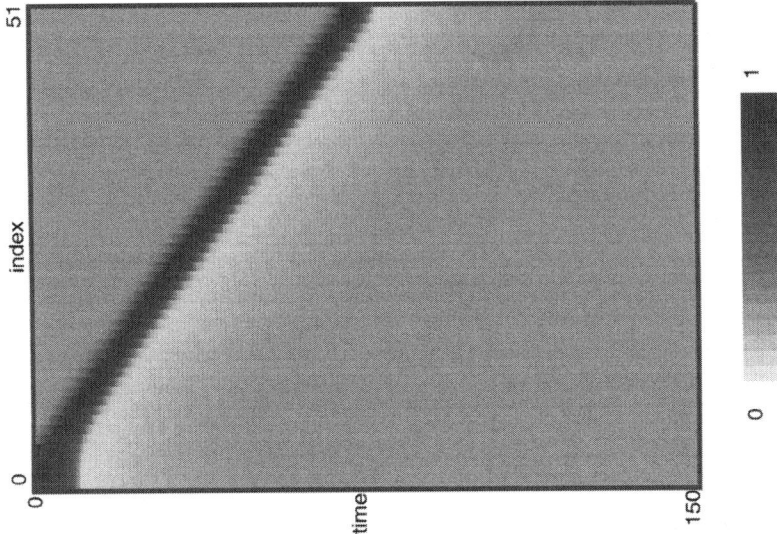

**Figure B.15** XPPAUT output of the spatial FitzHugh–Nagumo simulation.

routine with an output step size of 0.25. We integrate the equations for 200 time units. We plot v20 so that the appearance of an action potential will be clear.

Run XPPAUT with this file. Now we will give some initial conditions. Rather than type them in one by one, we will define $V_j(0)$ by a formula. Click on $\boxed{\texttt{Initial conds}}$ $\boxed{\texttt{formUla}}$. When prompted for the variable type in v[0..50] and type in heav(5-[j]) for the formula. Then tap the Enter key a few times. Note that the index is referred to as [j] in the formula rather than just j. You should see an action potential appear on the screen. Click on $\boxed{\texttt{Graphic stuff}}$ $\boxed{\texttt{Add curve}}$ and choose V40 for the y-axis and color 7 (green) for the color. The potentials of the 20th and 40th points will appear. In the **Initial Data Window**, click on the box next to V0. Scroll down and click on the box next to V50. Now click on the button labeled $\boxed{\texttt{arry}}$. A new window will appear with the space–time plot of the potential that should look something like Figure B.15. You can fool around with the parameters for this plot by clicking on the edit box in the window.

As a last bit of analysis, we can look at the spatial profile at a fixed point in time. To do this, we will transpose all the space–time data so that the 51 columns of the potential at a particular point in time become 51 rows in the second column; the first column will hold the indices. Click on $\boxed{\texttt{File}}$ Transpose. Edit the dialog box so that NCols=51 (the number of columns); Row 1=300 (the output time step is 0.25, so row 300 represents $t = 75$). Click $\boxed{\texttt{OK}}$ to complete the transpose. (You can undo this by clicking on $\boxed{\texttt{File}}$ $\boxed{\texttt{Transpose}}$ and then clicking $\boxed{\texttt{Cancel}}$ in the dialog box.) Once you have transposed the data, just plot V0 versus time. This is the spatial profile at $t = 75$.

# B.5   Stochastic Equations

## B.5.1   A Simple Brownian Ratchet

XPPAUT has many features useful for stochastic modeling. In particular, it can simulate Brownian motion and continuous Markov processes. Before turning to a sodium channel simulation, we first create an XPPAUT file for a ratchet that moves between $-1$ and $1$ and is not allowed to exit the boundaries. It is easiest to treat this in discrete time. The function normal(0,1) produces a normally distributed random number with mean 0 and standard deviation 1. Thus, the exercise can be written as the following simple ODE file:

```
par f1=-5,f2=5,h=.1,q=2
@ total=1000,meth=discrete
init x=-1
xp=x+h*.5*(f1*(sign(-x)+sign(x+1))+f2*(sign(x)+sign(1-x)))
+sqrt(q*h)*normal(0,1)
dx/dt=max(min(xp,1),-1)
done
```

The statement meth=discrete tells XPPAUT to treat this as a map rather than a continuous differential equation. In this mode, xp is the new value of x under the random dynamics. However we do not want $x$ to escape the boundaries of $\pm1$, so when we update the new value of x it is constrained by the function max(min(xp,1),-1).

## B.5.2   A Sodium Channel Model

In the next example we simulate a sodium channel model due to Joe Patlak. The functions $\alpha_m, \beta_m, \alpha_h$ are the usual voltage dependent functions for the Hodgkin–Huxley equations. XPPAUT can simulate a multi–state Markov process by defining a "Markov" variable (which has $N$ states, $0, 1, \ldots, N-1$) and the transition matrix. Each row of the transition matrix is given on a single line following the declaration of the Markov variable. Each entry is contained within the curly braces, { and }. For example, suppose that you had a two state process with transition rates $a$ from 0 to 1 and $b$ from 1 to 0. Then you would write

```
markov z 2
{0} {a}
{b} {0}
```

Note that you can put anything you want in the diagonals, since they are ignored. Here is a complete ODE file for the above process:

```
# two state markov model
par a=.2,b=.3
```

```
markov z 2
{0} {a}
{b} {0}
@ total=50,xhi=50,xp=z,yp=z,yhi=1.5,ylo=-.5
# ddummy/dt=0
done
```

The last line should be uncommented if your version of XPPAUT does not accept this file. Older versions require at least one differential equation. Run this and integrate the equations. See $z$ flip up and down.

Now with this trivial example in mind, we turn to the sodium channel model. Here is the XPPAUT file:

```
# model for the hh Na channel
# due to patlack
#
par vhold=-100,vnew=10
par ton=1,toff=16,ena=50
par k1=.24,k2=.4,k3=1.5
v=vhold+heav(t-ton)*heav(toff-t)*(vnew-vhold)
am=.1*(v+40)/(1-exp(-(v+40)/10))
bm=4*exp(-(v+65)/18)
ah=.07*exp(-(v+65)/20)
markov z 5
{0} {3*am} {0} {0} {0}
{bm} {0} {2*am} {0} {k1}
{0} {2*bm} {0} {am} {k2}
{0} {0} {3*bm} {0} {k3}
{0} {0} {0} {0} {ah}
aux cond=(z==3)
aux ina=(z==3)*(v-ena)/40
aux pot=v
@ meth=euler,dt=.01,total=16
@ yp=ina,ylo=-1.5,xlo=0,xhi=15,bound=100
done
```

The voltage is stepped from a value vhold to vnew. The Markov process has five states. Only state 3 is conducting. Thus, the auxiliary variable cond=(z==3) is zero if the channel is not conducting and 1 if it is. The current passed is given by the variable ina. Euler's method is used for this calculation, since that is usually the best method to use for any stochastic model. Integrate this a few times to see the channel open transiently. This is a transient channel, so even at high potentials it stays on only briefly. The Hodgkin–Huxley equations arise by assuming that there are many independent channels. Since they are assumed to be independent, we can simulate the effect of $m$

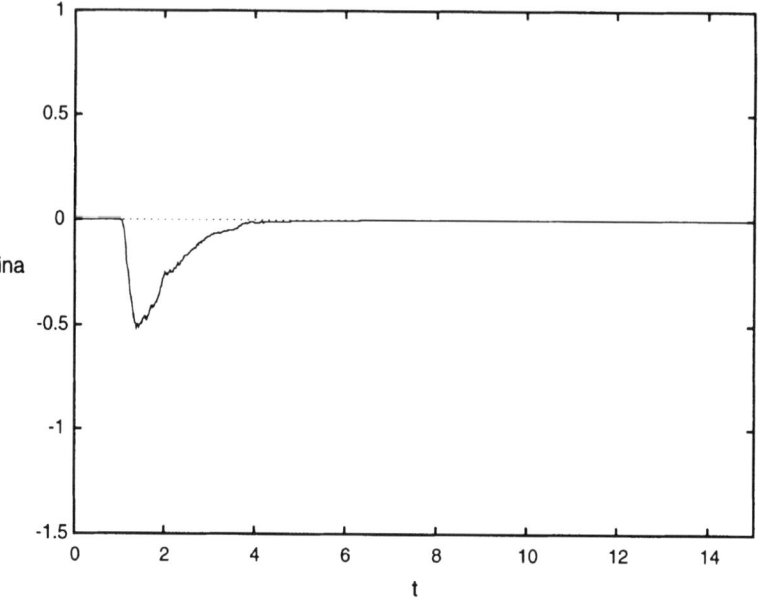

**Figure B.16**    XPPAUT results for the sodium channel simulation.

channels by just integrating the equations $m$ times and averaging the output. **XPPAUT** does this for you. Click on [nUmerics] [Stochastics] [Compute] to tell **XPPAUT** how many trials. Choose z as the variable to range over, 200 steps with Start=0 and End=0. Then click [Ok]. The equations will be integrated 200 times; this is equivalent to having 200 independent channels. Once **XPPAUT** has done this (you can keep track by looking at the bottom), then click on [stocHastics] [Mean] to load the data browser with the mean values of all its rows over the 200 trials. Click on [Escape] to get back to the main menu and then click on [Restore] to see the mean value shown in Figure B.16. This looks very similar to the deterministic solution. Try simulating fewer channels (e.g., 10) and more channels, (e.g., 1000). Why do we have to take such small steps?

### B.5.3   A Flashing Ratchet

As a final example of a stochastic equation, we simulate the "flashing ratchet" model: an asymmetric ratchet that flashes on and off at a particular rate according to a Markov process and subject to simple delta-correlated noise. Here is the model:

$$dx = -zf(x)dt + \sigma d\xi,$$

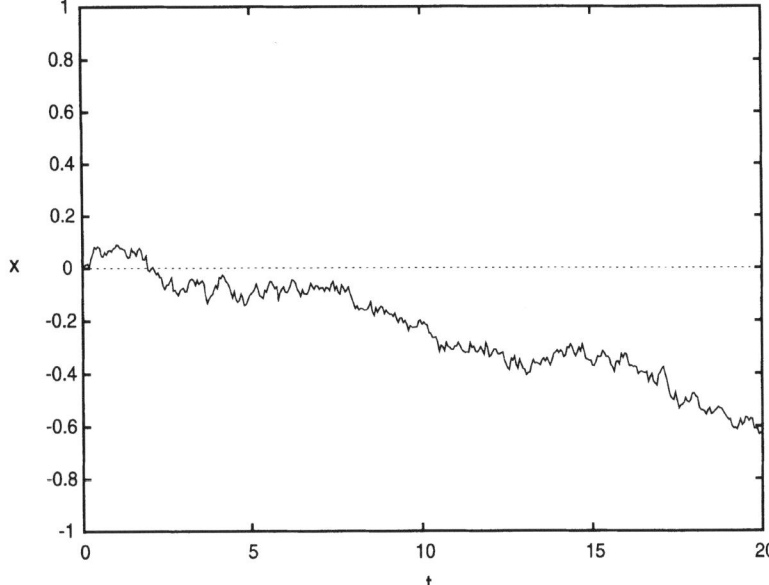

**Figure B.17** Stochastic simulation of a simple Brownian ratchet.

where $z$ is a two–state Markov process either off (0) or on (1) and $f$ is just the derivative of an asymmetric potential. The potential should be made periodic. Here is the **XPPAUT** file:

```
# ratchet
#   -1 for 0<x<1
#    1/a for 1<x<1+a
ff(x)=if(x<1)then(-1)else(1/a)
f(x)=ff(mod(x,1+a))
par a=.25
par alpha=.2,beta=.2
par sig=.5
wiener xi
dx/dt=f(x)*z+sig*xi
markov z 2
{0} {alpha}
{beta} {0}
@ total=200,bound=200,meth=euler
done
```

We first define the potential on the interval $[0, 1+a)$ and then extend it to the whole line modulo $1+a$. We define a two-state markov process that flips randomly between 0 and 1. The `wiener` `w` declaration tells XPPAUT that this is a normally distributed number scaled by the internal time step `dt`. Thus, if you change `dt`, the standard deviation is scaled accordingly by $\sqrt{dt}$. Integrate the equations a few times. Next, do 50 simulations and look at the mean trajectory. (Use nUmerics stocHastic Compute as was done in the channel model.) You will see in Figure B.17 that there is persistent downward drift. This is what is predicted by theory.

# Numerical Algorithms

This appendix contains samples of the different kinds of numerical constructs used to produce the figures in this book. Each of the files produces a figure in the book and is labeled accordingly. In general, each of the files contains a new construct not previously used. However, several are included for easy comparison to one that is very similar. We provide a one-line description of the intent of the code. It will be useful to look back at the figures these files produce as well.

The files presented here are working XPPAUT code. XPPAUT uses concise straightforward code, making it easy to translate to other languages and packages. The code as written can be saved as an .ode file and run using XPPAUT or Winpp. The code is commented heavily at the beginning and less so at the end, where the reader will be able to pick up on the more obvious features.

This appendix provides a good overview, and it provides those with programming experience a template for writing their own code. Furthermore, XPPAUT code for every computational figure in the book is available for download on the web site for the book, including the corresponding set files. As readers solve examples and exercises using other packages such as Berkeley Madonna, MATLAB, and Mathematica. These files will be included on the web site if they are submitted to us.

**Figure 1.10**. This is the first basic code to run. Once you have a simple XPPAUT model running, it is straightforward to modify it to increase its complexity.

```
#solves single channel model
#this is basically the same equation with the steady
#state level changed to ninf rather than zero
#the initial condition on the open fraction
fo(0)=1
#parameters
#finf is the steady state value
#tau is the the time scale
param finf=0.5,tau=3
#differential equation
dfo/dt=-(fo-finf)/tau
#auxiliary function
#rate is the instantaneous rate of change
aux rate=-(fo-finf)/tau
done
```

**Figure 2.6**. This program provides a template for plotting functions with XPPAUT using the aux functionality.

```
#.ode file for plotting functions
#x is the dummy variable for plotting the auxiliary functions
#initial value of x
x(0)=-80
#differential equation
#the solution to this equation is x=t+x(0)
#it lets us plot functions f(x)
dx/dt=1
#equilibrium activation and inactivation
#auxillary functions to plot
aux finfact=1/(exp(-(x+25)/5)+1)
aux finfinact=1/(exp((x+50)/2)+1)
#the characteristic times to plot
aux tauact=5/cosh((x+25)/2*5)
aux tauinact=5/cosh(-(x+50)/2*2)
done
```

**Figures 2.9, 2.10, 2.11, 2.12**. This is a simple two-variable model (the Morris–Lecar model) that includes function definitions outside of the differential equations.

```
#simulation of the Morris-Lecar equations
dv/dt = (-gca*minf*(V-Vca)-gk*w*(V-VK)-gl*(V-Vl)+Iapp)/C
dw/dt = phi*(winf-w)/tauw
```

```
#initial conditions
v(0)=-60.855
w(0)=0.014915
#functional forms for the equations
#keep the functions separate from the differential equations makes
#the differential equations less complicated to sort through,
#but is not necessary.
minf=0.5*(1+tanh((v-v1)/v2))
winf=0.5*(1+tanh((v-v3)/v4))
tauw=1/cosh((v-v3)/(2*v4))
#parameters
param Iapp=0,vk=-84,vl=-60,vca=120
param gk=8,gl=2,C=20
param v1=-1.2,v2=18
param v3=2,v4=30,phi=.04,gca=4.4
#numerical parameters:
#these are sometimes included to avoid needing to
#set them when the program is run
#total is the total time the simulation will run
#dt is the time step
#the rest of these parameters define the view in the graph
@ total=150,dt=0.25,ylo=-75,yhi=45,xlo=0,xhi=150,xp=t,yp=va=4.4
done
```

**Figures 5.8, 5.10.** This file shows that the ordering of the program elements in XPPAUT is not crucial.

```
#Keizer/Levine: reduced open-cell model
#Parameter values slightly modified from original paper.
#Two figs generated:
#(1) oscillating time series: jin=1.5 for 60<t<400 and jin=0.3 otherwise
#(2) bifurcation plot:   jin=0.1 to jin=4.5
#numerical parameters
@ meth=cvode, dtmax=1, dt=0.05, total=400, maxstor=1000000
@ bounds=100000000, xp=w,  yp=w, toler=1.0e-6, atoler=1.0e-6
@ xlo=0, xhi=400, ylo=0, yhi=2.5
#initial conditions
CAI(0)=0.2, CATOT(0)=1.
#parameters
params fi=0.01,Kserca=0.2
params Ka=0.4,Kb=0.6,Kc=0.1,kcm=0.1
params vpmca=5, Kpmca=0.6, jin=1.5
params vserca=100,vryr=5,vleak=0.2,sigma=0.02
```

```
#function definitions
winf=(1+(Ka/CAI)^4+(CAI/Kb)^3)/(1+(1/Kc)+(Ka/CAI)^4+(CAI/Kb)^3)
CAER=(CATOT-CAI)/sigma
tau=winf/kcm
Pryr=winf*(1+(CAI/Kb)^3)/(1+(Ka/CAI)^4 + (CAI/Kb)^3)
#differential equations
dCAI/dt=fi*((vryr*Pryr+vleak)*(CAER-CAI)-vserca*(CAI^2/(CAI^2+Kserca^2)))\
    -vpmca*CAI^2/(CAI^2+Kpmca^2)+jin)
dCATOT/dt=fi*(jin-vpmca*CAI^2/(CAI^2+Kpmca^2))
aux Pryr=winf*(1+(CAI/Kb)^3)/(1+(Ka/CAI)^4 + (CAI/Kb)^3)
done
```

**Figures 5.14, 5.15**. Note here that the keyword for parameters is flexible.

```
# Li-Rinzel open-cell model .ode file.
# Dimensional version
# The equations
dCAI/dt=fi/Vi*((L+Pip3*((IP3*CAI*h)/((IP3+Ki)*(CAI+Ka)))^3)*(CAER-CAI)\
  -Vserca*CAI^2/(Kserca^2+CAI^2)+epsilon*(Jin-Vpmca*CAI^2/(Kpmca*Kpmca+CAI^2)))
dh/dt=A*(Kd-(CAI+Kd)*h)
dCATOT/dt=fi/Vi*epsilon*(Jin-Vpmca*CAI^2/(Kpmca^2+CAI^2))
CAER=(CATOT-CAI)/sigma
# The parameters
# Jin=aMol/s
par Jin=1200,fi=0.01
# Vi=pL
par Vi=4
# L,Pip3=pL/s
par L=0.37,Pip3=26640
# IP3,CAI,CAER,CATOT,Ki,Ka,Kserca,Kd,Kpmca=uM
par IP3=0.9,Ki=1.0,Ka=0.4
# Vserca, Vp=aMol/s [sic]
par Vserca=400,Kserca=0.2,A=0.5,Kd=0.4
# sigma,epsilon,fi=unitless
par sigma=0.185,epsilon=0.01,Vpmca=2000,Kpmca=0.3
# The initial conditions
CAI(0)=0.2
h(0)=0.8
CATOT(0)=4.0
aux CAER=CAER
#parameters for graphing
@ TOTAL=300,DT=0.02,xlo=0,xhi=2,ylo=0,yhi=1.5,MAXSTOR=20000
@ xplot=t,yplot=CAI
#parameters for AUTO bifurcation analysis
```

```
@ dsmin=1e-5,dsmax=.1,parmin=-.5,parmax=.5,autoxmin=-.5,autoxmax=.5
@ autoymax=.4,autoymin=-.5
set vvst {xplot=c,yplot=h,xlo=0,xhi=1.5,ylo=0,yhi=1.5,total=100,\
dt=0.01,meth=qualrk}
done
```

**Figures 5.19, 5.20.** This simulation exhibits bursting behavior in the right parameter range. Note here an alternative method for declaring initial conditions.

```
#Morris-Lecar-like beta-cell
init V=-65.0, n=0.00016, CAI=0.2
# equations
dV/dt=(gL*(Vl-V)+gK*n*(Vk-V)-Ica-Ikca+Iapp)/Cm
dn/dt=(ninf-n)/tau
dCAI/dt=f*(-alpha*Ica - vlpm*CAI)
# where
minf=0.5*(1+tanh((v-v1)/v2))
ninf=0.5*(1+tanh((v-v3)/v4))
tau=1/(phi*cosh((v-v3)/(2*v4)))
# For bifurcation diagram, set auto=1 and use gkcastar
# as the bifurcation parameter:
param gkcastar=100.0, auto=0
param Vk=-75, Vl=-75, Vca=25
param Iapp=0, gK=2700, gL=150, gCa=1000, Cm=5300
param v1=-20.0, v2=24, v3=-16, v4=11.2, phi=0.035
param gkca=2000.0, Kkca=5.0
# Ikca
Ikca = auto*gkcastar*(V-Vk)+((1-auto)*gkca*CAI/(Kkca+CAI))*(V-Vk)
# Calcium Handling
par alpha=4.50e-6, vlpm=0.15, f=0.001
# Ikatp
par gkatp=0.0
Ica = gca*minf*(V-Vca)
@ meth=cvode, atol=1.0e-6, tol=1.0e-6, dt=10.0
@ total=40000, maxstor=10000
@ xp=t, yp=v, bound=100000000
@ xlo=0, xhi=40000, ylo=-70, yhi=-10
done
```

**Figures 6.2, 6.3.** This file provides a template for simple coupled dynamics. Note that by specifying the argument explicitly, functions (such as minf(v) here) need not be duplicated. Rather, one passes the appropriate variable as in the differential equation for V1.

```
#mlgap.ode
```

```
#gap-junction coupled cells.
#Morris-Lecar dynamics with modified parameters.
#Notably: vc=-5, vd=10, phi=0.5  (originally: 2, 30, 0.04, respectively)
#Use weak gc (=1) for antiphase and strong gc (=2) for inphase.
#differential equations
dV1/dt=(I-gca*minf(V1)*(V1-Vca)-gk*w1*(V1-VK)-gl*(V1-Vl)+\
   gc*heav(t-ton)*(V2-V1))/c
dw1/dt = phi*(winf(V1)-w1)/tauw(V1)
dV2/dt = (I-gca*minf(V2)*(V2-Vca)-gk*w2*(V2-VK)-gl*(V2-Vl)+\
   gc*heav(t-ton)*(V1-V2))/c
dw2/dt = phi*(winf(V2)-w2)/tauw(V2)
init V1=-20,w1=.2
init V2=-30,w2=0
minf(v)=0.5*(1+tanh((v-va)/vb))
winf(v)=0.5*(1+tanh((v-vc)/vd))
tauw(v)=1/cosh((v-vc)/(2*vd))
#parameters
param vk=-84,vl=-60,vca=120
param i=-10,gk=8,gl=1,c=20
param va=-1.2,vb=18,gc=2,ton=100
param vc=-5,vd=10,phi=0.5,gca=8
#numerical parameters
@ total=400,dt=.25,xhi=400,ylo=-40,yhi=30
done
```

**Figure 6.9**. This file includes a delay using a Heaviside function. This file also demonstrates alternative notation for the equations. In particular, note that the dv/dt notation may be substituted for v'.

```
# simple model for coincidence detection.
# ML cell receives two just-subthreshold inputs that are identical and
# excitatory -  tdel represents timing difference between them.
# this ML model uses the standard ML (Type II) params from Chapt 2 of this
# book (same as Rinzel/Ermentrout in  Koch and Segev's book.
# Note: treat tdel as a variable (with ode: tdel'=0) so can do Poincare
# map for response tuning curve
isyn2=gsyne*s2*(Vsyne-v1)
isyn1=gsyne*s1*(Vsyne-v1)
#differential equations
v1' = (I-gca*minf(v1)*(v1-vca)-gk*w1*(v1-vk)-gl*(v1-vl)\
      +isyn2+isyn1)/c
w1' = phi*(winf(v1)-w1)/tauw(v1)
s2' = alphae*sinf(vr2(t))*(1-s2)-betae*s2
s1' = alphae*sinf(vr1(t))*(1-s1)-betae*s1
```

```
tdel'= 0
init v1=-60.9,w1=.0149,s2=0.,s1=0.,tdel=0
minf(v)=.5*(1+tanh((v-va)/vb))
winf(v)=.5*(1+tanh((v-vc)/vd))
tauw(v)=1/cosh((v-vc)/(2*vd))
sinf(v)=1/(1+exp(-(v-thetasyn)/ksyn))
# fix the start time of excitation from R2 and delay the
# start of excitation from R1  by tdel (if tdel<0, R1 input precedes
# R2 input)
vr2(t)=100*heav(t-t0)*heav(teon+t0-t)+vrest
vr1(t)=100*heav(t-t0-tdel)*heav(teon+t0+tdel-t)+vrest
param t0=100,teon=5,vrest=-60
param gsyne=1.5,vsyne=100
param vk=-84,vl=-60,vca=120
param i=0,gk=8,gl=2,c=20
param va=-1.2,vb=18,vc=2,vd=30,phi=.04,gca=4.
param thetasyn=20,ksyn=2,alphae=1,betae=0.3
#  aux quantities
aux isyn2x=isyn2
aux isyn1x=isyn1
@ total=400,bound=100000,nout=2,dt=.25,xhi=400,ylo=-100,yhi=100
done
```

**Figure 7.6.** This is the basic diffusion equation. It demonstrates the two–dimensional graphing capabilities of XPPAUT.

```
param D=1
param dx=1
c0'=D*(c1-c0)/(dx*dx)
c[1..39]'=D*(c[j+1]-2*c[j]+c[j-1])/(dx*dx)
c40'=D*(c39-c40)/(dx*dx)
init c[0..19]=0
init c20=0.5
init c[21..29]=1
init c30=0.5
init c[31..40]=0
```

**Figure 7.8** This file provides a template for one-dimensional reaction–diffusion equations.

```
param dx=1
f(x)=x*(1-x)*(x-0.1)
V0'=(V1-V0)/(dx*dx)+f(V0)
V[1..39]'=(V[j+1]-2*V[j]+V[j-1])/(dx*dx)+f(V[j])
V40'=(V39-V40)/(dx*dx)+f(V40)
```

```
init V[0..30]=0
init V31=0.1
init V[32..40]=0.2
```

**Figure 8.7**. This file incorporates recovery, giving a traveling pulse with wave front and back.

```
# Calcium wave simulation
#parameters
param ip3=.7, a2=.2, Caer=1
param d1=0.1, d2=1, d3=0.2, d5=0.2
param v1=20, v2=0.004, v3=1.2, k3=0.15, tau=2
param deff=16,dx=10
# dx in units of um
# deff in units of um^2/sec
#the initial condition, equation and a function
Ca[0..4](0)=.1
Ca[5..100](0)=.01
w[0..100](0)=.8652
dCa0/dt=f(Ca0,w0)+deff*(Ca1-Ca0)/dx^2
dCa[1..99]/dt=f(Ca[j],w[j])+deff*(Ca[j-1]-2*Ca[j]+Ca[j+1])/dx^2
dCa100/dt = f(Ca100,w100)+deff*(Ca99-Ca100)/dx^2
dw[0..100]/dt=(winf(Ca[j])-w[j])/tau
f(Ca,w)=(v2+v1*(w*Ca/(Ca+d5))^3)*(Caer-Ca)-v3*Ca^2/(k3^2+Ca^2)
winf(Ca)=(d2*ip3/(ip3+d3))/(Ca+(d2*(ip3+d1)/(ip3+d3)))
#program end
@ total=200,trans=0,DT=.1,xlo=0,xhi=200,ylo=0,yhi=1
@ maxstore=1000000,bounds=10000
@ xplot=x,yplot=Ca0
done
# A forcing term can be added if you want
# capp = i1*(heav(mod(t,period))*heav(duty*period-mod(t,period)))
```

**Figure 8.12A,B**. The following file is interesting because although there are discrete release sites, one parameter set essentially results in continuous $Ca^{2+}$ wave propagation seen in Figure 8.12A while another results in discrete release and saltatory propagation as seen in Figure 8.12B. To replicate the discrete release and saltatory propagation seen in Figure 8.12B, set taur=0.01 and total=0.5. To replicate the continous wave of Figure 8.12A set taur=1 and total=2.0. Use the 2DArray option in XPPAUT to view the results.

```
# Fire-diffuse-fire model simulation
# time in s
# space in um
# d in um^2/s
# parameters (here for continuous wave)
```

```
param d=30,dx=0.2,sigma=5,taur=1,taud=10000000,cth=0.1
#the initial condition, equation and a function
c[0..100](0)=0
s[0..20](0)=1.0
s[21..100](0)=0
#global 1 c0-cth {s0=1}
global 1 c10-cth {s10=1}
global 1 c20-cth {s20=1}
global 1 c30-cth {s30=1}
global 1 c40-cth {s40=1}
global 1 c50-cth {s50=1}
global 1 c60-cth {s60=1}
global 1 c70-cth {s70=1}
global 1 c80-cth {s80=1}
global 1 c90-cth {s90=1}
global 1 c100-cth {s100=1}
dc0/dt = sigma*heav(s0)/2/taur-c0/taud+d*(c1-c0)/dx^2
dc[1..99]/dt = sigma*heav(s[j])/taur-c[j]/taud+d*(c[j-1]-2*c[j]+c[j+1])/dx^2
dc100/dt = sigma*heav(s100)/2/taur-c100/taud+d*(c99-c100)/dx^2
ds[0..100]/dt = -heav(s[j])/taur
aux logc[0..100] = c[j]
#numerical parameters (total for continuous wave)
@ total=2,trans=0,dt=0.0001,xlo=0,xhi=2000,ylo=0,yhi=1
@ maxstore=1000000,bounds=10000
@ xplot=x,yplot=Ca10
done
```

**Figure 9.10**. This file incorporates an active delay into the equation.

```
# Discrete time lag oscillator
p p=4, b=.5, tau=2.5
dx/dt = 1/(1+delay(x,tau)^p) - b*x
x(0)=1.1
@ delay=10
done
```

**Figure 11.3**. This file shows the basic two-state stochastic model.

```
# Example two state channel simulation
# parameters
params kp=1.5, km=0.5, tau=1000
#initial condition
po(0)=0.5
#stochastic variable
markov n 2
```

```
{0}{kp}
{km}{0}
#differential equation
po'=-(po-n)/tau
aux n=n
#numerical parameters
@ total=100,trans=0,DT=.001,xlo=0,xhi=100,ylo=-0.1,yhi=1.1
@ maxstore=1000000,bounds=10000
@ xplot=t,yplot=n
#@ njmp=100
done
```

**Figure 11.9**. This files introduces Wiener variables.

```
#weiner variables
wiener w[0..99]
#differential equations
x[0..99]'=w[j]
mean = sum(0,99)of(shift(x0,i'))/100
aux mean = mean
aux var = sum(0,99)of((shift(x0,i')-mean)^2)/100
aux w0 = w0
#numerical parameters
@ total=1000,trans=0,DT=0.1,xlo=0,xhi=1000,ylo=-100,yhi=100
@ maxstore=1000000,bounds=10000
@ xplot=t,yplot=x
@ njmp=1
done
```

**Figure 11.11A**. The next two files are used to explore the differences between twenty states and two states in a Markov process. Notice that the rows and columns of the transition probability matrix follow the  XPPAUT convention and are transposed relative to (11.5). Diagonal entries are not used by XPPAUT.

```
#parameters
params kp=0.5, km=0.5
params c=2, gl=0.5, gch=1.0, vl=-70, vch=20
#initial condition
v(0)=0
#stochastic variable
markov n 2
{0}{kp}
{km}{0}
#differential equation
v'=(-gl*(v-vl)-gch*n*(v-vch))/c
```

```
aux n=n
#numerical parameters
@ total=100,trans=0 ,DT=.001,xlo=0,xhi=100,ylo=-80,yhi=0
@ maxstore=1000000,bounds=10000
done
```

**Figure 11.11C**. This is the twenty–state Markov process. Again, the transition probability matrix follows the XPPAUT convention and is transposed relative to (11.13).

```
#parameters
params kp=0.5, km=0.5
params c=2, gl=0.5, gch=1.0, vl=-70, vch=20
#initial condition
v(0)=0
#stochastic variable
markov n 21
{0}{20*kp}{0}{0}{0}{0}{0}{0}{0}{0}{0}{0}{0}{0}{0}{0}{0}{0}{0}{0}
{km}{0}{19*kp}{0}{0}{0}{0}{0}{0}{0}{0}{0}{0}{0}{0}{0}{0}{0}{0}{0}
{0}{2*km}{0}{18*kp}{0}{0}{0}{0}{0}{0}{0}{0}{0}{0}{0}{0}{0}{0}{0}{0}
{0}{0}{3*km}{0}{17*kp}{0}{0}{0}{0}{0}{0}{0}{0}{0}{0}{0}{0}{0}{0}{0}
{0}{0}{0}{4*km}{0}{16*kp}{0}{0}{0}{0}{0}{0}{0}{0}{0}{0}{0}{0}{0}{0}
{0}{0}{0}{0}{5*km}{0}{15*kp}{0}{0}{0}{0}{0}{0}{0}{0}{0}{0}{0}{0}{0}
{0}{0}{0}{0}{0}{6*km}{0}{14*kp}{0}{0}{0}{0}{0}{0}{0}{0}{0}{0}{0}{0}
{0}{0}{0}{0}{0}{0}{7*km}{0}{13*kp}{0}{0}{0}{0}{0}{0}{0}{0}{0}{0}{0}
{0}{0}{0}{0}{0}{0}{0}{8*km}{0}{12*kp}{0}{0}{0}{0}{0}{0}{0}{0}{0}{0}
{0}{0}{0}{0}{0}{0}{0}{0}{9*km}{0}{11*kp}{0}{0}{0}{0}{0}{0}{0}{0}{0}
{0}{0}{0}{0}{0}{0}{0}{0}{0}{10*km}{0}{10*kp}{0}{0}{0}{0}{0}{0}{0}{0}
{0}{0}{0}{0}{0}{0}{0}{0}{0}{0}{11*km}{0}{9*kp}{0}{0}{0}{0}{0}{0}{0}
{0}{0}{0}{0}{0}{0}{0}{0}{0}{0}{0}{12*km}{0}{8*kp}{0}{0}{0}{0}{0}{0}
{0}{0}{0}{0}{0}{0}{0}{0}{0}{0}{0}{0}{13*km}{0}{7*kp}{0}{0}{0}{0}{0}
{0}{0}{0}{0}{0}{0}{0}{0}{0}{0}{0}{0}{0}{14*km}{0}{6*kp}{0}{0}{0}{0}
{0}{0}{0}{0}{0}{0}{0}{0}{0}{0}{0}{0}{0}{0}{15*km}{0}{5*kp}{0}{0}{0}
{0}{0}{0}{0}{0}{0}{0}{0}{0}{0}{0}{0}{0}{0}{0}{16*km}{0}{4*kp}{0}{0}
{0}{0}{0}{0}{0}{0}{0}{0}{0}{0}{0}{0}{0}{0}{0}{0}{17*km}{0}{3*kp}{0}{0}
{0}{0}{0}{0}{0}{0}{0}{0}{0}{0}{0}{0}{0}{0}{0}{0}{0}{18*km}{0}{2*kp}{0}
{0}{0}{0}{0}{0}{0}{0}{0}{0}{0}{0}{0}{0}{0}{0}{0}{0}{0}{19*km}{0}{kp}
{0}{0}{0}{0}{0}{0}{0}{0}{0}{0}{0}{0}{0}{0}{0}{0}{0}{0}{0}{20*km}{0}
#differential equation
v'=(-gl*(v-vl)-gch*n/20*(v-vch))/c
aux n=n
#numerical parameters
@ total=100,trans=0 ,DT=.001,xlo=0,xhi=100,ylo=-80,yhi=0
@ maxstore=1000000,bounds=10000
```

done

**Figure 11.14**. This file incorporates stochastic dynamics into the differential equations
for the Morris–Lecar model.

```
# Deterministic Morris-Lecar model follows Rinzel and Ermentrout's
# chapter in Koch & Segev.  Stochastic ion channel dynamics added
# using Langevin formulation.
# For stochastic excitability, oscillations, and bistability use
# iapp = 10, 12, 10 and v3 = 10, 10, 15
#initial conditions
v(0)=-40
w(0)=1
#weiner variable
wiener b
#parameters
params n=100
params v1=-1,v2=15,v3=10,v4=14.5,gca=1.33,phi=.333
params vk=-70,vl=-50,iapp=10,gk=2.0,gl=.5,om=1
minf(v)=.5*(1+tanh((v-v1)/v2))
# The 0.05 is a modification needed to lift w nullcline
# so that stochastic excitability can be realized
ninf(v)=.5*(1+tanh((v-v3)/v4))+0.05
lamn(v)= phi*cosh((v-v3)/(2*v4))
ica=gca*minf(v)*(v-100)
#differential equations
v'=  (iapp+gl*(vl-v)+gk*w*(vk-v)-ica)*om
w'= (lamn(v)*(ninf(v)-w))*om+sqrt(lamn(v)*((1-2*ninf(v))*w+ninf(v))/n)*b
aux I_ca=ica
#numerical parameters
@ total=500,trans=0,DT=.01,xlo=0,xhi=500,ylo=-60,yhi=50
@ maxstore=1000000,bounds=10000
@ xplot=t,yplot=v
done
```

# References

Alberts, B., Bray, D., Lewis, J., Raff, M., Roberts, K. and Watson, J. (1994). *Molecular Biology of the Cell*, 3rd edition, Garland Publishing, New York.

Allbritton, N., Meyer, T. and Stryer, L. (1992). Range of messenger action of calcium ion and inositol 1,4,5-trisphosphate, *Science* **258**, 1812.

Anderson, C. and Stevens, C. (1973). Voltage clamp analysis of acetylcholine produced end-plate current fluctuations at frog neuromuscular junction, *J. Physiol. (Lond.)* **235**(3), 655–691.

Asher, U. and Petzold, L. (1998). *Computer Methods for Ordinary Differential Equations and Differential-Algebraic Equations*, 1st edition, SIAM.

Atri, A., Amundson, J., Clapham, D. and Sneyd, J. (1993). A single pool model for intracellular calcium oscillations and waves in the *Xenopus laevis* oocyte, *Biophys. J.* **65**, 1727–1739.

Atwater, I., Dawson, C., Scott, A., Eddlestone, G. and Rojas, E. (1980). The nature of the oscillatory behavior in electrical activity for pancreatic beta cell., *Horm. Metab. Res. Suppl. 1980* **Suppl. 10**, 100–107.

Bell, G., Burant, C., Takeda, J. and Gould, G. (1993). Structure and function of mammalian facilitative sugar transporters, *J. Biol. Chem.* **268**, 19161–19164.

Berg, H. (1993). *Random Walks in Biology*, expanded edition, Princeton University Press, Princeton, N.J.

Berg, H. (2000). Constraints on models for the flagellar rotary motor, *Phil. Trans. R. Soc. Lond. B* **355**, 491–501.

Berridge, M. (1997). Elementary and global aspects of calcium signalling, *J. Physiol. (Lond.)* **499**, 291.

Berridge, M. (1998). Neuronal calcium signaling, *Neuron* **21**, 13.

Berridge, M. and Rapp, P. (1979). A comparative survey of the function, mechanism, and control of cellular oscillators, *J. Exp. Biol.* **81**, 217–279.

Berry, R. (1993). Torque and switching in the bacterial flagellar motor. An electrostatic model, *Biophys. J.* **64**(4), 961–973.

Bertram, R., Smolen, P., Sherman, A., Mears, D., Atwater, I., Martin, F. and Soria, B. (1995). A role for calcium release-activated current (CRAC) in cholinergic modulation of electrical activity in pancreatic $\beta$-cells, *Biophys. J.* **68**, 2323–2332.

Bezprozvanny, I., Watras, J. and Ehrlich, B. (1991). Bell-shaped calcium response curves of Ins(1,4,5)P$_3$- and calcium-gated channels from endoplasmic reticulum of cerebellum, *Nature* **351**, 751–754.

Billing, G. D. and Mikkelsen, K. V. (1996). *Introduction to Molecular Dynamics and Chemical Kinetics*, Wiley, New York.

Bliss, R., Painter, P. and Marr, A. (1982). Role of feedback inhibition in stabilizing the classical operon, *J. Theor. Biol.* **97**, 177–193.

Bohnensack, R. (1982). The role of the adenine nucleotide translocator in oxidative phosphorylation. A theoretical investigation on the basis of a comprehensive rate law of the translocator, *J. Bioenerg. Biomembr.* **14**, 45–61.

Boissonade, J. and De Kepper, P. (1980). Transitions from bistability to limit cycle oscillations. Theoretical analysis and experimental evidence in an open chemical system, *J. Phys. Chem.* **84**, 501–506.

Borisuk, M. and Tyson, J. (1998). Bifurcation analysis of a model of mitotic control in frog eggs, *J. Theor. Biol.* **195**, 69–85.

Carslaw, H. and Jaeger, J. (1959). *Conduction of Heat in Solids*, 2nd edition, Clarendon Press, Oxford.

Chance, B., Pye, P., Ghosh, A. and Hess, B. (1973). *Biological and Biochemical Oscillators*, Academic Press, New York.

Chauwin, J.-F., Oster, G. and Glick, B. (1998). Protein import into mitochondria: A comparison of the Brownian ratchet and power stroke models, *Biophys. J.* **74**(4), 1732–1743.

Chay, T. (1997). Effects of extracellular calcium on electrical bursting and intracellular and luminal calcium oscillations in insulin secreting pancreatic $\beta$-cells, *Biophys. J.* **73**, 1673–1688.

Chay, T. and Keizer, J. (1983). Minimal model for membrane oscillations in the pancreatic Beta cell, *Biophys. J.* **42**, 181–190.

Chen, K., Csikasz-Nagy, A., Gyorffy, B., Val, J., Novak, B. and Tyson, J. J. (2000). Kinetic analysis of a molecular model of the budding yeast cell cycle, *Mol. Biol. Cell* **11**, 369–391.

Cheng, H., Lederer, W. and Cannell, M. (1993). Calcium sparks: elementary events underlying excitation-contraction coupling in heart muscle, *Science* **262**, 740.

Cheng, H., Lederer, W. and Cannell, M. (1996). Calcium sparks and [Ca$^{2+}$]$_i$ waves in cardiac myocytes, *Am. J. Physiol.* **270**, C148–59.

Chow, C. and White, J. (1996). Spontaneous action potentials due to channel fluctuations, *Biophys. J.* **71**, 3013–21.

Cole, K. (1968). *A Quantitative Description of Membrane Current and Its Application to Conductance and Excitation in Nerve.*, University of California Press, Berkeley.

Cole, N., Smith, C., Sciaky, N., Terasaki, M. and Edidin, M. (1996). Diffusional mobility of golgi proteins in membranes of living cells, *Science* **273**, 797–801.

Cooper, G. M. (1997). *The Cell: A Molecular Approach*, ASM Press, Washington DC.

Crank, J. (1975). *The Mathematics of Diffusion*, 2nd edition, Clarendon Press, Oxford.

Cuthbertson, K. and Chay, T. (1991). Modelling receptor-controlled intracellular calcium oscillators, *Cell Calcium* **12**, 97–109.

Dayan, P. and Abbott, L. (2001). *Theoretical Neuroscience: Computational and Mathematical Modeling of Neural Systems*, MIT Press, Cambridge, MA.

delCastillo, J. and Moore, J. (1959). On increasing the velocity of a nerve impulse, *J. Physiol. (London)* **148**, 665–670.

Derenyi, I. and Vicsek, T. (1996). The kinesin walk: a dynamic model with elastically coupled heads, *Proc. Natl. Acad. Sci.* **93**, 6775–6779.

Destexhe, A., Mainen, Z. and Sejnowski, T. (1998). Kinetic models of synaptic transmission, *in Methods in Neuronal Modeling*, C. Koch and I. Segev (eds.), The MIT Press, Cambridge.

DeYoung, G. and Keizer, J. (1992). A single-pool inositol 1,4,5-trisphosphate-receptor-based model for agonist-stimulated oscillations in $Ca^{2+}$ concentration, *Proc. Natl. Acad. Sci. USA* **89**, 9895–9899.

Dimroth, P., Wang, H., Grabe, M. and Oster, G. (1999). Energy transduction in the sodium F-ATPase of *Propionigenium modestum*, *Proc. Natl. Acad. Sci. USA* **96**(9), 4924–4929.

Doering, C. (1990). Modeling complex systems: Stochastic processes, stochastic differential equations, and Fokker-Planck equations, *in 1990 Lectures in Complex Systems*, L. Nadel and D. Stein (eds.), Vol. III of *Sante Fe Institute Studies in the Sciences of Complexity*, Addison-Wesley, Redwood City, CA, pp. 3–51.

Doering, C. (1995). Randomly rattled ratchets, *Nuovo Cimento D* **17**, 685–97.

Doering, C. (1998). Stochastic ratchets, *Physica A* **254**, 1–6.

Dogterom, M. and Yurke, B. (1997). Measurement of the force-velocity relationship for growing microtubules, *Science* **278**, 856–860.

Doi, M. and Edwards, S. (1986). *The Theory of Polymer Dynamics*, Oxford University Press, New York.

Dolmetsch, R. and Lewis, R. (1994). Signaling between intracellular $Ca^{2+}$ stores and depletion-activated $Ca^{2+}$ channels generates $[Ca^{2+}]_i$ oscillations in T lymphocytes, *J Gen Physiol.* **103**(3), 365–388.

Dunlap, J. (1999). Molecular bases for circadian clocks, *Cell* **96**, 271–290.

Dupont, G. and Goldbeter, A. (1994). Properties of intracellular $Ca^{2+}$ waves generated by a model based on $Ca^{2+}$-induced $Ca^{2+}$ release, *Biophys. J.* **67**(6), 2191–204.

Eckert, R. and Tillotson, D. (1981). Calcium-mediated inactivation of the calcium conductance in caesium-loaded giant neurons of *Aplysia*, *J. Physiol. (Lond)* **314**, 265–280.

Edelstein-Keshet, L. (1988). *Mathematical Models in Biology*, 1st edition, McGraw-Hill, New York.

Edwards, Jr., L. (1988). *Cellular and Molecular Bases of Biological Clocks*, 1st edition, Springer, New York.

Elston, T. (2000). Models of post-translational protein translocation, *Biophys. J.* **79**(5), 2235–2251.

Elston, T. and Doering, C. (1995). Numerical and analytical studies of nonequilibrium fluctuation-induced transport processes, *J. Stat. Phys.* **83**, 359–383.

Elston, T. and Oster, G. (1997). Protein turbines I: The bacterial flagellar motor, *Biophys. J.* **73**(2), 703–721.

Elston, T., Wang, H. and Oster, G. (1998). Energy transduction in ATP synthase, *Nature* **391**, 510–514.

Ermentrout, B. (1998). Neural networks as spatio-temporal pattern-forming systems, *Rep. Prog. Phys.* **61**, 353–430.

Ermentrout, B. (2002). *Simulating, Analyzing, and Animating Dynamical Systems: A Guide to XPPAUT for Researchers and Students*, 1st edition, SIAM, New York.

Feynman, R., Leighton, R. and Sands, M. (1963). *The Feynman Lectures on Physics*, Addison-Wesley, Reading, MA. Ratchet and Pawl: Vol. 1, Chap 46.

Finch, E., Turner, T. and Goldin, S. (1991). Calcium as a coagonist of inositol 1,4,5-trisphosphate-induced calcium release, *Science* **252**, 443–446.

FitzHugh, R. (1961). Impulses and physiological states in models of nerve membrane, *Biophys J.* **1**, 445–466.

Fox, R. (1997). Stochastic versions of the Hodgkin–Huxley equations, *Biophys. J.* **72**, 2068–2074.

Fox, R. and Choi, M. (2001). Rectified Brownian motion and kinesin motion along microtubules, *Phys. Rev. E.*

Friel, D. (1995). $[Ca^{2+}]_i$ oscillations in sympathetic neurons: an experimental test of a theoretical model, *Biophys. J.* **68**(5), 1752–1766.

Friel, D. and Tsien, R. W. (1992). Phase-dependent contributions from $Ca^{2+}$ entry and $Ca^{2+}$ release to caffeine-induced $Ca^{2+}$ oscillations in bullfrog sympathetic neurons, *Neuron* **8**, 1109–1125.

Fygenson, D., Marko, J. and Libchaber, A. (1997). Mechanics of microtubule-based membrane extension, *Phys. Rev. Lett.* **79**(22), 4497–500.

Gantmacher, F. (1959). *Theory of Matrices, Volume II*, Chelsea Publishing, New York.

Gardiner, C. W. (1997). *Handbook of Stochastic Methods*, 3rd edition, Springer Verlag, New York.

Gillespie, D. (1977). Exact stochastic simulation of coupled chemical reactions, *J Phys. Chem.* **81**, 2340–2361.

Girard, S., Luckhoff, A., Lechleiter, J., Sneyd, J. and Clapham, D. (1992). Two-dimensional model of calcium waves reproduces the patterns observed in *Xenopus* oocytes, *Biophys. J.* **61**(2), 509–17.

Goldbeter, A. (1995). A model for circadian oscillations in the *Drosophila* period protein (PER), *Proc. R. Soc. Lond. B* **261**, 319–324.

Goldbeter, A. (1996). *Biochemical Oscillations and Cellular Rhythms*, Cambridge Univ. Press, Cambridge, UK.

Goldbeter, A. and Koshland, D. (1981). An amplified sensitivity arising from covalent modification in biological systems, *Proc. Natl. Acad. Sci. U.S.A.* **78**, 6840–6844.

Goldbeter, A., Dupont, G. and Berridge, M. (1990). Minimal model for signal-induced $Ca^{2+}$ oscillations and for their frequency encoding through protein phosphorylation, *Proc. Natl. Acad. Sci. USA* **87**, 1461–1465.

Goodwin, B. (1965). Oscillatory behavior in enzymatic control processes, *Adv. Enz. Regul.* **3**, 425–438.

Goodwin, B. (1966). An entrainment model for timed enzyme synthesis in bacteria, *Nature* **209**, 479–481.

Griffith, J. (1968a). Mathematics of cellular control processes. I. Negative feedback to one gene, *J. Theor. Biol.* **20**, 202–208.

Griffith, J. (1968b). Mathematics of cellular control processes. II. Positive feedback to one gene, *J. Theor. Biol.* **20**, 209–216.

Grimes, A. (1980). *Human Red Cell Metabolism*, 1st edition, Blackwell Sciences, New York.

Grodsky, G. (1972). A threshold distribution hypothesis for packet storage of insulin and its mathematical modeling, *J. Clin. Invest.* **51**, 2047–2059.

Grynkiewicz, G., Poenie, M. and Tsien, R. Y. (1985). A new generation of $Ca^{2+}$ indicators with greatly improved fluorescence properties, *J. Biol. Chem.* **260**(6), 3440–50.

Guckenheimer, J. (1986). Multiple bifurcation problems for chemical reactors, *Physica D* **20**, 1–20.

Györke, S. and Fill, M. (1993). Ryanodine receptor adaptation: control mechanism of $Ca^{2+}$-induced $Ca^{2+}$ release in heart, *Science* **260**, 807–809.

Hanggi, P., Talkner, P. and Borkovec, M. (1990). Reaction-rate theory: 50 years after Kramers, *Rev. Mod. Phys.* **62**, 254–341.

Hansel, D. and Sompolinsky, H. (1998). Modeling feature selectivity in local cortical circuits, *in Methods in Neuronal Modeling*, C. Koch and I. Segev (eds.), The MIT Press, Cambridge.

Happel, J. and Brenner, H. (1986). *Low Reynolds Number Hydrodynamics*, Vol. 1 of *Mechanics of fluids and transport processes*, Nijhoff, The Hague.

Higgins, J. (1967). The theory of oscillating reactions, *Ind. Eng. Chem.* **59**, 18–62.

Hill, T. (1977). *Free Energy Transduction in Biology*, Academic Press, New York.

Hille, B. (2001). *Ionic Channels of Excitable Membranes*, 3rd edition, Sinauer Associates, Inc., Sunderland.

Hodgkin, A. and Huxley, A. (1939). Action potentials recorded from inside a nerve fibre, *J. Physiol (Lond.)* **144**, 710–711.

Hodgkin, A. and Huxley, A. (1952). A quantitative description of membrane current and its application to conduction and excitation in nerve., *J. Physiol (Lond.)* **117**, 500–544.

Honda, M., Takiguchi, K., Ishikawa, S. and Hotani, H. (1999). Morphogenesis of liposomes encapsulating actin depends on the type of actin-crosslinking, *J. Mol. Biol.* **287**, 293–300.

Howard, J. (2001). *Mechanics of Motor Proteins and the Cytoskeleton*, 1st edition, Sinauer Associates, Sunderland, MA.

Huxley, A. (1957). Muscle structure and theories of contraction, *Prog. Biophys. Biophys. Chem.* **7**, 255–318.

Huxley, A. and Simmons, R. (1971). Proposed mechanism of force generation in striated muscle, *Nature* **233**, 533–538.

Iino, M. (1990). Biphasic $Ca^{2+}$-dependence of inositiol 1,4,5-trisphosphate-induced $Ca^{2+}$ release in smooth muscle cells of the guinea pig *Taenia caeci* gonadotrophs, *J. Gen. Physiol.* **95**, 1103–1122.

Israelachvili, J. (1992). *Intermolecular and Surface Forces*, 2nd edition, Academic Press, New York.

Jafri, M. and Keizer, J. (1994). Diffusion of inositol 1,4,5-trisphosphate but not $Ca^{2+}$ is necessary for a class of inositol 1,4,5-trisphosphate-induced $Ca^{2+}$ waves, *Proc. Natl. Acad. Sci. USA* **91**, 9485.

Jafri, M. and Keizer, J. (1997). Agonist-induced calcium waves in oscillatory cells: a biological example of Burgers' equation, *Bull Math Biol.* **59**(6), 1125–44.

Johnston, D. and Wu, S. (1995). *Foundations of Cellular Neurophysiology*, 1st edition, MIT Press, Cambridge.

Julicher, F. and Bruinsma, R. (1998). Motion of RNA polymerase along DNA: A stochastic model, *Biophys. J.* **74**(3), 1169–1185.

Kandel, E., Schwartz, J. and Jessell, T. (eds.) (2000). *Principles of Neural Science*, 4th edition, McGraw-Hill, New York.

Kaplan, D. and Glass, L. (1995). *Understanding Nonlinear Dynamics*, 1st edition, Springer Verlag, New York.

Keener, J. (1999). *Principles of Applied Mathematics*, Perseus Books, New York.

Keener, J. (2000). Propagation of waves in an excitable medium with discrete release sites, *SIAM J. Appl. Math* **61**, 317–334.

Keener, J. and Sneyd, J. (1998). *Mathematical Physiology*, 1st edition, Springer, New York.

Keizer, J. (1987). *Statistical Thermodynamics of Nonequilibrium Processes*, Springer-Verlag, New York.

Keizer, J. and Levine, L. (1996). Ryanodine receptor adaptation and $Ca^{2+}$-induced $Ca^{2+}$ release-dependent $Ca^{2+}$ oscillations, *Biophys. J.* **71**, 3477–3487.

Keizer, J. and Magnus, G. (1989). ATP-sensitive potassium channel and bursting in the pancreatic beta-cell, *Biophys. J.* **56**, 229–242.

Keizer, J., Li, Y.-X., Stojilković, S. and Rinzel, J. (1995). InsP$_3$-induced $Ca^{2+}$ excitability of the endoplasmic reticulum, *Molecular Biology of the Cell* **6**, 945–951.

Keizer, J., Smith, G., Ponce-Dawson, S. and Pearson, J. (1998). Saltatory propagation of $Ca^{2+}$ waves by $Ca^{2+}$ sparks, *Biophys. J.* **75**, 595.

Keller, A. (1995). Model genetic circuits encoding autoregulating transcription factors, *J. Theor. Biol.* **172**, 169–85.

Keller, D. and Bustamante, C. (2000). The mechanochemistry of molecular motors, *Biophys. J.* **78**, 541–556.

Kirk, J., Orr, J. and Hope, C. (1968). A mathematical analysis of red blood cell and bone marrow stem cell control mechanisms, *Brit. J. Haematology* **15**, 35–46.

Koch, C. (1999). *Biophysics of Computation : Information Processing in Single Neurons*, Oxford University Press, New York.

Koch, C. and Segev, I. (2000). The role of single neurons in information processing, *Nat. Neurosci* **3**, 1171–1177.

Koch, C. and Segev, I. (eds.) (1998). *Methods in Neuronal Modeling*, 2nd edition, MIT Press, Cambridge.

Kohn, K. (1999). Molecular interaction map of the mammalian cell cycle control and DNA repair systems, *Mol. Biol. Cell* **10**, 2703–2734.

Konopka, R. and Benzer, S. (1971). Clock mutants in *Drosophila melanogaster*, *Proc. Natl. Acad. Sci. USA* **68**, 2112–2116.

Kuba, K. and Nishi, S. (1976). Rhythmic hyperpolarizations and depolarization of sympathetic ganglion cells induced by caffeine, *J. Neurophysiol.* **39**, 547–563.

Kuba, K. and Takeshita, S. (1981). Simulation of intracellular $Ca^{2+}$ oscillation in a sympathetic neuron, *J. Theor. Biol.* **93**, 1009–1031.

Kukuljan, M., Rojas, E., Catt, K. and Stojilković, S. (1994). Membrane potential regulates inositol 1,4,5-trisphosphate-controlled cytoplasmic $Ca^{2+}$ oscillations in pituitary gonadotrophs, *J. Biol. Chem.* **269**, 4860–4865.

Kuznetsov, Y. (1998). *Elements of Applied Bifurcation Theory*, 2nd edition, Springer-Verlag, New York.

Landau, L., Lifshitz, E. and Pitaevskii, L. (1980). *Statistical Physics*, 3rd edition, Pergamon Press, Oxford.

Läuger, P. (1990). Microscopic models of the bacterial flagellar motor, *Comments Theoret. Biol.* **2**, 99–123.

Läuger, P. (1991). F0F1-ATPases, *in Electrogenic Ion Pumps*, L. Nadel and D. Stein (eds.), Addison-Wesley, pp. 252–269.

Lawrence, P. (1992). *The Making of a Fly: The Genetics of Animal Design*, Blackwell Science, Oxford.

Lechleiter, J. and Clapham, D. (1992). Molecular mechanisms of intracellular calcium excitability in *X. laevis* oocytes, *Cell.* **69**, 283–294.

Leinhard, G., Slot, J., James, D. and Mueckler, M. (1992). How cells absorb glucose, *Scientific American* **January**, 89.

Leloup, J. and Goldbeter, A. (1998). A model for circadian rhythms in *Drosophila* incorporating the formation of a complex between the PER and TIM proteins, *J. Biol. Rhythms* **13**, 70–87.

Leloup, J. and Goldbeter, A. (1999). Limit cycle models for circadian rythms based on transcriptional regulation in *Drosophila* and *neurospora*, *J. Biol. Rhythms* **14**, 433–448.

Lema, M., Golombek, D. and Echave, J. (2000). Delay model of the circadian pacemaker, *J. Theor. Biol.* **204**, 565–573.

Li, Y.-X. and Rinzel, J. (1994). Equations for $InsP_3$ receptor-mediated $Ca^{2+}$ oscillations derived from a detailed kinetic model: A Hodgkin–Huxley like formalism, *J. Theor. Biol.* **166**, 461–473.

Li, Y.-X., Keizer, J., Stojilković, S. and Rinzel, J. (1995b). $Ca^{2+}$ excitability of the ER membrane: an explanation for $IP_3$-induced $Ca^{2+}$ oscillations, *Am. J. Phys.* **269**, C1079–C1092.

Li, Y.-X., Rinzel, J., Vergara, L. and Stojilković, S. (1995a). Spontaneous electrical and calcium oscillations in unstimulated pituitary gonadotrophs, *Biophys. J.* **69**, 785–795.

Li, Y.-X., Stojilković, S., Keizer, J. and Rinzel, J. (1997). Sensing and refilling calcium stores in an excitable cell, *Biophys. J.* **72**, 1080–1091.

Lindenberg, K. and Seshadri, V. (1979). Analytic theory of extrema. I. Asymptotic theory for Fokker-Planck processes, *J. Chem. Phys.* **71**(10), 4075–4084.

Logan, D. (1997). *Applied Mathematics*, John Wiley and Sons, New York.

Mackey, M. (1996). Periodic hemolytic anemia, *in Case Studies in Mathematical Modeling: Ecology, Physiology and Cell Biology*, H. Othmer, F. Adler, J. Dallon and M. Lewis (eds.), Prentice Hall, Upper Saddle River, NJ, pp. 149–178.

Mahadevan, L. and Matsudaira, P. (2000). Motility powered by supramolecular springs and ratchets, *Science* **288**, 95–99.

Maki, L. and Keizer, J. (1995). Mathematical analysis of a proposed mechanism for oscillatory insulin secretion in perifused HIT-15 cells, *Bull. Math. Biology* **57**, 569–591.

McCormick, D. (1999). Membrane potential and action potential, *in Fundamental Neuroscience*, M. J. Zigmond, F. E. Bloom, S. C. Landis, J. L. Roberts and L. R. Squire (eds.), Academic Press, New York.

McLaughlin, D., Shapley, R., Shelley, M. and Wielaard, D. (2000). A neuronal network model of macaque primary visual cortex (v1): orientation selectivity and dynamics in the input layer 4calpha, *Proc. Natl. Acad. Sci. USA.* **97**(14), 8087–8092.

McMahon, T. (1984). *Muscles, Reflexes, and Locomotion*, Princeton University Press, Princeton.

Meinhardt, H. (1998). *The Algorithmic Beauty of Sea Shells*, Springer Verlag, New York.

Meyer, T. and Stryer, L. (1988). Molecular model for receptor-stimulated calcium spiking, *Proc. Natl. Acad. Sci. USA* **85**, 5051–5055.

Mogilner, A. and Oster, G. (1996). Cell motility driven by actin polymerization, *Biophys. J.* **71**(6), 3030–3045.

Mogilner, A. and Oster, G. (1999). The polymerization ratchet model explains the force-velocity relation for growing microtubules, *Eur. J. Biophys.* **28**(3), 235–242.

Morris, C. and Lecar, H. (1981). Voltage oscillations in the barnacle giant muscle, *Biophys J.* **35**, 193–213.

Murray, A. and Hunt, T. (1993). *The Cell Cycle. An Introduction*, W.H. Freeman and Co., New York.

Murray, J. (1981). A pre-pattern formation mechanism for animal coat markings, *J. Theor. Biol.* **88**, 161–99.

Murray, J. (1989). *Mathematical Biology*, 1st edition, Springer-Verlag, New York.

Nagumo, J., Arimoto, S. and Yoshizawa, S. (1962). An active pulse transmission line simulating nerve axon, *Proc. IRE.* **50**, 2061–2070.

Naraghi, M. and Neher, E. (1997). Linearized buffered $Ca^{2+}$ diffusion in microdomains and its implications for calculation of $[Ca^{2+}]$ at the mouth of a calcium channel, *J. Neurosci.* **17**, 6961.

Naray-Szabo, G. and Warshel, A. (1997). *Computational Approaches to Biochemical Reactivity*, Vol. 19 of *Understanding chemical reactivity*, Kluwer Academic, Boston.

Nasmyth, K. (1996). At the heart of the budding yeast cell cycle, *Trends Genet.* **12**, 405–412.

Neher, E. (1986). Concentration profiles of intracellular $Ca^{2+}$ in the presence of diffusible chelator, *Exp. Brain Res.* **14**, 80.

Neher, E. (1998). Usefulness and limitations of linear approximations to the understanding of $Ca^{2+}$ signals, *Cell Calcium* **24**, 345.

Nohmi, M., Hua, S.-Y. and Kuba, K. (1992). Basal $Ca^{2+}$ and the oscillation of $Ca^{2+}$ in caffeine-treated bullfrog sympathetic neurones, *J. Physiol. (Lond.)* **450**, 513–528.

Nuccitelli, R., Yim, D. L. and Smart, T. (1993). The sperm-induced $Ca^{2+}$ wave following fertilization of the *Xenopus* egg requires the production of Ins(1,4,5)P3, *Dev. Biol.* **158**(1), 200–12.

Nykamp, D. and Tranchina, D. (2000). A population density approach that facilitates large scale modeling of neural networks: Analysis and an application to tuning, *J. Comput. Neurosci.* **8**(1), 19–50.

O'Donovan, M., Wenner, P., Chub, N., Tabak, J. and Rinzel, J. (1998). Mechanisms of spontaneous activity in the developing spinal cord and their relevance to locomotion, *Ann. N.Y. Acad. Sci. USA* **860**, 130–141.

Omurtag, A., Knight, B. and Sirovich, L. (2000). On the simulation of large populations of neurons, *J. Comput. Neurosci.* **8**(1), 51–64.

Oosawa, F. and Hayashi, S. (1986). The loose coupling mechanism in molecular machines of living cells, *Adv Biophys* **22**, 151–183.

Oster, G. and Wang, H. (2000a). Reverse engineering a protein: The mechanochemistry of ATP synthase, *Bioc. Biophys. Acta (Bioenergetics)* **1458**(1-2), 482–510.

Oster, G. and Wang, H. (2000b). Why is the efficiency of the F1 ATPase so high?, *J. Bioenerg. Biomembr.* **32**, 459–469.

Oster, G., Perelson, A. and Tilney, L. (1982). A mechanical model for acrosomal extension in *Thyone*, *J. Math. Biol* **15**, 259–65.

Othmer, H. and Tang, Y. (1993). Oscillations and waves in a model of InsP$_3$-controlled calcium dynamics, *in Experimental and Theoretical Advances in Biological Pattern Formation*, H. G. Othmer, P. Maini and J. Murray (eds.), Plenum Press, New York, pp. 277–299.

Pape, P., Jong, D. and Chandler, W. (1998). Effects of partial sarcoplasmic reticulum calcium depletion on calcium release in frog cut muscle fibers equilibrated with 20 mM EGTA, *J. Gen. Physiol.* **112**(3), 263–95.

Parent, L., Supplisson, S., Loo, D. and Wright, E. (1992a). Electrogenic properties of the cloned Na$^+$/glucose cotransporter: I. Voltage clamp studies, *J. Membrane Biol.* **125**, 49–62.

Parent, L., Supplisson, S., Loo, D. and Wright, E. (1992b). Electrogenic properties of the cloned Na$^+$/glucose cotransporter: II. A transport model under nonrapid equilibrium conditions, *J. Membrane Biol.* **125**, 63–79.

Parker, I. and Ivorra, I. (1990). Inhibition by Ca$^{2+}$ of inositol trisphosphate-mediated Ca$^{2+}$ liberation - a possible mechanism for oscillatory release of Ca$^{2+}$, *Proc. Natl. Acad. Sci. USA* **87**(1), 260–264.

Parker, I., Choi, J. and Yao, Y. (1996). Elementary events of InsP3-induced Ca$^{2+}$ liberation in *Xenopus* oocytes: hot spots, puffs and blips, *Cell Calcium* **20**, 105.

Pearson, J. (1993). Complex patterns in a simple system, *Science* **261**, 189–192.

Perez Velasquez, J. and Carlen, P. (2000). Gap junctions, synchrony and seizures, *Trends Neurosci.* **23**(2), 68–74.

Pernarowski, M., Miura, R. and Kevorkian, J. (1992). Perturbation techniques for models of bursting electrical activity in the pancreatic Beta cells, *SIAM J. Appl. Math.* **52**, 1627–1650.

Peskin, C. and Oster, G. (1995). Coordinated hydrolysis explains the mechanical behavior of kinesin, *Biophys. J.* **68**(4), 202s–210s.

Peskin, C., Odell, G. and Oster, G. (1993). Cellular motions and thermal fluctuations: the Brownian ratchet, *Biophys. J.* **65**, 316–324.

Pietrobon, D. and Caplan, S. (1985). Flow-force relationships for a six-state proton pump model: intrinsic uncoupling, kinetic equivalence of input and output forces, and domain of approximate linearity, *Biochemistry* **24**, 5764–5778.

Pinto, D., Brumberg, J., Simons, D. and Ermentrout, G. (1996). A quantitative population model of whisker barrels: re-examining the Wilson-Cowan equations, *J. Comput. Neurosci.* **3**(3), 5247–264.

Ponce-Dawson, S., Keizer, J. and Pearson, J. (1999). Fire-diffuse-fire model of dynamics of intracellular calcium waves, *Proc. Natl. Acad. Sci. USA* **96**, 6060.

Porte, Jr., D. (1990). Banting lecture 1990. Beta-cells in type II diabetes mellitus., *Diabetes* **40**(2), 166–180.

Pratusevich, V. and Balke, C. (1996). Factors shaping the confocal image of the calcium spark in cardiac muscle cells, *Biophys. J.* **71**(6), 2942–57.

Pye, E. (1971). Periodicities in intermediary metabolism, *in Biochronometry*, M. Menaker (ed.), National Acad. Sci. Press, Washington, DC, pp. 623–636.

Quon, M. and Campfield, L. (1991a). A mathematical model and computer simulation study of insulin sensitive glucose transporter regulation, *J Theor. Biol.* **150**, 93–107.

Quon, M. and Campfield, L. (1991b). A mathematical model and computer simulation study of insulin receptor regulation, *J Theor. Biol.* **150**, 59–72.

Rapp, P. (1979). An atlas of cellular oscillators, *J. Exp. Biol* **81**, 281–306.

Reif, F. (1965). *Fundamentals of Statistical and Thermal Physics*, McGraw Hill, New York.

Rieke, F., Warland, D., de Ruyter van Steveninck, R. and Bialek, W. (1997). *Spikes: Exploring the Neural Code*, MIT Press, Cambridge, MA.

Rinzel, J. and Ermentrout, B. (1998). Analysis of neural excitability and oscillations, *in Methods in Neuronal Modeling*, C. Koch and I. Segev (eds.), 2nd edition, The MIT Press, Cambridge, pp. 251–291.

Risken, H. (1989). *The Fokker-Planck Equation*, 2nd edition, Springer-Verlag, New York.

Roberts, W. (1994). Localization of calcium signals by a mobile calcium buffer in frog saccular hair cells, *J. Neurosci.* **14**, 3246.

Rubinow, S. (1973). *Mathematical Problems in the Biological Sciences*, S.I.A.M., Philadelphia.

Rubinow, S. (1980). Biochemical reaction theory, *in Mathematical models in molecular and cellular biology*, L. A. Segel (ed.), Cambridge Univ. Press, Cambridge, UK, pp. 13–111.

Ruoff, P. and Rensing, L. (1996). The temperature-compensated goodwin model simulates many circadian clock properties, *J. Theor. Biol.* **179**, 275–285.

Sanchez-Andres, J., Gomis, A. and Valdeolmillos, M. (1995). The electrical activity of mouse pancreatic beta-cells recorded in vivo show glucose-dependent oscillations, *J. Physiol. (London)* **486**, 223–228.

Scharf, B., Fahrner, K., Turner, L. and Berg, H. (1998). Control of direction of flagellar rotation in bacterial chemotaxis, *Proc. Nat. Acad. Sci. USA* **95**, 201–206.

Schnitzer, M. and Block, S. (1995). Statistical kinetics of processive enzymes, *Cold Spring Harbor Symposia on Quantitative Biology* **Vol. LX**, 793–802.

Segel, L. (1984). *Modeling dynamic phenomena in molecular and cellular biology*, 1st edition, Cambridge University Press, New York.

Segel, L., Chet, I. and Henis, Y. (1977). A simple quantitative assay for bacterial motility, *J. Gen. Microbiol.* **98**, 329–337.

Segev, I., Rinzel, J. and Shepard, G. (1995). *The Theoretical Foundation of Dendritic Function: Selected Papers of Wilfrid Rall with Commentaries*, MIT Press, Cambridge, MA.

Shadlen, M. and Newsome, W. (1998). The variable discharge of cortical neurons: implications for connectivity, computation, and information coding, *J. Neurosci.* **18**(10), 3870–3896.

Sherman, A. (1996). Contributions of modeling to understanding stimulus-secretion coupling in pancreatic $\beta$-cells, *Am. J. Physiol.* **271**, E362–E372.

Sherman, A. (1997). Calcium and membrane potential oscillations in pancreatic $\beta$-cells, *in Case Studies in Mathematical Modeling—Ecology, Physiology, and Cell Biology*, H. G. Othmer, F. R. Adler, M. A. Lewis and J. C. Dallon (eds.), Prentice-Hall, New Jersey, pp. 199–217.

Sherman, A., Rinzel, J. and Keizer, J. (1988). Emergence of organized bursting in clusters of pancreatic Beta cells by channel sharing., *Biophys. J.* **54**, 411–425.

Siffre, M. (1975). Six months alone in a cave, *National Geographic* **147**(3), 426–435.

Sigworth, F. (1980). The variance of sodium current fluctuations at the node of Ranvier, *J. Physiol (Lond)* **307**, 97–129.

Simon, S., Peskin, C. and Oster, G. (1992). What drives the translocation of proteins?, *Proc. Natl. Acad. Sci. USA* **89**(9), 3770–3774.

Smith, D. and Geeves, M. (1995). Strain-dependent cross-bridge cycle for muscle, *Biophys. J.* **69**(2), 524–537.

Smith, G. (1996). Analytical steady-state solution to the rapid buffering approximation near an open $Ca^{2+}$ channel, *Biophys. J.* **71**, 3064.

Smith, G. (2001). Modeling local and global $Ca^{2+}$ signals using reaction-diffusion equations, *in Computational Neuroscience: Realistic Modeling for Experimentalists.*, E. D. Schutter (ed.), CRC Press.

Smith, G., Dai, L., Miura, R. and Sherman, A. (2001). Asymptotic analysis buffered $Ca^{2+}$ diffusion near a point source, *SIAM J. Appl. Math.* **61**(5), 1816–1838.

Smith, G., Keizer, J., Stern, M., Lederer, W. and Cheng, H. (1998). A simple numerical model of $Ca^{2+}$ spark formation and detection in cardiac myocytes, *Biophys. J.* **75**, 15.

Smith, G., Wagner, J. and Keizer, J. (1996). Validity of the rapid buffering approximation near a point source for $Ca^{2+}$ ions, *Biophys. J.* **70**, 2527.

Sneyd, J., Dale, P. and Duffy, A. (1998). Traveling waves in buffered systems: applications to calcium waves, *SIAM J. Applied Math.* **58**, 1178.

Stein, W. and Lauger, P. (1990). Kinetic properties of F0F1-ATPases, *Biophys. J.* **57**, 255–267.

Stern, B. and Nurse, P. (1996). A quantitative model for the cdc2 control of S phase and mitosis in fission yeast, *Trends Genet.* **12**, 345–350.

Stern, M. (1992). Buffering of $Ca^{2+}$ in the vicinity of a channel pore, *Cell Calcium* **13**, 183.

Strogatz, S. (1994). *Nonlinear Dynamics and Chaos*, 1st edition, Perseus Books, Cambridge.

Surana, U., Robitsch, H., Price, C., Schuster, T., Fitch, I., Futcher, A. and Naysmyth, K. (1991). The role of CDC28 and cyclins during mitosis in the budding yeast *S. cerevisiae*, *Cell* **65**, 145–161.

Tabak, J., Senn, W., O'Donovan, M. and Rinzel, J. (2000). Modeling of spontaneous activity in the developing spinal cord using activity dependent depression in an excitatory network, *J. Neurosci.* **20**(8), 3041–3056.

Tang, Y. and Othmer, H. (1994). A model of calcium dynamics in cardiac myocytes based on the kinetics of ryanodine-sensitive calcium channels, *Biophys. J.* **67**, 2223–2235.

Tang, Y., Stephenson, J. and Othmer, H. (1996). Simplification and analysis of models of calcium dynamics based on $IP_3$-sensitive calcium channel kinetics, *Biophys. J.* **70**, 246–263.

Taylor, S. (1999). Deconstructing type 2 diabetes, *Cell* **97**, 9–12.

Thomas, A., Bird, G., Hajnoczky, G., Robb-Gaspers, L. and Putney Jr., J. (1996). Spatial and temporal aspects of cellular calcium signaling, *FASEB J.* **13**, 1505–1517.

Topp, B., Promislow, K., de Vries, G., Miura, R. and Finegood, D. (2000). A model of $\beta$-cell mass, insulin and glucose kinetics: Pathways to diabetes, *J. Theor. Biol.* **206**, 605–619.

Traub, R., Jefferys, J. and Whittington, M. (1999). *Fast Oscillations in Cortical Circuits*, MIT Press, Cambridge, MA.

Traub, R., Whittington, M., Stanford, I. and Jefferys, J. (1996). A mechanism for generation of long-range synchronous fast oscillations in the cortex, *Nature* **383**(6601), 621–624.

Troyer, T., Krukowski, A., Priebe, N. and Miller, K. (1998). Contrast-invariant orientation tuning in cat visual cortex: thalamocortical input tuning and correlation-based intracortical connectivity, *J. Neurosci.* **18**(15), 5908–5927.

Trussell, L. (1999). Synaptic mechanisms for coding timing in auditory neurons, *Annu. Rev. Physiol.* **61**, 477–496.

Tse, A., Tse, F. and Hille, B. (1993). Rhythmic exocytosis stimulated by GnRh-induced calcium oscillations in rat gonadotropes, *Science* **260**, 82–84.

Tuckwell, H. (1995). *Elementary Applications of Probability Theory*, 2nd edition, Chapman and Hall, New York.

Tyers, M., Tokiwa, G. and Futcher, B. (1993). Comparison of the *Saccharomyces cerevisiae* G1 cyclins: Cln3 may be an upstream activator of Cln1, Cln2 and other cyclins, *EMBO J.* **12**, 1955–1968.

Tyson, J., Hong, C., Thron, C. and Novak, B. (1999). A simple model of circadian rhythms based on dimerization and proteolysis of PER and TIM, *Biophys. J.* **77**, 2411–2417.

Tyson, J., Novak, B., Odell, G., Chen, K. and Thron, C. (1996). Chemical kinetic theory: understanding cell-cycle regulation, *Trends Biochem. Sci.* **21**, 89–96.

Valdeolmillos, M., Santos, R., Contreras, D., Soria, B. and Rosario, L. (1989). Glucose-induced oscillations of intracellular $Ca^{2+}$ concentration resembling bursting electrical activity in single mouse islets of Langerhans, *FEBS Lett.* **259**, 19–23.

Vale, R. (2000). Millennial musings on molecular motors, *Trends Bioc. Sci.* **24**(12), M38–M42.

Van Vreeswijk, C., Abbott, L. and Ermentrout, G. (1994). When inhibition not excitation synchronizes neural firing, *J. Comput. Neurosci.* 1(4), 313–321.

Wagner, J. and Keizer, J. (1994). Effects of rapid buffers on $Ca^{2+}$ oscillations, *Biophys. J.* **67**, 447–456.

Wagner, J., Li, Y.-X., Pearson, J. and Keizer, J. (1998). Simulation of the fertilization calcium wave in *Xenopus laevis* eggs, *Biophys. J.* **75**, 2088–2097.

Wang, H. and Oster, G. (1998). Energy transduction in the F1 motor of ATP synthase, *Nature* **396**, 279–282.

Wang, H., Elston, T., Mogilner, A. and Oster, G. (1998). Force generation in RNA polymerase, *Biophys. J.* **74**(3), 1186–1202.

Wang, X. and Buzsaki, G. (1996). Gamma oscillation by synaptic inhibition in a hippocampal interneuronal network model, *J. Neurosci.* **16**(20), 6402–6413.

Wang, X. and Rinzel, J. (1992). Alternating and synchronous rhythms in reciprocally inhibitory model neurons, *Neural Computation* **4**, 84–97.

Warshel, A. (1991). *Computer Modeling of Chemical Reactions in Enzymes and Solutions*, Wiley, New York.

Weiss, G. (1967). First passage time problems in chemical physics, *Adv. Chem. Phys.* **13**, 1–18.

Weiss, T. (1996). *Cellular Biophysics*, 1st edition, MIT Press, Cambridge.

Welch, G. and Kell, D. (1986). Not just catalysts-molecular machines in bioenergetics, *in The Fluctuating Enzyme*, G. R. Welch (ed.), Wiley Interscience, New York, pp. 252–269.

White, J., Budde, T. and Kay, A. (1995). A bifurcation analysis of neuronal subthreshold oscillations, *Biophys. J.* **69**(4), 1203–17.

White, J., Rubinstein, J. and Kay, A. (2000). Channel noise in neurons, *TINS* **23**(3), 131–137.

Whitesell, R., Powers, A., Regen, D. and Abumrad, N. (1991). Transport and metabolism of glucose in an insulin-secreting cell line $\beta$ TC-1, *Biochem.* **30**, 11560–11566.

Wilson, H. and Cowan, J. (1972). Excitatory and inhibitory interactions in localized populations of model neurons, *Biophys. J.* **12**(1), 1–24.

Winfree, A. (1987). *When Time Breaks Down*, 1st edition, Princeton University Press, Princeton.

Wong, T.-F. (1975). *Kinetics of Enzyme Mechanisms*, Academic Press, New York.

Yamada, W., Koch, C. and Adams, P. (1998). Multiple channels and calcium dynamics, *in Methods in Neuronal Modeling*, C. Koch and I. Segev (eds.), 2nd edition, The MIT Press, Cambridge, Massachusetts, pp. 251–291.

Yao, Y., Choi, J. and Parker, I. (1995). Quantal puffs of intracellular $Ca^{2+}$ evoked by inositol trisphosphate in *Xenopus* oocytes, *J. Physiol. (Lond.)* **482**, 533.

Yasuda, R., Noji, H., Kinosita, K. and Yoshida, M. (1998). F1-ATPase is a highly efficient molecular motor that rotates with discrete $120°$ steps, *Cell* **93**, 1117–1124.

Zigmond, M., Bloom, F., Landis, S., Roberts, J. and Squire, L. (1999). *Fundamental Neuroscience*, Academic Press, New York.

# Index

# Interdisciplinary Applied Mathematics

Printed in Great Britain
by Amazon.co.uk, Ltd.,
Marston Gate.

1605599R0